YOUR TRUSTED SOURCE FOR LUXURY WATCH

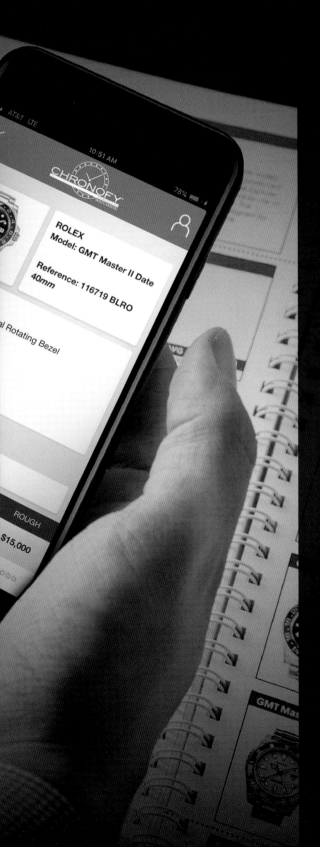

- Purchase Guarantee
- Theft Check
- Real Time Pricing Updates
- Access to Industry Experts
- Multiple Language and Currency Options
- Web Portal
  Mac + PC Access
- Mobile App
  Apple Store + Google Play

CHRONOFY®
POWERED BY WATCHBOX

THE WATCH GUIDE

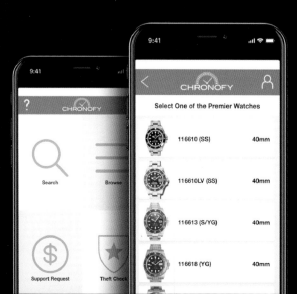

CONTENTS

A

A. Lange & Söhne 40
Alexander Shorokhoff 46
Alpina 48
Anonimo 49
Aquadive 50
Aristo 51
Armin Strom 52
Arnold & Son 54
ArtyA 56
Audemars Piguet 58
Azimuth 62

B

Ball Watch Co. 64
Baume & Mercier 68
Bell & Ross 70
Blancpain 72
Borgward 78
Bovet 80
Breguet 82
Breitling 86
Bremont 90
BRM 92
Carl F. Bucherer 98
Bulgari 94

C

Carl F. Bucherer 98
Cartier 102
Casio 106
Chanel 107
Chopard 108
Christophe Claret 114
Chronoswiss 115
Claude Meylan 118
Frédérique Constant 141
Corum 119
Cuervo y Sobrinos 122
Cvstos 124
Czapek & Cie. 125

D

Davosa 126
Detroit Watch Co. 128
Roger Dubuis 270
duManège 129

E

Eberhard & Co. 130
Eterna 132

F

Fabergé 134
F.P. Journe 136
Franck Muller 138
Frédérique Constant 141

G

Paul Gerber 254
Girard-Perregaux 144
Glashütte Original 148
Graham 154
Grand Seiko 156
Greubel Forsey 158

H

H. Moser & Cie. 160
Habring² 162
Hager Watches 164
Hamilton 165
Hanhardt 167
Harry Winston 168
Hermès 170
Hublot 172
HYT 176

I

Itay Noy 178
IWC 180

J

Jaeger-LeCoultre 186
Jaquet-Droz 192
Jörg Schauer 193
Junghans 194
Urban Jürgensen & Sønner 310

K

Kobold 196
Kudoke 198

L

Maurice Lacroix 208
Laurent Ferrier 199
Longines 200
Louis Moinet 203
Louis Vuitton 204
Luminox 206

M

Manufacture Royale 207
Maurice Lacroix 208
MB&F 211
MeisterSinger 212
Mido 214
Richard Mille 268
Mk II 216
Louis Moinet 203
Montblanc 217
H. Moser & Cie. 160
Mühle-Glashütte 221
Franck Muller 138

N

Ulysse Nardin 306
Nivrel 224
Nomos 226
Itay Noy 178

O

Omega 230
Oris 235

P

Panerai 238
Parmigiani 242
Patek Philippe 246
Paul Gerber 254
Peter Speake-Marin 255
Piaget 256
Porsche Design 260
Pramzius 262

R

Rado 263
Ressence 265
RGM 266
Richard Mille 268
Roger Dubuis 270
Rolex 274

S

Jörg Schauer 193
Schaumburg Watch 280
Schwarz Etienne 281
Seiko 282
Alexander Shorokhoff 46
Sinn 284
Peter Speake-Marin 255
Stowa 288
Armin Strom 52

T

TAG Heuer 290
Temption 294
Tissot 296
Towson Watch Company 298
Tudor 300
Tutima 303

U

Ulysse Nardin 306
Urban Jürgensen & Sønner 310
Urwerk 311
UTS 312

V

Vacheron Constantin 314
Van Cleef & Arpels 320
Vortic 321
Vostok-Europe 322
Louis Vuitton 204

W

Wempe Glashütte 324

Z

Zeitwinkel 326
Zenith 327

Movement manufacturers

Concepto 332
ETA 334
Ronda 338
Sellita 340

Editorial

Letter to the Readers 8
Independent Watchmaking:
The Independent Scene 2018 14
Masters and Mavericks:
The Outsiders 20
Watch Tech:
The Watch Aftermarket 30
Watch Your Watch 342
Glossary 344
Masthead 352

Advertisers

Anonimo
Armin Strom
Briller
Carl F. Bucherer
Casio
Chris Aire
Chronofy
Claude Meylan
Deep Blue
Franck Muller
Greubel Forsey
Hager
Island Watch
Itay Noy
Luminox
Meistersinger
Mk II
Mühle Glashütte
Orbis
Orbita
Paul Forrest
Quill & Pad
Rolex
Shorokhoff
Sturmanskie
Swiss Kubik
Towson Watch Company
Tutima
Vortic
Watchbox
Wempe Glashütte
William Henry Knives
Wolf

WEMPE
GLASHÜTTE I/SA

Is there a more appropriate place than an observatory to begin a stellar new chapter in the history of watchmaking?

Inaugurated in 1910 in Germany's renowned watchmaking town of Glashütte, the neglected ruin of the observatory building was rebuilt by the company WEMPE some 100 years later. The establishment of the only testing facility for chronometers according to the German industrial standard and of the WEMPE watchmaking school make Glashütte Observatory nowadays the perfect production site for the wristwatch chronometers of the WEMPE GLASHÜTTE I/SA collection.

DOES THE WORD "CHRONOMETER" HAVE A SUPERLATIVE FORM? ZEITMEISTER.

NEW YORK, 700 FIFTH AVENUE AT 55TH STREET, T 212.397.9000

AT THE BEST ADDRESSES IN GERMANY AND IN LONDON, PARIS, MADRID, VIENNA AND NEW YORK. WWW.WEMPE.COM

WEMPE Zeitmeister
GLASHÜTTE I/SA

A milestone in fine German watchmaking: The first German wristwatch chronometers that underwent the strict testing procedures at Glashütte Observatory. WEMPE ZEITMEISTER Chronograph in stainless steel with self-winding movement. Available exclusively at Wempe for $2,515.

LETTER TO THE READER

VORTIC WATCH COMPANY
2013

PRESERVING
AMERICAN HISTORY
ONE WATCH AT A TIME

FORT COLLINS, CO

www.vorticwatches.com

Dear Reader, Thomas Kuhn, who wrote *The Structure of Scientific Revolutions* and coined the term "paradigm shift," would probably agree that change begins slowly, almost unnoticeably, and then gains momentum and traction. After that, all bets are off as to where whatever is changing will go.

It's a process that applies to all areas of our lives these days, it would seem, be it the political world, where whole-cloth facts and a somewhat sophomoric tone once confined to the soapbox have now gone mainstream, or slews of industries, which are having to deal with digitalization, Big Data, Blockchain, and sundry bit-stuff.

IT and the Internet have, for better or for worse, been transforming the world in which we live and especially the processes—in most industries. For some, of course, especially those in the advertising field, it is the Great Killer Application per se, while for others, it is simply a bane, in spite of the advantages, a time killer that has put paid to vacations and weekends, a job killer, an immense cesspool where one can find gems but only after a hard search.

How is this affecting the watch world, an industry whose product is still based on the strict rationality of engineering? Increasingly, brands that once considered the Internet a necessary evil are now featuring neat little shopping carts on their sites, where the curious browser can become a real customer. This, of course, is putting pressure on retailers, who are not happy. Yet the move is understandable, since brands need to attract people who live through their phones and have neither the time nor the inclination to visit a shop.

The other aspect is best captured by the term "disruptive." We can't really define it, but we sense what it means: counterintuition, a *je ne sais quoi* of nihilism, a reminder that there is no negative feedback. Like ads and taglines that stay with you because they are irritating rather than inspiring. Without mentioning names, let me refer to one that is supposed to awaken visions of cars racing around a circuit at insane speeds, but for most people will conjure images of a frenetic life consumed by deadlines that are impossible to keep and the boss sending text messages on Sunday afternoons. And if you are under too much pressure, the first thing you should get rid of is, in fact, the time-tyrant on your wrist.

In fact, the entire world of communication is being stood on its head: Baselworld, once the biggest annual watch confab, was hit by an exodus. A number of brands went MIA in 2017, but the trickle suddenly became a torrent, with the cancellation of the Swatch Group and its eighteen brands in July 2018. Price, it seems, is the problem. Communication budgets are shrinking. And there are so many options.

WRISTWATCH ANNUAL
2019

THE CATALOG

of

PRODUCERS, PRICES, MODELS,

and

SPECIFICATIONS

BY PETER BRAUN

WITH MARTON RADKAI

ABBEVILLE PRESS PUBLISHERS

New York London

LETTER TO THE READER

What is the alternative? Where will the serious journalists go to see the collections and meet the people behind them? Events? Other trade fairs? The press releases uploaded by the brands, which anyone can read? Will everything be relegated to the Internet, with its frantic scrambling for "who's to be the first" and it's immanent evanescence?

A word about paper, then: Paper is a solid value. Paper tends to survive. Paper appeals to the senses and is therefore a memorable experience. Images still look far better on paper than on any screen. That is why *Wristwatch Annual*—and many magazines—still have appeal and are being produced, in spite of harsh headwinds. And it's why brands still advertise on paper: Because it has lasting value. Like a beautiful watch.

Once again, *Wristwatch Annual* has surveyed over one hundred brands carefully selected to balance the known with the lesser known, the crazy with the sober, the elegant with the trendy, the sportive with the philosophical. Among the newcomers this year is Colorado-based Vortic, because founder RT Custer had a superb idea to salvage old movements from the Golden Age of American watchmaking. Casio has long been knocking at the door with its muscular G-shock, too. In contrast, there is Claude Meylan, which has earned a spot for its ethereal, colorful skeleton watches.

More detail can be found in the first section of the book: Elizabeth Doerr, from the Quill & Pad site, takes an expert's look at the independent scene (page 14). Our Masters and Mavericks (page 20) section looks at the industry outsiders who are making waves, like Fiona Krüger and Itay Noy. For a different esthetic perspective, I have also added a brief interview with Indra Kupferschmid (page 29), a typeface specialist who wandered through Baselworld and looked at numerals. . . . Finally, rather than delve into an esoteric technical aspect of the watch, we'll explore the watch aftermarket, as it's called in the automotive industry, that is, some of the tools and articles you might need or want after purchasing a watch, notable winders for the automatics (page 30) . . .

To close, a few notes of thanks. First, to Elizabeth Doerr for her yearly contribution in the independent scene. Thanks, too, to Ginny Carroll for excellent proofing and steering production of the book. Any errors are, regretfully, mine. Gratitude is owed to our advertisers, who make *Wristwatch Annual* possible, and who thereby help keep an industry booster in people's hands. And finally, thanks to you, the reader, who has purchased this book and is hopefully enjoying it. As I mentioned already, the book you have in your hand will last a long time. Your contribution helps us maintain our independence and publish this yearly, durable tribute to beautiful timepieces.

By the way: Any prices mentioned in the book are subject to change.

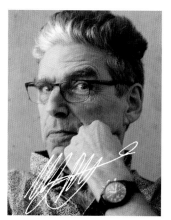

Marton Radkai

The timepiece portrayed on this edition of *Wristwatch Annual* is the **Heritage Tourbillon DoublePeripheral Limited Edition**. It comes in a 42.5-millimeter rose gold case with a height of 11.9 millimeters. The silver dial, under a double-domed sapphire crystal with antireflective coating, has a sunburst finish with milled rings on the periphery and applique rose gold indices. Inside is the automatic CFB T3000 manufacture caliber (36.5-millimeter diameter, height 4.60 millimeters), which drives the hour, minute, and small seconds hands, and the tourbillon. The rotor is of rose gold, and the engraved bridge of white gold. The mechanism features a stop-seconds function and has a power reserve of sixty-five hours.

ADVERTISEMENT

TREASURING VALUES
SINCE 1888

Swiss watch maker Carl F. Bucherer just celebrated its 130th anniversary. Its origins lie in the beautiful city of Lucerne in the heart of Switzerland, where founder Carl Friedrich Bucherer opened his first boutique for fine jewelry and watches in 1888. Being part of the Bucherer AG, the brand yet is very proud of having its own history as a watch manufacture, producing ingenious in-house made timepieces, born out of a rich heritage, driven by innovation.

130 Years of Sophistication and a bright future

___ A new limited masterpiece emphasizes this unique symbiosis of tradition and pioneering spirit, that Carl F. Bucherer is proud of: The **Heritage Tourbillon DoublePeripheral**, just launched in October 2018 in New York, not only pays tribute to the past, it marks the start of a new product range set to carry on Carl F. Bucherer's legacy: the Heritage collection.

Hallmark of Lucerne, the Chapel Bridge

"The Heritage Tourbillon DoublePeripheral Limited Edition is the embodiment of 130 years of sophistication at Carl F. Bucherer."

SASCHA MOERI, CEO CARL F. BUCHERER

___ The first model from the new collection is limited to 88 pieces in a tribute to the year in which the Carl F. Bucherer brand was founded. The new timepiece reflects the golden luster of the baroque city Lucerne, with an elaborately finished 42.50 mm, 18-karat rose gold case. The shape and the design of the dial, were inspired by various vintage watch models from the 1960s. Another homage to the brand's birthplace is the movement bridge, crafted in 18-karat white gold, which features a hand-engraved cityscape of Lucerne. It covers the entire back of the movement, with the exception of the tourbillon.

Inside the Carl F. Bucherer ateliers

ADVERTISEMENT

Heritage Tourbillon DoublePeripheral Limited Edition

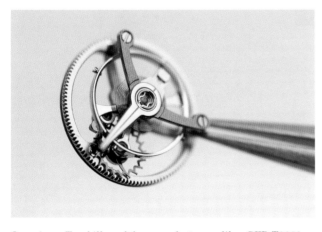

Ingenious: Tourbillon of the manufacture caliber CFB T3000

*"We are inspired by the past,
yet always look ahead into the future.
We are bound to tradition and driven
by innovation and passion.
"Made of Lucerne".
These are our core values"*

SASCHA MOERI, CEO CARL F. BUCHERER

A Pioneer of Peripheral Technology

___At the heart of the new masterpiece is the in-house CFB T3000 caliber, featuring a tourbillon and an automatic winding system that are both mounted peripherally. The tourbillon is regarded as a highlight of the art of watchmaking. In the T3000 the designers and watchmakers at Carl F. Bucherer have succeeded in giving it a floating appearance. The cage of the minute tourbillon is supported peripherally by three ball bearings – invisible to the viewer, giving it a floating appearance. The Heritage Tourbillon DoublePeripheral Limited Edition has a power reserve of 65 hours and is a certified chronometer.

___The brand thus combines retro design elements with state-of-the-art manufacturing technologies, thereby successfully bridging the gap between the past and the present, into a bright future.

#CARLFBUCHERER

THE INDEPENDENT SCENE 2018

ELIZABETH DOERR

The world of independent watchmakers continued to gain traction in 2018 as evidenced by the large spaces allotted to them at the major fairs. This community of extremely engaged and creative watchmakers continues to attract newcomers, and to enlarge its field of vision.

The watch industry's recovery seems to have really raised the spirits and energy of the independent scene. Baselworld's buzzing Les Ateliers hall was the place to be seen for the second year running, while the 2018 SIHH's Carré des Horlogers jumped to a record sixteen independent brands and creators. And a traveling exhibition of independent timepieces organized by Rome's Maxima Gallery of contemporary arts entitled "Watchmakers: The Masters of Art Horology" featured the work of the pantheon of the scene: Philippe Dufour, Kari Voutilainen, Jean Daniel Nicolas (Mr. Daniel Roth), Vianney Halter, Christophe Claret, Denis Flageollet (De Bethune), Hajime Asaoka, Romain Gauthier, Roger Smith,

The tourbillon becomes an essential esthetic focal point on Beat Haldimann's H1 Stars.

Ludovic Ballouard, Laurent Ferrier, Christian Klings, and George Daniels.

FOR WOMEN: A NEW MARKET FOR INDEPENDENT CREATORS

Romain Gauthier introduced his first automatic wristwatch in 2017 and, as one might expect, it was anything but ordinary: a 39.5 mm timepiece boasting a visible movement wound by a 22-karat-gold microrotor. For 2018, Gauthier redesigned it for women in two variations, both housed in red gold and featuring black Tahitian or Australian extra-white mother-of-pearl on the dials and as inlay on some of the visible movement components. Christened **Insight Micro-Rotor Lady**, the inner workings of the mechanics are fully on view. Gauthier has designed an impeccably finished watch whose style is efficient and elegant at the same time. But we expect no less of this brilliant engineer.

Another unexpected 2018 watch for the female wrist (but not only) comes from **Beat Haldimann**. The **H1 Stars** features a redesigned visual for the Swiss independent's famous central tourbillon watch. Haldimann is known for his elegant style of complicated watchmaking that is firmly rooted in the principles of creative, philosophically inclined, yet minimalist, haute horlogerie. The H1 Stars conforms perfectly with its two

High-end mechanics for women, by Romain Gauthier.

ARMIN STROM
SWISS WATCH MANUFACTURE

MANUFACTURE CALIBRE
ARF15 16½'''

MIRRORED FORCE RESONANCE
GUILLOCHÉ DIAL

Miami FL: King Jewelers (305) 935 4900
Denver CO: Oster Jewelers (303) 572 1111
Costa Mesa CA: Watch Connection (714) 432 8200
Toronto Canada: Ebillion Watches (416) 960 5500

Hartford CT: Armstrong Rockwell Watches & Fine Jewelry (860) 246 9858
New York NY: Analog Shift operated by Chronovillage (212) 989 1252
Wellington FL: Provident Jewelry (561) 798 0777
Jupiter, FL: Provident Jewelry (561) 747 4449
Naples, FL: Provident Jewelry (239) 649 7200

INDEPENDENT WATCHMAKING

Remi Maillat (right) and his Everywhere timepiece, which tells you when the sun is rising or setting throughout the planet.

Touch of mystery on the Sundance dial, by Louis Moinet.

dainty diamonds placed on the hands so that they glow against the now-dark dial—all of which never takes away from the real star of this show even for an instant: the central flying tourbillon.

And **Louis Moinet** entered the high-end market for women's watches in 2018 with the ultra-beautiful **Sundance** sporting a dial crafted using a still-secret method of hand-applying fanciful, bright colors that make each dial of this 60-piece edition unique.

Evocative of the Milky Way, the exceptional dials form a backdrop to the "Dewdrop" hands emanating from an openworked sun at their arbors. Bezel-set diamonds form the hour markers. With so much beauty on the dial side, it may be difficult to remember there is an automatic movement on the inside of the structural case to keep accurate time.

NEWCOMERS

Krayon is the brain child of engineer Rémi Maillat, formerly of Cartier. Maillat worked on the **Everywhere** timepiece for four years before finally introducing it: Two years were needed just for mathematical calculations and two for the development and production of the first piece. The result is a watch whose claim to fame is the ability to display the times of sunrise and sunset anywhere in the world. This is revolutionary, as the previous timepieces boasting this display could do so only in a specific location for which it was set by its maker. No wonder the Everywhere won the Innovation Prize at the 2018 GPHG.

Geneva-based **Akrivia** is the new shooting star of the independent scene. Founded and run by Rexhep Rexhepi, the boutique brand has thus far been known for watches with a distinct "open" look that showcases their beautifully finished mechanics. But in 2018, Rexhepi introduced a brand-new style for his little company, one that veers off the path he has already beaten. The **Chronomètre Contemporain** is the first timepiece of the new Rexhep Rexhepi collection of chronometers outfitted with a manual-wind movement and an enamel dial, and available in two variations. As the name says, its design is also quite pleasingly contemporary.

Akrivia's Chronomètre Contemporain hints at classicism.

INDEPENDENT WATCHMAKING

Independents are free from brand constraints.

Czapek's Aqua Blue model features hand-made "ricochet" guilloché.

Czapek & Cie. (see page 125), the product of three impassioned watch industry professionals, has mastered the art of great quality and fair pricing in the luxury segment. And, as the variety of extensions and variations coming out in 2018 shows, the brand has found its audience with grand feu enamel dials, manufacture movements, and moderately sized (practically unisex) cases. The latest is the addition of hand-guilloché dials for the **33s model** in a variety of funky colors with cool names like "guilloché sea salt," "black prince," and "rhubarb red."

A.H.C.I.

The heart piece of the independent scene remains the A.H.C.I. (Académie Horlogère des Créateurs Indépendents/Horological Academy of Independent Creators), a group of independent creators founded in 1985. The way of these individualistic watchmaker-inventors is indeed anachronistic, and though the products that emerge may not be everyone's cup of tea all of the time, they do attract the attention of collectors of rare taste who follow not only the horological escapades of these 30-odd extraordinary men—

AHCI's booth at Baselworld, where the creativity breaks through the norms.

MARTIN M - 130 THE MOST ADVANCED AIRCRAFT OF ITS TIME

MADE IN MARYLAND

MARTIN M-130 THE MOST DISTINCTIVE TIME PIECE OF OUR TIME

MADE IN MARYLAND

Towson Watch Company

www.towsonwatchcompany.com

INDEPENDENT WATCHMAKING

Lang & Heyne's Anton could be worn upside down.

with a few women—of varying age and nationality, but also the passion and personality that goes into each extremely limited timepiece.

Marco Lang and his boutique brand **Lang & Heyne**, at home in the cradle of Saxon watchmaking, Dresden, came out with an absolutely stunning rectangular watch with a new shaped caliber in 2017: Georg measures 40 × 32 mm and is as svelte as can be at 9.4 mm in height, thanks to beautifully finished, manual winding Caliber VIII. Its beauty is capped by a simple enamel dial as the cherry on top. Baselworld 2018 saw the creative watchmaker returning with a version of Georg called **Anton**, featuring a flying tourbillon at 6 o'clock in a case with precisely the same measurements despite Caliber XI's very visible and complicated addition.

Self-taught Japanese horological virtuoso **Hajime Asaoka** has been associated with the A.H.C.I. for several years, but it is his appropriately named **Chronograph** that catapulted him into the limelight of the select circle of connoisseurs who appreciate high-quality work by independent watchmakers. After working for years developing multiple tourbillon movements as well as Tsunami, a time-only movement with a massive 16 mm balance wheel, Asaoka set his sights on the column wheel chronograph. Chronograph is based on the Tsunami movement with its oversized balance wheel, though much of the new caliber needed to be redesigned for new hand locations and to make space

INDEPENDENT WATCHMAKING

for the chronograph mechanism. Asaoka's main goal—aside from creating his own example of the beloved complication—was to place the chronograph mechanism on full view on the dial side of the 38 mm stainless steel case, a size that perfectly complements today's feel for vintage.

President of the A.H.C.I., **Konstantin Chaykin**, resides in Moscow. But that does not stop him from creating some of the world's best and most complicated timepieces. However, it was 2017's fun **Joker** that has brought him a great deal of attention: It is an ETA-driven wristwatch with an inventive and funky clown face whose eyes and mouth move along with the displays. Topping that, Chaykin released not only several more Joker versions in 2018 but also a "joint venture" automaton watch that features Chaykin's Joker on one side and co-A.H.C.I. founder Svend Andersen's famous dogs-playing-poker *montre à tact* automaton on the other side. Simply called the Automaton Joker, Chaykin's clever press release exclaims: "The Joker had to play poker with one of the dogs!"

Elizabeth Doerr is a freelance journalist specializing in watches and was senior editor of Wristwatch Annual *until the 2010 edition. She is now the editor in chief of* Quill & Pad, *an online magazine that keeps a watch on time (www.quillandpad.com).*

Asoaka's compact chronograph movement is visible from the dial side.

Konstantin Chaykin Joker has received a movement-side automaton by Svend Andersen.

MASTERS AND MAVERICKS

THE OUTSIDERS

MARTON RADKAI

Diversity is one of Nature's ways of strengthening its many ecosystems by increasing what could be called the collective intelligence. The same could be said of many industries like watchmaking, where outsiders arrive and pollinate the system with new ideas and challenges, preventing it from becoming too complacent.

Ask any cross-section of specialists in the industry about their favorite personalities and you will inevitably hear a host of great names, like Wiederrecht, Prescher, Sarpaneva, McGonigle, Gerber, Beat Haldimann. . . . Everyone has their favorites. And then there are the likes of Maximilian Büsser, successful founder of, and facilitator at, MB&F, which has been pushing and pulling the envelope of watchmaking for over ten years now. Büsser himself, while an engineer, is not a watchmaker. Ever since his days at Harry Winston, though, he's been challenging young watchmakers to excel.

Büsser and friends, and many others, steadfastly follow a personal creative vision. In doing so, they venture into spaces that defy conventions and invite others to do the same. One person from outside the industry who took the step is Fiona Krüger. As a Master's art student at Lausanne's ECAL, she had been tasked with designing a watch. One of the jury members for her project was Max Büsser. "He asked pointed questions and seemed genuinely interested," she remembers. "I didn't know what he meant to the industry, otherwise I would have been terribly nervous."

Krüger was new to watches, but a visit to the Patek Philippe Museum in Geneva opened her eyes to the myriad possibilities

Together with watchmakers, Max Büsser creates kinetic art that tells time and sparks the imagination.

MASTERS AND MAVERICKS

buried in such tiny mechanisms. "I saw watches of all different shapes and sizes," she recalls. "I saw that I was not just limited to a watch being round and flat, it could be anything that I thought of." It was the birth of what would become her trademark skull-shaped watch.

HEAD START

The skull project quickly became the Skull Watch, which has given birth to all sorts of variations large and small, and ultimately led to a career creating watches. The first was somewhat austere, in steel ("for the mechanical esthetics," she notes) and the second in black and white, adding graphical elements that highlighted the shapes. Then came the larger colorful pieces, for which she channeled her youthful years in Mexico. "All the skulls are inspired by the Santa Calavera (the holy skull), because none of them look mean or sad," she points out. "They actually celebrate life, with thoughts of time and mortality."

While developing her brand, Krüger was also exploring the relationship with a powerful industry and its many opportunities and

Notebooks full of ideas.

coming to grips with what is essentially a different palette. "There's the added aspect of movement in the watch, which is like a toy, and with the animation you can tell a story," she points out. "And there are all these creative techniques, the métiers d'art, enameling, engraving, guilloché. It was like being a kid in Willy Wonka's candy factory."

Like many watchmakers, Krüger also begins with an idea and a blank page that gradually fills up. Her notepads are filled with the genesis of creations, quickly sketched, but remarkably complete. The hard part, however, is transferring a large drawing to the tiny surface of a watch. It's only then that she goes to meet the artisans who are going to realize the work in 3D. "I tell them what I would like, we have collaborative discussions," she says. "My job is to bring them to the table and together we create something new."

Trust the remarkable Aurélie Picaud, director of watches at Fabergé, to pick out Fiona Krüger and enameler Anita Porchet for the design of the Lady Libertine III, depicting Mozambique's red soil (in rubies) where it meets the blue sea. It was a fruitful confrontation between the artist with an affinity for fuzzy brush strokes, and the enameler striving for razor-sharp contours. "Wouldn't it be amazing if you could turn an enamel artisan into Jackson Pollock?" Krüger thought. The white caps on the sea demanded daubs of silver enamel. "For Anita, it was like an imperfection," Krüger remembers. "She had to figure out how to do those brush marks without making it look messy."

Sea, sand, and red earth on a Fabergé dial: Painter Fiona Krüger and enameler Anita Porchet joined forces and broke down barriers.

SEIKO *Laco* ORIENT

CITIZEN  TIMEX

LONGISLANDWATCH.COM
AFFORDABLE QUALITY TIMEPIECES ONLINE
SINCE 2003

MASTERS AND MAVERICKS

Time explodes: the Mechanical Entropy model from the Chaos collection.

The Skull is itself a symbol of finite time.

THE ARROW OF TIME

Perhaps unconsciously, she was already triangulating her next project. There was an almost spiritual encounter with physics on viewing *The Arrow of Time*, a BBC documentary by Brian Cox. Add a few ideas from Dada artists, who would throw random items into the air and create a kind of aleatory art out of where they fell. And finally, the key moment, seeing Cornelia Parker's installation "Cold Dark Matter," the reproduction of the instant when a garden shed explodes. "Everyone will recognize something exploding, from comic books, or the Big Bang," she says, "so maybe I could blow up time?" It was the birth of the Crash.

She chatted about the idea with Jean-Marc Wiederrecht, founder of Agenhor, who enthusiastically endorsed the project. The key was to make a skeleton watch, but to cut away the dial so as to see the movement of the balance wheel. Krüger worked with an engineer at the company's premises in the Geneva suburb of Meyrin. "It was like two brains working together," she recalls, "his was the technical one, mine the design."

The result was Entropy I of the Chaos collection. It's a skeleton with jagged breaks in the bridge. The hour-wheel may as well be cracking up, the indices are splinters, each one different. Colors have been added to the metal to give the models a youthful, trendy look. The esthetics hint at comic books, which is what Fiona Krüger was looking for. "I needed some sort of symbolism or imagery that is universally recognizable," she says. "If somebody looks at the watch without going any deeper they will already have a relationship with the object, but if they want to discover more they can figure out the different layers behind the watch."

TIME SHARE

One lesson from Fiona Krüger's watches, be it the Entropy or the Skull, is that bolting an attractive dial to an industrially produced movement may not be enough. A watch must in some way strike up a conversation with its owner, or tell a story. And this is precisely what another outsider has sought and found in making watches . . .

Itay Noy, independent timepiece maker, as he calls himself, makes watches that connect with their owners at various levels. The City Squares, one of his earliest pieces, features a dial made to order using a section of whatever city the owner chose. The Mask is a face you'll look at when reading the time.

From an idea to reality: Outsider Fiona Krüger begins with visuals.

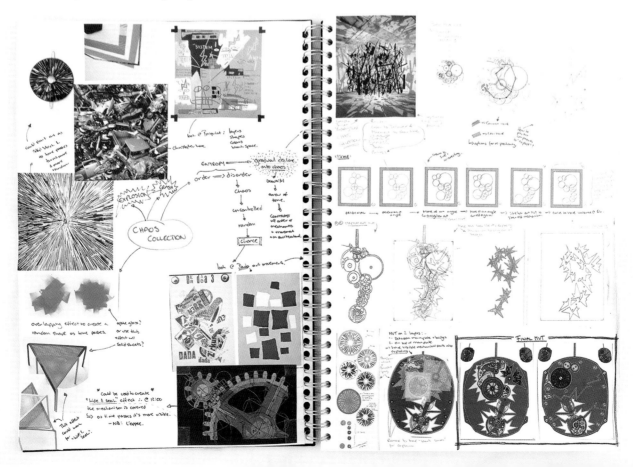

MASTERS AND MAVERICKS

On the more recent pieces, Time Tone, the hours are arranged around the dial on a disk and appear as circles of slightly different tone. Again, the owner chooses which color will represent the hour hand.

Noy, who hails from Israel, studied jewelry and object design, and did a master's in industrial design, all of which provided him with the tools to give physical expression to his artistic ideas. And these are not only legion but, more often than not, very original. Many of his watches have been featured in special exhibitions and museums, and in 2007 he clinched the coveted Andrea M. Bronfman Prize for Contemporary Crafts in Israel.

His love of watches and clocks was sparked during a job at a watch store when he was just 21. Like Fiona Krüger, he too noted the fascinating dichotomy between the outward appearance and the technical aspects: "I have designed many different kinds of objects," he points out, "but watches and clocks are unique, because they contain the heart inside them, they can come alive and stop." It's a vision of watchmaking that reveals itself in different ways, often by a mysterious interaction between mechanism and the dial. It's quite obvious in the Open Mind, where an opening in the skull on the dial reveals the ticking escapement. The Time Tone mentioned above also plays with the idea of moving layers, quite literally, and so does his latest, the Full Moon. Here the moon's or the days' progress through the month can be physically seen through tiny cuts in the surface of the dial.

THE ANTI-TRENDER

Over time, Noy has become adept at creating his own modules and particularly what he calls his "dynamic dials." Barring his association with a small Swiss movement maker, IsoProg, to create his own IN.IP13 caliber, his involvement with watch industry insiders, unlike Fiona Krüger's, has been fairly limited. This may account for the singularity of his pieces. There was something immanently dialectical in his art, as if two watchmakers, each with his own specific vision, were at work, each confronting and challenging the other to find a synthesis. "I feel like I am an artist that makes functional objects, and a watchmaker making conceptual art," Itay Noy told me. "I go through a long process of searching until I find something that intrigues me. I try to think of things that will be relevant in many years' time." And one way to catalyze ideas is to check what big brands are doing at Baselworld, and then go in a different direction. "I am not a follower," he says, tersely.

SHEER ENERGY

La Chaux-de-Fonds, Switzerland, was destroyed by fire in 1794 and rebuilt for watchmakers along the best lines for natural lighting. This particularly impressive combination of form and function earned the city and neighboring Le Locle a place on the UNESCO World Heritage list. But you'll be hard-pressed to find that lumbering name on a watch dial. Unless you purchase a duManège.

Founder and CEO, for lack of a better word, is Julien Fleury, a young *chaudefonnier*, a graduate of the local art school, who studied graphic design, trained as a jeweler, and one day decided to put together a watch honoring his hometown. It started as a little idea, and grew into a clear and present goal. And because it was to be assembled locally using local components and local suppliers, people he calls "des copains," buddies, he decided to put "made in La Chaux-de-Fonds" on the dial as a unique proposition. As for the brand name duManège, it refers to an old riding school built in the city in 1855, which later provided families with low-cost housing and became an architectural ideal of community living.

Fleury is not the complete maverick, since his own DNA is steeped in watchmaking on both of his parents' sides. One grandfather owned a dial factory; the other ran a watch manufactory. Still, he actually comes from outside the box and has managed to develop some special ideas, like regional patriotism, to which he adds his graphical sense. "I am mainly inspired by architecture and by luxury cars, especially the old ones, and by what has been done in watchmaking," he noted during an interview. "I am particularly drawn to the 1930s, '40s, and '50s, especially military watches that have a very readable display and clean dials."

His first timepiece, the Exploration, had all those elements, including a fairly edgy 44.5-mm, multipart steel case (titanium and gold also available) with black ceramic inserts. The coolness was in the details and revealed his close relationship to the horological world—the satin-brushed surfaces, contrasting polished beveling, openworked hands, markers that shorten on the left to

Itay Noy watches dialogue with their owners.

Time Tone: What color is your hour?

Julien Fleury, founder of duManège watches.

GREUBEL FORSEY
ART *of* INVENTION

GMT EARTH

Limited edition
33 Timepieces in white gold

WWW.GREUBELFORSEY.COM

FOR INFORMATION · TIME ART DISTRIBUTION LLC
Phone +1 212 221-8041 · info@timeartdistribution.com

MASTERS AND MAVERICKS

Customizing dials of the duManège Heritage line.

The Manège, La Chaux-de-Fonds' old riding school and apartment block.

make way for the subdials. He even added a power reserve indicator—the two spring barrels of the customized Technotime automatic movement can drive the watch for up to 120 hours—and a retrograde date hand that takes up about one-third of the dial.

The Exploration, which was crowdfunded, has become his mainstay in some ways. Yet anyone can strike the right chord with buyers the first time around. That is beginner's luck. The important thing is the next step, and even the one after that. For his second chapter, Fleury shifted gears by turning to esthetics and métiers d'art, the various decorative crafts that he could source in his region. As a tribute to his region, he came out with the Heritage line, which is divided into several categories. One is a simple, affordable watch at a very affordable price. The other wanders into the fields of art and jewelry. The "Sapin" features *champlevé* enamel fir trees, a unique, Swiss Art Nouveau element developed by art students in La Chaux-de-Fonds. Some dials are totally pure, made of colored stone or meteorite. Some of his clients request unique, customized dials that he can have hand-painted, while others are simply sold off the shelf.

"My success is quite modest," he points out. "I'm not trying to sell quantities but rather quality. So I'm very proud of each and every timepiece that is released by my workshop.

Fleury is young, and looks even younger. His innate determination is written on his face. In his free time, he is a marathon runner, which may explain the wiry body, but also many of his other abilities, such as resilience and organization. Not all of his projects have been successful, so he has experienced defeats. What sportsman hasn't? "Most people were rather impressed that I should throw myself into such an adventure, especially my friends who were in watchmaking, since they know of the constraints and difficulties that one has to start a watch brand," he told me, adding as an afterthought, "but my family was very proud."

SYNTHESIS

The French have a saying: "Impossible is not French." In reality it may be wishful thinking, but it does ring true for the watch industry. Think MB&F, or HYT, Urwerk, or even some of the big brands with armies of top watchmakers who produce almost impossibly complicated pieces. The point is this: No matter how off-the-wall, an idea will, at some point, be greeted with a "let's figure out how to do this." Watchmakers, the best ones, are like that. And so the outsider aiming to realize his or her horological vision will find assistance. "One always comes across new and interesting stories, and we can learn from the specific competencies of the 'externals,'" says Marc Jenni, one of the very talented independents, who joined Austrian Robert Punkenhofer to rehabilitate Carl Suchy & Söhne, an old Viennese brand that once delivered watches to the Habsburg court.

Max Büsser, who has been creating watches for twenty-seven years, has come to grips with his being an outsider and what he, an engineer by training, describes as a feeling of inferiority. He's hands-on with anything that touches the creative process, he points out, and as hands-off as possible for the engineering. The outsider status is, in his words, one of his greatest assets."

"I am not bogged down by the 'dos and don'ts' taught in watchmaking schools," he told me recently. "Take the flying balance wheel on our Legacy Machine. Why never show the most beautiful part of a movement? Because a watchmaker's first reflex will be to keep the escapement and balance wheel protected in the movement. Of course, our flying balance is much more complicated to engineer, machine, and regulate but it is so beautiful . . ." And that sense of beauty, in engineering as in esthetics, is what truly drives the industry.

Julien Fleury's first product: the DM Exploration.

An MB&F-L'Epée collaboration: The Fifth Element, including an extraterrestrial on the balance wheel.

TO THE LETTER

There are many different ways of looking at a watch and appreciating it. The details that catch the eye, though, can be very subjective. Indra Kupferschmid, a typographer and professor of communication design, has little to do with watches. But she was involved in the creation of the Helvetica Watch by Mondaine, and once strolled through Baselworld noticing things. An interview was a must . . .

MR: First and foremost: Do you have favorite typefaces?

IK: Yes, but for different things. It's like choosing a piece of clothing for some activity. So there's a typeface for a poster, and there's another for, say, a watch or a bar of chocolate.

MR: Typeface should be discreet, but in the field of marketing, it must also attract. Isn't that something of a contradiction.

IK: Everything moves between those two poles. On one hand, you want optimal readability, with a low-impact font, but you also want to stimulate awareness. And between the two there are a whole bunch of shadings. There are typefaces that have such a personal identity they cannot appear small, because you will lose all the detail. Then there are the boring typefaces that will not attract anyone.

MR: So what about the dial of a watch, a tiny little thing that has to do a little bit of both?

IK: On the dial, you're not dealing with long-distance copy that has to be read. It's more about identity, design ideas, branding, or how the brand is being positioned vis-à-vis the customer. This all has to be contained within the numerals on a watch.

Avante Garde

MR: What about the logos?

IK: Numerals and logos are not usually created from a font family, they are drawn as logos so that only those words have to look good, and the rest doesn't matter. Even readability is not that important, because you recognize a logo as an image. You do have to make sure the numbers are all separate, though.

MR: How could you tell whether the numbers have been well drawn or not? Is it in the eye of the beholder?

IK: A numeral can suggest an austere watch, or a cool watch, or a vigorous macho type. Or am I more friendly and accessible? It's often unconscious. What I did notice was that many of the companies imitate each other. Also, they sometimes use typefaces that are not as well drawn or in character as they could be. Maybe they should commission a typeface designer to make twelve numbers that they would then put on their watches.

Helvetica

MR: What about the vintage trend? Watch brands are sometimes using typefaces that are no longer really contemporary.

IK: In typography, there's a tradition of updating typefaces. But some typefaces really reference a certain epoch, like Avant Garde for the 1970s. They appeared on advertising, on packaging, all over the place. If you use these typefaces today you will recall the period in which they were used . . . Helvetica recalls the 1950s and 1960s, and that Swiss style. Futura, however, was popular in the 1920s and 1930s, but was used a lot in the 1960s, for instance, in the iconic VW Beetle advertisements.

MR: Would you have any advice for the watch industry?

IK: It would be nice if typefaces were not considered as a kind of afterthought. It's the same thing with big, beautiful buildings: Millions invested, ten years' construction work, and then the writing on the wall is in Arial . . .

Futura

WATCH TECH

THE WATCH AFTERMARKET

MARTON RADKAI

For the past few years, *Wristwatch Annual* has taken a look at technical aspects of the watch, such as crowns, shock absorbers, chronographs, mechanisms, and so on. This year, it's the turn of the aftermarket, a term usually reserved for all the frills, bells, and whistles that are offered to the car owner after the sale has been made.

"There were guys there wearing gold Hublots and asking if we were selling speakers," chuckles Simon Wolf, CEO of Wolf 1834, recalling a trade fair where he presented his wares. His mirth is not meant as disparagement to luxury watch owners, but rather to express the sheer size of his potential market. And in almost Bernie Sanders fashion, he begins rapidly crunching the numbers: 22 million watches made annually by the Swiss, many more million made in Hong Kong and China and elsewhere, and most of them are *automatics*, leaving millions

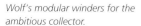

Wolf's modular winders for the ambitious collector.

Simon Wolf, minding the winding around the world.

of watches that need to be kept wound and on time.

Wolf sells watch winders. They come in many shapes and sizes, from single units for the small-time collector to modular systems that can be daisy-chained while the collector's collection grows and grows. Some have batteries; others can be plugged into the grid. He also sells chic sets with drawers or trays for the nonautomatic watches and jewelry, and has a range of cases, sheaths, and other containers for those traveling with precious cargo, including high-end neckties.

The main reason for having winders, of course, is if you own automatic watches, because, according to Wolf, we sedentary folk simply don't walk around enough to keep an automatic properly wound. The consequences can be quite serious. "Modern watches are very complicated," he explains. "If you have a perpetual calendar and you make a mistake setting it after it stopped,

STURMANSKIE
THE FIRST IN SPACE

Only one watch company holds the distinction of being the first watch in space - Sturmanskie

It is rare when the phrase "own a piece of history" has real meaning. In the case of the Sturmanskie Yuri Gagarin commemorative edition it is not hyperbole.

Inspired by the original watch Gagarin wore during his historic 1961 flight, the watch is the only timepiece in the world authorized to use Gagarin's likeness. Hand assembled in Moscow, the watch uses the same Poljot movement as the original and truly brings the history of space travel to your wrist.

Detente Watch Group, 244 Upton Road, Suite 4, Colchester, CT 06415.
877-486-7865. sales@detentewatches.com

WATCH TECH

Wolf's roll case for your best pieces . . .

If you've got it, show it!

it can easily cost you $5,000 or $10,000 to fix it."

While the winders may look simple on the outside, they are fairly complex gadgets, with parts and algorithms often protected by patents to keep out the competition. The Wolf winders, for example, have a special patented rotation counter, which ensures that each watch is getting the right number of turns. This information is often provided by brands, since each watch model or movement is a bit different. Achieving the ideal tension involves not only tightening the mainspring but also periods of rest and unwinding. So winders generally come as programmable or preprogrammed units. What the uninitiated think looks like a loudspeaker in the Wolf portfolio is actually the knobs framing an LCD display to set the winding rate. There are apps, as well, that allow the user to program the number of turns to give their watches. "It allows you to choose from one turn per day to 1,999, which you would never use," says Wolf. "Why do we do it? Because if someone wants to turn their watch 1,999 per day, they can."

SAFETY FIRST

Not surprisingly, a company offering watch winders also has an interest in selling safes, since watch owners will need a place to store their valuable collection. At Baselworld 2018, Wolf showed me the latest in this

The Wolf safe: solid insurance.

To see the poetic beating of Heart's Passion®
please visit paulforrestco.com.

IT'S NOT ABOUT TIME, IT'S ABOUT PASSION
MANUFACTURED IN SWITZERLAND.

828 W. Indiantown Road · Jupiter, FL 33458
(561) 747-4449 · www.providentjewelry.com

WATCH TECH

Erwin Sattler's safe with a weather station and clock.

Sattler's cool school winder and display case.

technology, a blunt black exhibition unit—custom-made, of course, to be fitted into a person's home or office. This one, he explained, had been certified by the American Underwriters Laboratories. It was subjected to 1700°F fire (temperature inside doesn't get beyond 300°F), dropped from a certain height, baked again.... It is bulletproof and waterproof. Entry-level price is about $1,200.

It wasn't the first time I had found the combination safe-watch winder, by the way. Visiting the *manufacture* Erwin Sattler in Munich, besides the company's remarkable standing clocks and table clocks, all very modern, I was shown a range of very chic, contemporary watch cabinets the company makes in collaboration with safe-maker Stockinger and the watch-winder manufacturer Beluwo. Some, I was told, have doors made of 18-millimeter bulletproof glass—in case you are attacked by armed robbers, your collection is safe.

The needs of collectors, though, go beyond just mechanics, and they are as diverse as human society itself. Like any item of value, watches are meant to be displayed; hence, companies selling watch winders must pay attention to interior decoration. Wolf, for instance, has a very modern, almost trendy feel, as if some of the products are extensions of the electronic devices that occupy human space. Sattler, on the other hand, has opted for techno cool, with lots of glass and metal, like a skyscraper.

Then there is Orbita, with a portfolio of cases and cabinets with a more classical flavor that emphasize solid craftsmanship quite in tune with the vintage trend that is sweeping the industry. One of its more recent products is a three-watch wooden trunk made in collaboration with Gerstner Tool Chests of Dayton, Ohio—definitely a piece you can leave in any living room without fear of stylistic clash, since it is timelessly natural.

And if that appears too organic, Orbita also has a fairly large portfolio of cases that will appeal to more contemporary tastes.

COLLECTOR'S ITEMS

Watch winders may be beyond the average wallet's grasp, of course, and if your collection of perpetual calendars or moon phases is not that extensive, might be avoidable. But asking around in the community of watch journalists and collectors and makers generated a list of items that one should have and some thoughts about owning a watch. Friend of *Wristwatch Annual*, journalist, watch owner, and expert Elizabeth Doerr was the first to respond: "A safe and a good watch case with a soft bed and [that] separates the watches so they don't scratch," she told me in a terse but impactful email. And, she added, almost as an afterthought, a strap-changing tool, too (see separate box on following page).

The Orbita winders: from rural charm to hipster chic.

Retro Zeitgeist

Known by those who dare depths.

If you're looking for an automatic dive watch without having to go overboard, consider the Hager Aquamariner.

The Adventurer's Watch...
in or out of the water

HAGER

www.HagerWatches.com

WATCH TECH

STRAPPING WATCHES

One of the more frequent reasons to visit the watch shop is to change straps. A spring bar tool (the official name for the strap changer) costs in the region of $20 and will save time and money. The colleagues at *Hodinkee* suggest Bergeon, a venerable *horlogerie* company from Le Locle, as a source, and it is indeed quite a treasure trove of watch paraphernalia, including special kits for the watch collector starting at around $200.

Having a spring bar tool might also encourage the watch collector to experiment with his or her watches by changing straps. If you own one of those 45- to 48-millimeter behemoths, for example, you may want to harmonize the look by purchasing an extra bund strap, where the watch rests comfortably on a leather flap rather than directly on your wrist (try StrapsCo in Champlain, New York, for a broad selection).

For an especially rugged look, as befits some of the more muscular watches on the market, there is always SNPR Leather Works, where you'll find straps for a range of major brands. Designer and leather craftsman Joe d'Agostino will be happy to custom make one for you, which is one way to turn a regular timepiece into a personal experience. He also gives his straps a special treatment: "My leather has oils and waxes introduced during the tanning process so they are pretty impervious to the elements," he told me. "They will develop a nice natural patina over time."

The Fin Watch Strap Company in Helsinki also has a very creative approach to watchstrap making. For a feeling of robustness, for example, they make 4-millimeter-thick straps that might impart a certain Viking feel. Also, in addition to leather from the venerable Horween tannery in Chicago, they use reindeer leather and a very attractive military-grade leather known as "ammo leather," which is taken from ammunition pouches, belts, and bags, and has a natural patina.

And if you feel like turning your watch into a more trendy experience, there are many options in any shop selling straps. But the most chic and trendy place must be ABP in Paris, which operates out of a little shop in the 4th Arrondissement. For the past 20 years, they have been exploring all sorts of avenues for the watch fan, with leathers of different colors and other materials. Their catalog is designed to turn any wristflower into a radiant prince or princess at the ball.

A final word about leather straps: Avoid getting them wet, and every two months, use a leather care conditioner, such as Obenauf's or Leather Honey conditioner, to revitalize them and keep them supple. An alternative is simply olive oil.

Watches need attention, and it's not only the leather straps. Rubber straps, too, can be in need of some cleaning and reinvigorating. And there is the watch itself, which needs polishing and cleaning, especially if you are going to exhibit it at an event or a trade fair, or just show it off to friends. If the watch case has scratches that are quite noticeable, please visit a specialist, who will have the right pastes and sands and technique. On the other hand, no one is preventing you from doing some general maintenance on your timepieces with a soft toothbrush, special cloths, and gentle products. But you must be careful, as gold, for instance, is quite a soft metal. The place to go is Wrist-Clean, a hub of watch surface maintenance.

Top: Ammo leather in its original form.
Below: The spring bar tool: for DIY strap-changes.

The color of leather: ABP in Paris boosts even the most banal timepiece.

WILLIAM HENRY

WILLIAMHENRY.COM

WATCH TECH

Use gloves when manipulating watches.

A $70 demagnetizer will get your watch working again.

GOING PUBLIC

Whenever I'm preparing to go to a trade fair, or a flea market, or a brand visit, or some other horological jamboree, there are two items I never forget to take along. The first is a pair of light cotton jeweler's gloves. They are not expensive, so buying a few pairs is a good idea in case you are showing your collection to a friend at home as well. They prevent you from leaving your fingerprints on the watches, especially the crystal, which is not pleasant for the next person. Furthermore, the oils in your skin are not entirely harmless and can, over time, etch the metal a bit. Anyone showing you a watch will be thankful.

The other tool is a jeweler's magnifying glass. They are cheap as well and readily available. Some can be easily clamped into your eye socket, leaving your hands free to manipulate the watch. This will also make you look like a serious watch fan. I always bring mine along with me, even though brands always provide one. What you stare at for long minutes, while the owner of the watch counts to 100 and back, is up to you.

I tend to look at the finishing, the clean lines, the balance of colors, the typography of the numerals, the quality of an enamel dial, the incredible ballet of a complicated watch's workings. You're also looking for flaws that might suggest sloppy manufacturing. Staring too long at the pallet fork going back and forth can have a hypnotic effect, so remember to detach your eye at some point.

THE WATCHMAKER'S TIP

Finally, one of the more unusual devices you may want to keep handy to avoid costly and useless repairs is a demagnetizer. "Super strong magnets have never been more common, whether it's bag closures or electronic devices," says Bill Yao. A sudden change in the accuracy of your timepiece could result from a number of issues, but magnetism has been the first we check for and one of the few things you can remedy yourself." Yao, who runs the company Mk II (see page 216), knows what he is talking about and suggested spontaneously purchasing the K&D demagnetizer, available

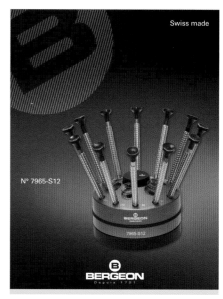
These screwdrivers are always handy.

for about $70 or so. And like many of the products and tools suggested here (except the winders), it, too, can be used for other items in your home, such as accidentally magnetized screwdrivers.

A magnified look at a watch will reveal both the beauty and the flaws.

A. LANGE & SÖHNE

In summer 2015, A. Lange & Söhne inaugurated a new manufacture in Glashütte. It was a big enough event for German chancellor Angela Merkel to attend. It was a particularly nice capstone for the work of Walter Lange, who died in January 2017 after a life of outstanding entrepreneurship. That has not stopped the brand from sticking with its unique esthetic and mechanical codes.

Lange, as the company is known for short, exemplifies the steady, careful, and effective way Germans do business. On December 7, 1990, on the exact day 145 years after the firm was founded by his great-grandfather Ferdinand Adolph Lange, Walter Lange re-registered the brand A. Lange & Söhne in its old hometown of Glashütte. Ferdinand Adolph had originally launched the company as a way to provide work to the local population. And shortly after German reunification in 1990, that is exactly what Glashütte needed as well.

The company quickly regained its outstanding reputation as a robust innovator and manufacturer of classically beautiful watches. A. Lange & Söhne uses only mechanical, manually wound *manufacture* calibers or automatic winders finished according to the highest Glashütte standards. The movements are decorated and assembled by hand with the fine adjustment done in five positions. The typical three-quarter plate and all the structural parts of the movement are made of undecorated German silver; the balance cock is engraved freehand. The movements combine equal parts traditional elements and patented innovations, like the Lange large date, the SAX-O-MAT with an automatic "zero reset" for the seconds hand, or the patented constant force escapement (Lange 31, Lange Zeitwerk).

The entry-level family is the Saxonia, while the Lange 1, introduced in 1994, is considered the company leader. Two leading novelties in 2017 were the stunning Tourbograph "Pour le Mérite," combining a flyback chronograph, a perpetual calendar, and a tourbillon, and the perpetual calendar in the 1815 line. This tough bar was surpassed again with the Triple Split chronograph presented in 2018.

Lange Uhren GmbH
Ferdinand-A.-Lange-Platz 1
D-01768 Glashütte
Germany

Tel.:
+49-35053-44-0

Fax:
+49-35053-44-5999

E-mail:
info@lange-soehne.com

Website:
www.lange-soehne.com

Founded:
1990

Number of employees:
750 employees, almost half of whom are watchmakers

U.S. distributor:
A. Lange & Söhne
645 Fifth Avenue
New York, NY 10022
800-408-8147

Most important collections/price range:
Lange 1 / $34,700 to $335,800; Saxonia / $14,800 to $62,100; 1815 / $24,800 to $236,900; Richard Lange / $33,200 to $231,500; Zeitwerk / $77,700 to $128,100

Lange 1
Reference number: 191.039
Movement: manually wound, Lange Caliber L121.1; ø 30.6 mm, height 5.7 mm; 43 jewels; 21,600 vph; swan-neck fine adjustment, hand-engraved balance cock, 8 screw-mounted gold chatons, parts finished and assembled by hand; 72-hour power reserve
Functions: hours, minutes, subsidiary seconds; power reserve indicator; large date
Case: white gold, ø 38.5 mm, height 9.8 mm; sapphire crystal; transparent case back; water-resistant to 3 atm
Band: reptile skin, buckle
Price: $34,700
Variations: yellow gold ($34,700); pink gold ($34,700); platinum ($49,500)

Lange 1 Moon Phase
Reference number: 192.029
Movement: manually wound, Lange Caliber L121.3; ø 30.6 mm, height 6.3 mm; 47 jewels; 21,600 vph; 8 screw-mounted gold chatons, swan-neck fine adjustment, hand-engraved balance cock, parts finished and assembled by hand; 72-hour power reserve
Functions: hours, minutes, subsidiary seconds; power reserve indicator; large date, moon phase
Case: white gold, ø 38.5 mm, height 10.2 mm; sapphire crystal; transparent case back; water-resistant to 3 atm
Band: reptile skin, buckle
Price: $41,700
Variations: pink gold ($41,700); platinum ($54,700)

Grosse Lange 1
Reference number: 117.028
Movement: manually wound, Lange Caliber L095.1; ø 34.1 mm, height 4.7 mm; 42 jewels; 21,600 vph; 7 screw-mounted gold chatons, swan-neck fine adjustment, hand-engraved balance cock, parts finished and assembled by hand; 72-hour power reserve
Functions: hours, minutes, subsidiary seconds; power reserve indicator; large date
Case: white gold, ø 40.9 mm, height 8.8 mm; sapphire crystal; transparent case back; water-resistant to 3 atm
Band: reptile skin, buckle
Price: $42,300
Variations: pink gold with light dial ($42,300); platinum ($56,400)

A. LANGE & SÖHNE

Lange 1 Time Zone
Reference number: 116.032
Movement: manually wound, Lange Caliber L031.1; ø 34.1 mm, height 6.7 mm; 54 jewels; 21,600 vph; 4 screw-mounted gold chatons; zone time (hours, minutes) with day/night indicator and city ring, forward-switchable; 72-hour power reserve
Functions: hours, minutes, subsidiary seconds; 2nd time zone; large date; power reserve indicator; day/night indicator for both time zones
Case: pink gold, ø 41.9 mm, height 11 mm; sapphire crystal; transparent case back; water-resistant to 3 atm
Band: reptile skin, buckle
Price: $49,400
Variations: white gold with light dial ($49,400); platinum ($64,300)

Lange 1 Tourbillon Perpetual Calendar
Reference number: 720.038F
Movement: automatic, Lange Caliber L082.1; ø 34.1 mm, height 7.8 mm; 76 jewels; 21,600 vph; 1-minute tourbillon with movement side stop function, off-center balance, 6 screw-mounted gold chatons; 50-hour power reserve
Functions: hours, minutes, subsidiary seconds; day/night indicator; perpetual calendar with large date, weekday, month, moon phase, leap year
Case: white gold, ø 41.9 mm, height 12.2 mm; sapphire crystal; transparent case back; water-resistant to 3 atm
Band: reptile skin, folding clasp
Price: $335,800
Variations: pink gold ($335,800)

Small Lange 1
Reference number: 181.037
Movement: manually wound, Lange Caliber L121.1; ø 30.6 mm, height 5.7 mm; 43 jewels; 21,600 vph; swan-neck fine adjustment, hand-engraved balance cock, 8 screw-mounted gold chatons, parts finished and assembled by hand; 72-hour power reserve
Functions: hours, minutes, subsidiary seconds; power reserve indicator; large date
Case: pink gold, ø 36.8 mm, height 9.5 mm; sapphire crystal; transparent case back; water-resistant to 3 atm
Band: reptile skin, buckle
Remarks: guilloché dial
Price: $34,400
Variations: white gold with gray dial ($34,400)

Saxonia Large Date
Reference number: 381.029
Movement: automatic, Lange Caliber L086.8; ø 30.4 mm, height 5.2 mm; 40 jewels; 21,600 vph; hand-engraved balance cock, screw balance, swan-neck fine adjustment; 72-hour power reserve
Functions: hours, minutes, subsidiary seconds; large date
Case: white gold, ø 38.5 mm, height 9.6 mm; sapphire crystal; transparent case back; water-resistant to 3 atm
Band: reptile skin, buckle
Price: $25,900
Variations: pink gold ($25,900)

Saxonia Moon Phase
Reference number: 384.031
Movement: automatic, Lange Caliber L086.5; ø 30.4 mm, height 5.2 mm; 40 jewels; 21,600 vph; hand-engraved balance cock, screw balance, swan-neck fine adjustment; 72-hour power reserve
Functions: hours, minutes, subsidiary seconds; large date, moon phase
Case: pink gold, ø 40 mm, height 9.8 mm; sapphire crystal; transparent case back; water-resistant to 3 atm
Band: reptile skin, buckle
Price: $29,600
Variations: white gold ($29,600)

Saxonia Thin
Reference number: 205.086
Movement: manually wound, Lange Caliber L093.1; ø 28 mm, height 2.9 mm; 21 jewels; 21,600 vph; screw balance, swan-neck fine adjustment, hand-engraved balance cock, 3 gold chatons; 72-hour power reserve
Functions: hours, minutes
Case: white gold, ø 39 mm, height 6.2 mm; sapphire crystal; transparent case back
Band: reptile skin, buckle
Remarks: copper-blue gold flow dial
Price: $22,000
Variations: 37-mm case in white and pink gold, silver-colored dial ($15,200)

A. LANGE & SÖHNE

Saxonia
Reference number: 219.043
Movement: manually wound, Lange Caliber L941.1; ø 25.6 mm, height 3.2 mm; 21 jewels; 21,600 vph; 4 gold chatons, screw balance, swan-neck fine adjustment, hand-engraved balance cock; 45-hour power reserve
Functions: hours, minutes, subsidiary seconds
Case: pink gold, ø 35 mm, height 7.3 mm; sapphire crystal; transparent case back
Band: reptile skin, buckle
Remarks: mother-of-pearl dial (mop)
Price: $16,600
Variations: pink or white gold with silver-colored dial ($14,800); white gold with blue dial ($14,800); white gold with mother-of-pearl dial (16,600)

Datograph Auf/Ab
Reference number: 405.035
Movement: manually wound, Lange Caliber L951.6; ø 30.6 mm, height 7.9 mm; 46 jewels; 18,000 vph; 4 screw-mounted gold chatons, swan-neck fine adjustment, hand-engraved balance cock; 60-hour power reserve
Functions: hours, minutes, subsidiary seconds; power reserve indicator; flyback chronograph with precisely jumping minute counter; large date
Case: platinum, ø 41 mm, height 13.1 mm; sapphire crystal; transparent case back; water-resistant to 3 atm
Band: reptile skin, buckle
Price: $92,500
Variations: pink gold ($71,600)

Triple Split
Reference number: 424.038F
Movement: manually wound, Lange Caliber L132.1; ø 30.6 mm, height 9.4 mm; 46 jewels; 18,000 vph; off-center balance, in-house hairspring, column-wheel control or chronograph; 5 screw-mounted gold chatons; 50-hour power reserve
Functions: hours, minutes; subsidiary seconds; large date; power reserve indicator; flyback chronograph with triple flyback hand for reference measurements up to 12 hours, precisely jumping minute counter with double hand, continuous chrono hours
Case: white gold, ø 43.2 mm, height 15.6 mm; sapphire crystal; transparent case back; water-resistant to 3 atm
Band: reptile skin, folding clasp
Price: on request; limited to 100 pieces

Lange 31
Reference number: 130.039
Movement: manually wound, Lange Caliber L034.1; ø 37.3 mm, height 9.6 mm; 62 jewels; 21,600 vph; double spring barrel, 3 screw-mounted gold chatons; constant force escapement (remontoir), stop-seconds mechanism, hand-engraved balance cock, parts finished and assembled by hand; 744-hour power reserve
Functions: hours, minutes, subsidiary seconds; power reserve indicator; large date
Case: white gold, ø 45.9 mm, height 15.9 mm; sapphire crystal; transparent case back; water-resistant to 3 atm
Band: reptile skin, folding clasp
Remarks: comes with winding key; limited to 100 pieces
Price: $144,700

Zeitwerk
Reference number: 140.029
Movement: manually wound, Lange Caliber L043.1; ø 33.6 mm, height 9.3 mm; 68 jewels; 18,000 vph; 2 screw-mounted gold chatons, constant force escapement (remontoir), hand-engraved balance cock; 36-hour power reserve
Functions: hours, minutes (digital, jumping), subsidiary seconds; power reserve indicator
Case: white gold, ø 41.9 mm, height 12.6 mm; sapphire crystal; transparent case back; water-resistant to 3 atm
Band: reptile skin, buckle
Price: $77,700
Variations: pink gold with silver-colored dial ($77,700)

Zeitwerk Decimal Strike
Reference number: 143.050
Movement: manually wound, Lange Caliber L043.7; ø 36 mm, height 10 mm; 78 jewels; 18,000 vph; three-quarter plate, 2 screw-mounted gold chatons, constant force escapement (remontoir), hand-engraved balance cock; 36-hour power reserve
Functions: hours, minutes (digital, jumping), subsidiary seconds; power reserve indicator, automatic 10-minute chimes
Case: "honey gold," ø 44.2 mm, height 13.1 mm; sapphire crystal; transparent case back; water-resistant to 3 atm
Band: reptile skin, buckle
Price: $128,100; limited to 100 pieces

A. LANGE & SÖHNE

1815 Annual Calendar
Reference number: 238.026
Movement: manually wound, Lange Caliber L051.3; ø 30.6 mm, height 5.7 mm; 26 jewels; 21,600 vph; 3 screw-mounted gold chatons, hand-engraved balance cock, parts finished and assembled by hand; 72-hour power reserve
Functions: hours, minutes, subsidiary seconds; annual calendar with date, weekday, month, moon phase
Case: white gold, ø 40 mm, height 10.1 mm; sapphire crystal; transparent case back; water-resistant to 3 atm
Band: reptile skin, buckle
Price: $41,200
Variations: pink gold ($41,200)

1815 Chronograph
Reference number: 414.031
Movement: manually wound, Lange Caliber L951.5; ø 30.6 mm, height 6.1 mm; 34 jewels; 21,600 vph; 4 screw-mounted gold chatons, hand-engraved balance cock, parts finished and assembled by hand; 60-hour power reserve
Functions: hours, minutes, subsidiary seconds; flyback chronograph
Case: pink gold, ø 39.5 mm, height 11 mm; sapphire crystal; transparent case back; water-resistant to 3 atm
Band: reptile skin, buckle
Price: $52,300
Variations: silver-colored dial ($50,000); white gold ($52,300)

1815 "Hommage to Walter Lange"
Reference number: 297.026
Movement: manually wound, Lange Caliber L1924; ø 31.6 mm, height 6.1 mm; 36 jewels; 21,600 vph; 3 screw-mounted gold chatons, hand-engraved balance cock, parts finished and assembled by hand; 60-hour power reserve
Functions: hours, minutes, subsidiary seconds, jumping sweep seconds with start/stop function
Case: white gold, ø 40 mm, height 10.1 mm; sapphire crystal; transparent case back; water-resistant to 3 atm
Band: reptile skin, buckle
Price: $49,800; limited to 145 pieces
Variations: pink gold, limited to 90 pieces ($49,800); yellow gold, limited to 27 pieces ($49,800)

Tourbograph Perpetual "Pour le Mérite"
Reference number: 706.025F
Movement: manually wound, Lange Caliber L133.1; ø 32 mm, height 10.9 mm; 52 jewels; 21,600 vph; 1-minute tourbillon, chain and fusée transmission, 6 screw-mounted gold chatons, with two diamond capstones, screw balance; 36-hour power reserve
Functions: hours, minutes, subsidiary seconds; split-seconds chronograph; calendar with date, weekday, month, moon phase, leap year
Case: platinum, ø 43 mm, height 16.6 mm; sapphire crystal; transparent case back; water-resistant to 3 atm
Band: reptile skin, folding clasp
Price: on request; limited to 50 pieces

Richard Lange
Reference number: 232.032
Movement: manually wound, Lange Caliber L041.2; ø 30.6 mm, height 6 mm; 26 jewels; 21,600 vph; hand-engraved balance cock, 2 screw-mounted gold chatons, parts finished and assembled by hand, in-house balance spring with patent-pending anchoring clip; 38-hour power reserve
Functions: hours, minutes, sweep seconds
Case: pink gold, ø 40.5 mm, height 10.5 mm; sapphire crystal; transparent case back; water-resistant to 3 atm
Band: reptile skin, buckle
Price: $33,200
Variations: platinum with silver-colored dial ($47,300)

Richard Lange Jumping Seconds
Reference number: 252.032
Movement: manually wound, Lange Caliber L094.1; ø 33.6 mm, height 6 mm; 50 jewels; 21,600 vph; zero-reset mechanism, constant force escapement (remontoir); 42-hour power reserve
Functions: hours (off-center), minutes (off-center), large seconds (jumping); winding reminder
Case: pink gold, ø 39.9 mm, height 10.6 mm; sapphire crystal; transparent case back; water-resistant to 3 atm
Band: reptile skin, buckle
Price: $78,600; limited to 100 pieces
Variations: platinum with silver-colored dial, limited to 100 pieces ($84,200)

A. LANGE & SÖHNE

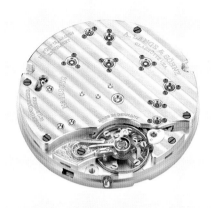

Caliber L121.3
Manually wound; stop-seconds mechanism, 8 screw-mounted gold chatons, swan-neck fine adjustment; double spring barrel, 72-hour power reserve
Functions: hours, minutes, subsidiary seconds; power reserve indicator; moon phase; day/night indicator
Diameter: 30.6 mm
Height: 6.3 mm
Jewels: 47
Balance: glucydur with eccentric adjustment cams
Frequency: 21,600 vph
Balance spring: in-house manufacture
Shock protection: Kif
Remarks: plates and bridges of untreated German silver, engraved balance cock

Caliber L031.1
Manually wound; stop-seconds mechanism; double spring barrel, 72-hour power reserve
Functions: hours, minutes, subsidiary seconds; 2nd time zone (world time display with city reference ring); power reserve indicator, day/night indicator for both time zones; large date
Diameter: 34.1 mm; **Height:** 6.7 mm
Jewels: 54, including 4 screw-mounted gold chatons
Balance: glucydur with weighted screws
Frequency: 21,600 vph;
Balance spring: Nivarox 1 with special terminal curve and swan-neck fine adjustment
Shock protection: Kif
Remarks: hand-engraved balance and intermediate wheel cocks

Caliber L082.1
Automatic, 1-minute tourbillon with a patented stop-seconds mechanism; one-way gold rotor with platinum mass; single barrel, 50-hour power reserve
Functions: hours, minutes, subsidiary seconds; day/night indicator; perpetual calendar with large date, weekday, month, moon phase, leap year
Diameter: 34.1 mm
Height: 7.8 mm
Jewels: 76, including 6 screw-mounted gold chatons and 1 diamond capstone
Balance: glucydur with eccentric regulating cams
Frequency: 21,600 vph;
Balance spring: in-house manufacture
Remarks: hand-engraved balance cock

Caliber L086.5
Automatic; single barrel, 72-hour power reserve
Functions: hours, minutes, subsidiary seconds; large date, moon phase
Diameter: 30.4 mm
Height: 5.2 mm
Jewels: 40
Balance: glucydur
Frequency: 21,600 vph
Balance spring: in-house manufacture
Remarks: plates of untreated German silver, hand-engraved balance cock, moon-phase display decorated with 852 stars, calculated to remain accurate for 122.6 years

Caliber L133.1
Manually wound; chain and fusée drive, 1-minute tourbillon; 2 diamond capstones, column wheel control of chronograph functions, German silver mainplate and bridges; single spring barrel, 36-hour power reserve
Functions: hours, minutes, subsidiary seconds; split-seconds chronograph; perpetual calendar with date, weekday, month, moon phase, leap year
Diameter: 32 mm
Height: 10.9 mm
Jewels: 52, including 6 screw-mounted gold chatons
Balance: glucydur with weighted screws
Balance spring: in-house manufacture
Remarks: hand-engraved balance cock and chronograph bridges

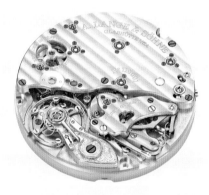

Caliber L094.1
Manually wound; constant force escapement (remontoir), swan-neck fine adjustment, stop-seconds mechanism; single barrel, 42-hour power reserve
Functions: hours (off-center), minutes (off-center), large seconds (jumping); winding reminder
Diameter: 33.6 mm
Height: 6 mm
Jewels: 50, including 8 screw-mounted gold chatons
Balance: glucydur with weighted screws
Frequency: 21,600 vph
Balance spring: in-house manufacture
Remarks: plates and bridges of untreated German silver, hand-engraved balance cock

A. LANGE & SÖHNE

Caliber L1924

Manually wound; swan-neck fine adjustment, stop-seconds mechanism; single barrel, 60-hour power reserve
Functions: hours, minutes, subsidiary jumping seconds.
Diameter: 31.6 mm
Height: 6.1 mm
Jewels: 36, including 3 screw-mounted gold chatons
Balance: glucydur with weighted screws
Frequency: 21,600 vph
Balance spring: in-house manufacture
Remarks: three-quarter plate of untreated German silver, assembled and decorated according to highest quality criteria

Caliber L951.6

Manually wound; stop-seconds mechanism, jumping minute counter; single barrel, 60-hour power reserve
Functions: hours, minutes, subsidiary seconds; power reserve indicator; flyback chronograph; large date
Diameter: 30.6 mm
Height: 7.9 mm
Jewels: 46
Balance: glucydur with weighted screws
Frequency: 18,000 A/h
Balance spring: in-house manufacture
Shock protection: Incabloc
Remarks: three-quarter plate of untreated German silver, hand-engraved balance cock

Caliber L132.1

Manually wound; column wheel chronograph control; single barrel, 55-hour power reserve
Functions: hours, minutes, subsidiary seconds; power reserve indicator; flyback chronograph with triple flyback function for references up to 12 hours; precisely jumping chrono and flyback minute counter, continuous chrono and flyback hour counter
Diameter: 30.6 mm
Height: 9.4 mm
Jewels: 46, including 5 screw-mounted gold chatons
Balance: glucydur with eccentric adjustment cams
Frequency: 21,600 vph
Balance spring: in-house manufacture
Remarks: hand-engraved balance cock

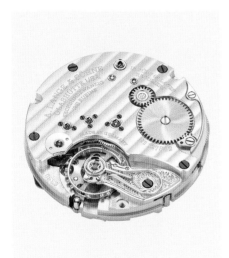

Caliber L051.3

Manually wound; swan-neck fine adjustment, stop-seconds mechanism, plates and bridges of German silver; single barrel, 72-hour power reserve
Functions: hours, minutes, subsidiary seconds; annual calendar with date, weekday, month, moon phase
Diameter: 30.6 mm
Height: 5.7 mm
Jewels: 26, including 3 screw-mounted gold chatons
Balance: glucydur with weighted screws
Frequency: 21,600 vph
Balance spring: in-house manufacture
Remarks: finished and assembled by hand, hand-engraved balance cock

Caliber L043.7

Manually wound; jumping minutes, constant force escapement (remontoir), stop-seconds mechanism; single spring barrel, 36-hour power reserve
Functions: hours and minutes (digital, jumping), subsidiary seconds; power reserve indicator, signal every 10 minutes and top of the hour (can be switched off)
Diameter: 36 mm
Height: 10 mm
Jewels: 78
Balance: glucydur with eccentric adjustment cams
Frequency: 18,000 A/h
Balance spring: in-house manufacture, patent pending on hairspring clamp
Shock protection: Incabloc

Caliber L034.1

Manually wound; key winding with torque brake; constant force mechanism (remontoir), stop-seconds mechanism; two mainsprings, each 185 cm long; 744-hour power reserve (31 days) with switch-off mechanism
Functions: hours, minutes, subsidiary seconds; power reserve indicator; large date
Diameter: 37.3 mm
Height: 9.6 mm
Jewels: 62
Balance: glucydur with weighted screws
Frequency: 21,600 vph
Balance spring: Nivarox 1 with special terminal curve and swan-neck fine adjustment
Shock protection: Kif

ALEXANDER SHOROKHOFF

The ultimate goal for the watch connoisseur may be realizing one's own ideas for timepieces. In the first stages of his life, Alexander Shorokhoff, born in Moscow in 1960, was an engineer and then an architect with his own construction company. These turned out to be an excellent platform to begin expanding into the field of fine timepieces. In 1992, shortly after the demise of the Soviet Union, Shorokhoff founded a distribution company in Germany to market Russia's own Poljot watches. This gave him the insight and practice needed to launch phase two of his plan: establishing his own manufacturing facilities for an independent watch brand under his own name.

At Shorokhoff Watches, three main creative lines are bundled under the general concept "Art on the Wrist": Heritage, Avantgarde, and Vintage. The three lines share a design with a distinctly artistic orientation. They all focus on technical quality, sophisticated hand-engraving, and the cultural backdrop. "We consider watches not only as timekeepers, but also as works of art," says Alexander Shorokhoff. It's a statement that can be seen in the watches. The brand is at home in the world of international and Russian art and culture. Each dial is designed down to the smallest detail. The engraving and finishing of the movements is unique as well.

All of them are taken apart in Alzenau, reworked, and then reassembled with great care, which is why the brand has stamped each watch with "Handmade in Germany." Some of the modules used in these timepieces have been developed by the company itself. Before a watch leaves the *manufacture,* it is subjected to strict quality control. The timepiece's functionality must be given the cleanest bill of health before it can be sent out to jewelers around the world.

Alexander Shorokhoff Uhrenmanufaktur
Hanauer Strasse 25
63755 Alzenau
Germany

Tel.:
+49-6023-919-93

Fax:
+49-6023-919-949

E-mail:
info@alexander-shorokhoff.de

Website:
www.alexander-shorokhoff.de

Founded:
2003

Number of employees:
15

Annual production:
approx. 900 watches

Distribution:
About Time Luxury Group
210 Bellevue Avenue
Newport, RI 02840
401-846-0598

Most important collections/price range
Heritage / approx. $4,500; Avantgarde / approx. $1,500

Vintage 5
Reference number: AS.V5-C
Movement: automatic, Poljot Caliber 2416; ø 24.6 mm, height 4.55 mm; 29 jewels; 18,000 vph; hand-engraved and finished movement; 36-hour power reserve
Functions: hours, minutes, sweep seconds; date
Case: stainless steel, ø 40 mm, height 10.1 mm; sapphire crystal; transparent case back; water-resistant to 5 atm
Band: calfskin, buckle
Price: $1,750; limited to 35 pieces
Variations: stainless steel Milanese bracelet; silver-colored dial (limited to 35 pieces)

Babylonian I
Reference number: AS.BYL01
Movement: manually wound, Caliber 2609.AS (base Poljot 2609); ø 25.6 mm, height 4.05 mm; 17 jewels; 21,600 vph; finely finished movement; 42-hour power reserve
Functions: hours, minutes, sweep seconds
Case: stainless steel, ø 46.5 mm, height 11.5 mm; sapphire crystal; transparent case back; water-resistant to 5 atm
Band: ostrich leather, buckle
Price: $2,835; limited to 500 pieces

Regulator R02
Reference number: AS.R02-1
Movement: manually wound, Caliber 3105.AS (base Poljot 3105); ø 31 mm, height 5.38 mm; 23 jewels; 21,600 vph; hand-engraved and finished movement; 40-hour power reserve
Functions: hours (off-center), minutes, subsidiary seconds; date
Case: stainless steel, ø 43.5 mm, height 11.55 mm; sapphire crystal; transparent case back; water-resistant to 5 atm
Band: calfskin, buckle
Remarks: partially skeletonized dial
Price: $2,400; limited to 98 pieces
Variations: stainless steel Milanese bracelet; various dial colors

Crossing
Reference number: AS.JH01-4
Movement: automatic, Dubois Dépraz Caliber 14400A; 25 jewels; 28,800 vph; blued screws, hand-engraved rotor; 40-hour power reserve
Functions: hours (digital, jumping), minutes (off-center), subsidiary seconds
Case: stainless steel, ø 43.5 mm, height 11.3 mm; sapphire crystal; transparent case back; water-resistant to 5 atm
Band: reptile skin, buckle
Price: $4,650; limited to 25 pieces
Variations: various dials; stainless steel Milanese bracelet

Winter
Reference number: AS.LA-WIN-3
Movement: automatic, Caliber 2824.AS (base ETA 2824-2); ø 25.6 mm, height 4.6 mm; 25 jewels; 28,800 vph; hand-engraved rotor; 38-hour power reserve
Functions: hours, minutes, sweep seconds; date
Case: stainless steel, ø 39 mm, height 10.6 mm; sapphire crystal; transparent case back; water-resistant to 5 atm
Band: galuchat, buckle
Remarks: handmade dial set with diamond dust
Price: $1,850

Miss Avantgarde Diamant
Reference number: AS.AVG02-WGG-D
Movement: automatic, ETA Caliber 2824-2; ø 25.6 mm, height 4.6 mm; 25 jewels; 28,800 vph; hand-engraved rotor, blued screws; 39-hour power reserve
Functions: hours, minutes, sweep seconds; date
Case: white gold, ø 39 mm, height 12.75 mm; bezel set with diamonds; sapphire crystal; transparent case back
Band: galuchat, buckle
Price: $9,790; limited to 20 pieces
Variations: various strap colors

Chrono-Regulator CR02
Reference number: AS.CR02-1
Movement: manually wound, Caliber 31679.AS (base Poljot 3133); ø 31 mm, height 7.35 mm; 23 jewels; 21,600 vph; blued screws, skeletonized, hand-engraved and finished movement; 42-hour power reserve
Functions: hours (off-center), minutes, subsidiary seconds; chronograph; date
Case: stainless steel, ø 43.5 mm, height 13.65 mm; sapphire crystal; transparent case back; water-resistant to 5 atm
Band: reptile skin, buckle
Remarks: partially skeletonized dial
Price: $7,100; limited to 68 pieces

Fedor Dostoevsky Chronograph
Reference number: AS.FD41
Movement: manually wound, Soprod Caliber 7750SORM-3H; ø 30 mm, height 7.9 mm; 25 jewels; 28,800 vph; blued screws, hand-engraved and finished movement; 50-hour power reserve
Functions: hours, minutes, subsidiary seconds; power reserve indicator; chronograph; date
Case: stainless steel, 43 × 43 mm; height 14.2 mm; sapphire crystal; transparent case back; water-resistant to 3 atm
Band: reptile skin, buckle
Price: $9,700

Regulator R01
Reference number: AS.R01-3
Movement: manually wound, Caliber 3104.AS (base Poljot 3104); ø 31 mm, height 5.38 mm; 17 jewels; 21,600 vph; blued screws, hand-engraved and finished movement; 42-hour power reserve
Functions: hours (off-center), minutes, subsidiary seconds; date
Case: stainless steel, ø 43.5 mm, height 11.55 mm; sapphire crystal; transparent case back; water-resistant to 5 atm
Band: calfskin, buckle
Price: $1,400
Variations: folding clasp; stainless steel Milanese bracelet

ALPINA

The brand Alpina essentially grew out of a confederation of watchmakers known as the Alpina Union Horlogère, founded by Gottlieb Hauser. The group expanded quickly to reach beyond Swiss borders into Germany, where it opened a factory in Glashütte. For a while in the 1930s, it even merged with Gruen, one of the most important watch companies in the United States.

After World War II, the Allied Forces decreed that the name Alpina could no longer be used in Germany, and so that branch was renamed "Dugena" for Deutsche Uhrmacher-Genossenschaft Alpina, or the German Watchmaker Cooperative Alpina.

Today, Geneva-based Alpina is no longer associated with that watchmaker cooperative of yore. Now a sister brand of Frédérique Constant, it has a decidedly modern collection enhanced with a series of movements designed, built, and assembled in-house: the Tourbillon AL-980, the World Timer AL-718, the Automatic Regulator AL-950, the Small Date Automatic AL-710, and, more recently, the Flyback-Chronograph Automatic AL-760, which features the patented "Direct-Flyback" technology. Owners Peter and Aletta Stas have built up an outstanding business over the years, and in 2016, they sold it to Citizen Group, but continue managing the brands until 2020.

Alpina likes to call itself the inventor of the modern sports watch. Its iconic Block-Uhr of 1933 and the Alpina 4 of 1938, with an in-house automatic movement, set the pace for all sports watches, with a waterproof stainless steel case, an antimagnetic system, and shock absorbers. But beyond a target group engaged in water and air sports, the brand is now looking at the twenty-first-century hipsters, whose lives are electronic. The Horological Smartwatch, equipped with a quartz movement, connects with mobile phones and other electronic devices and can display the data on an analog dial. The idea of producing a watch for the active person with a liking for hip electronics and traditional mechanics has proven itself a good recipe for the brand.

Alpina Watch International SA
Route de la Galaise, 8
CH-1228 Plan-les-Ouates, Geneva
Switzerland

Tel.:
+41-0-22-860-87-40

Fax:
+41-0-22-860-04-64

E-mail:
info@alpina-watches.com

Website:
www.alpina-watches.com

Founded:
1883

Number of employees:
100

Annual production:
10,000 watches

U.S. distributor:
Alpina Frederique Constant USA
350 5th Avenue, 29th Fl.
New York, NY 10118
646-438-8124
lmellor@usa.frederique-constant.com

Most important collections/price range:
AlpinerX / from approx. $995 to $1,295; Seastrong Diver Quartz GMT / from approx. $795 to $995; Startimer Pilot Automatic / from approx. $995 to $1,295; Startimer Pilot / from approx. $895

Startimer Pilot Heritage Automatic GMT

Reference number: AL-555DGS4H6
Movement: automatic, Caliber AL-555 (Basis Sellita SW200-1); ø 25.6 mm, height 4.6 mm; 26 jewels; 28,800 vph; 38-hour power reserve
Functions: hours, minutes, sweep seconds; additional 24-hour display (2nd time zone) with crown-activated inner scale; date
Case: stainless steel, 40.75 × 42 mm, height 12.65 mm; sapphire crystal; water-resistant to 10 atm
Band: calfskin, buckle
Price: $1,395

Startimer Pilot Automatic Shadow Line

Reference number: AL-525GG4S4
Movement: automatic, Caliber AL-525 (Basis Sellita SW200-1); ø 25.6 mm, height 4.6 mm; 26 jewels; 28,800 vph; 38-hour power reserve
Functions: hours, minutes, sweep seconds; date
Case: stainless steel with rose gold PVD coating, ø 44 mm, height 10.74 mm; sapphire crystal; screw-in crown; water-resistant to 10 atm
Band: calfskin, buckle
Price: $1,295

Alpiner 4 Shadow Line

Reference number: AL-525BB5FBAQ6
Movement: automatic, Caliber AL-525 (Basis Sellita SW200-1); ø 25.6 mm, height 4.6 mm; 26 jewels; 28,800 vph; 38-hour power reserve
Functions: hours, minutes, sweep seconds; date
Case: stainless steel with black PVD coating, ø 44 mm, height 13.15 mm; bidirectional 12-hour bezel; sapphire crystal; screw-in crown; water-resistant to 10 atm
Band: calfskin, buckle
Price: $1,495

ANONIMO SA
Chemin des Tourelles 4
CH-2400 Le Locle
Switzerland

Tel.:
+41-22-566-06-06

E-mail:
info@anonimo.com

Website:
www.anonimo.com

Founded:
1997; moved to Switzerland in 2013

Distributor:
Duber Time
131 4th Street North
St. Petersburg, FL 33701
For international inquiries contact the brand directly or go to anonimo.com/storelocator.

Most important collections/price range:
Militare / $2,000 to $5,500

ANONIMO

The brand Anonimo was launched on the banks of the Arno, in Florence. Watchmaking has a long history in the capital of Tuscany, going back to one Giovanni de Dondi (1318–1389), who built his first planetarium around 1368. Then came the Renaissance man Lorenzo della Volpaia (1446–1512), an architect and goldsmith, who also worked with calendars and astronomical instruments. And finally, there were the likes of the mathematician Galileo and the incomparable Leonardo da Vinci.

In more recent times, the Italian watchmaking industry was busy equipping submarine crews and frogmen with timepieces. The key technology came from Switzerland, but the specialized know-how for making robust, water-resistant watches sprang from small enterprises with competencies in building cases. The founders of Anonimo understood this strength and decided to put it in the service of their "anonymous" brand—a name chosen to "hide" the fact that many small, discreet companies are involved in their superbly finished watches.

In 2013, armed with some fresh capital and a new management team, Anonimo came out with three watch families running on Swiss technology: the mechanical movements are partially complemented with Dubois Dépraz modules or ETA and Sellita calibers. On the whole, though, the collections reflect exquisite conception and manufacturing, and the quality of the materials is unimpeachable: corrosion-resistant stainless steel, fine bronze, and titanium, appreciated for its durability and hypoallergenic properties. The design is definitely vintage, a bit 1960s with a hint of a cushion case, but use of only three numerals—4, 8, and 12, which sketch an A on the dial—is quite modern. On the military models, the crown has been placed in a protected area between the two upper lugs. Thanks to a clever hinge system, that crown can be pressed onto the case for an impermeable fit or released for time-setting. The latest collection is, as the name Epurato suggests, more streamlined, slightly smaller. The color-conscious buyer has a number of different dials to choose from.

Epurato Blu Intenso

Reference number: AM-1120.02.003.A03
Movement: automatic, Sellita Caliber SW200-1; ø 25.6 mm, height 4.6 mm; 25 jewels; 28,800 vph; 38-hour power reserve
Functions: hours, minutes, sweep seconds; date
Case: stainless steel, ø 42 mm, height 14 mm; sapphire crystal; water-resistant to 5 atm
Band: calfskin, buckle and interchangeable strap
Price: $2,480
Variations: different dial colors and case materials (bronze and DLC)

Militare Vintage Classic

Reference number: AM-1020.01.001.A02
Movement: automatic, Sellita Caliber SW260-1; ø 25.6 mm, height 5.6 mm; 31 jewels; 28,800 vph; oscillating mass with côtes de Genève; 42-hour power reserve
Functions: hours, minutes, subsidiary seconds; date
Case: stainless steel, ø 43.4 mm, height 14.5 mm; sapphire crystal; screw-in crown; transparent case back; water-resistant to 12 atm
Band: calfskin, buckle
Remarks: crown pressed onto case by upper lug for impermeable seal
Price: $2,800
Variations: various dial colors and cases

Nautilo NATO Automatic Bicolor

Reference number: AM-1002.11.007.A16
Movement: automatic, Sellita Caliber SW200-1; ø 25.6 mm, height 4.6 mm; 25 jewels; 28,800 vph; 38-hour power reserve
Functions: hours, minutes, sweep seconds; date
Case: stainless steel with DLC treatment, ø 44.4 mm, height 14.2 mm; stainless steel bezel with ceramic inlay, unidirectional bezel with 0-60 scale; sapphire crystal; screw-in crown; water-resistant to 20 atm
Band: NATO strap and additional rubber strap complimentary, buckle
Price: $2,370
Variations: comes with additional rubber strap

AQUADIVE

Aquadive USA
P.O. Box 113
Wicomico, VA 23184

Tel.:
888-397-9363

E-mail:
info@aquadive.com

Website:
www.aquadive.com

Founded:
1962

Number of employees:
18

Distribution:
Retail and direct online sales

Most important collections/price range:
Bathyscaphe / $1,290 to $3,990, NOS Model 77

According to Laver's Law, a style that shows up again at the fifty-year mark is "quaint." So much for the outward impact, maybe. But what about the intrinsic long-term personal and ephemeral value, the sometimes collective memories associated with a particular moment in our lives? Today, the very sight of a watch from times past might bring forth images of a different era, much like hearing the songs of Procol Harum or touching Naugahyde in an old Dodge Dart. Nostalgia is a powerful impulse, especially in an era like ours, which appears enamored by its own frenetic pace and refuses categorically to stop and reflect.

So when a group of watch experts decided to revive an iconic watch of the sixties and seventies, they were bound to strike a positive note. In its day, the Aquadive was considered a solidly built and reliable piece of equipment seriously coveted by professional divers. It might still be around had it not been put out to pasture during the quartz revolution.

In its twenty-first-century incarnation, the Aquadive bears many hallmarks of the original. The look is unmistakable: the charmingly awkward hands, the puffy cushion case, the sheer stability it exudes. In fact, some of the components, like the 200 NOS case and sapphire crystal, are leftovers from the old stock. The Swiss-made automatic movements and the gaskets, of course, are new.

Modern technologies, like DLC, and advances in CNC machining have transformed the older concepts. And to ensure reliability, the watches are assembled in Switzerland. The top of the current line is the Bathyscaphe series, machined from a block of stainless steel, and featuring new shock absorbers and an automatic helium release valve.

Bathyscaphe 100
Reference number: 1002.11.36211
Movement: automatic, ETA Caliber 2836-2; ø 25.6 mm, height 4.6 mm; 25 jewels; 28,800 vph; 42-hour power reserve; regulated in 5 positions
Functions: hours, minutes, sweep seconds; date
Case: stainless steel, ø 43 mm, height 14 mm; unidirectional bezel with 0-60 scale; antimagnetic soft iron inner case; sapphire crystal; screw-in crown; automatic helium release valve; water-resistant to 100 atm
Band: Isofrane, buckle
Price: $1,690; limited to 500 pieces
Variations: mesh bracelet ($1,890); gun metal DLC-coated version ($1,890)

Bathyscaphe 100 Bronze
Reference number: 1006.13.365311
Movement: automatic, ETA Caliber 2836-2; ø 25.6 mm, height 4.6 mm; 25 jewels; 28,800 vph; 42-hour power reserve; regulated in 5 positions
Functions: hours, minutes, sweep seconds; date
Case: German bronze alloy, ø 43 mm, height 15 mm; unidirectional bezel with 0-60 scale; sapphire crystal; screw-in crown; automatic helium release valve; water-resistant to 100 atm
Band: Isofrane, buckle
Price: $1,690; limited to 100 pieces

Bathysphere 100 GMT
Reference number: 1001.13.935113
Movement: automatic, ETA Caliber 2893-2; ø 25.6 mm, height 4.2 mm; 21 jewels; 28,800 vph; 42-hour power reserve; regulated in 5 positions
Functions: hours, minutes, sweep seconds; date, GMT hand for 24 hours indication
Case: stainless steel case, ø 43 mm, height 15 mm; unidirectional bezel with 0-60 scale; sapphire crystal; screw-in crown; automatic helium release valve; water-resistant to 100 atm
Band: Isofrane, buckle
Price: $1,990; limited to 300 pieces
Variations: mesh bracelet ($2,150); DLC-coated gun metal ($2,090); 2 additional dial colors

ARISTO

Aristo Vollmer GmbH
Erbprinzenstr. 36
D-75175 Pforzheim
Germany

Tel.:
+49-7231-17031

Fax:
+49-7231-17033

E-mail:
info@aristo-vollmer.de

Website:
www.aristo-vollmer.de

Founded:
1907/1998

Number of employees:
16

Annual production:
9,000 watches and 9,000 bracelets

Distribution:
retail

U.S. distributor:
Long Island Watch, Marc Frankel
273 Walt Whitman Road, Suite 217
Huntington Station, NY 11746
631-470-0762; 888-673-1129 (fax)
www.longislandwatch.com

Most important collections/price range:
Aristo watches starting at $400 up to Vollmer watches at $1,900; Erbprinz watches up to $1,600

"If you lie down with dogs . . ." goes the old saying. And if you work closely with watchmakers . . . you may catch their more beneficial bug and become one yourself. That at any rate is what happened to the watch case and metal bracelet manufacturer Vollmer, Ltd, established in Pforzheim, Germany, by Ernst Vollmer in 1922. Third-generation president Hansjörg Vollmer decided he was interested in producing watches as well.

Vollmer, who studied business in Stuttgart, had the experience, but also the connections with manufacturers in Switzerland. He speaks French fluently, another asset. He acquired Aristo and launched a series of pilot watches in 1998 housed in sturdy titanium cases with bold onion crowns and secured with Vollmer's own light and comfortable titanium bracelets. Bit by bit, thanks to affordable prices and no-nonsense design—reviving some classic dials from World War II—Vollmer's watches caught hold. The collection grew with limited editions and a few chronometers.

In October 2005, Vollmer GmbH and Aristo Watches finally consolidated for a bigger impact. Besides their own lines, they produce quartz watches, automatics, and chronographs under the names Messerschmitt and Aristella. The Aristo brand has been trademarked worldwide and is sold mainly in Europe, North America, and Asia. The collection is divided up into Classic, Design, and Sports, with the mechanical segment further split based on elements Land, Water, and Air. The timepieces range from quality wristwatches with historical movements from older Swiss production to attractive ladies' watches or replicas of classic military watches, all assembled in Pforzheim. The company also has an established name as a manufacturer of classic pilot watches. And it took another step toward the higher end of the market by launching the "Erbprinz" series, named after the street where the company also has a workshop for manufacturing metal bracelets.

Me262 "Rote B"
Reference number: 3H262-RBM
Movement: automatic, ETA Caliber 2824-2; ø 25.6 mm, height 4.6 mm; 25 jewels; 28,800 vph; 38-hour power reserve
Functions: hours, minutes, sweep seconds
Case: stainless steel, ø 44 mm, height 11 mm; sapphire crystal; transparent case back; water-resistant to 5 atm
Band: stainless steel Milanese mesh, folding clasp
Remarks: dial from the aluminum of an original Messerschmitt Me262
Price: $790

Vintage 41
Reference number: 3H190
Movement: automatic, ETA Caliber 2824-2; ø 25.6 mm, height 4.6 mm; 25 jewels; 28,800 vph; 38-hour power reserve
Functions: hours, minutes, sweep seconds; date
Case: stainless steel, ø 40.5 mm, height 10 mm; mineral glass; water-resistant to 5 atm
Band: calfskin, buckle
Price: $590
Variations: stainless steel Milanese mesh ($540)

Erbprinz "Goldstadt" Chrono
Reference number: GS-4
Movement: automatic, ETA Caliber 7750; ø 30 mm, height 7.9 mm; 25 jewels; 28,800 vph; 42-hour power reserve
Functions: hours, minutes, subsidiary seconds; chronograph; date, weekday
Case: stainless steel, ø 44 mm, height 14.5 mm; gold-plated bezel; sapphire crystal; transparent case back; water-resistant to 5 atm
Band: stainless steel Milanese mesh, folding clasp
Price: $1,590

ARMIN STROM

For more than thirty years, Armin Strom's name was associated mainly with the art of skeletonizing. But this "grandmaster of skeletonizers" then decided to entrust his life's work to the next generation, which turned out to be the Swiss industrialist and art patron Willy Michel.

Michel had the wherewithal to expand the one-man show into a full-blown *manufacture* able to conceive, design, and produce its own mechanical movements. The endeavor attracted Claude Geisler, a very skilled designer, and Michel's own son, Serge, who became business manager. When this triumvirate joined forces, it was able to come up with a technically fascinating movement at the quaint little *manufacture* in the Biel suburb of Bözingen within a brief period of time.

The new movement went on to grow into a family of ten, which forms the backbone of a new collection, including a tourbillon with microrotor—no mean feat for a small firm. The ARF15 caliber of the Mirrored Force Resonance, for example, features two balance wheels placed close enough to influence each other (resonance) and give the movement greater stability. The two oscillating systems are connected by a clutch spring.

This essential portfolio has given the *manufacture* the industrial autonomy to implement its projects quickly and independently. Armin Strom has additionally created an online configurator (on its homepage), giving fans and collectors the opportunity to personalize their watches. All components can be selected individually and combined, from the dial, hands, and finishing, to the straps. The finished product can be picked up at a local dealership or at the manufacturer in Biel/Bienne, including a tour of the place.

Armin Strom AG
Bözingenstrasse 46
CH-2502 Biel/Bienne
Switzerland

Tel.:
+41-32-343-3344

Fax:
+41-32-343-3340

E-mail:
info@arminstrom.com

Website:
www.arminstrom.com

Founded:
2006 (first company 1967)

Number of employees:
22

Annual production:
approx. 400 watches

U.S. representative:
Jean Marc Bories
Head of North America
+1 (929) 353 5395
Jean-marc@arminstrom.com

Most important collections/price range:
Offers an online configurator for individual design using six in-house movements (manual, power reserve, automatic with or without date, tourbillon, resonance, and skeletons): $9,900 to $100,000 plus.

Mirrored Force Resonance Fire
Reference number: RG15-RF.5N
Movement: manually wound, Caliber ARF15; ø 36.6 mm, height 7.7 mm; 43 jewels; 25,200 vph; two separate regulating systems are connected by a resonance clutch spring and mutually stabilize each other; finely finished movement; 48-hour power reserve
Functions: hours, minutes (off-center), two subsidiary seconds
Case: rose gold, ø 43.4 mm, height 13 mm; sapphire crystal; transparent case back; water-resistant to 5 atm
Band: reptile skin, buckle
Price: $67,100; limited to 50 pieces
Variation: stainless steel ($54,100)

Pure Resonance Water
Reference number: ST17-RW.05.AL.L.14
Movement: manually wound, Caliber ARF16; ø 34.4 mm, height 7.05 mm; 38 jewels; 25,200 vph; two separate regulating systems are connected by a resonance clutch spring and mutually stabilize each other; fine decoration on mainplate and bridges; 48-hour power reserve
Functions: hours, minutes (off-center), subsidiary seconds
Case: stainless steel, ø 42 mm, height 12 mm; sapphire crystal; transparent case back; water-resistant to 5 atm
Band: reptile skin, buckle
Remarks: comes with extra rubber strap
Price: $49,000

Tourbillon Skeleton Earth
Reference number: ST15-TE.90
Movement: manually wound, Caliber ATC11-S; ø 36.6 mm, height 6.2 mm; 24 jewels; 18,000 vph; 1-minute tourbillon, double spring barrel, skeletonized plates, wheels, and bridges, Breguet hairspring, screw balance with gold weight screws, crown wheels visible on dial side; 240-hour power reserve
Functions: hours, minutes, subsidiary seconds
Case: stainless steel with black PVD coating, ø 43.4 mm, height 13 mm; sapphire crystal; transparent case back; water-resistant to 5 atm
Band: reptile skin, double folding clasp
Price: $75,000
Variations: Air ($77,000); Fire ($84,000); Water ($75,000)

ARMIN STROM

Skeleton Pure Water
Reference number: ST15-PW.05
Movement: manually wound, Caliber ARM09-S; ø 36.6 mm, height 6.2 mm; 34 jewels; 18,000 vph; 2 spring barrels, screw balance with gold weight screws, Breguet hairspring, crown wheels visible on dial side; skeletonized plate, gearwheels, and spring barrel bridges, mainplate with blue PVD coating; 168-hour power reserve
Functions: hours, minutes, subsidiary seconds; power reserve indicator
Case: stainless steel, ø 43.4 mm, height 13 mm; sapphire crystal; transparent case back; water-resistant to 5 atm
Band: reptile skin, buckle
Remarks: comes with extra rubber strap
Price: $32,400; limited to 100 pieces
Variations: Fire ($45,400); Air ($35,400)

Edge Double Barrel Rose Gold
Reference number: RG16-EB.5N
Movement: manually wound, Caliber ARM16; ø 36.6 mm, height 7.7 mm; 34 jewels; 18,000 vph; double spring barrel, winding wheels visible on dial side; 192-hour power reserve
Functions: hours, minutes, subsidiary seconds; power reserve indicator
Case: rose gold, ø 46.8 mm, height 13.2 mm; sapphire crystal; transparent case back; water-resistant to 5 atm
Band: reptile skin, buckle
Remarks: comes with extra rubber strap
Price: $39,900; limited to 100 pieces
Variations: stainless steel with black PVD coating ($26,900)

Gravity Earth
Reference number: ST13-GE.90
Movement: automatic, Caliber AMR13; ø 36.6 mm, height 6 mm; 32 jewels; 18,000 vph; screw balance with gold weight screws, Breguet hairspring; microrotor on dial side; 120-hour power reserve
Functions: hours, minutes, subsidiary seconds
Case: stainless steel with black PVD coating, ø 43.4 mm, height 13 mm; sapphire crystal; transparent case back; water-resistant to 5 atm
Band: reptile skin, buckle
Remarks: comes with extra rubber strap
Price: $13,900; limited to 100 pieces
Variations: Air ($16,900); Fire ($26,900); Water ($13,900)

Caliber ATC11-S
Manually wound; 1-minute tourbillon, gold escape wheel and pallet lever with hardened functional surfaces, fully skeletonized movement, mainplate with black PVD coating; double mainspring barrel, 240-hour power reserve
Functions: hours, minutes, subsidiary seconds
Diameter: 36.6 mm
Height: 6.2 mm
Jewels: 24
Balance: screw balance with variable inertia
Frequency: 18,000 vph
Balance spring: Breguet hairspring
Shock protection: Incabloc
Remarks: finely finished movement

Caliber ARF15
Manually wound 25,200 vph; two separate regulating systems are connected by a resonance clutch spring and mutually stabilize each other; single spring barrel, 48-hour power reserve
Functions: hours, minutes (off-center), two independent symmetrically mirrored subsidiary seconds
Diameter: 36.6 mm
Height: 7.7 mm
Jewels: 43
Balance: two balance wheels oscillating in opposite directions on a single hairspring
Frequency: 25,200 vph
Remarks: 226 components; fine, hand-decorated movement

Caliber ARM16
Manually wound gold escape wheel and pallet lever with hardened functional surfaces; double mainspring barrel, 192-hour power reserve
Functions: hours, minutes, subsidiary seconds; power reserve indicator
Diameter: 36.6 mm
Height: 7.7 mm
Jewels: 34
Balance: screw balance with variable inertia
Frequency: 18,000 vph
Balance spring: Breguet hairspring
Remarks: finely finished movement

ARNOLD & SON

John Arnold holds a special place among the British watchmakers of the eighteenth and nineteenth centuries because he was the first to organize the production of his chronometers along industrial lines. He developed his own standards and employed numerous watchmakers. During his lifetime, he is said to have manufactured around 5,000 marine chronometers, which he sold at reasonable prices to the Royal Navy and the West Indies merchant fleet. Arnold chronometers were packed in the trunks of some of the greatest explorers, from John Franklin and Ernest Shackleton to Captain Cook and Dr. Livingstone.

As Arnold & Son was once synonymous with precision timekeeping on the high seas, it stands to reason, then, that the modern brand should also focus its design policies on the interplay of time and geography as well as the basic functions of navigation. Independence from The British Masters Group has meant that the venerable English chronometer brand has been reorienting itself, setting its sights on classic, elegant watchmaking. With the expertise of watch manufacturer La Joux-Perret behind it (and the expertise housed in the building behind the complex on the main road between La Chaux-de-Fonds and Le Locle), it has been able to implement a number of new ideas.

There are two main lines: The Royal Collection celebrates John Arnold's art, with luxuriously designed models inspired from past creations with delicate complications, tourbillons, or world-time displays, or unadorned manual windings featuring the new Caliber A&S 1001 by La Joux-Perret. The Instrument Collection is dedicated to exploring the seven seas and offers a sober look reflecting old-fashioned meters. Typically, these timepieces combine two displays on a single dial: a chronograph with jumping seconds, for example, between the off-center displays of time and the date hand or separate escapements driving a dual time display—left, the sidereal time; right, the solar time; and between the two, the difference. Perhaps the most remarkable timepiece in the collection is the skeletonized Time Pyramid, with a dual power reserve, a crown between the lugs, and an overall modern look.

Arnold & Son
38, boulevard des Eplatures
CH-2300 La Chaux-de-Fonds
Switzerland

Tel.:
+41-32-967-9797

Fax:
+41-32-968-0755

E-mail:
info@arnoldandson.com

Website:
www.arnoldandson.com

Founded:
1995

Number of employees:
approx. 30

U.S. distributor:
Arnold & Son USA
510 West 6th Street, Suite 309
Los Angeles, CA 90014
213-622-1133

Most important collections/price range:
Globetrotter, Tourbillon Chronometer No. 36, Time Pyramid, Nebula, TB88, TBR, TE8 (Tourbillon), Time Pyramid, UTTE / from approx. $10,000 to $325,000

Tourbillon Chronometer No. 36 Tribute Edition

Reference number: 1ETAS.B01A.C113S
Movement: manually wound, Arnold & Son Caliber 8600; ø 37.8 mm, height 5.9 mm; 33 jewels; 28,800 vph; 1-minute tourbillon; mainplate black DLC-coated, NAC-coated bridges and cocks, finely finished movement; 90-hour power reserve; COSC-certified chronometer
Functions: hours, minutes, subsidiary seconds
Case: stainless steel, ø 46 mm, height 12.66 mm; sapphire crystal; transparent case back; water-resistant to 3 atm
Band: reptile skin, buckle
Price: $37,400; limited to 28 pieces
Variations: pink gold ($55,400)

Globetrotter

Reference number: 1WTAS.S01A.D137S
Movement: automatic, Arnold & Son Caliber 6022; ø 38 mm, height 6.55 mm (14 mm includes arched bridge and hemisphere); 29 jewels; 28,800 vph; 3D world time display in sculptural hemisphere design; 45-hour power reserve
Functions: hours, minutes; world time indicator (2nd time zone)
Case: stainless steel, ø 45 mm, height 17.2 mm; sapphire crystal; transparent case back; water-resistant to 3 atm
Band: calfskin, buckle
Price: $16,995

Nebula Steel

Reference number: 1NEAS.B01A.D134A
Movement: manually wound, Arnold & Son Caliber 5101; ø 31.5 mm, height 4.04 mm; 24 jewels; 21,600 vph; skeletonized and finely finished movement; 90-hour power reserve
Functions: hours, minutes, subsidiary seconds
Case: stainless steel, ø 41.5 mm, height 8.73 mm; sapphire crystal; transparent case back; water-resistant to 3 atm
Band: calfskin, buckle
Remarks: skeletonized dial
Price: $14,500
Variations: pink gold ($25,950)

DBG Steel
Reference number: 1DGAS.S01AC121S
Movement: manually wound, Arnold & Son Caliber 1209; ø 35 mm, height 3.9 mm; 42 jewels; 21,600 vph; double spring barrel, 2 independent gearwheels and escapement systems; 40-hour power reserve
Functions: hours, minutes (double, 2 time zones), sweep seconds; day/night indicator (per time zone)
Case: stainless steel, ø 44 mm, height 9.89 mm; sapphire crystal; transparent case back; water-resistant to 3 atm
Band: reptile skin, buckle
Price: $27,900

Eight-Day Royal Navy
Reference number: 1EDAS.U01A.D136A
Movement: manually wound, Arnold & Son Caliber 1016; ø 33 mm, height 4.7 mm; 33 jewels; 21,600 vph; 192-hour power reserve
Functions: hours, minutes, subsidiary seconds; power reserve indicator; date
Case: stainless steel, ø 43 mm, height 10.7 mm; sapphire crystal; transparent case back; water-resistant to 3 atm
Band: reptile skin, buckle
Price: $12,950
Variations: black and silver-white dial

HM Perpetual Moon
Reference number: 1GLAR.I01A.C122A
Movement: manually wound, Arnold & Son Caliber 1512; ø 34 mm, height 5.35 mm; 27 jewels; 21,600 vph; astronomically precise 122-year moon phase; 90-hour power reserve
Functions: hours, minutes; moon phase
Case: pink gold, ø 42 mm, height 11.43 mm; sapphire crystal; transparent case back; water-resistant to 3 atm
Band: reptile skin, buckle
Remarks: 3D hand-engraved moon
Price: $29,950
Variations: blue guilloché dial ($29,950); stainless steel with black or blue dial ($16,300 and $16,950)

DSTB
Reference number: 1ATAS.U01A.C121S
Movement: automatic, Arnold & Son Caliber 6003; ø 38 mm, height 7.39 mm; 32 jewels; 28,800 vph; true-beat escapement on dial, finely finished movement; 45-hour power reserve
Functions: hours, minutes (off-center), subsidiary seconds (jumping)
Case: stainless steel, ø 43.5 mm, height 13 mm; sapphire crystal; transparent case back; water-resistant to 3 atm
Band: reptile skin, buckle
Price: $30,750
Variations: black dial ($29,995); pink gold with anthracite dial ($48,550)

Time Pyramid Black
Reference number: 1TPBS.R01A
Movement: manually wound, Arnold & Son Caliber 1615; ø 37 mm, height 4.4 mm; 27 jewels; 21,600 vph; skeletonized movement; double spring barrel; 90-hour power reserve
Functions: hours, minutes, subsidiary seconds; double power reserve indicator
Case: stainless steel with black DLC coating, ø 44.6 mm, height 10 mm; sapphire crystal; transparent case back; water-resistant to 3 atm
Band: reptile skin, buckle
Remarks: pyramid-shaped movement inspired from table clocks by John and Roger Arnold
Price: $31,900; limited to 50 pieces; **Variations:** stainless steel ($31,900); pink gold $43,200)

Time Pyramid
Reference number: 1TPAR.S01A.C125A
Movement: manually wound, Arnold & Son Caliber 1615; ø 37 mm, height 4.4 mm; 27 jewels; 21,600 vph; skeletonized movement; double spring barrel; 90-hour power reserve
Functions: hours, minutes, subsidiary seconds; double power reserve indicator
Case: pink gold, ø 44.6 mm, height 10 mm; sapphire crystal; transparent case back; water-resistant to 3 atm
Band: reptile skin, buckle
Remarks: pyramid-shaped movement inspired from table clocks by John and Roger Arnold
Price: $43,200
Variations: stainless steel ($31,900); with black DLC coating ($31,900)

ARTYA

Shaking up the staid atmosphere of watchmaking can be achieved many ways. The conservative approach is to make some small engineering advance and then talk loudly of tradition and innovation. Yvan Arpa, founder of ArtyA watches, enjoys "putting his boot in the anthill," in his own words.

This refreshingly candid personality arrived at watchmaking because, after spending his *Wanderjahre* crossing Papua New Guinea on foot and practicing Thai boxing in its native land, any corporate mugginess back home did not quite cut it for him. Instead he turned the obscure brand Romain Jerome into the talk of the industry with novel material choices: "I looked for antimatter to gentrify common matter," he reflects, "like the rust: proof of the passage of time and the sworn enemy of watchmaking."

Leaving Romain Jerome liberated Arpa from brand constraints. He founded ArtyA, where he could get his "monster" off the slab as it were, with a divine spark. "I had worked with water, rust, dust, and other elements, and then I really caught fire," says Arpa. From cases hit with an electrical arc to cut-up Euro bills, ArtyA's watches hit nerves and drew a gamut of emotional responses. That's his aim, to surprise and amaze. His dials shake up the owner, and are often genuinely unique. They can include real butterfly wings, exquisite engravings by Bram Ramon, or mysterious mother-of-pearl crafting—for 2017, he even revived the snipped money idea, this time with dollars. He never shuns a crazy idea, like the Son of Sounds, with their guitar-shaped cases and chrono pushers designed like guitar pegs—Alice Cooper owns one, obviously.

Arpa wants us not only to wear a watch, but to reflect on aspects of our world and society, the meaning of money, bullets, skulls, our love-hate relationship with electronics, the passage of time, love and violence, the beauty of nature frozen in death, and the significance of music. His provocations, though, do not arise from a sophomoric need to be contrarian, but rather from his long and rich experience of an industry that tends to play it safe. No wonder Samsung recruited him to design their Gear 3 hybrid pocket watch.

Luxury Artpieces SA
Route de Thonon 146
CH-1222 Vésenaz
Switzerland

Tel.:
+41-22-752-4940

Website:
www.artya.com

Founded:
2010

Number of employees:
12

Annual production:
at least 365 (one a day)

U.S. distributor:
Contact headquarters for all enquiries.

Most important collections/price range:
Son of a Gun / $8,800 to $167,000; Son of Art / $3,800 to $21,000; Son of Earth / $4,300 to $183,000; Son of Love / $4,300 to $54,500; Son of Sound / $4,300 to $22,110; Son of Gears / $6,550 to $16,550

Artya 3 Gongs Minute Repeater, Regulator, and Double Axis Tourbillon

Movement: manually wound, by MHC, design by ArtyA; ø 13.2 mm, height 6.6 mm; 21,600 vph; 46 jewels; double-axis tourbillon, 30-second in one direction and 60-second in the other; minute repeater; 64-hour power reserve
Functions: hours (off-center), sweep minutes, seconds on tourbillon
Case: titanium and ArtyOr with PVD treatment, ø 64.6 × 47.3 mm, height 18.1 mm; sapphire crystal; transparent case back; water-resistant to 5 atm
Band: reptile skin, buckle
Remarks: special gongs, customizable
Price: $480,000; unique piece

Son of Sound Purple Rain

Movement: automatic, ArtyA; ø 33 mm, height 12 mm; 19 jewels; 21,600 vph; 48-hour power reserve
Functions: sliding hours, minutes, subsidiary seconds; date; patented active "tuning pegs" system for chronograph functions/date setting; 30-minute counter
Case: stainless steel, ø 37 × 49 mm, height 15 mm; transparent engraved and screwed-down case back; protected against humidity and dust, but not water-resistant
Band: calfskin, buckle
Price: $20,100; unique piece

Son of a Gun Russian Roulette Extreme

Movement: manually wound, ArtyA; ø 32.6 mm, height 5.7 mm; 19 jewels; 52-hour power reserve
Functions: hours, minutes
Case: stainless steel, ArtyOr inserts, 44 mm, height 12 mm; engraved and screwed-down case back; water-resistant to 5 atm
Remarks: multilayered, fast spinning dial with hand-set real bullet
Band: calfskin, buckle
Price: $9,900; limited to 99 pieces

ARTYA

Son of Gears Snowflakes Megève Edition

Movement: manually wound ArtyA movement; ø 33 mm, height 12 mm; 19 jewels; 21,600 vph; 52-hour power reserve
Functions: hours, minutes
Case: stainless steel with brushed lateral inserts, ø 44 mm, height 18 mm; engraved and screwed-down transparent case back; water-resistant to 3 atm
Band: calfskin, buckle
Remarks: artfully skeletonized dial in shape of snowflake
Price: $3,990; limited to 99 pieces

Son of Gears Shams

Movement: manually wound ArtyA movement; ø 33 mm, height 12 mm; 19 jewels; 21,600 vph; 52-hour power reserve
Functions: hours, minutes
Case: stainless steel with black PVD, ø 44 mm, height 18 mm; engraved and screwed-down transparent case back; water-resistant to 3 atm
Band: calfskin, buckle
Remarks: artfully skeletonized dial shaped like the sun with luminescent material
Price: $7,400; unique piece
Variations: without luminescent material; limited to 99 pieces ($5,400)

Son of a Gun Shuriken

Movement: automatic, ArtyA Aion; ø 26.20 mm, height 3.60 mm; 25 jewels; 28,800 vph; with côtes de Genève; rhodium-plated gold oscillator; COSC-certified; 52-hour power reserve
Functions: hours, minutes, seconds
Case: stainless steel, ø 44 mm, height 18 mm; engraved and screwed-down transparent case back; water-resistant to 3 atm
Band: calfskin, buckle
Remarks: dial animation with spinning shuriken (star-shaped throwing weapon)
Price: $11,900; 9 unique pieces
Variation: case and shuriken with black PVD

Son of Sea Tourbillon

Movement: exclusive ArtyA flying tourbillon manual winding; 100-hour power reserve
Functions: hours, minutes, seconds
Case: stainless steel with exclusive blue carbon fiber lateral inserts, ø 44 mm, height 18 mm; engraved and screwed-down transparent case back; water-resistant to 3 atm
Band: reptile skin, buckle
Remarks: dial made with natural pigments, butterfly wings, fish scales, gold leaf, by artist D. Arpa-Cirpka
Price: $137,000; unique piece
Variations: all unique pieces in the collection; bezel set with double row of diamonds ($158,000)

Butterfly Delicacy Set

Movement: automatic ArtyA; 50-hour power reserve
Functions: hours, minutes, seconds
Case: stainless steel with lateral inserts, ø 44 mm, height 18 mm; engraved and screwed-down transparent case back; water-resistant to 3 atm
Band: calfskin, buckle
Remarks: dial decorated with genuine butterfly wings and gold leaf by artist D. Arpa-Cirpka; bezel set with double row of diamonds
Price: $21,700; unique piece
Variations: comes in 38-mm or 47-mm case, with or without PVD and diamonds; different, hand-made, unique

Son of Art Horse Enamel

Movement: automatic ArtyA; 50-hour power reserve
Functions: hours, minutes, seconds
Case: stainless steel, ø 47 mm, height 18 mm; engraved and screwed-down case back; water-resistant to 3 atm
Band: calfskin, buckle
Remarks: hand-engraved and sculpted dial, gold inlaying and enameling by artist Bram Ramon
Price: $24,900; unique piece

AUDEMARS PIGUET

The history of Audemars Piguet is one of the most engaging stories of Swiss watchmaking folklore: Ever since their school days together in the Vallée de Joux, Jules-Louis Audemars (b. 1851) and Edward-Auguste Piguet (b. 1853) knew they would follow in the footsteps of their fathers and grandfathers and become watchmakers. They were members of the same sports association, sang in the same choir, attended the same vocational school—and both became outstandingly talented watchmakers. The *manufacture* founded over 140 years ago by these two is still in family hands, and it has become one of the leading names in the industry.

In the history of watchmaking, only a handful of watches really achieved cult status. One of them is the Royal Oak by Audemars Piguet. It was born as a radical answer to the global invasion of the quartz watch. Audemars Piguet contacted the designer Gérald Genta to create a watch for a new generation of customers, a sportive luxury timepiece with a modern look, which could be worn every day. The result was a luxurious watch of stainless steel. The octagonal bezel held down with boldly "industrial" hexagonal bolts onto a 39-millimeter case was almost provocative. The watch, big for its time, was nicknamed "Jumbo." It ran on the then thinnest automatic movement, a slice 3.05 millimeters high.

Thanks to the ongoing success of the sporty Royal Oak collection, which is still its flagship, the company was able to expand its portfolio. It also sold off its shares in Jaeger-LeCoultre in 2000 and, in August 2009, opened the Manufacture des Forges in Le Brassus.

The second key to the brand's enduring success was no doubt the acquisition of the atelier Renaud et Papi in 1992. APRP, as it is known, specializes in creating and executing complex complications, a skill it lets other brands share in as well. At the SIHH, like clockwork, CEO François-Henry Bennahmias always presents a range of new models in the Royal Oak family with some remarkable complications, chief among them being the Supersonnerie, a minute repeater with a special architecture to maximally amplify the sound. Another is the openworked Royal Oak with two balance wheels operating in opposite directions on a single shaft and, in 2018, a gritty, technoid machine with a tourbillon.

Manufacture d'Horlogerie
Audemars Piguet
Route de France 16
CH-1348 Le Brassus
Switzerland

Tel.:
+41-21-642-3900

E-mail:
info@audemarspiguet.com

Website:
www.audemarspiguet.com

Founded:
1875

Number of employees:
approx. 1,300

Annual production:
40,000 watches

U.S. distributor:
Audemars Piguet (North America) Inc.
Service Center of the Americas
3040 Gulf to Bay Boulevard
Clearwater, FL 33759

Most important collection/price range:
Royal Oak / from approx. $17,800; Millenary / from approx. $28,400; special concept watches

Royal Oak Offshore Tourbillon Chronograph

Reference number: 26421ST.OO.A002CA.01
Movement: manually wound, AP Caliber 2947; ø 39.78 mm, height 11.6 mm; 30 jewels; 21,600 vph; minutes-tourbillon; skeletonized movement; 173-hour power reserve
Functions: hours, minutes; chronograph
Case: stainless steel, ø 45 mm, height 16.1 mm; bezel screwed to case back with 8 white gold screws; sapphire crystal; transparent case back; ceramic crown and pushers, screw-in crown; water-resistant to 10 atm
Band: rubber, buckle
Price: on request; limited to 50 pieces

Royal Oak Offshore Tourbillon Chronograph

Reference number: 26421OR.OO.A002CA.01
Movement: manually wound, AP Caliber 2947; ø 39.78 mm, height 11.6 mm; 30 jewels; 21,600 vph; minutes-tourbillon; skeletonized movement; 173-hour power reserve
Functions: hours, minutes; chronograph
Case: rose gold, ø 45 mm, height 16.1 mm; bezel screwed to case back with 8 white gold screws; sapphire crystal; transparent case back; ceramic crown and pushers, screw-in crown; water-resistant to 10 atm
Band: rubber, buckle
Price: on request; limited to 50 pieces

Royal Oak Offshore Diver

Reference number: 15710ST.OO.A052CA.01
Movement: automatic, AP Caliber 3120; ø 26.6 mm, height 4.26 mm; 40 jewels; 21,600 vph; hand-decorated movement; 60-hour power reserve
Functions: hours, minutes, sweep seconds; date
Case: stainless steel, ø 42 mm, height 14.1 mm; bezel screwed to case back with 8 white gold screws, crown-activated scale ring, with 0-60 scale; sapphire crystal; screw-in crown; water-resistant to 30 atm
Band: rubber, buckle
Price: $19,900
Variations: various colors

AUDEMARS PIGUET

Royal Oak Offshore Chronograph
Reference number: 26470ST.OO.A030CA.01
Movement: automatic, AP Caliber 3126/3840; ø 29.92 mm, height 7.16 mm; 59 jewels; 21,600 vph; 50-hour power reserve
Functions: hours, minutes, subsidiary seconds; chronograph; date
Case: stainless steel, ø 42 mm, height 14.6 mm; bezel screwed to case back with 8 white gold screws; sapphire crystal; transparent case back; ceramic crown and pushers, screw-in crown; water-resistant to 10 atm
Band: rubber, buckle
Price: $25,600
Variations: various straps, cases, and dials

Royal Oak Offshore Chronograph
Reference number: 26470OR.OO.A125CR.01
Movement: automatic, AP Caliber 3126/3840; ø 29.92 mm, height 7.16 mm; 59 jewels; 21,600 vph; 50-hour power reserve
Functions: hours, minutes, subsidiary seconds; chronograph; date
Case: rose gold, ø 42 mm, height 14.6 mm; bezel screwed to case back with 8 white gold screws; sapphire crystal; transparent case back; ceramic crown and pushers, screw-in crown; water-resistant to 10 atm
Band: reptile skin, buckle
Price: $40,700
Variations: various straps, cases, and dials

Royal Oak Concept Flying Tourbillon GMT
Reference number: 26589IO.OO.D002CA.01
Movement: manually wound, AP Caliber 2954; ø 35.6 mm, height 9.9 mm; 24 jewels; 21,600 vph; flying 1-minute tourbillon, finely hand-finished movement; 237-hour power reserve
Functions: hours, minutes; additional 24-hour display (2nd time zone), displays crown position for changing functions
Case: titanium, ø 44 mm, height 16.1 mm; ceramic bezel, screwed to case back with 8 white gold screws; sapphire crystal; transparent case back; ceramic crown and pushers
Band: rubber, folding clasp
Price: on request

Royal Oak Perpetual Calendar
Reference number: 26579CE.OO.1225CE.01
Movement: automatic, AP Caliber 5134; ø 29 mm, height 4.31 mm; 38 jewels; 19,800 vph; 40-hour power reserve
Functions: hours, minutes; perpetual calendar with date, weekday, calendar week, month, moon phase, leap year
Case: ceramic, ø 41 mm, height 9.5 mm; screwed to case back with 8 white gold screws; sapphire crystal; transparent case back; screw-in crown
Band: ceramic, folding clasp
Price: $93,900

Royal Oak Double Balance Skeleton
Reference number: 15467OR.OO.1256OR.01
Movement: automatic, AP Caliber 3132; ø 26.59 mm, height 5.57 mm; 38 jewels; 21,600 vph; double balance with 2 counter-wound hairsprings; skeletonized movement; 45-hour power reserve
Functions: hours, minutes, sweep seconds
Case: rose gold, ø 37 mm, height 10 mm; bezel screwed to case back with 8 white gold screws; sapphire crystal; transparent case back; water-resistant to 5 atm
Band: rose gold, folding clasp
Remarks: skeletonized dial
Price: $68,500
Variations: white gold ($76,000)

Royal Oak Double Balance Skeleton
Reference number: 15466BC.GG.1259BC.01
Movement: automatic, AP Caliber 3132; ø 26.59 mm, height 5.57 mm; 38 jewels; 21,600 vph; double balance with 2 counter-wound hairsprings; skeletonized movement; 45-hour power reserve
Functions: hours, minutes, sweep seconds
Case: white gold, ø 37 mm, height 10 mm; bezel screwed to case back with 8 white gold screws; sapphire crystal; transparent case back; water-resistant to 5 atm
Band: white gold, folding clasp
Remarks: skeletonized dial; "frosted gold" surface treatment
Price: $76,000
Variations: rose gold ($68,500)

AUDEMARS PIGUET

Royal Oak Automatic
Reference number: 15450ST.OO.1256ST.03
Movement: automatic, AP Caliber 3120; ø 26.6 mm, height 4.25 mm; 40 jewels; 21,600 vph; hand-decorated movement; 60-hour power reserve
Functions: hours, minutes, sweep seconds; date
Case: stainless steel, ø 37 mm, height 9.8 mm; bezel screwed to case back with 8 white gold screws; sapphire crystal; transparent case back; water-resistant to 5 atm
Band: stainless steel, folding clasp
Price: $16,500

Royal Oak "Jumbo" Extra Thin
Reference number: 15202BA.OO.1240BA.02
Movement: automatic, AP Caliber 2121; ø 28.4 mm, height 3.05 mm; 36 jewels; 19,800 vph; finely hand-finished movement; 40-hour power reserve
Functions: hours, minutes; date
Case: yellow gold, ø 39 mm, height 8.1 mm; bezel screwed to case back with 8 white gold screws; sapphire crystal; transparent case back; water-resistant to 5 atm
Band: yellow gold, folding clasp
Price: $55,400
Variations: rose gold; white gold; stainless steel

Royal Oak Chronograph
Reference number: 26331BA.OO.1220BA.01
Movement: automatic, AP Caliber 2385; ø 26.2 mm, height 5.5 mm; 37 jewels; 21,600 vph; hand-decorated movement; 40-hour power reserve
Functions: hours, minutes, subsidiary seconds; chronograph; date
Case: yellow gold, ø 41 mm, height 11 mm; bezel screwed to case back with 8 white gold screws; sapphire crystal; screw-in crown and pusher; water-resistant to 5 atm
Band: yellow gold, folding clasp
Price: $61,900

Royal Oak Tourbillon Chronograph Skeleton
Reference number: 26343CE.OO.1247CE.01
Movement: automatic, AP Caliber 2936; ø 29.9 mm, height 8.07 mm; 28 jewels; 21,600 vph; minutes-tourbillon; skeletonized movement; 72-hour power reserve
Functions: hours, minutes, subsidiary seconds; chronograph
Case: ceramic, ø 44 mm, height 13.2 mm; bezel screwed to case back with 8 white gold screws; sapphire crystal; transparent case back
Band: ceramic, folding clasp
Price: on request; limited to 4 × 25 pieces (by band color)

Ladies Millenary
Reference number: 77244OR.GG.1272OR.01
Movement: manually wound, AP Caliber 5201 × 32.74 × 28.59 mm, height 4.16 mm; 19 jewels; 21,600 vph; inverted movement design with balance and escapement on dial side; 49-hour power reserve
Functions: hours, minutes, subsidiary seconds
Case: white gold, 39.5 × 35.4 mm; bezel and lugs of "frosted gold"; sapphire crystal; transparent case back
Band: rose gold Milanese mesh, folding clasp
Remarks: mother-of-pearl dials
Price: $53,000

Ladies Millenary
Reference number: 77247BC.ZZ.1272BC.01
Movement: manually wound, AP Caliber 5201 × 32.74 × 28.59 mm, height 4.16 mm; 19 jewels; 21,600 vph; inverted movement design with balance and escapement on dial side; 49-hour power reserve
Functions: hours, minutes, subsidiary seconds
Case: white gold, 39.5 × 35.4 mm; bezel and lugs set with diamonds; sapphire crystal; transparent case back
Band: white gold Milanese mesh, folding clasp
Remarks: mother-of-pearl dials
Price: $47,000

AUDEMARS PIGUET

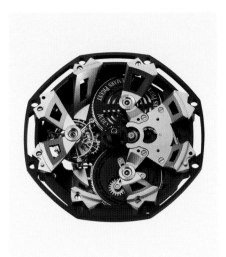

Caliber 2947

Manually wound, 1-minute tourbillon; column wheel control of chronograph functions; double spring barrel, 173-hour power reserve
Functions: hours, minutes, chronograph
Diameter: 39.78 mm
Height: 11.6 mm
Jewels: 30
Frequency: 21,600 vph
Remarks: inverted movement construction; cocks affixed to a circular plate projected toward center of movement; 353 parts

Caliber 2951

Manually wound; flying 1-minute tourbillon; single spring barrel, 72-hour power reserve
Functions: hours, minutes; power reserve indicator
Diameter: 29.5 mm
Height: 7.07 mm
Jewels: 17
Frequency: 21,600 vph
Remarks: fully in harmony with Royal Oak Concept Flying Tourbillon: decorated movement set with precious stones, painted sections on plates, bridges, and cocks; 255 parts

Caliber 2954

Manually wound; flying 1-minute tourbillon; double spring barrel, 237-hour power reserve
Functions: crown position display for changing functions
Diameter: 35.6 mm
Height: 9.9 mm
Jewels: 24
Frequency: 21,600 vph
Remarks: 348 parts

Caliber 2936

Manually wound; 1-minute tourbillon; single spring barrel, 72-hour power reserve
Functions: hours, minutes, subsidiary seconds; chronograph
Diameter: 29.9 mm
Height: 8.07 mm
Jewels: 28
Frequency: 21,600 vph
Remarks: skeletonized movement; 299 parts

Caliber 5201

Manually wound; inverted movement design with balance and escapement on dial; single spring barrel; 49-hour power reserve
Functions: hours, minutes, subsidiary seconds
Measurements: 32.74 × 28.59 mm
Height: 4.16 mm
Jewels: 19
Balance: with variable inertia
Frequency: 21,600 vph
Shock protection: Kif Elastor
Remarks: all parts decorated by hand; dial side of mainplate decorated with horizontal côtes de Genève and perlage on movement side; 157 parts

Caliber 3126-3840

Automatic; bidirectionally winding gold rotor; single spring barrel, 50-hour power reserve
Base caliber: 3120
Functions: hours, minutes, subsidiary seconds; chronograph; date
Diameter: 29.92 mm
Height: 7.16 mm
Jewels: 59
Balance: with variable inertia
Frequency: 21,600 vph
Shock protection: Kif Elastor
Remarks: beveled and polished steel parts, perlage on mainplate, bridges with côtes de Genève

AZIMUTH

Creativity can take on all forms and accept all forms as well. This appears to be the philosophy behind Azimuth, an independent watch brand that has sprouted an eclectic and surprising bouquet of watch designs. For the company, the path is by no means well-beaten: Azimuth always guarantees a raised eyebrow with avant-garde designs for luxury timepieces.

The company has produced several iconic models, like the Mr. Roboto, which looks, indeed, like a robot and is perfectly in tune with our times. Then there is the self-explanatory spaceship series and the automobile series, like the TT and GT, which all enjoy cult status. The Gran Turismo takes its cue from the racetracks of a generation ago.

For 2019, Azimuth once again started pushing the envelope, always a good idea for an unconventional company. It dipped into its Spaceship collection to create the Predator 2.0, which might have been the product of sci-fi writers and scientists dreaming of space travel. Cyber-robotics and interplanetary exploration are currently woven into the fabric of everyday life; the idea of centuries-old horological traditions being progressive and disruptive is now embraced by legions of watch lovers all over the world.

The Predator 2.0 is powered by a hand-wound caliber. The watch's bridges are crafted in aluminum, which makes the movement extra-light. "The Predator 2.0 is yet another expression of our desire to always be bold and adventurous in the field of mechanical watchmaking. The watch's inspiration, design and technical qualities are driven by our never-ending quest in the field of progressive horology," says Azimuth's CEO and technical director, Giuseppe Picchi.

Azimuth Watch Co. Sàrl
Rue des Draizes no. 5
CH-2000 Neuchâtel
Switzerland

Tel.:
+41-79-765-1466

E-mail:
gpi@azimuthwatch.com
sales@azimuthwatch.com

Website:
www.azimuthwatch.com

Founded:
2003

Number of employees:
6

U.S. distributor:
About Time Luxury Group
210 Bellevue Avenue
Newport RI 02840
401-952-4684

Most important collections/price range:
SP-1 / from $4,850

SP-1 Gran Turismo

Reference number: SP.SS.GT.L003
Movement: automatic winding, ETA 2671; 28,800 vph; ø 17.2 mm, height 4.80 mm
Functions: hours, minutes, seconds
Case: stainless steel with black PVD treatment, 50 × 45 mm; water-resistant to 3 atm
Band: calfskin strap, folding clasp
Price: $4,850; limited to 100 pieces
Variations: top in high-gloss polished stainless steel, gold PVD coating or urban camouflage PVD coating

SP-1 Gran Turismo Pavé Diamonds

Reference number: SP.SS.GT.N006
Movement: automatic winding, ETA 2671; 28,800 vph; ø 17.2 mm, height 4.80 mm
Functions: hours, minutes, seconds
Case: stainless steel, 50 × 45 mm; bezel set with diamonds; water-resistant to 3 atm
Band: calfskin strap, folding clasp
Remarks: available in full pavé setting
Price: $26,000

SP-1 Twin Turbo

Reference number: SP.SS.TT.N002
Movement: manual winding, ETA 2512-1; 21,600 vph; ø 17.2 mm, height 2.85 mm
Functions: hours, minutes, 2 time zones
Case: stainless steel and aluminum, 51 × 50 mm, water-resistant to 3 atm
Band: calfskin strap, folding clasp
Remarks: 2 vintage movements
Price: $6,000; limited to 88 pieces
Variations: top hood in silver, yellow, red (red limited to 50 pieces)

AZIMUTH

SP-1 Predator 2.0
Reference number: SP.Ti.PR.N001
Movement: manual winding, AZM 769 modified and skeletonized, 21,600 vph; ø 36.6 mm, height 4.5 mm
Functions: jumping hours, minutes, seconds
Case: titanium and stainless steel, diameter 44 mm, domed sapphire crystal; water-resistant to 3 atm
Band: rubber, folding clasp
Remarks: 3D titanium minute hand
Price: $5,700
Variations: case in bronze

SP-1 King Casino
Reference number: SP.KC.SS.N001
Movement: automatic, in-house modified (base ETA), 21,600 vph; ø 25.6 mm, height 6.0 mm
Functions: casino game function via crown; hours, minutes, seconds
Case: stainless steel, 45 × 45 mm, domed sapphire crystal, water-resistant to 3 atm
Band: calfskin, folding clasp
Remarks: roulette and baccarat game functions
Price: $3,650
Variations: chocolate color–plated or yellow gold–plated

SP-1 Crazy Rider
Reference number: SP.SS.CR.N004
Movement: automatic, in-house modified, 28,800 vph; ø 47.7 mm, height 4.35 mm
Functions: 24-hour chain drive hour system, minutes
Case: stainless steel with PVD treatment, titanium bezel with black PVD treatment, 55 × 36 mm, sapphire crystal, water-resistant to 3 atm
Band: calfskin, folding clasp
Price: $5,250

SP-1 Mr. Roboto Bronzo
Reference number: SP.BR.MRB.L001
Movement: automatic, in-house modified, 28,800 vph; ø 32.5 mm, height 6.7 mm
Functions: regulator hours, retrograde minutes, GMT
Case: bronze, 43 mm × 50 mm, sapphire crystal, water-resistant to 3 atm
Band: calfskin, bronze tang buckle
Price: $7,000; limited to 100 pieces

SP-1 Mr. Roboto R2
Reference number: SP.SS.ROT.N001
Movement: automatic, in-house modified, sapphire rotor, 28,800 vph; ø 32.5 mm, height 6.7 mm
Functions: regulator hours, retrograde minutes, GMT
Case: stainless steel, 47 × 55 mm, sapphire crystal, water-resistant to 3 atm
Band: calfskin, folding clasp
Price: $6,000
Variations: mid-case in titanium with blue PVD treatment

SP-1 Twin Barrel Tourbillon
Reference number: SP.TB.TI.L001
Movement: manual winding tourbillon, in-house modified, 5-day power reserve, twin barrels, 28,800 vph; 36.3 × 32.0 mm, height 6.4 mm
Functions: jumping hours, minutes, specially modified twin-disk jumping hour system on 3D minute hand
Case: titanium with carbon fiber side inserts, 45 × 50 mm, height 18 mm, domed sapphire crystal, water-resistant to 5 atm
Band: calfskin, folding clasp
Price: $89,000; limited to 25 pieces

BALL WATCH CO.

BALL Watch Company SA
Rue du Châtelot 21
CH-2300 La Chaux-de-Fonds
Switzerland

Tel.:
0041-32-724-53-00

Fax:
0041-32-724-53-01

E-mail:
info@ballwatch.ch

Website:
www.ballwatch.com

Founded:
1891

U.S. distributor:
BALL Watch USA
1920 Dr. Martin Luther King Jr. St. N
Suite D
St. Petersburg, FL 33704
727-896-4278

Most important collections/price range:
Engineer, Fireman, Trainmaster / $1,300 to $6,500

Engineer, Fireman, Trainmaster, Conductor . . . these names for the Ball Watch Co. collections trace back to the company's origins and evoke the glorious age when trains puffing smoke and steam crisscrossed America. Back then, the pocket watch was a necessity to maintain precise rail schedules. By 1893, many companies had adopted the General Railroad Timepiece Standards, which included such norms as regulation in at least five positions, precision to within thirty seconds per week, Breguet balance springs, and so on. One of the chief players in developing the standards was Webster Clay Ball. This farmboy-turned-watchmaker from Fredericktown, Ohio, decided to leave the homestead for a more lucrative occupation. He apprenticed as a watchmaker, became a salesperson for Dueber watch cases, and finally opened the Webb C. Ball Company in Cleveland. In 1891, he added the position of chief inspector of the Lake Shore Lines to his CV. When a hogshead's watch stopped, resulting in an accident, Ball decided to establish quality benchmarks for watches that included antimagnetic technology, and he set up an inspection system for the timepieces.

Today, Ball Watch Co. has maintained its lineage, although now producing in Switzerland. These rugged, durable watches aim to be "accurate in adverse conditions," so the company tagline says—and at a very good price. Functionality is therefore a priority, so Ball does have a growing list of technologies it works into its timepieces. Top billing goes to the Amortizer and Springlock shock absorption systems that increase accuracy of the watches and protect them from bumps and shakes. The A-Proof system fends off magnetic fields up to 80,000 A/m while allowing a transparent case back. Ball has also developed special oils for cold temperatures, and it is one of few brands to use tritium gas tubes to light up dials, hands, and markers.

Finally, the company has also developed an in-house movement named RRM7309. It's an automatic that boasts COSC certification, with eighty hours of power. The quality seal is built into the name: "RR" stands simply for railroad, a look back to the company's origins.

Engineer Hydrocarbon AeroGMT II

Reference number: DG2018C-S3C-BK
Movement: automatic, Ball Caliber RR1201-C; ø 25.6 mm, height 4.1 mm; 21 jewels; 28,800 vph; 42-hour power reserve; COSC-certified chronometer
Functions: hours, minutes, sweep seconds; date; 2nd time zone indication
Case: stainless steel, ø 42 mm, height 13.85 mm; sapphire bidirectional bezel with micro gas tube illumination; sapphire crystal; crown protection; water-resistant to 10 atm
Band: stainless steel, folding buckle and extension
Remarks: micro gas tube illumination
Price: $3,499
Variations: rubber strap

Engineer Hydrocarbon Airborne

Reference number: DM2076C-S1CAJ-BK
Movement: automatic, Ball Caliber RR1102-CSL; ø 25.6 mm, height 5.05 mm; 25 or 26 jewels; 28,800 vph; 38-hour power reserve; COSC-certified chronometer; Springlock antishock system
Functions: hours, minutes, sweep seconds; day, date
Case: stainless steel, ø 42 mm, height 13.85 mm; unidirectional ceramic bezel; sapphire crystal; crown protection; water-resistant to 12 atm
Band: stainless steel, folding buckle and extension
Remarks: micro gas tube illumination
Price: $4,399

Engineer Hydrocarbon NEDU

Reference number: DC3026A-SC-BK
Movement: automatic, Ball Caliber RR1402-C; ø 30 mm, height 7.9 mm; 25 jewels; 28,800 vph; 48-hour power reserve; COSC-certified chronometer
Functions: hours, minutes, subsidiary seconds; day, date; 12-hour chronograph operable underwater
Case: stainless steel, ø 42 mm, height 17.30 mm; patented helium system; ceramic unidirectional bezel; sapphire crystal; crown protection system; water-resistant to 60 atm
Band: titanium/stainless steel, folding buckle and extension
Remarks: micro gas tube illumination
Price: $5,099
Variations: blue dial; rubber strap

Engineer Hydrocarbon DEVGRU
Reference number: NM3200C-SJ-BK
Movement: automatic, Ball Caliber RR1102-SL; ø 25.6 mm, height 5.05 mm; 25 or 26 jewels; 28,800 vph; 38-hour power reserve; Springlock antishock system
Functions: hours, minutes, sweep seconds; day, date; power reserve indication
Case: stainless steel, ø 42 mm, height 13.4 mm; patented shock absorption system; sapphire crystal with outer protective flange; patented crown protection; water-resistant to 10 atm
Band: stainless steel, folding buckle and extension
Remarks: micro gas tube illumination
Price: $2,299
Variations: blue dial; rubber strap

Engineer Master II Diver
Reference number: DM3020A-SAJ-BK
Movement: automatic, Ball Caliber RR1102; ø 25.6 mm, height 5.05 mm; 25 or 26 jewels; 28,800 vph; 38-hour power reserve
Functions: hours, minutes, sweep seconds; day, date
Case: stainless steel, ø 42 mm, height 14.55 mm; inner bezel with micro gas tube illumination; sapphire crystal; screw-in crown; water-resistant to 30 atm
Band: stainless steel, folding buckle
Remarks: micro gas tube illumination
Price: $2,499
Variations: rubber strap

Engineer Master II Diver Worldtime
Reference number: DG2022A-S3AJ-BK
Movement: automatic, Ball Caliber RR1501; ø 31.4 mm, height 6.95 mm; 25 jewels; 28,800 vph; 38-hour power reserve
Functions: hours, minutes, sweep seconds; day, date; world time display
Case: stainless steel, ø 45 mm, height 15.4 mm; luminous bidirectional rotating inner bezel; sapphire crystal; screw-in crown; water-resistant to 30 atm
Band: stainless steel bracelet, folding buckle
Remarks: micro gas tube illumination
Price: $2,999
Variations: rubber strap

Engineer Master II Skindiver Heritage
Reference number: DM3208C-SC-BK
Movement: automatic, Ball Caliber RR1102-C; ø 25.6 mm, height 5.05 mm; 25 or 26 jewels; 28,800 vph; 38-hour power reserve; COSC-certified chronometer
Functions: hours, minutes, sweep seconds; date
Case: stainless steel, ø 41 mm, height 14.9 mm; unidirectional bezel; mu-metal shield; sapphire crystal; screw-in crown; water-resistant to 10 atm
Band: stainless steel, folding buckle
Remarks: micro gas tube illumination
Price: $2,799

Engineer Master II Aviator
Reference number: NM1080C-L14A-BK
Movement: automatic, Ball Caliber RR1102; ø 25.6 mm, height 5.05 mm; 25 or 26 jewels; 28,800 vph; 38-hour power reserve
Functions: hours, minutes, sweep seconds; day, date
Case: stainless steel, ø 46 mm, height 11.55 mm; mu-metal shield; antireflective convex sapphire crystal; screw-in crown; water-resistant to 10 atm
Band: calfskin, standard buckle
Remarks: micro gas tube illumination
Price: $1,999
Variations: stainless steel bracelet; rubber strap

Engineer II Moon Phase
Reference number: NM2282C-LLJ-BK
Movement: automatic, Ball Caliber RR1801; ø 25.6 mm, height 5.05 mm; 25 jewels; 28,800 vph; 42-hour power reserve
Functions: hours, minutes, sweep seconds; date, moon phase
Case: stainless steel, ø 41 mm; luminous moon phase indication; sapphire crystal; screw-in crown; water-resistant to 10 atm; Amortiser antishock system
Band: reptile skin strap, standard buckle
Remarks: micro gas tube illumination
Price: $1,799
Variations: stainless steel bracelet; blue, gray dial

BALL WATCH CO.

Engineer II Volcano
Reference number: NM3060C-PCJ-GY
Movement: automatic, Ball Caliber RR1102-C; ø 25.6 mm, height 5.05 mm; 25 or 26 jewels; 28,800 vph; 38-hour power reserve; COSC-certified chronometer
Functions: hours, minutes, sweep seconds; day, date
Case: patented mu-metal and carbide composite, ø 45 mm, height 12.4 mm; sapphire crystal; screw-in crown; water-resistant to 10 atm
Band: rubber, standard buckle
Remarks: micro gas tube illumination
Price: $3,499
Variations: canvas NATO strap

Engineer II Magneto S
Reference number: NM3022C-N1CJ-BK
Movement: automatic, Ball Caliber RR1103-CSL; ø 25.6 mm, height 4.6 mm; 25 or 26 jewels; 28,800 vph; 38-hour power reserve; COSC-certified chronometer; Springlock antishock system
Functions: hours, minutes, sweep seconds; date
Case: stainless steel, ø 42 mm, height 12.9 mm; A-Proof antimagnetic system; sapphire crystal; screw-in crown; transparent case back; water-resistant to 10 atm
Band: cordura fabric, standard buckle
Remarks: micro gas tube illumination
Price: $3,399

Trainmaster Eternity
Reference number: NM2080D-S1J-BE
Movement: automatic, Ball Caliber RR1102; ø 25.6 mm, height 5.05 mm; 25 or 26 jewels; 28,800 vph; 38-hour power reserve
Functions: hours, minutes, sweep seconds; day, date
Case: stainless steel, ø 39.5 mm, height 11.8 mm; sapphire crystal; screw-in crown; transparent case back; water-resistant to 3 atm
Band: stainless steel, folding buckle
Remarks: micro gas tube illumination
Price: $2,299
Variations: black dial; reptile skin strap

Trainmaster Kelvin
Reference number: NT3888D-LL1J-GYF
Movement: automatic, Ball Caliber RR1601; ø 25.6 mm, height 5.1 mm; 21 jewels; 28,800 vph; 42-hour power reserve
Functions: hours, minutes, sweep seconds; date; patented mechanical thermometric indication
Case: stainless steel, ø 39.5 mm, height 11.8 mm; sapphire crystal; transparent case back; water-resistant to 3 atm
Band: reptile skin strap, folding buckle
Remarks: micro gas tube illumination
Price: $3,599
Variations: silver dial; stainless steel bracelet; TMT Celsius scale

Trainmaster Manufacture 80 Hours
Reference number: NM3280D-S1CJ-BK
Movement: automatic, Ball Caliber RRM7309-C; ø 34.24 mm, height 5.16 mm; 25 jewels; 28,800 vph; 80-hour power reserve; COSC-certified chronometer
Functions: hours, minutes, sweep seconds; date
Case: stainless steel, ø 40 mm, height 12.25 mm; sapphire crystal; transparent case back; screw-in crown; water-resistant to 5 atm
Band: stainless steel bracelet, folding buckle
Remarks: micro gas tube illumination
Price: $2,799

Trainmaster Worldtime
Reference number: GM2020D-S1CJ-SL
Movement: automatic, Ball Caliber RR1501-C; ø 31.4 mm, height 6.95 mm; 25 jewels; 28,800 vph; 38-hour power reserve; COSC-certified chronometer
Functions: hours, minutes, sweep seconds; day, date; world time display
Case: stainless steel, ø 41 mm, height 12.5 mm; sapphire crystal; transparent case back; screw-in crown; water-resistant to 5 atm
Band: stainless steel bracelet, folding buckle
Remarks: micro gas tube illumination
Price: $2,699
Variations: black dial; reptile skin strap

BALL WATCH CO.

Trainmaster Worldtime Chronograph
Reference number: CM2052D-LL1J-SLBE
Movement: automatic, Ball Caliber RR1502; ø 30 mm, height 7.9 mm; 25 jewels; 28,800 vph; 48-hour power reserve
Functions: hours, minutes, subsidiary seconds; day, date; chronograph with accumulated measurement up to 12 hours; world time display
Case: stainless steel, ø 42 mm, height 13.7 mm; sapphire crystal; transparent case back; screw-in crown; water-resistant to 5 atm
Band: reptile skin strap, standard buckle
Remarks: micro gas tube illumination
Price: $4,299; **Variations:** black with red dial, silver with red dial; stainless steel bracelet

Fireman Enterprise
Reference number: NM2188C-S5J-BK
Movement: automatic, Ball Caliber RR1103; ø 25.6 mm, height 4.6 mm; 25 or 26 jewels; 28,800 vph; 38-hour power reserve
Functions: hours, minutes, sweep seconds; magnified date
Case: stainless steel, ø 40 mm, height 11.3 mm; sapphire crystal; screw-in crown; water-resistant to 10 atm
Band: stainless steel bracelet, folding buckle
Remarks: micro gas tube illumination
Price: $1,199
Variations: white dial; NATO strap

Fireman NECC
Reference number: DM3090A-P5J-BK
Movement: automatic, Ball Caliber RR1103; ø 25.6 mm, height 4.6 mm; 25 or 26 jewels; 28,800 vph; 38-hour power reserve
Functions: hours, minutes, sweep seconds; magnified date
Case: stainless steel with titanium carbide coating, ø 42 mm, height 13.2 mm; stainless steel carbide rotating bezel; sapphire crystal; transparent case back; screw-in crown; water-resistant to 30 atm
Band: rubber strap, standard buckle
Remarks: micro gas tube illumination
Price: $1,599
Variations: white or blue dial; stainless steel case; stainless steel bracelet

Fireman Storm Chaser DLC Glow II
Reference number: CM2192C-L4A-GY
Movement: automatic, Ball Caliber RR1402; ø 30 mm, height 7.9 mm; 25 jewels; 28,800 vph; 48-hour power reserve
Functions: hours, minutes, subsidiary seconds; day, date; 12-hour chronograph; tachymeter
Case: stainless steel with DLC coating, ø 43 mm, height 15.45 mm; sapphire crystal; screw-in crowns; water-resistant to 10 atm
Band: calfskin, standard buckle
Remarks: micro gas tube illumination
Price: $2,999; limited to 1,999 pieces
Variations: white dial

Engineer III Bronze
Reference number: NM2186C-L3J-BK
Movement: automatic, Ball Caliber RR1102-SL; ø 25.6 mm, height 5.05 mm; 25 or 26 jewels; 28,800 vph; 38-hour power reserve; Springlock antishock system
Functions: hours, minutes, sweep seconds; day, date
Case: bronze, ø 43 mm, height 13.45 mm; mu-metal shield; sapphire crystal; screw-in crown; water-resistant to 10 atm; Amortiser antishock system
Band: calfskin, standard buckle
Remarks: micro gas tube illumination
Price: $2,300

Engineer M Marvelight
Reference number: NM2128C-S1C-BE
Movement: automatic, Ball Caliber RRM7309-C; ø 34.24 mm, height 5.16 mm; 25 jewels; 28,800 vph; 80-hour power reserve; COSC-certified chronometer
Functions: hours, minutes, sweep seconds; date
Case: stainless steel, ø 43 mm, height 12.85 mm; sapphire crystal; transparent case back; screw-in crown; water-resistant to 10 atm; Amortiser antishock system
Band: stainless steel bracelet, folding buckle
Remarks: micro gas tube illumination
Price: $2,499
Variations: 40-mm case; black or gray dial; calfskin strap

BAUME & MERCIER

Baume & Mercier and its elite watchmaking peers Cartier and Piaget make up the quality timepiece nucleus in the Richemont Group's impressive portfolio. The tradition-rich brand counts among the most accessible and most affordable watches of the Genevan luxury brands. In the past decade, it has created a number of remarkable—and often copied—classics. The twelve-sided Riviera and the Catwalk have had to step off the stage, but the classic rectangular Hampton continues to evolve. In recent years, the company has worked hard to gain acceptance in the men's market for its Classima Executives line and to build on watchmaking glory of days gone by, when Baume & Mercier was celebrated as a chronograph specialist.

Though the brand has taken up residence in Geneva, most of the watches are produced in a reassembly center built a few years ago in Les Brenets near Le Locle. Individual parts are made by specialized suppliers according to the strictest of quality guidelines. Some of these manufacturers are sister companies within the Richemont Group.

Keeping the brand abreast of trends in both male and female fashions is key to its design strategy and is ensured by the brand's integrated design studio within the Richemont Luxury Group in Geneva. The newly interpreted, iconic Classima picked up on the modern minimalist zeitgeist, which could be developed into more luxurious versions, one with a window on the balance wheel, the other with a second time zone. Economic realities also meant staying on the lower end of the price scales and targeting a younger age group with the Petite Promesse line, which falls somewhere between being jewelry and a timekeeper at the same time, thanks to the graceful, double-looped strap in bright colors or cool stainless steel. Meanwhile, the brand revived the name of an old movement from the 1960s, the Baumatic, using a silicon hairspring, escape wheel, and anchor that together deliver five days of power. The movement is used in the Clifton collection.

Baume & Mercier
rue André de Garrini 4
CH 1217 Meyrin
Switzerland

Tel.:
+41-022-580-2948

Fax:
+41-44-972-2086

Website:
www.baume-et-mercier.com
Register on the website to contact via e-mail

Founded:
1830

Annual production:
100,000 (estimated)

U.S. distributor:
Baume & Mercier
Richemont North America
New York, NY 10022
800-637-2437

Most important collections/price range:
Clifton (men) / $2,700 to $13,950; Capeland (men) / $4,350 to $19,990; Hampton (men and women) / $3,450 to $15,000; Linea (women) / $1,950 to $15,750; Classima / $1,750 to $5,950; Petite Promesse (women) / $2,450 to $3,300

Clifton Baumatic
Reference number: 10436
Movement: automatic, Caliber Baumatic BM12.1975A; ø 28.2 mm, height 4.2 mm; 21 jewels; 28,800 vph; silicon hairspring, anchor, and escape wheel; balance with variable inertia; 120-hour power reserve; COSC-certified chronometer
Functions: hours, minutes, sweep seconds; date
Case: stainless steel, ø 40 mm, height 10.3 mm; sapphire crystal; transparent case back; water-resistant to 5 atm
Band: reptile skin, buckle
Price: $2,990

Clifton Baumatic
Reference number: 10400
Movement: automatic, Caliber Baumatic BM12.1975A; ø 28.2 mm, height 4.2 mm; 21 jewels; 28,800 vph; silicon hairspring, anchor, and escape wheel; balance with variable inertia; 120-hour power reserve
Functions: hours, minutes, sweep seconds; date
Case: stainless steel, ø 40 mm, height 10.3 mm; sapphire crystal; transparent case back; water-resistant to 5 atm
Band: stainless steel, triple folding clasp
Price: $2,950

Clifton Baumatic
Reference number: 10401
Movement: automatic, Caliber Baumatic BM12.1975A; ø 28.2 mm, height 4.2 mm; 21 jewels; 28,800 vph; silicon hairspring, anchor, and escape wheel; balance with variable inertia; 120-hour power reserve
Functions: hours, minutes, sweep seconds; date
Case: stainless steel, ø 40 mm, height 10.3 mm; bezel in rose gold; sapphire crystal; transparent case back; rose gold crown; water-resistant to 5 atm
Band: reptile skin, buckle
Price: $3,500

BAUME & MERCIER

Clifton Club Indian "Burt Munro" Limited Edition

Reference number: 10404
Movement: automatic, ETA Caliber 7750; ø 30 mm, height 7.9 mm; 25 jewels; 28,800 vph; skeletonized rotor; 48-hour power reserve
Functions: hours, minutes, subsidiary seconds; chronograph; date
Case: stainless steel, ø 44 mm, height 14.95 mm; bezel with black DLC coating; sapphire crystal; water-resistant to 5 atm
Band: calfskin, buckle
Remarks: named for the record-breaking New Zealand motorcyclist
Price: $3,900; limited to 1,967 pieces (for the year the record was set)

Clifton Club Indian Legend Tribute Limited Edition

Reference number: 10402
Movement: automatic, ETA Caliber 7750; ø 30 mm, height 7.9 mm; 25 jewels; 28,800 vph; skeletonized rotor; 48-hour power reserve
Functions: hours, minutes, subsidiary seconds; chronograph; date
Case: stainless steel, ø 44 mm, height 13.3 mm; bezel with black DLC coating; sapphire crystal; water-resistant to 5 atm
Band: calfskin, buckle
Remarks: partially skeletonized dial
Price: $3,900; limited to 1,901 pieces

Clifton Club

Reference number: 10339
Movement: automatic, Sellita Caliber SW200-1; ø 25.6 mm, height 4.6 mm; 26 jewels; 28,800 vph; 48-hour power reserve
Functions: hours, minutes, sweep seconds; date
Case: stainless steel with black DLC coating, ø 42 mm, height 10.3 mm; unidirectional bezel, with 0-60 scale; sapphire crystal; screw-in crown; water-resistant to 10 atm
Band: rubber, buckle
Price: $2,250

Classima

Reference number: 10441
Movement: quartz
Functions: hours, minutes; date
Case: stainless steel with rose gold PVD coating, ø 40 mm, height 5.95 mm; sapphire crystal; water-resistant to 5 atm
Band: calfskin, buckle
Price: $1,150

Classima Automatic

Reference number: 10453
Movement: automatic, Sellita Caliber SW200-1; ø 25.6 mm, height 4.6 mm; 26 jewels; 28,800 vph; 38-hour power reserve
Functions: hours, minutes, sweep seconds; date
Case: stainless steel, ø 42 mm, height 8.85 mm; sapphire crystal; water-resistant to 5 atm
Band: calfskin, buckle
Price: $1,650

Clifton Club Quartz

Reference number: 10413
Movement: quartz
Functions: hours, minutes, sweep seconds; date
Case: stainless steel, ø 42 mm, height 9.55 mm; sapphire crystal; water-resistant to 10 atm
Band: stainless steel, triple folding clasp
Price: $1,490

BELL & ROSS

If there is such a class as "military chic," Bell & Ross is undoubtedly one of the leaders. The Paris-headquartered brand develops, manufactures, assembles, and regulates its timepieces in a modern factory in La Chaux-de-Fonds in the Jura mountains of Switzerland. The early models had a certain stringency that one might associate with soldierly life, but in the past years, working with outside specialists, the company has ventured into even more complicated watches such as tourbillons and wristwatches with uncommon shapes. This kind of ambitious innovation has only been possible since perfume and fashion specialist Chanel—which also maintains a successful watch line in its own right—became a significant Bell & Ross shareholder and brought the watchmaker access to the production facilities where designers Bruno Belamich and team can create more complicated, more interesting designs for their esthetically unusual "instrument" watches. And to prove perhaps that watchmakers are not riding the coattails (or fenders) of iconic cars, in 2016 Belamich and his team presented the AeroGT at the Geneva International Auto Show, a super–sports car that can stand on its own next to a series of same-class Italians.

Belamich continues to prove his skills where technical features and artful proportions are concerned, and what sets Bell & Ross timepieces apart from those of other, more traditional professional luxury makers is their special, roguish look—a delicate balance between striking, martial, and poetic. And it is this beauty for the eye to behold that makes the company's wares popular with style-conscious "civilians" as well as with the pilots, divers, astronauts, sappers, and other hard-riding professionals drawn to Bell & Ross timepieces for their superior functionality. And the brand is also capable of producing more feminine timepieces as well, or at least objects that will stimulate the inner warrior that slumbers in everyone.

Bell & Ross Ltd.
8 rue Copernic
F-75116 Paris
France

Tel.:
+33-1-73-73-93-00

Fax:
+33-1-73-73-93-01

E-mail:
sav@bellross.com

Website:
www.bellross.com

Founded:
1992

U.S. distributor:
Bell & Ross, Inc.
605 Lincoln Road, Suite 300
Miami Beach, FL 33139
888-307-7887; 305-672-3840 (fax)
information@bellross.com
www.bellross.com

Most important collections/price range:
Instrument BR-X1, BR 01, and BR 03 / approx. $3,100 to $200,000

BR V2-94 Racing Bird

Reference number: BRV294-BB-ST/SST
Movement: automatic, Caliber BR-CAL.301 (base ETA 2894); ø 28.6 mm, height 6.1 mm; 37 jewels; 28,800 vph; 42-hour power reserve
Functions: hours, minutes, subsidiary seconds; chronograph; date
Case: stainless steel, ø 41 mm, height 13.9 mm; bezel with aluminum insert, sapphire crystal; transparent case back; screw-in crown and pusher; water-resistant to 10 atm
Band: stainless steel, folding clasp
Price: $5,700; limited to 999 pieces
Variations: calfskin strap ($5,400)

BR V2-94 Steel Heritage

Reference number: BRV294-HER-ST/SRB
Movement: automatic, Caliber BR-CAL.301 (base ETA 2894); ø 28.6 mm, height 6.1 mm; 37 jewels; 28,800 vph; 42-hour power reserve
Functions: hours, minutes, subsidiary seconds; chronograph; date
Case: stainless steel, ø 41 mm, height 13.9 mm; unidirectional bezel with aluminum insert, with 0-60 scale; sapphire crystal; transparent case back; screw-in crown and pusher; water-resistant to 10 atm
Band: rubber, folding clasp
Price: $4,300
Variations: stainless steel bracelet ($4,600)

BR V2-93 GMT

Reference number: BRV293-BL-ST
Movement: automatic, Caliber BR-CAL.303 (base ETA 2893-2); ø 25.6 mm, height 4.1 mm; 21 jewels; 28,800 vph; 42-hour power reserve
Functions: hours, minutes, sweep seconds; additional 24-hour display (2nd time zone); date
Case: stainless steel, ø 41 mm, height 12.15 mm; bidirectional bezel with 0-24 scale; sapphire crystal; screw-in crown; water-resistant to 10 atm
Band: stainless steel, folding clasp
Price: $3,500
Variations: rubber strap ($3,200)

BELL & ROSS

BR03-92 Diver Black
Reference number: BR0392-D-BL-ST/SRB
Movement: automatic, Caliber BR-CAL.302 (base ETA 2892-A2); ø 25.6 mm, height 3.6 mm; 21 jewels; 28,800 vph; 42-hour power reserve
Functions: hours, minutes, sweep seconds; date
Case: stainless steel, 42 × 42 mm, height 12.3 mm; unidirectional bezel screwed to monocoque case with 4 screws, with 0-60 scale; sapphire crystal; screw-in crown; water-resistant to 30 atm
Band: rubber, buckle
Price: $3,700

BR03-92 Diver Blue
Reference number: BR0392-D-BU-ST/SRB
Movement: automatic, Caliber BR-CAL.302 (base ETA 2892-A2); ø 25.6 mm, height 3.6 mm; 21 jewels; 28,800 vph; 42-hour power reserve
Functions: hours, minutes, sweep seconds; date
Case: stainless steel, 42 × 42 mm, height 12.3 mm; unidirectional bezel screwed to monocoque case with 4 screws, with 0-60 scale; sapphire crystal; screw-in crown; water-resistant to 30 atm
Band: rubber, buckle
Price: $3,700

BR03-92 Diver Bronze
Reference number: BR0392-D-BL-BR/SCA
Movement: automatic, Caliber BR-CAL.302 (base ETA 2892-A2); ø 25.6 mm, height 3.6 mm; 21 jewels; 28,800 vph; 42-hour power reserve
Functions: hours, minutes, sweep seconds; date
Case: bronze, 42 × 42 mm, height 12.3 mm; unidirectional bezel screwed to monocoque case with 4 screws, with 0-60 scale; sapphire crystal; screw-in crown; water-resistant to 30 atm
Band: calfskin, buckle
Price: $3,990; limited to 999 pieces

BR03-94 R.S.18
Reference number: BR0394-RS18
Movement: automatic, Caliber BR-CAL.301 (base ETA 2894-2); ø 28.6 mm, height 6.1 mm; 37 jewels; 28,800 vph; 42-hour power reserve
Functions: hours, minutes, subsidiary seconds; chronograph; date
Case: titanium, 42 × 42 mm, height 12.5 mm; bezel screwed to monocoque case with 4 screws; sapphire crystal; water-resistant to 10 atm
Band: rubber, buckle
Remarks: homage to Renault F1 team
Price: $6,500; limited to 999 pieces

BR-X1 R.S.18
Reference number: BRX1-RS18
Movement: automatic, Caliber BR-CAL.313 (base ETA Caliber 2892-A2 with Dubois Dépraz module); ø 25.6 mm, height 6.5 mm; 56 jewels; 28,800 vph; skeletonized movement; 42-hour power reserve
Functions: hours, minutes, subsidiary seconds; chronograph; date
Case: titanium, 45 × 45 mm, height 14.8 mm; bezel screwed to monocoque case with 4 screws; sapphire crystal; transparent case back; water-resistant to 10 atm
Band: rubber, buckle
Remarks: homage to Renault F1 Team; color-coded in F1 steering wheel style
Price: $21,500; limited to 250 pieces

BR-X1 Tourbillon R.S.18
Reference number: BRX1-CHTB-RS18
Movement: manually wound, Caliber BR-CAL.283; ø 34 mm, height 7.4 mm; 35 jewels; 21,600 vph; flying 1-minute tourbillon; skeletonized movement; 96-hour power reserve
Functions: hours, minutes; power reserve indicator; chronograph
Case: titanium, 45 × 45 mm, height 14.85 mm; bezel screwed to monocoque case with 4 screws; sapphire crystal; transparent case back; water-resistant to 10 atm
Band: rubber, buckle
Remarks: homage to Renault F1 team
Price: $189,000; limited to 20 pieces

BLANCPAIN

In its advertising, the Blancpain watch brand has always proudly declared that, since 1735, the company has never made quartz watches and never will. Indeed, Blancpain is Switzerland's oldest watchmaker, and by sticking to its ideals, the company was put out of business by the "quartz boom" of the 1970s.

The Blancpain brand we know today came into being in the mid-eighties, when Jean-Claude Biver and Jacques Piguet purchased the venerable name. The company was subsequently moved to the Frédéric Piguet watch factory in Le Brassus, where it quickly became largely responsible for the renaissance of the mechanical wristwatch. This success caught the attention of the Swatch Group—known at that time as SMH. In 1992, it swooped in and purchased both companies to add to its portfolio. Movement fabrication and watch production were melded to form the Blancpain Manufacture in mid-2010.

But being quartzless does not mean being old-fashioned. Over the past several years, Blancpain president Marc A. Hayek has put a great deal of energy into developing the company's technical originality. He is frank about the fact that making new calibers did harness most of Blancpain's creative potential, leaving little to apply to its existing collection of watches. Still, in terms of complications, Blancpain watches have always been in a class of their own. Furthermore, the farsighted move now means that other brands in the family have outstanding movements at their disposal, notably the Z9 from Harry Winston.

The Blancpain portfolio has been growing and subtly modernizing with each new model. The watches feature the company's own basic movement and a choice of manual or automatic winding, like the new collection, the Fifty Fathoms Bathyscaphe, a modern interpretation of the classic diver's watch of 1953. As part of the planned consolidation of the entire collection, the other major families—the Villeret, Le Brassus, L'Evolution, and Sport—are being reworked over time.

Blancpain SA
Le Rocher 12
CH-1348 Le Brassus
Switzerland

Tel.:
+41-21-796-3636

Website:
www.blancpain.com

Founded:
1735

U.S. distributor:
Blancpain
The Swatch Group (U.S.), Inc.
1200 Harbor Boulevard
Weehawken, NJ 07086
201-271-4680

Most important collections/price range:
L'Evolution, Villeret, Fifty Fathoms, Le Brassus, Women / $9,800 to $400,000

Le Brassus Carrousel Répétition Minutes

Reference number: 00235-3631-55B
Movement: automatic, Blancpain Caliber 235; ø 32.8 mm, height 9.1 mm; 54 jewels; 21,600 vph; escapement with 1-minute flying carrousel
Functions: hours, minutes; minute repeater
Case: pink gold, ø 45 mm, height 15.35 mm; sapphire crystal; transparent case back
Band: reptile skin, folding clasp
Remarks: enamel dial
Price: $412,100

Le Brassus Tourbillon Carrousel

Reference number: 2322-3631-55B
Movement: manually wound, Blancpain Caliber 2322; ø 35.3 mm, height 5.85 mm; 70 jewels; 21,600 vph; escapement system with flying 1-minute tourbillon and 1-minute carrousel with differential compensation; 3 spring barrels, 168-hour power reserve
Functions: hours, minutes; power reserve indicator (on rear); date
Case: pink gold, ø 44.6 mm, height 11.94 mm; sapphire crystal; transparent case back; water-resistant to 3 atm
Band: reptile skin, folding clasp
Remarks: enamel dial
Price: $319,000

Villeret Tourbillon Volant Une Minute 12 Jours

Reference number: 66240-3431-55B
Movement: automatic, Blancpain Caliber 242; ø 30.6 mm, height 6.1 mm; 43 jewels; 28,800 vph; 1-minute tourbillon; 288-hour power reserve
Functions: hours, minutes; power reserve indicator (on rear)
Case: platinum, ø 42 mm, height 11.65 mm; sapphire crystal; transparent case back; water-resistant to 3 atm
Band: reptile skin, folding clasp
Remarks: enamel dial
Price: $148,800; limited to 188 pieces
Variations: pink gold ($127,400)

BLANCPAIN

Villeret Carrousel Moonphase
Reference number: 6622L-3631-55B
Movement: automatic, Blancpain Caliber 225L; ø 31.9 mm, height 6.86 mm; 40 jewels; 28,800 vph; flying 1-minute carrousel; 120-hour power reserve
Functions: hours, minutes; date, moon phase
Case: pink gold, ø 42 mm, height 12.74 mm; sapphire crystal; transparent case back; water-resistant to 3 atm
Band: reptile skin, folding clasp
Remarks: enamel dial
Price: $129,600
Variations: platinum (limited to 88 pieces, $151,000)

Villeret Quantième Perpétuel 8 Jours
Reference number: 6659-3631-55B
Movement: automatic, Blancpain Caliber 5939A; ø 32 mm, height 7.25 mm; 42 jewels; 28,800 vph; 192-hour power reserve
Functions: hours, minutes, subsidiary seconds; perpetual calendar with date, weekday, month, moon phase, leap year
Case: pink gold, ø 42 mm, height 13.5 mm; sapphire crystal; transparent case back; water-resistant to 3 atm
Band: reptile skin, folding clasp
Remarks: enamel dial
Price: $54,430
Variations: pink gold Milanese meshband ($72,240)

Villeret Tourbillon Volant Heure Sautante Minute Rétrograde
Reference number: 66260-3633-55B
Movement: manually wound, Blancpain Caliber 260MR; ø 32 mm, height 5.85 mm; 39 jewels; 21,600 vph; flying 1-minute tourbillon; 144-hour power reserve
Functions: hours (digital, jumping), minutes (retrograde)
Case: pink gold, ø 42 mm, height 11 mm; sapphire crystal; transparent case back; water-resistant to 3 atm
Band: reptile skin, folding clasp
Remarks: enamel dial
Price: $148,800

Villeret Quantième Complet 8 Days
Reference number: 6639A-3631-55B
Movement: automatic, Blancpain Caliber 6639; ø 32 mm, height 7.6 mm; 35 jewels; 28,800 vph; 192-hour power reserve
Functions: hours, minutes, subsidiary seconds; date, weekday, month, moon phase
Case: pink gold, ø 42 mm, height 13.03 mm; sapphire crystal; transparent case back; water-resistant to 3 atm
Band: reptile skin, folding clasp
Remarks: enamel dial
Price: $41,900
Variations: pink gold Milanese mesh band ($61,200)

Villeret Quantième Complet
Reference number: 6654A-1127-55B
Movement: automatic, Blancpain Caliber 6654; ø 32 mm, height 5.5 mm; 28 jewels; 28,800 vph; 2 spring barrels, 72-hour power reserve
Functions: hours, minutes, sweep seconds; date, weekday, month, moon phase
Case: stainless steel, ø 40 mm, height 10.94 mm; sapphire crystal; transparent case back; water-resistant to 3 atm
Band: reptile skin, folding clasp
Price: $14,900
Variations: stainless steel Milanese mesh band ($17,300)

Villeret Grande Date Jour Rétrograde
Reference number: 6668-3642-55B
Movement: automatic, Blancpain Caliber 6950GJ; ø 32 mm, height 5.27 mm; 40 jewels; 28,800 vph; 72-hour power reserve
Functions: hours, minutes, sweep seconds; large date, weekday (retrograde)
Case: pink gold, ø 40 mm, height 11.1 mm; sapphire crystal; transparent case back; water-resistant to 3 atm
Band: reptile skin, folding clasp
Remarks: opalin dial
Price: $24,500

BLANCPAIN

Fifty Fathoms Automatic
Reference number: 5015-12B40-O52A
Movement: automatic, Blancpain Caliber 1315; ø 30.6 mm, height 5.65 mm; 35 jewels; 28,800 vph; silicon hairspring; 120-hour power reserve
Functions: hours, minutes, sweep seconds; date
Case: titanium, ø 45 mm, height 15.4 mm; unidirectional bezel with sapphire crystal inlay, with 0-60 scale; sapphire crystal; transparent case back; water-resistant to 30 atm
Band: textile, buckle
Price: $15,700

Fifty Fathoms Bathyscaphe Jour Date
Reference number: 5052-1110-063A
Movement: automatic, Blancpain Caliber 1315DD; ø 30.6 mm, height 5.65 mm; 37 jewels; 28,800 vph; 3 spring barrels, 120-hour power reserve
Functions: hours, minutes, sweep seconds; date, weekday
Case: stainless steel, ø 43 mm, height 14.25 mm; unidirectional bezel with ceramic inlay, with 0-60 scale; sapphire crystal; transparent case back; screw-in crown; water-resistant to 30 atm
Band: leather, buckle
Price: $12,700; limited to 500 pieces

Fifty Fathoms Automatique Grande Date
Reference number: 5050-12B30-B52A
Movement: automatic, Blancpain Caliber 6918B; ø 32 mm, height 7.15 mm; 44 jewels; 28,800 vph; silicon hairspring; 3 spring barrels, 120-hour power reserve
Functions: hours, minutes, sweep seconds; large date
Case: titanium, ø 45 mm, height 16.27 mm; unidirectional bezel with sapphire crystal inlay, with 0-60 scale; sapphire crystal; transparent case back; water-resistant to 30 atm
Band: textile, buckle
Price: $17,500

Fifty Fathoms Bathyscaphe
Reference number: 5000-0240-O52A
Movement: automatic, Blancpain Caliber 1315; ø 30.6 mm, height 5.65 mm; 35 jewels; 28,800 vph; silicon hairspring; 120-hour power reserve
Functions: hours, minutes, sweep seconds; date
Case: ceramic, ø 43.6 mm, height 13.83 mm; unidirectional bezel, with 0-60 scale; sapphire crystal; transparent case back; screw-in crown; water-resistant to 30 atm
Band: textile, buckle
Price: $12,860
Variations: various textile straps

Fifty Fathoms Bathyscaphe Chronograph Flyback
Reference number: 5200-0130-NABA
Movement: automatic, Blancpain Caliber F385; ø 31.8 mm, height 6.65 mm; 37 jewels; 36,000 vph; silicon hairspring; 50-hour power reserve
Functions: hours, minutes, subsidiary seconds; flyback chronograph; date
Case: ceramic, ø 43.6 mm, height 15.25 mm; unidirectional bezel, with 0-60 scale; sapphire crystal; water-resistant to 30 atm
Band: textile, buckle
Price: $17,200
Variations: sailcloth strap ($17,200)

Fifty Fathoms Bathyscaphe Quantième Annuel
Reference number: 5071-1110-B52A
Movement: automatic, Blancpain Caliber 6054.P; ø 32 mm, height 5.73 mm; 34 jewels; 28,800 vph; 72-hour power reserve
Functions: hours, minutes, sweep seconds; annual calendar with date, weekday, month
Case: stainless steel, ø 43 mm, height 13.46 mm; unidirectional bezel with ceramic inlay, with 0-60 scale; sapphire crystal; transparent case back; screw-in crown; water-resistant to 30 atm
Band: textile, buckle
Price: $26,100

X Fathoms
Reference number: 5018-1230-64A
Movement: automatic, Blancpain Caliber 9918B (base Blancpain 1315); ø 36 mm, height 13 mm; 48 jewels; 28,800 vph; 3 spring barrels, 120-hour power reserve
Functions: hours, minutes, sweep seconds; mechanical depth gauge (two-part scale) with maximum depth indicator, 5-minute short-time counter (countdown)
Case: titanium, ø 55.65 mm, height 24 mm; unidirectional bezel, with 0-60 scale; sapphire crystal; helium valve; water-resistant to 30 atm
Band: rubber, buckle
Price: $40,700

L'Evolution Tourbillon Carrousel
Reference number: 92322-34B39-55B
Movement: manually wound, Blancpain Caliber 2322V2; ø 35.3 mm, height 5.85 mm; 70 jewels; 21,600 vph; escapement system with flying 1-minute tourbillon and 1-minute carrousel with differential compensation; 3 spring barrels, 168-hour power reserve
Functions: hours, minutes; power reserve indicator, (on rear)
Case: platinum, ø 47.4 mm, height 11.66 mm; sapphire crystal; transparent case back; water-resistant to 3 atm
Band: reptile skin, folding clasp
Remarks: skeletonized mainplate and dial
Price: $373,130; limited to 50 pieces

L'Evolution Flyback Chronograph
Reference number: 8886F-1503-52B
Movement: automatic, Blancpain Caliber 69F9; ø 32 mm, height 8.4 mm; 44 jewels; 21,600 vph; 40-hour power reserve
Functions: hours, minutes; chronograph with flyback function; large date
Case: white gold, ø 43 mm, height 16.04 mm; carbon bezel; sapphire crystal; transparent case back; water-resistant to 10 atm
Band: textile, folding clasp
Remarks: carbon dial
Price: $55,700
Variations: pink gold ($51,460)

Villeret Women Ultraplate
Reference number: 6104-3642-55A
Movement: automatic, Blancpain Caliber 913; ø 21 mm, height 3.28 mm; 20 jewels; 28,800 vph; 40-hour power reserve
Functions: hours, minutes, sweep seconds
Case: pink gold, ø 29.2 mm, height 9.2 mm; sapphire crystal; transparent case back; water-resistant to 10 atm
Band: reptile skin, buckle
Price: $12,700

Villeret Women Quantième Moonphase
Reference number: 6106-1127-55A
Movement: automatic, Blancpain Caliber 913QL; ø 23.7 mm, height 4.5 mm; 20 jewels; 28,800 vph; 40-hour power reserve
Functions: hours, minutes, sweep seconds; date, moon phase
Case: stainless steel, ø 29.2 mm, height 10.36 mm; sapphire crystal; transparent case back; water-resistant to 3 atm
Band: reptile skin, buckle
Price: $10,500

Villeret Women Quantième Moonphase
Reference number: 6126-4628-55B
Movement: automatic, Blancpain Caliber 913QL; ø 23.7 mm, height 4.5 mm; 20 jewels; 28,800 vph; 40-hour power reserve
Functions: hours, minutes, sweep seconds; date, moon phase
Case: pink gold, ø 29.2 mm, height 10.36 mm; bezel set with diamonds; sapphire crystal; transparent case back; water-resistant to 3 atm
Band: reptile skin, buckle
Remarks: dial set with 8 diamonds
Price: $15,900

BLANCPAIN

Caliber F385
Automatic; column-wheel control of chrono functions; single spring barrel, 50-hour power reserve
Functions: hours, minutes, subsidiary seconds; flyback chronograph; date
Diameter: 31.8 mm
Height: 6.65 mm
Jewels: 37
Balance: silicon
Frequency: 36,000 vph
Balance spring: flat hairspring
Shock protection: Kif
Remarks: finely worked movement, bridges with côtes de Genève

Caliber 2358
Automatic; escapement with 1-minute carrousel; single spring barrel, 65-hour power reserve
Functions: hours, minutes; minute repeater with cathedral chimes; flyback chronograph, sweep 30-minute counter
Diameter: 32.8 mm
Height: 11.7 mm
Jewels: 59
Balance: glucydur with golden regulating screws
Frequency: 28,800 vph
Balance spring: flat hairspring
Shock protection: Kif
Remarks: hand-engraved bridges and rotor; 546 parts

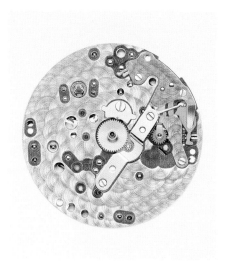

Caliber 913
Automatic; single spring barrel, 40-hour power reserve
Functions: hours, minutes, sweep seconds
Diameter: 21 mm
Height: 3.28 mm
Jewels: 20
Balance: glucydur
Frequency: 28,800 vph
Remarks: 174 parts

Caliber 152B
Manually wound; inverted movement with time indication on movement side, bridges with black ceramic inlay; single spring barrel, 40-hour power reserve
Functions: hours, minutes
Diameter: 35.64 mm
Height: 2.95 mm
Jewels: 21
Balance: screw balance
Frequency: 21,600 vph
Balance spring: flat hairspring
Shock protection: Kif

Caliber 225L
Automatic; flying 1-minute carrousel, two separate gear works; single spring barrel, 120-hour power reserve
Functions: hours, minutes; date, moon phase
Diameter: 31.9 mm
Height: 6.86 mm
Jewels: 40
Balance: glucydur with screw balance
Frequency: 28,800 vph
Balance spring: silicon
Shock protection: Kif
Remarks: 281 parts

Caliber 242
Automatic; flying 1-minute tourbillon with silicon balance and pallet fork horns; peripheral rotor at edge of movement; quadruple spring barrel, 288-hour power reserve
Functions: hours, minutes; power reserve indicator (on rear)
Diameter: 30.6 mm
Height: 6.1 mm
Jewels: 43
Balance: silicon
Frequency: 21,600 vph
Remarks: finely finished movement, hand-guillochéed bridges; 243 parts

DAYNIGHT RESCUE GMT T-100

Luminous Ceramic Bezel • 44mm Stainless Steel Case • 300M/1,000 FT Water Resistant
Swiss Made ETA 2893 Automatic Movement • Scratch Resistant Sapphire Crystal
16 Flat Tubes on Dial • 21 Total Tritium Tubes

www.deepbluewatches.com   718-484-7717   info@deepbluewatches.com

BORGWARD

It is not unusual for prospective watch brand founders to search for the name of a dormant or even defunct horological company to connect their business with a glorious past. Watchmaker Jürgen Betz looked elsewhere when he launched a series of watches under the name Borgward. This former automobile company had a reputation for outstanding quality, reliability, and durability. For connoisseurs and fans, Borgward meant technical prowess, perfect styling, and precision engineering.

Carl F. Borgward began his career as an automobile designer in 1924, when he built a small three-wheeled van. In the early 1930s, his budding company took over the Hansa-Lloyd automobile factory and went on to conquer a global market with the Lloyd, Goliath, and Borgward brands. The real Borgward legend, however, began in the 1950s with the "Goddess," the famed Isabella Coupé, whose elegant lines and state-of-the-art technology literally heralded a new era in automotive design in Germany. In 1961, the company went bankrupt due to poor management. But the legend lives on and became the inspiration for Betz when building his Borgward watch B511. Support came from his friend Eric Borgward, grandson of Carl, who confirmed: "My grandfather would have liked it." Since the Borgward Zeitmanufaktur was founded in July 2010, it has produced three collections: the B511 limited to 511 pieces, the P100 limited to 1,890 pieces, and the B2300 limited to 1,942 pieces. They are all "made in Germany" but based on Swiss technology. At the heart of each watch is either an ETA 2824 with three hands and calendar or the famous ETA 7750 Valjoux chronograph automatic. The decoration of the movements with various stripes and perlage is the bailiwick of Betz himself. For production, Borgward relies on space in the northern city of Bremen.

Jürgen Betz also has some special offers for watch enthusiasts: a workshop in the pretty town of Efringen-Kirchen just north of Basel, where participants learn how to finish a mainplate, take apart and reassemble a watch, and print the dial.

BORGWARD
Zeitmanufaktur GmbH & Co. KG
Markgrafenstrasse 16
D-79588 Efringen-Kirchen
Germany

Tel.:
01149-7628-805-7840

Fax:
01149-7628-805-7841

E-mail:
manufaktur@borgward.ag

Website:
www.borgward.ag

Founded:
2010

Number of employees:
3

Annual production:
approx. 180

Distribution:
Please contact Borgward directly for enquiries.

Most important collections:
P100, B2300, 1957, New Heritage Steam

New Heritage Steam Chronograph
Reference number: NHS.CL.01
Movement: automatic, ETA Caliber 7750-2; ø 30 mm, height 7.9 mm; 25 jewels; 28,800 vph; blackened oscillating mass; movement decorated with perlage and côtes de Genève; 42-hour power reserve
Functions: hours, minutes, subsidiary seconds; chronograph
Case: stainless steel, ø 44 mm, height 15 mm; bezel in bronze; sapphire crystal; transparent case back; water-resistant to 5 atm
Band: calfskin, buckle
Price: $2,995; limited to 98 pieces
Variations: stainless steel bracelet ($3,190)

P100 Retrospective
Reference number: P100R.AL.02
Movement: automatic, ETA Caliber 2824-2; ø 25.6 mm, height 4.6 mm; 25 jewels; 28,800 vph; blackened oscillating mass; movement decorated with perlage and côtes de Genève; 42-hour power reserve
Functions: hours, minutes, sweep seconds; date
Case: stainless steel, ø 40 mm, height 12 mm; sapphire crystal; transparent case back; water-resistant to 5 atm
Band: calfskin, buckle
Price: $1,450; limited to 1,890 pieces
Variations: stainless steel Milanese mesh bracelet ($1,650); white or brown dial

Fiftyseven
Reference number: B57.CL.06.MIL
Movement: automatic, ETA Caliber 7750; ø 30 mm, height 7.9 mm; 25 jewels; 28,800 vph; blackened oscillating mass; movement decorated with perlage and côtes de Genève; 42-hour power reserve
Functions: hours, minutes; chronograph; date, weekday
Case: stainless steel, ø 40 mm, height 15.9 mm; sapphire crystal; transparent case back; water-resistant to 5 atm
Band: stainless steel Milanese mesh, folding clasp
Price: $3,140
Variations: calfskin strap ($2,940); various dial colors

BORGWARD

Big Fiftyseven
Reference number: BIG57.CL.09
Movement: automatic, ETA Caliber 7750; ø 30 mm, height 7.9 mm; 25 jewels; 28,800 vph; perlage on plate, bridges with côtes de Genève, partly skeletonized galvano-blackened oscillating mass; 42-hour power reserve
Functions: hours, minutes; chronograph; date, weekday
Case: stainless steel, ø 44 mm, height 15.9 mm; sapphire crystal; transparent case back; water-resistant to 5 atm
Band: calfskin, buckle
Price: $3,050
Variations: stainless steel Milanese mesh ($3,250)

P100 Diamond Chrono Medium
Reference number: P100.CMZDIA.05.L01
Movement: automatic, ETA Caliber 7750; ø 30 mm, height 7.9 mm; 25 jewels; 28,800 vph; blackened oscillating mass; 38-hour power reserve
Functions: hours, minutes, subsidiary seconds; chronograph; date
Case: stainless steel, ø 36 mm, height 15.5 mm; bezel set with 70 diamonds; sapphire crystal; transparent case back; water-resistant to 5 atm
Band: calfskin, buckle
Remarks: mother-of-pearl dial
Price: $5,530
Variations: white dial; stainless steel Milanese mesh ($5,730)

B2300 Medium
Reference number: B2300.AM.05.MIL
Movement: automatic, ETA Caliber 2824-2; ø 25.6 mm, height 4.6 mm; 25 jewels; 28,800 vph; blackened oscillating mass; movement decorated with perlage and côtes de Genève; 42-hour power reserve
Functions: hours, minutes, sweep seconds; date
Case: stainless steel, ø 38 mm, height 11 mm; sapphire crystal; transparent case back; water-resistant to 5 atm
Band: stainless steel Milanese mesh, folding clasp
Remarks: mother-of-pearl dial
Price: $1,790; limited to 1,942 pieces
Variations: white dial; calfskin strap ($1,590)

Deluxe
Reference number: DELUXE.CL.09
Movement: automatic, ETA Caliber 7751; ø 30 mm, height 7.9 mm; 25 jewels; 28,800 vph; personalized oscillating mass, movement decorated with perlage and côtes de Genève; 46-hour power reserve
Functions: hours, minutes, subsidiary seconds; additional 24-hour display; chronograph; date, weekday, month, moon phase
Case: stainless steel, ø 42.5 mm, height 14.5 mm; sapphire crystal; transparent case back; water-resistant to 10 atm
Band: calfskin, buckle
Price: $7,830
Variations: stainless steel Milanese mesh ($8,030); various personalization options

New Heritage Steam Automatic
Reference number: NHS.AL.01
Movement: automatic, ETA Caliber 2824-2; ø 25.6 mm, height 4.6 mm; 25 jewels; 28,800 vph; blackened movement, movement with perlage and côtes de Genève; 42-hour power reserve
Functions: hours, minutes, sweep seconds; date
Case: stainless steel, ø 44 mm, height 12.5 mm; bezel in bronze; sapphire crystal; transparent case back; water-resistant to 5 atm
Band: calfskin, buckle
Price: $2,145; limited to 98 pieces
Variations: stainless steel ($2,345)

New Heritage Steam Manually Wound
Reference number: NHS.HA.01
Movement: manually wound, Borgward Caliber B02 (base ETA 6498); ø 36.6 mm, height 4.5 mm; 17 jewels; 18,000 vph; blackened movement with côtes de Genève; 46-hour power reserve
Functions: hours, minutes, subsidiary seconds
Case: stainless steel, ø 44 mm, height 12.5 mm; bezel in bronze; sapphire crystal; transparent case back; water-resistant to 5 atm
Band: calfskin, buckle
Price: $1,935; limited to 98 pieces
Variations: stainless steel ($2,135)

BOVET

If any brand can claim real connections to China, it is Bovet, founded by Swiss businessman Edouard Bovet. Bovet emigrated to Canton, China, in 1818 and sold four watches of his own design there. On his return to Switzerland in 1822, he set up a company for shipping his Fleurier-made watches to China. The company name, pronounced "Bo Wei" in Mandarin, became a synonym for "watch" in Asia, and at one point it had offices in Canton. For more than eighty years, Bovet and his successors supplied the Chinese ruling class with valuable timepieces.

In 2001, the brand was bought by entrepreneur Pascal Raffy. He ensured the company's industrial independence by acquiring several other companies as well, notably the high-end watchmaker Swiss Time Technology (STT) in Tramelan, which he renamed Dimier 1738. In addition to creating its own line of watches, this *manufacture* produces complex technical components such as tourbillons for Bovet watches. Assembly of Bovet creations takes place at the headquarters in the thirteenth-century Castle of Môtiers in Val-de-Travers not far from Fleurier.

Bovet is an equal opportunity manufacturer of fine watches. The same careful approach to the craft goes into the watches for men as those for women. These high-end timekeepers do have several distinctive features. The first is intricate dial work, featuring not only complex architecture in the men's series, but also very fine enameling using a grand-feu technique, and intricate guilloché patterns. The latter include some remarkable unique pieces that transcend the world of timekeeping and enter the realm of art.

The second special feature is placement of the lugs and crown at 12 o'clock, recalling Bovet's tasteful pocket watches of the nineteenth century. On some models, the wristbands are made to be easily removed so the watch can be worn on a chain or cord. Other watches convert to table clocks, and the Amadeo Fleurier Miss Audrey series can even be worn as a necklace.

Bovet Fleurier S.A.
109 Pont-du-Centenaire
CP109
CH-1228 Plan-les-Ouates
Switzerland

Tel.:
+41-22-731-4638

Fax:
+41-22-884-1450

E-mail:
info@bovet.com

Website:
www.bovet.com

Founded:
1822

Annual production:
around 2,000 timepieces

Bovet LLC U.S.A.:
888-909-1822

Most important collections/price range:
Amadeo Fleurier, Dimier, Pininfarina, Sportster / $18,500 to $1,000,000

Amadeo Fleurier Virtuoso VIII
Reference number: T10GD003
Movement: manually wound, Bovet Caliber 17BM03-GD; ø 38 mm, height 5.70 mm, 18,000 vph; 1-minute flying tourbillon; 10-day power reserve
Functions: hours, minutes, seconds (on tourbillon cage); big date; power reserve indicator
Case: red gold, ø 44 mm, height 13.45 mm; sapphire crystal; transparent case back; water-resistant to 3 atm
Band: reptile skin, buckle
Remarks: flexible lugs to turn the watch into a table clock or pocket watch
Price: $210,600; limited edition of 39 pieces
Variations: platinum ($275,400, limited to 20 pieces); white gold, ($221,400, limited to 39 pieces)

Récital 22 Grand Récital
Reference number: R220001-USA
Movement: manually wound, Bovet Caliber 17DM03-TEL; ø 38 mm, height 15.70 mm; 18,000 vph; 1-minute double-sided flying tourbillon; 9-day power reserve
Functions: hours (24), minutes (retrograde), seconds (on double-sided tourbillon), double-sided date, moon phase; power reserve indicator, perpetual calendar (retrograde)
Case: platinum, ø 46.30 mm, height 19.60 mm; sapphire crystal; transparent case back; water-resistant to 3 atm
Band: reptile skin, buckle
Price: $502,200; limited to 60 movements
Variations: red gold ($469,800)

Pininfarina OttantaSei
Reference number: TPINS002
Movement: manually wound, Bovet Caliber 17BM03MM; ø 38 mm, height 6.60 mm; 18,000 vph; flying 1-minute double-face tourbillon; 240-hour power reserve
Functions: hours, minutes, (off-center); power reserve indicator
Case: titanium with blue PVD, ø 44 mm, height 12 mm; sapphire crystal; transparent case back; water-resistant to 3 atm
Band: rubber, folding clasp
Price: $194,400; limited to 86 movements
Variations: titanium; red gold ($189,000); with white lacquer or blue circular brushed dial

Château de Motiers 40— "Butterfly"

Reference number: HMS5048-SD12
Movement: automatic, Bovet Caliber 11BA15; ø 26.20 mm, height 3.60 mm; 28,800 vph; 42-hour power reserve
Functions: hours, minutes
Case: red gold, ø 40 mm, height 8.43 mm; bezel set with 109 diamonds; sapphire crystal; transparent case back; water-resistant to 3 atm
Band: reptile skin, buckle
Remarks: mother-of-pearl dial with miniature painting of butterfly using special luminescent material and paint
Price: $50,350; unique piece

Récital 19 Miss Dimier

Reference number: R19S001-SD1
Movement: automatic, Bovet Caliber 11DA17; ø 26.20 mm, height 5.15 mm; 28,800 vph; 42-hour power reserve
Functions: hours, minutes
Case: stainless steel, ø 35.40 × 30.40 mm, height 10.35 mm; sapphire crystal; transparent case back; water-resistant to 3 atm
Band: synthetic satin, buckle
Remarks: blue lacquered dial with circular brushing
Price: $14,600
Variations: black mother-of-pearl dial

Amadeo Fleurier 39— "Blue Burdocks"

Reference number: AF39559-SD23
Movement: automatic, Bovet Caliber 11BA13; ø 26.20 mm, height 5.07 mm; 28,800 vph; 72-hour power reserve
Functions: hours, minutes
Case: pink gold, ø 39 mm, height: 5.07 mm; sapphire crystal; water-resistant to 3 atm
Band: reptile skin, buckle
Remarks: enamel miniature painting of blue burdocks on gold base plate
Price: $101,600; unique piece

Amadeo Fleurier Virtuoso V

Reference number: ACHS016
Movement: automatic, Bovet Caliber 13BM11AIHSMR; ø 31.00 mm, height 5.09 mm; 21,600 vph; 5-day power reserve
Functions: jumping hours, minutes (retrograde); hours, minutes, seconds on rear
Case: white gold, ø 43.5 mm, height 15.70 mm; sapphire crystal; transparent case back; water-resistant to 3 atm
Band: reptile skin, buckle
Remarks: reversed hand fitting
Price: $73,500

19Thirty Fleurier

Reference number: NTS0001
Movement: manually wound, Bovet Caliber 15BM04; ø 35 mm, height 3.80 mm; 21,600 vph; 7-day power reserve
Functions: hours, minutes, subsidiary seconds; power reserve indicator
Case: stainless steel, ø 42 mm, height 9.05 mm; sapphire crystal; transparent case back; water-resistant to 3 atm
Band: reptile skin, buckle
Price: $18,200
Variations: blue or black dial; with Arabic, Roman, or Chinese numerals

Amadeo Fleurier 36 Miss Audrey

Reference number: AS36007-SD12
Movement: automatic, Bovet Caliber 11BA13; ø 26.20 mm, height 5.09 mm; 28,800 vph; 42-hour power reserve
Functions: hours, minutes
Case: stainless steel, ø 36 mm, height 11.25 mm; sapphire crystal; bezel set with 60 round diamonds; water-resistant to 3 atm
Band: synthetic with calfskin lining, buckle
Remarks: 4 diamond indices; necklace bow set with diamonds
Price: $16,800;
Variations: turquoise guilloché dial

BREGUET

We never quite lose that attachment to the era in which we were born and grew up, nor do some brands. Abraham-Louis Breguet (1747–1823), who hailed from Switzerland, brought his craft to Paris in the *Sturm und Drang* atmosphere of the late eighteenth century. It was fertile ground for one of the most inventive watchmakers in the history of horology, and his products soon found favor with the highest levels of society.

Little has changed two centuries later. After a few years of drifting, in 1999 the brand carrying this illustrious name became the prize possession of the Swatch Group and came under the personal management of Nicolas G. Hayek, CEO. Hayek worked assiduously to restore the brand's roots, going as far as rebuilding the legendary Marie Antoinette pocket watch and contributing to the restoration of the Petit Trianon at Versailles.

Breguet is a full-fledged *manufacture*, and this has allowed it to forge ahead uncompromisingly with upscale watches and even jewelry. In modern facilities on the shores of Lake Joux, traditional craftsmanship still plays a significant role in the production of its fine watches, but at the same time, Breguet is one of the few brands to work with modern materials for its movements. This is not just a PR trick, but rather a sincere attempt to improve quality and rate precision. Many innovations have debuted at Breguet, for instance pallet levers and balance wheels made of silicon, the first Breguet hairspring with the arched terminal curve made of this glassy material, or even a mechanical high-frequency balance beating at 72,000 vph. Other innovations include the electromagnetic regulation of a minute repeater or the use of two micromagnets to achieve contactless anchoring of a balance wheel staff.

Breguet, now under the auspices of Nicolas G. Hayek's grandson, Marc A. Hayek, continues to explore the edges of the technologically possible in watchmaking, while maintaining the brand's particular connection to traditional processes and esthetic codes. Even when creating a sports watch, like the new Type XXI Flyback, Breguets always give off a scent of luxury.

Montres Breguet SA
CH-1344 L'Abbaye
Switzerland

Tel.:
+41-21-841-9090

Fax:
+41-21-841-9084

Website:
www.breguet.com

Founded:
1775 (Swatch Group since 1999)

U.S. distributor:
Breguet
The Swatch Group (U.S.), Inc.
1200 Harbor Boulevard, 7th Floor
Weehawken, NJ 07087
201-271-1400

Most important collections:
Classique, Tradition, Héritage, Marine, Reine de Naples, Type XX, Type XXI, Type XXII

Classique Phase de Lune Dame
Reference number: 9088BR 29 RC0 DD00
Movement: automatic, Breguet Caliber 537L; ø 20 mm; 26 jewels; 25,200 vph; silicon hairspring and escapement; 45-hour power reserve
Functions: hours, minutes, subsidiary seconds; moon phase
Case: rose gold, ø 30 mm, height 9.8 mm; bezel and lugs set with 66 diamonds; sapphire crystal; transparent case back; water-resistant to 3 atm
Band: rose gold, folding clasp
Remarks: enamel dial
Price: $51,200
Variations: white gold ($51,200)

Classique Hora Mundi
Reference number: 5717PT EU 9ZU
Movement: automatic, Breguet Caliber 77F0; ø 27.1 mm, height 6.15 mm; 43 jewels; 28,800 vph; silicon pallet lever, escape wheel and hairspring; 55-hour power reserve
Functions: hours, minutes, sweep seconds; world time display, day/night indicator (2nd time zone); date
Case: platinum, ø 43 mm, height 13.55 mm; sapphire crystal; transparent case back; screw-in crown; water-resistant to 3 atm
Band: reptile skin, folding clasp
Price: $94,200
Variations: rose gold ($78,900)

Marine Équation Marchante
Reference number: 5887PT Y2 9WV
Movement: automatic, Breguet Caliber 581DPE; ø 37.2 mm; 57 jewels; 28,800 vph; 1-minute tourbillon; silicon pallet lever, escape wheel, and hairspring; 80-hour power reserve
Functions: hours, minutes, subsidiary seconds (on tourbillon cage), running equation of time; perpetual calendar with date (retrograde), weekday, month
Case: platinum, ø 43.9 mm, height 11.75 mm; sapphire crystal; transparent case back; water-resistant to 10 atm
Band: reptile skin, folding clasp
Price: $230,400
Variations: rose gold ($215,000)

Classique Extra-Thin
Reference number: 5157BR 11 9V6
Movement: automatic, Breguet Caliber 502.3; ø 27.1 mm, height 2.4 mm; 35 jewels; 21,600 vph; silicon pallet fork horns and hairspring; 45-hour power reserve
Functions: hours, minutes
Case: rose gold, ø 38 mm, height 5.4 mm; sapphire crystal; transparent case back; water-resistant to 3 atm
Band: reptile skin, buckle
Price: $18,800
Variations: white gold ($18,800)

Classique 7787
Reference number: 7787BB 29 9V6
Movement: automatic, Breguet Caliber 591 DRL; ø 26 mm; 25 jewels; 28,800 vph; silicon hairspring, pallet lever, and escape wheel; 38-hour power reserve
Functions: hours, minutes, sweep seconds; power reserve indicator; moon phase
Case: white gold, ø 39 mm, height 10.2 mm; sapphire crystal; transparent case back; water-resistant to 3 atm
Band: reptile skin, folding clasp
Remarks: enamel dial
Price: $30,200
Variations: rose gold ($29,700)

Classique 7147
Reference number: 7147BR 29 9WU
Movement: automatic, Breguet Caliber 502.3SD; ø 24.9 mm, height 2.4 mm; 35 jewels; 21,600 vph; silicon pallet lever and hairspring; 45-hour power reserve
Functions: hours, minutes, subsidiary seconds
Case: rose gold, ø 40 mm, height 6.1 mm; sapphire crystal; transparent case back; water-resistant to 3 atm
Band: reptile skin, buckle
Remarks: enamel dial
Price: $21,000
Variations: white gold ($21,500)

Classique Extra-Thin Tourbillon
Reference number: 5367BR 29 9WU
Movement: automatic, Breguet Caliber 581; ø 36 mm, height 3 mm; 33 jewels; 28,800 vph; 1-minute tourbillon, silicon pallet lever and hairspring; hubless peripheral rotor for winding; 80-hour power reserve
Functions: hours, minutes, subsidiary seconds (on tourbillon cage)
Case: rose gold, ø 42 mm, height 7.45 mm; sapphire crystal; transparent case back; water-resistant to 3 atm
Band: reptile skin, folding clasp
Remarks: enamel dial
Price: $147,500
Variations: platinum ($161,800)

Tradition Tourbillon Fusée
Reference number: 7047PT 11 9ZU
Movement: manually wound, Breguet Caliber 569; ø 35.7 mm, height 10.82 mm; 43 jewels; 18,000 vph; silicon Breguet hairspring, chain and worm screw torque regulator; 1-minute tourbillon; 50-hour power reserve
Functions: hours, minutes; power reserve indicator (on movement side)
Case: platinum, ø 41 mm, height 15.95 mm; sapphire crystal; transparent case back; water-resistant to 3 atm
Band: reptile skin, folding clasp
Price: $189,700
Variations: rose gold ($175,600)

Tradition Seconde Rétrograde
Reference number: 7097BR G1 9WU
Movement: automatic, Breguet Caliber 505 SR1; ø 33 mm; 38 jewels; 21,600 vph; silicon Breguet hairspring and lever pallets; 50-hour power reserve
Functions: hours, minutes (off-center), subsidiary seconds (retrograde)
Case: rose gold, ø 40 mm, height 11.65 mm; sapphire crystal; transparent case back; water-resistant to 3 atm
Band: reptile skin, folding clasp
Price: $32,700
Variations: white gold ($33,500)

BREGUET

Tradition
Reference number: 7057BR R9 9W6
Movement: manually wound, Breguet Caliber 507DR1; ø 33 mm; 34 jewels; 21,600 vph; silicon Breguet hairspring and lever pallets; 50-hour power reserve
Functions: hours, minutes (off-center), subsidiary seconds (retrograde)
Case: rose gold, ø 40 mm, height 11.65 mm; sapphire crystal; transparent case back; water-resistant to 3 atm
Band: reptile skin, buckle
Price: $27,600

Tradition Dame
Reference number: 7038BB 1T 9V6 D00D
Movement: automatic, Breguet Caliber 505 SR; ø 33 mm; 38 jewels; 21,600 vph; silicon Breguet hairspring and lever pallets; 50-hour power reserve
Functions: hours, minutes (off-center), subsidiary seconds (retrograde)
Case: white gold, ø 37 mm, height 11.85 mm; bezel set with 68 diamonds; sapphire crystal; transparent case back; crown with ruby cabochon; water-resistant to 3 atm
Band: reptile skin, buckle set with 19 diamonds
Price: $38,900
Variations: rose gold ($38,100)

Héritage Phases de Lune Rétrograde
Reference number: 8861BR 11 386 D000
Movement: manually wound, Breguet Caliber 586L; ø 20 mm; 38 jewels; 21,600 vph; silicon Breguet hairspring, silicon pallet lever and escape wheel; 36-hour power reserve
Functions: hours, minutes; moon phase
Case: rose gold, 25 × 35 mm, height 9.75 mm; bezel set with 140 diamonds; sapphire crystal; water-resistant to 3 atm
Band: calfskin, folding clasp
Price: $33,300
Variations: rose gold bracelet ($51,700); white gold ($34,300)

Héritage Tourbillon
Reference number: 5497PT 12 9V6
Movement: manually wound, Breguet Caliber 187H; ø 26 mm; 21 jewels; 18,000 vph; 1-minute tourbillon; 50-hour power reserve
Functions: hours, minutes (off-center), subsidiary seconds (on tourbillon cage)
Case: platinum, 35 × 42 mm; sapphire crystal; water-resistant to 3 atm
Band: reptile skin, folding clasp
Price: $142,900

Type XXI Chrono Cadran Vintage
Reference number: 3817ST X2 3ZU
Movement: automatic, Breguet Caliber 584 Q/2; ø 30 mm; 26 jewels; 28,800 vph; silicon hairspring and escapement, central minute totalizer; 48-hour power reserve
Functions: hours, minutes, subsidiary seconds; additional 24-hour display (2nd time zone); flyback chronograph; date
Case: stainless steel, ø 42 mm, height 15.2 mm; unidirectional bezel with 0-60 scale; sapphire crystal; transparent case back; water-resistant to 10 atm
Band: calfskin, folding clasp
Price: $13,900

Type XXII
Reference number: 3880BR Z2 9XV
Movement: automatic, Breguet Caliber 589F; ø 30 mm, height 8.3 mm; 28 jewels; 72,000 vph; high-frequency silicon escapement, sweep minute counter; 40-hour power reserve
Functions: hours, minutes, subsidiary seconds; additional 24-hour display (2nd time zone); flyback chronograph; date
Case: rose gold, ø 44 mm, height 18.05 mm; bidirectional 60-minute bezel; sapphire crystal; transparent case back; screw-in crown; water-resistant to 10 atm
Band: reptile skin, folding clasp
Price: $35,500
Variations: stainless steel ($20,100)

Marine Date

Reference number: 5517TI G2 5ZU
Movement: automatic, Breguet Caliber 777A; ø 33.8 mm; 26 jewels; 28,800 vph; silicon pallets and hairspring; 55-hour power reserve
Functions: hours, minutes, sweep seconds; date
Case: titanium, ø 40 mm, height 11.5 mm; sapphire crystal; transparent case back; screw-in crown; water-resistant to 10 atm
Band: rubber, folding clasp
Price: $18,500
Variations: reptile skin band ($18,500); rose gold ($28,700); white gold ($28,700)

Marine Chronograph

Reference number: 5527BB Y2 9WV
Movement: automatic, Breguet Caliber 582QA; ø 32.7 mm; 28 jewels; 28,800 vph; silicon pallets and hairspring; 48-hour power reserve
Functions: hours, minutes, subsidiary seconds; chronograph; date
Case: white gold, ø 42.3 mm, height 13.85 mm; sapphire crystal; transparent case back; screw-in crown; water-resistant to 10 atm
Band: reptile skin, folding clasp
Price: $33,800
Variations: rubber strap ($33,800); rose gold ($33,800); titanium ($22,600)

Marine Alarme Musicale

Reference number: 5547BR 12 9ZU
Movement: automatic, Breguet Caliber 518F/1; ø 27.1 mm; 36 jewels; 28,800 vph; silicon pallets and hairspring; 45-hour power reserve
Functions: hours, minutes, sweep seconds; additional 24-hour display (2nd time zone), power reserve indicator for chimes; alarm (adjustable to the minute); date
Case: rose gold, ø 40 mm, height 13.05 mm; sapphire crystal; transparent case back; water-resistant to 5 atm
Band: reptile skin, folding clasp
Price: $40,900
Variations: rubber strap ($40,900); white gold ($40,900); titanium ($28,600)

Reine de Naples

Reference number: 8908BR 5T 864 D00D
Movement: automatic, Breguet Caliber 537 DRL2; ø 19.7 mm; 45 jewels; 25,200 vph; silicon Breguet balance wheel, escapement, and hairspring; 45-hour power reserve
Functions: hours, minutes, subsidiary seconds; power reserve indicator; moon phase
Case: rose gold, 28.45 × 36.5 mm, height 10.05 mm; bezel and flange set with 128 diamonds; sapphire crystal; transparent case back; crown with ruby cabochon
Band: satin, folding clasp with diamonds
Price: $36,100
Variations: rose gold bracelet ($61,800)

Reine de Naples Mini

Reference number: 8918BR 58 864 D00D
Movement: automatic, Breguet Caliber 537/1; ø 19.7 mm; 20 jewels; 21,600 vph; 40-hour power reserve
Functions: hours, minutes
Case: rose gold, 24.95 × 33 mm, height 13.05 mm; bezel and flange set with 117 diamonds; sapphire crystal; transparent case back; crown with diamond cabochon; water-resistant to 3 atm
Band: satin, folding clasp set with 26 diamonds
Remarks: mother-of-pearl dial with drop-shaped diamonds
Price: $35,100
Variations: white gold ($36,100)

Reine de Naples Mini

Reference number: 8928BR 5W 844 DD0D
Movement: automatic, Breguet Caliber 586/1; ø 15 mm; 29 jewels; 21,600 vph; silicon hairspring; 38-hour power reserve
Functions: hours, minutes
Case: pink gold, 24.95 × 33 mm, height 8.5 mm; bezel, flange, and lugs set with 139 diamonds; sapphire crystal; crown with diamond cabochon; water-resistant to 3 atm
Band: satin, folding clasp set with 26 diamonds
Remarks: mother-of-pearl dial
Price: $35,100
Variations: white gold ($36,100)

BREITLING

In 1884, Léon Breitling opened his workshop in St. Imier in the Jura mountains and immediately began specializing in integrated chronographs. His business strategy was to focus consistently on instrument watches with a distinctive design. High quality standards and the rise of aviation completed the picture.

Today, Breitling's relationship with air sports and commercial and military aviation is clear from its brand identity. The watch company hosts a series of aviation days, owns an aerobatics team, and sponsors several aviation associations.

The unveiling of its own, modern chronograph movement at Basel in 2009 was a major milestone in the company's history and also a return to its roots. The new design was to be "100 percent Breitling" and industrially produced in large numbers at a reasonable cost. Although Breitling's operations in Grenchen and in La Chaux-de-Fonds both boast state-of-the-art equipment, the contract for the new chronograph was awarded to a small team in Geneva. By 2006, the brand-new Caliber B01 had made the COSC grade with flying colors, and it has enjoyed great popularity ever since. For the team of designers, the innovative centering system on the reset mechanism that requires no manual adjustment was one of the great achievements. Since then, the in-house caliber has evolved, but the cost for the company was immense. Ultimately, owner Théodore Schneider, in the second generation, decided to put management in the hands of Georges Kern of IWC fame. Together with the new owners, the investment company CVC Capital Partners, Kern decided to expand the brand beyond the pilot watch niche and look to the untapped markets in the Far East. The new collections were streamlined and given starker profiles, a recipe he brought in from his IWC days. Breitling is now highly athletic and elegant. Prices range from around $3,500 to the $10,000 region, and that includes ladies' watches. No more quartz, only mechanical. The company will also be making its own chronographs using the B1 movement and movements from outside vendors.

Breitling
Schlachthausstrasse 2
CH-2540 Grenchen
Switzerland

Tel.:
+41-32-654-5454

Fax:
+41-32-654-5400

E-mail:
sales@breitlingusa.com

Website:
www.breitling.com

Founded:
1884

Annual production:
700,000 (estimated)

U.S. distributor:
Breitling U.S.A. Inc.
206 Danbury Road
Wilton, CT 06897
203-762-1180
www.breitling.com

Most important collections:
Navitimer 1, Navitimer 8, Avenger, Premier, Chronomat, Superocean Heritage, Superocean, Professional

Navitimer 8 B01 Chronograph 43
Reference number: AB0117131B1A1
Movement: automatic, Breitling Caliber B01; ø 30 mm, height 7.2 mm; 47 jewels; 28,800 vph; column wheel control of chronograph functions; 70-hour power reserve; COSC-certified chronometer
Functions: hours, minutes, subsidiary seconds; chronograph; date
Case: stainless steel, ø 43 mm, height 13.97 mm; bidirectional bezel with reference marks; sapphire crystal; screw-in crown; water-resistant to 10 atm
Band: stainless steel, folding clasp
Price: $8,080

Navitimer 8 Chronograph 43
Reference number: M13314101B1X1
Movement: automatic, Breitling Caliber 13 (base ETA 7750); ø 30 mm, height 7.9 mm; 25 jewels; 28,800 vph; 42-hour power reserve; COSC-certified chronometer
Functions: hours, minutes, subsidiary seconds; chronograph; date, weekday
Case: stainless steel with black DLC coating, ø 43 mm, height 14.17 mm; bidirectional bezel with reference marks; sapphire crystal; screw-in crown; water-resistant to 10 atm
Band: calfskin, buckle
Price: $6,850

Navitimer 8 Automatic 41
Reference number: A17314101C1A1
Movement: automatic, Breitling Caliber 17 (base ETA 2824-2); ø 25.6 mm, height 4.6 mm; 25 jewels; 28,800 vph; 40-hour power reserve; COSC-certified chronometer
Functions: hours, minutes, sweep seconds; date
Case: stainless steel, ø 41 mm, height 10.74 mm; bidirectional bezel with reference marks; sapphire crystal; screw-in crown; water-resistant to 10 atm
Band: stainless steel, folding clasp
Price: $4,310

Navitimer 1 Automatic 38

Reference number: A17325211C1P1
Movement: automatic, Breitling Caliber 17 (base ETA 2824-2); ø 25.6 mm, height 4.6 mm; 25 jewels; 28,800 vph; 40-hour power reserve; COSC-certified chronometer
Functions: hours, minutes, sweep seconds; date
Case: stainless steel, ø 38 mm, height 10.1 mm; bidirectional bezel with integrated slide rule and tachymeter scale; sapphire crystal; water-resistant to 3 atm
Band: reptile skin, buckle
Price: $4,310

Navitimer 1 Chronograph 41

Reference number: A13324121G1X1
Movement: automatic, Breitling Caliber 13 (base ETA 7750); ø 30 mm, height 7.9 mm; 25 jewels; 28,800 vph; 42-hour power reserve; COSC-certified chronometer
Functions: hours, minutes, subsidiary seconds; chronograph; date
Case: stainless steel, ø 41 mm, height 14.44 mm; bidirectional bezel with integrated slide rule and tachymeter scale; sapphire crystal; water-resistant to 3 atm
Band: calfskin, buckle
Price: $6,040
Variations: stainless steel bracelet ($6,690)

Navitimer 1 Chronograph GMT 46

Reference number: A2432212/B726/441X/A20BA.1
Movement: automatic, Breitling Caliber 24 (base ETA 7751); ø 30 mm, height 7.9 mm; 25 jewels; 28,800 vph; 42-hour power reserve; COSC-certified chronometer
Functions: hours, minutes, subsidiary seconds; additional 24-hour display; chronograph; date
Case: stainless steel, ø 46 mm, height 15.5 mm; bidirectional bezel with integrated slide rule and tachymeter scale; sapphire crystal; water-resistant to 3 atm
Band: calfskin, buckle
Price: $6,535
Variations: stainless steel bracelet ($7,480)

Navitimer 1 B01 Chronograph 46

Reference number: RB0127121F1P1
Movement: automatic, Breitling Caliber B01; ø 30 mm, height 7.2 mm; 47 jewels; 28,800 vph; column wheel control of chronograph functions; 70-hour power reserve; COSC-certified chronometer
Functions: hours, minutes, subsidiary seconds; chronograph; date
Case: pink gold, ø 46 mm, height 14.51 mm; bidirectional bezel with integrated slide rule and tachymeter scale; sapphire crystal; transparent case back; water-resistant to 3 atm
Band: reptile skin, buckle
Price: $24,270

Navitimer 1 B01 Chronograph 43

Reference number: AB0121211B1X1
Movement: automatic, Breitling Caliber B01; ø 30 mm, height 7.2 mm; 47 jewels; 28,800 vph; column wheel control of chronograph functions; 70-hour power reserve; COSC-certified chronometer
Functions: hours, minutes, subsidiary seconds; chronograph; date
Case: stainless steel, ø 43 mm, height 14.22 mm; bidirectional bezel with integrated slide rule and tachymeter scale; sapphire crystal; water-resistant to 3 atm
Band: calfskin, buckle
Price: $8,215
Variations: stainless steel bracelet ($9,160)

Chronomat B01 Chronograph 44

Reference number: AB0115101F1A1
Movement: automatic, Breitling Caliber B01; ø 30 mm, height 7.2 mm; 47 jewels; 28,800 vph; column wheel control of chronograph functions; 70-hour power reserve; COSC-certified chronometer
Functions: hours, minutes, subsidiary seconds; chronograph; date
Case: stainless steel, ø 44 mm, height 16.95 mm; unidirectional bezel with 0-60 scale; sapphire crystal; screw-in crown and pusher; water-resistant to 50 atm
Band: stainless steel, folding clasp
Price: $8,720
Variations: reptile skin strap and buckle ($8,150)

BREITLING

Superocean Héritage II B20 Automatic 42

Reference number: UB2010121B1S1
Movement: automatic, Breitling Caliber B20 (base Tudor MT 5612); ø 31.8 mm, height 6.5 mm; 28 jewels; 28,800 vph; 70-hour power reserve; COSC-certified chronometer
Functions: hours, minutes, sweep seconds; date
Case: stainless steel, ø 42 mm, height 14.35 mm; unidirectional bezel in rose gold with ceramic inlay; sapphire crystal; screw-in crown; water-resistant to 20 atm
Band: rubber, folding clasp
Price: $5,715

Superocean Héritage II B20 Automatic 46

Reference number: AB2020121B1S1
Movement: automatic, Breitling Caliber B20 (base Tudor MT 5612); ø 31.8 mm, height 6.5 mm; 28 jewels; 28,800 vph; 70-hour power reserve; COSC-certified chronometer
Functions: hours, minutes, sweep seconds; date
Case: stainless steel, ø 46 mm, height 15 mm; unidirectional bezel with ceramic insert; sapphire crystal; screw-in crown; water-resistant to 20 atm
Band: rubber, folding clasp
Price: $4,560
Variations: stainless steel bracelet ($4,885)

Superocean Héritage II Chronograph

Reference number: A1331233/Q616/295S/A20D.2
Movement: automatic, Breitling Caliber 13 (base ETA 7750); ø 30 mm, height 7.9 mm; 25 jewels; 28,800 vph; 42-hour power reserve; COSC-certified chronometer
Functions: hours, minutes, subsidiary seconds; chronograph; date
Case: stainless steel, ø 46 mm, height 16.35 mm; unidirectional bezel with ceramic insert; sapphire crystal; screw-in crown; water-resistant to 20 atm
Band: rubber with calfskin overlay, folding clasp
Price: $5,930
Variations: stainless steel bracelet ($6,240)

Superocean Héritage II Chronograph 44

Reference number: A13313121B1A1
Movement: automatic, Breitling Caliber 13 (base ETA 7750); ø 30 mm, height 7.9 mm; 25 jewels; 28,800 vph; 42-hour power reserve; COSC-certified chronometer
Functions: hours, minutes, subsidiary seconds; chronograph; date, weekday
Case: stainless steel, ø 44 mm, height 15.65 mm; unidirectional bezel with ceramic insert, with 0-60 scale; sapphire crystal; screw-in crown; water-resistant to 20 atm
Band: stainless steel Milanese mesh bracelet, folding clasp
Price: $6,240

Superocean Héritage II B01 Chronograph 44

Reference number: AB0162121B1S1
Movement: automatic, Breitling Caliber B01; ø 30 mm, height 7.2 mm; 47 jewels; 28,800 vph; 70-hour power reserve; COSC-certified chronometer
Functions: hours, minutes, subsidiary seconds; chronograph; date
Case: stainless steel, ø 44 mm, height 15.5 mm; unidirectional bezel with ceramic insert; sapphire crystal; screw-in crown; water-resistant to 20 atm
Band: rubber, folding clasp
Price: $7,665
Variations: stainless steel bracelet ($7,990)

Superocean II 44

Reference number: A17392D7/BD68/162A
Movement: automatic, Breitling Caliber 17 (base ETA 2824-2); ø 25.6 mm, height 4.6 mm; 25 jewels; 28,800 vph; 40-hour power reserve; COSC-certified chronometer
Functions: hours, minutes, sweep seconds; date
Case: stainless steel, ø 44 mm, height 14.2 mm; unidirectional bezel, with 0-60 scale; sapphire crystal; screw-in crown; helium valve; water-resistant to 100 atm
Band: stainless steel, folding clasp
Price: $4,150
Variations: rubber strap and buckle ($3,650)

BREITLING

Avenger Blackbird 44
Reference number: V1731110/BD74/109W/M20BASA.1
Movement: automatic, Breitling Caliber 17 (base ETA 2824-2); ø 25.6 mm, height 4.6 mm; 25 jewels; 28,800 vph; 40-hour power reserve; COSC-certified chronometer
Functions: hours, minutes, sweep seconds; date
Case: titanium with black DLC coating, ø 44 mm, height 12.7 mm; unidirectional bezel with 0-60 scale; sapphire crystal; screw-in crown; water-resistant to 20 atm
Band: textile, buckle
Price: $5,105

Super Avenger II
Reference number: A1337111/BC28/154S/A20S.1
Movement: automatic, Breitling Caliber 13 (base ETA 7750); ø 30 mm, height 7.9 mm; 25 jewels; 28,800 vph; 42-hour power reserve; COSC-certified chronometer
Functions: hours, minutes, subsidiary seconds; chronograph; date
Case: stainless steel, ø 48 mm, height 17.75 mm; unidirectional bezel with 0-60 scale; sapphire crystal; screw-in crown; water-resistant to 30 atm
Band: rubber, buckle
Price: $5,875
Variations: stainless steel bracelet ($5,835)

Colt Lady
Reference number: A7738811/BD46/175A
Movement: quartz
Functions: hours, minutes, sweep seconds; date
Case: stainless steel, ø 33 mm, height 9.1 mm; unidirectional bezel with 0-60 scale; sapphire crystal; water-resistant to 20 atm
Band: stainless steel, folding clasp
Price: $3,670

Caliber B01
Automatic; column wheel control of chronograph functions; vertical clutch; single spring barrel, 70-hour power reserve; COSC-certified chronometer
Functions: hours, minutes, subsidiary seconds; chronograph; date
Diameter: 30 mm
Height: 7.2 mm
Jewels: 47
Balance: glucydur
Frequency: 28,800 vph

Caliber B04
Automatic; column wheel control of chronograph functions; vertical clutch; single spring barrel, 70-hour power reserve; COSC-certified chronometer
Functions: hours, minutes, subsidiary seconds; additional 24-hour display (2nd time zone); chronograph; date
Diameter: 30 mm
Height: 7.4 mm
Jewels: 47
Balance: glucydur
Frequency: 28,800 vph

Caliber B05
Automatic; column wheel control of chronograph functions; vertical clutch; time zone disk connected to hand mechanism by planetary transmission; single spring barrel, 70-hour power reserve; COSC-certified chronometer
Functions: hours, minutes, subsidiary seconds; world time display (crown-set 2nd time zone); chronograph; date
Diameter: 30 mm
Height: 8.1 mm
Jewels: 56
Balance: glucydur
Frequency: 28,800 vph

BREMONT

At the 2012 Olympic Games in London, stuntman Gary Connery parachuted into the stadium wearing an outfit that made him look suspiciously like the Queen. He was also the first to jet suit out of a helicopter. On both occasions he was wearing a Bremont watch. And so do many other adventurous types, like polar explorer Ben Saunders or Levison Wood, who was the first person to walk the length of the Nile.

Bremonts are the brainchild of brothers Nick and Giles English, themselves dyed-in-the-wool pilots and restorers of vintage airplanes. They understand that flying safety relies on outstanding mechanics, so they took their time engineering their watches. Naming their brand required some thought, however. The solution came when they remembered an adventure they had had in southern France when they were forced to land their vintage biplane in a field to avoid a storm. The farmer, a former World War II pilot and just as passionate about aircraft as Nick and Giles, was more than happy to put them up. His name: Antoine Bremont.

Ever since the watches hit the market in 2007, the brand has grown sharply. These British-made timepieces use a sturdy, COSC-certified automatic movement, especially hardened steel, a patented shock-absorbing system, and a rotor whose design recalls a flight of planes. The brand has sought its inspiration from various cultural icons in British or international history, including the Spitfire, Bletchley Park (where the German codes were broken during World War II), or Jaguar sports cars and, most recently, Norton motorcycles. It also partnered with Boeing to produce an elegant range of watches on an organic polymer strap. Water sport is another area Bremont has explored, with models inspired by the legendary J-Class yachts, like the ladies' model AC I 32, and a special set devoted to the America's Cup. In 2017, Bremont became the first official timekeeper at the Henley Royal Regatta, one of Great Britain's top rowing events.

Bremont Watch Company
P.O. Box 4741
Henley-on-Thames
RG9 9BZ
Great Britain

Tel.:
+44-845-094-0690

Fax:
+44-870-762-0475

E-mail:
info@bremont.com

Website:
www.bremont.com

Founded:
2002

Number of employees:
100+

Annual production:
several thousand watches

U.S. distributor:
Mike Pearson
1-855-BREMONT
Anthony.kozlowsky@bremont.com

Most important collections/price range:
ALT1, Bremont Boeing, Bremont Jaguar, MB, SOLO, Supermarine, U-2, and limited editions / $3,695 to $42,495

Norton V4

Movement: automatic, modified Caliber BE-50AE; ø 29.89 mm, height 16.5 mm; 28 jewels; 28,800 vph; COSC-certified chronometer; 42-hour power reserve
Functions: hours, minutes, subsidiary seconds; chronograph; date, world time zone
Case: stainless steel, DLC-treated case barrel, ø 43 mm, height 16.5 mm; transparent case back; sapphire crystal; water-resistant to 10 atm
Band: reptile skin, rose gold buckle
Remarks: Norton VR Rim inspired rotor; silver with tachymeter dial ring
Price: $7,295

AC-R-11

Movement: automatic, customized Caliber BE-36AE; ø 28 mm, height 7.5 mm; 25 jewels; 28,800 vph; 42-hour power reserve; rotor with America's Cup decoration; COSC-certified chronometer
Functions: hours, minutes, subsidiary seconds; date; chronograph with 15-minute regatta timer and 5-minute countdown at 12 o'clock
Case: hardened stainless steel and rose gold; ø 43 mm, height 16 mm; bidirectional bezel; DLC-treated case barrel; transparent case back; sapphire crystal; water-resistant to 10 atm
Band: rubber, titanium buckle
Price: $7,095
Variations: titanium case with black dial ($6,459); leather strap of various colors, or NATO strap

ALT1-P2/CR

Movement: automatic, modified Caliber BE-53AE; ø 29.89 mm, height 16.5 mm; 27 jewels; 28,800 vph; Bremont molded and decorated skeletonized rotor; COSC-certified chronometer; 42-hour power reserve
Functions: hours, minutes, subsidiary seconds; chronograph; date
Case: stainless steel, DLC-treated case barrel, ø 43 mm, height 16 mm; transparent case back; sapphire crystal; water-resistant to 10 atm
Band: leather with stainless steel deployment buckle and security clasp
Price: $4,995

BREMONT

Alt1-C
Reference number: ALT1-C/WH-BK
Movement: automatic, Caliber BE-50AE (base ETA "Valjoux" 7750-SO BI AC); ø 30 mm, height 7.9 mm; 25 jewels; 28,800 vph; 42-hour power reserve; COSC-certified chronometer
Functions: hours, minutes, subsidiary seconds; chronograph; date
Case: stainless steel, barrel with black DLC treatment, ø 43 mm, height 16 mm; sapphire crystal; transparent case back; water-resistant to 10 atm
Band: calfskin, buckle
Price: $6,495
Variations: stainless steel bracelet

Martin-Baker MBII
Reference number: MBII-BK/OR
Movement: automatic, Caliber BE-36AE (base ETA 2836-2); ø 25.6 mm, height 5.05 mm; 25 jewels; 28,800 vph; soft iron cage for antimagnetic protection; 38-hour power reserve; COSC-certified chronometer
Functions: hours, minutes, sweep seconds; date, weekday
Case: stainless steel, barrel with orange DLC treatment, ø 43 mm, height 12 mm; sapphire crystal; transparent case back; water-resistant to 10 atm
Band: calfskin, buckle
Price: $4,995
Variations: stainless steel bracelet

Supermarine S501
Reference number: S501/BK
Movement: automatic, Caliber BE-92AE (base ETA 2824-2); ø 25.6 mm, height 4.6 mm; 25 jewels; 28,800 vph; soft iron cage for antimagnetic protection; 38-hour power reserve; COSC-certified chronometer
Functions: hours, minutes, sweep seconds; date
Case: stainless steel, barrel with black DLC treatment, with crown protection, ø 43 mm, height 16 mm; unidirectional bezel, with 0-60 scale; sapphire crystal; transparent case back; water-resistant to 50 atm
Band: calfskin, buckle
Price: $4,775
Variations: stainless steel bracelet; various dials and bezels

Jaguar MKI
Movement: automatic, Bremont BWC/01-10; ø 33.4 mm; 25 jewels; 28,800 vph; 50-hour power reserve
Functions: hours, minutes, subsidiary seconds; date
Case: stainless steel; ø 43 mm, height 16 mm; transparent case back; sapphire crystal; water-resistant to 10 atm
Remarks: miniaturized Jaguar E-Type steering wheel rotor with Growler emblem
Price: $11,395

Solo-32-AJ
Movement: automatic, modified Caliber BE-10AE; ø 20.3 mm; 18 jewels; 28,800 vph; 40-hour power reserve; Bremont molded and decorated rotor; COSC-certified chronometer
Functions: hours, minutes, sweep seconds; date
Case: stainless steel; ø 32 mm, height 9.65 mm; transparent screw-in case back; sapphire crystal; water-resistant to 5 atm
Price: $3,695
Variations: various straps

AC35
Movement: automatic, BWC/01-10; ø 33.4 mm; 25 jewels; 28,800 vph; 50-hour power reserve; bidirectional Bremont molded and decorated rotor; COSC-certified chronometer
Functions: hours, minutes, subsidiary seconds, date
Case: rose gold with stainless steel case barrel; ø 43 mm, height 14.1 mm; sapphire crystal; water-resistant to 200 atm
Band: rubber, buckle
Remarks: America's Cup "Auld Mug" embossing and "2017 America's Cup Bermuda"
Price: $22,495

BRM

Is luxury on the outside or the inside? The answer to this question can tear the veil from the hype and reveal the true craftsman. For Bernard Richards, the true sign of luxury lies in "technical skills and perfection in all stages of manufacture." The exterior of the product is of course crucial, but all of BRM's major operations for making a wristwatch—such as encasing, assembling, setting, and polishing—are performed by hand in his little garage-like factory located outside Paris in Magny-sur-Vexin.

BRM is devoted to the ultra-mechanical look with the *haute-horlogerie* feel of high-end materials. His inspiration at the start came from the 1940s, when internal combustion engines meant business, the age of axle grease, pinups, real pilots, and a can-do attitude. The design: three dimensions visible to the naked eye, big mechanical landscapes. The inside: custom-designed components, fitting perfectly into Richards's automotive ideal. Gradually, though, Richards has been modernizing. Since the beginning of 2009, BRM aficionados have been able to engage in this process to an even greater degree: When visiting the BRM website, the client can now construct his or her own V12-44-BRM model.

BRM's unusual timepieces have mainly been based on the tried and trusted Valjoux 7750. But Richards has set lofty goals for himself and his young venture, for he intends to set up a true *manufacture* in his French factory. His BiRotor model is thus outfitted with the Precitime, an autonomous caliber conceived and manufactured on French soil. The movement features BRM's own shock absorbers mounted on the conical springs of its so-called Isolastic system. Plates and bridges are crafted in ARCAP, rotors are made of Fortale and tantalum. The twin rotors, found at 12 and 6 o'clock, are mounted on double rows of ceramic bearings that require no lubrication.

BRM
(Bernard Richards Manufacture)
2 Impasse de L'Aubette
ZA des Aulnaies
F-95420 Magny en Vexin
France

Tel.:
+33-1-61-02-00-25

Fax:
+33-1-61-02-00-14

Website:
www.brm-manufacture.com

Founded:
2003

Number of employees:
20

Annual production:
approx. 2,000 pieces

U.S. distributor:
BRM Manufacture North America
25 Highland Park Village, Suite 100-777
Dallas, TX 75205
214-231-0144
usa@brm-manufacture.com

Price range:
$3,000 to $150,000

TR1 Tourbillon

Movement: automatic, Precitime Caliber; ø 30 mm, height 7.9 mm; 26 jewels; 28,800 vph; 46-hour power reserve; 105-second tourbillon in ARCAP with reversed cage for visible escapement and suspended by 2 micro springs; patented isolastic system with 4 shock absorbers; automatic assembly with ceramic ball bearings
Functions: hours, minutes, sweep seconds
Case: titanium, ø 52 mm; sapphire crystal; antireflective on both sides; transparent case back; water-resistant to 10 atm
Band: leather, buckle
Price: $145,350
Variations: 48 mm ($136,150)

BiRotor

Movement: automatic, Precitime Caliber BiRotor; 24 × 32 mm; 35 jewels; 28,800 vph; 45-hour power reserve; Fortale HR and tantalum double rotors on ceramic ball bearings; patented isolastic system with 4 shock absorbers, ARCAP plates, bridges
Functions: hours, minutes, subsidiary seconds
Case: titanium with rose gold crown and strap lugs, 40 × 48 mm, height 9.9 mm; domed sapphire crystal; antireflective on both sides; domed sapphire crystal transparent case back; water-resistant to 30 m
Band: Nomex, buckle
Price: $68,500

R50 MK

Movement: automatic, heavily modified ETA Caliber 2161; ø 38 mm; 35 jewels; 28,800 vph; 48-hour power reserve; patented isolastic system with 3 shock absorbers; Fortale HR, tantalum, and aluminum rotor; hand-painted Gulf colors
Functions: hours, minutes, sweep seconds; power reserve indication
Case: Makrolon (polycarbonate) with rose gold crown and strap lugs, ø 50 mm, height 13.2 mm; sapphire crystal; antireflective on both sides; exhibition case back; water-resistant to 3 atm
Band: leather, buckle
Price: $28,550; limited to 30 pieces
Variations: rose gold ($65,000)

R46
Movement: automatic, heavily modified ETA Caliber 2161; ø 38 mm; 35 jewels; 28,800 vph; 48-hour power reserve; patented isolastic system with 3 shock absorbers; Fortale HR, tantalum and aluminum rotor; hand-painted Gulf colors
Functions: hours, minutes, sweep seconds; power reserve indication
Case: Makrolon with rose gold crown and strap lugs, ø 46 mm, height 10 mm; sapphire crystal; antireflective on both sides; exhibition case back; water-resistant to 3 atm
Band: leather, buckle
Price: $24,750; limited to 30 pieces

R12-46
Movement: automatic, ETA Valjoux Caliber 7753 modified in-house; ø 30 mm, height 7.90 mm; 27 jewels; 28,800 vph; skeletonized dial; shock absorbers connected to block; 42-hour power reserve
Functions: hours, minutes, subsidiary seconds; chronograph; date
Case: bronze, ø 46 mm, height 14 mm; stainless steel lugs and crown with black PVD; crystal sapphire; transparent case back; water-resistant to 10 atm
Band: leather, bronze buckle
Price: $13,550

MK 44 Green
Movement: automatic, ETA Valjoux Caliber 7753; ø 30 mm, height 7.90 mm; 27 jewels; 28,800 vph; 42-hour power reserve
Functions: hours, minutes, subsidiary seconds; date; chronograph
Case: Makrolon (polycarbonate), ø 45 mm; pushers, lugs, crown from single titanium block; sapphire crystal; exhibition case back; water-resistant to 10 atm
Band: technical fabrics for extra lightness
Remarks: lightest automatic chronograph ever made; skeleton dial with green hands
Price: $13,450
Variations: many options with configurator

DDF12-44-AR
Movement: automatic, ETA Valjoux Caliber 7753 modified in-house; ø 30 mm, height 7.90 mm; 27 jewels; 28,800 vph; skeletonized dial; shock absorbers connected to block; 42-hour power reserve
Functions: hours, minutes, subsidiary seconds; chronograph; date
Case: titanium with black PVD coating, ø 44 mm; stainless steel lugs and pushers; sapphire crystal; transparent case back; water-resistant to 10 atm
Band: leather, buckle
Remarks: skeletonized dial with red hands
Price: $12,750

V12-46-TSAABL
Movement: automatic, ETA Valjoux Caliber 7753 modified in-house; ø 30 mm, height 7.90 mm; 27 jewels; 28,800 vph; skeletonized dial; shock absorbers connected to movement, 3 vertical, 3 horizontal; 42-hour power reserve
Functions: hours, minutes, subsidiary seconds; chronograph; date with 10-hour corrector
Case: titanium with rose gold crown and strap lugs, ø 46 mm, height 12 mm; sapphire crystal; antireflective on both sides; transparent case back; screw-in crown; water-resistant to 10 atm
Band: calfskin, folding clasp
Price: $16,700
Variations: hands and springs in different colors on request

R6-46-CORV-GS Corvette Grand Sport
Movement: automatic, ETA Valjoux Caliber 2824 modified in-house; ø 25.6 mm, height 4.6 mm; 25 jewels; 28,800 vph; skeletonized dial; 38-hour power reserve
Functions: hours, minutes, seconds; date
Case: aluminum, ø 46 mm, height 10 mm; crystal sapphire; transparent case back; screw-in crown; water-resistant to 5 atm
Band: leather, buckle
Price: $5,700; limited to 163 pieces

BULGARI

Although Bulgari is one of the largest jewelry manufacturers in the world, watches have always played an important role for the brand. The purchase of Daniel Roth and Gérald Genta in the Vallée de Joux opened new perspectives for its timepieces, thanks to specialized production facilities and the watchmaking talent in the Vallée de Joux—especially where complicated timepieces are concerned. In March 2011, luxury goods giant Louis Vuitton Moët Hennessy (LVMH) secured all the Bulgari family shares in exchange for 16.5 million LVMH shares and a say in the group's future. The financial backing of the megagroup boosted the company's strategy to become fully independent.

In mid-2013, Jean-Christophe Babin, the man who turned TAG Heuer into a leading player in sports watches, was chosen to head the venerable brand. He had also managed to build up a manufacturing structure from scratch at TAG Heuer, which is exactly the direction Bulgari's watch division is headed in. The company has been building complete watches, including a number of outstanding calibers, like the 168 automatic based on a design by the great nineteenth-century watchmaker Jean Frédéric Leschot. In a modern building in the industrial zone of La Chaux-de-Fonds, the company manufactures its BVL 191 and BVL 193 calibers. Bulgari even manufactures its ultra-thin BVL 128.

The company has been releasing a series of increasingly thin and complicated automatics that suggest some very clever engineering is going on behind the scenes: a tourbillon, a minute repeater (2016), which is 3.12 millimeters high and whose dial features slotted indices for better sound transmission. The Octo Finissimo Automatic, with the 2.23-millimeter-high Caliber BVL 138, was the talk of Baselworld in 2017: It measures a mere 5.15 millimeters from case back to sapphire crystal. And 2018 saw the release of another tourbillon, this one "standing" at 3.95 millimeters.

Bulgari Horlogerie SA
rue de Monruz 34
CH-2000 Neuchâtel
Switzerland

Tel.:
+41-32-722-7878

Fax:
+41-32-722-7933

E-mail:
info@bulgari.com

Website:
www.bulgari.com

Founded:
1884 (Bulgari Horlogerie was founded in the early 1980s as Bulgari Time)

U.S. distributor:
Bulgari Corporation of America
555 Madison Avenue
New York, NY 10022
212-315-9700

Most important collections/price range:
Bulgari-Bulgari / from approx. $4,700 to $30,300; Diagono / from approx. $3,200; Octo / from approx. $9,500 to $690,000 and above; Daniel Roth and Gérald Genta collections

Octo Finissimo Tourbillon Automatic

Reference number: BGO42C14TTBXTSKAUTO
Movement: automatic, Bulgari Caliber BVL 288; ø 36 mm, height 1.95 mm; 24 jewels; 21,600 vph; flying 1-minute tourbillon; hubless peripheral rotor; 55-hour power reserve
Functions: hours, minutes
Case: titanium, ø 42 mm, height 3.95 mm; sapphire crystal; transparent case back; water-resistant to 3 atm
Band: titanium, double folding clasp
Remarks: currently thinnest wristwatch with automatic movement and tourbillon
Price: $118,000

Octo Finissimo Tourbillon Skeleton

Reference number: BGO40PLTBXTSK
Movement: manually wound, Bulgari Caliber BVL 268 Finissimo Squelette; ø 32.6 mm, height 1.95 mm; 26 jewels; 21,600 vph; flying 1-minute tourbillon; skeletonized and finely finished movement; 62-hour power reserve
Functions: hours, minutes
Case: platinum, ø 40 mm, height 5 mm; sapphire crystal; transparent case back; water-resistant to 3 atm
Band: reptile skin, buckle
Price: $132,000

Octo Finissimo Minute Repeater

Reference number: BGO40BTLMRXT
Movement: manually wound, Bulgari Caliber BVL 362; ø 28.5 mm, height 3.12 mm; 21,600 vph; finely hand-finished movement; 42-hour power reserve
Functions: hours, minutes, subsidiary seconds; minute repeater
Case: titanium, ø 40 mm, height 6.85 mm; sapphire crystal; transparent case back; water-resistant to 3 atm
Band: reptile skin, double folding clasp
Remarks: resonance dial with slot indices
Price: $156,000; limited to 50 pieces

BULGARI

Octo Finissimo Automatic
Reference number: BGO40C14TTXTAUTO
Movement: automatic, Bulgari Caliber BVL 138 Finissimo; ø 36 mm, height 2.23 mm; 23 jewels; 21,600 vph; platinum microrotor; finely finished with côtes de Genève; 60-hour power reserve
Functions: hours, minutes, subsidiary seconds
Case: titanium, ø 40 mm, height 5.15 mm; sapphire crystal; transparent case back
Band: titanium, folding clasp
Price: $13,900

Octo Finissimo Automatic
Reference number: BGO40C14TLXTAUTO
Movement: automatic, Bulgari Caliber BVL 138 Finissimo; ø 36 mm, height 2.23 mm; 23 jewels; 21,600 vph; platinum microrotor; finely finished with côtes de Genève; 60-hour power reserve
Functions: hours, minutes, subsidiary seconds
Case: titanium, ø 40 mm, height 5.15 mm; sapphire crystal; transparent case back
Band: reptile skin, buckle
Price: $12,800

Octo Finissimo Automatic
Reference number: BGOPGXTAUTO
Movement: automatic, Bulgari Caliber BVL 138 Finissimo; ø 36 mm, height 2.23 mm; 23 jewels; 21,600 vph; platinum microrotor; finely finished with côtes de Genève; 60-hour power reserve
Functions: hours, minutes, subsidiary seconds
Case: rose gold, ø 40 mm, height 5.15 mm; sapphire crystal; transparent case back
Band: rose gold, folding clasp
Price: $43,400

Octo Finissimo Skeleton
Reference number: BGO40TLXTSK (SAP code 102714)
Movement: manually wound, Bulgari Caliber BVL 128SK; ø 36 mm, height 2.35 mm; 28,800 vph; skeletonized bridges and plates; 65-hour power reserve
Functions: hours, minutes, subsidiary seconds; power reserve indicator
Case: stainless steel, ø 40 mm, height 5.37 mm; sapphire crystal; transparent case back; screw-in crown; water-resistant to 3 atm
Band: reptile skin, buckle
Price: $21,500

Octo Finissimo Skeleton
Reference number: BGO40TLXTSK/LE
Movement: manually wound, Bulgari Caliber BVL 128SK; ø 36 mm, height 2.35 mm; 28,800 vph; skeletonized bridges and plates; 65-hour power reserve
Functions: hours, minutes, subsidiary seconds; power reserve indicator
Case: titanium, ø 40 mm, height 5.37 mm; sapphire crystal; transparent case back; screw-in crown; water-resistant to 3 atm
Band: reptile skin, buckle
Price: $22,500

Octo Finissimo Ultranero Skeleton
Reference number: BGO40BSPGLXT/SK
Movement: manually wound, Bulgari Caliber BVL 128SK; ø 36 mm, height 2.35 mm; 28,800 vph; skeletonized bridges and plates; 65-hour power reserve
Functions: hours, minutes, subsidiary seconds; power reserve indicator
Case: stainless steel with black DLC coating, ø 40 mm, height 5.37 mm; bezel in rose gold; sapphire crystal; transparent case back; screw-in crown, in rose gold; water-resistant to 3 atm
Band: reptile skin, buckle
Remarks: skeletonized dial
Price: $26,600

BULGARI

Octo Ultranero Solotempo
Reference number: BGO41C9BSVD
Movement: automatic, Bulgari Caliber BVL 193; ø 25.6 mm, height 3.7 mm; 28 jewels; 28,800 vph; 50-hour power reserve
Functions: hours, minutes, sweep seconds; date
Case: stainless steel with black DLC coating, ø 41 mm, height 10.5 mm; sapphire crystal
Band: rubber, buckle
Price: $6,950

Octo Ultranero Solotempo
Reference number: BGO41BBSVD/N
Movement: automatic, Bulgari Caliber BVL 193; ø 25.6 mm, height 3.7 mm; 28 jewels; 28,800 vph; 50-hour power reserve
Functions: hours, minutes, sweep seconds; date
Case: stainless steel with black DLC coating, ø 41 mm, height 10.5 mm; sapphire crystal
Band: rubber, buckle
Price: $6,950

Octo Finissimo Tourbillon Ultranero
Reference number: BGO40BTLTBXT (SAP code 102560)
Movement: manually wound, Bulgari Caliber BVL 268 Finissimo Tourbillon; ø 32.6 mm, height 1.95 mm; 26 jewels; 21,600 vph; flying 1-minute tourbillon; finely finished movement; 52-hour power reserve
Functions: hours, minutes
Case: titanium with black DLC coating, ø 40 mm, height 5 mm; sapphire crystal; transparent case back; screw-in crown, in rose gold; water-resistant to 3 atm
Band: reptile skin, buckle
Price: $99,000

Octo L'Originale Velocissimo
Reference number: BGO41C14TVDCH (SAP code 102859)
Movement: automatic, Bulgari Caliber BVL 328 Velocissimo (base Zenith "El Primero"); ø 30 mm, height 6.62 mm; 31 jewels; 36,000 vph; column wheel control of chronograph functions, silicon escapement; 50-hour power reserve
Functions: hours, minutes, subsidiary seconds; chronograph; date
Case: titanium, ø 41 mm, height 13.07 mm; sapphire crystal; transparent case back; water-resistant to 10 atm
Band: rubber, folding clasp
Price: $10,200

Octo Ultranero Velocissimo
Reference number: BGO41BBSVDCH (SAP code 102630)
Movement: automatic, Bulgari Caliber Velocissimo (base Zenith "El Primero"); ø 30 mm, height 6.62 mm; 31 jewels; 36,000 vph; column wheel control of chronograph functions, silicon escapement; 50-hour power reserve
Functions: hours, minutes, subsidiary seconds; chronograph; date
Case: stainless steel with black DLC coating, ø 41 mm, height 13 mm; sapphire crystal; transparent case back; screw-in crown, in rose gold
Band: rubber, buckle
Price: $10,600

Octo Ultranero Velocissimo
Reference number: BGO41BBSPGVDCH
Movement: automatic, Bulgari Caliber Velocissimo (base Zenith "El Primero"); ø 30 mm, height 6.62 mm; 31 jewels; 36,000 vph; column wheel control of chronograph functions, silicon escapement; 50-hour power reserve
Functions: hours, minutes, subsidiary seconds; chronograph; date
Case: stainless steel with black DLC coating, ø 41 mm, height 13 mm; bezel in pink gold; sapphire crystal; transparent case back; screw-in crown, in rose gold
Band: rubber, buckle
Price: $13,400

BULGARI

Octo Roma
Reference number: OC41C3SSD
Movement: automatic, Bulgari Caliber BVL 191; ø 26.2 mm, height 3.8 mm; 26 jewels; 28,800 vph; finely finished with côtes de Genève; 42-hour power reserve
Functions: hours, minutes, sweep seconds; date
Case: stainless steel, ø 41 mm, height 10.5 mm; sapphire crystal; transparent case back; water-resistant to 10 atm
Band: stainless steel, folding clasp
Price: $6,500
Variations: comes with different cases, straps, and dials

Papillon Voyageur
Reference number: BRRP46C14GLGMTP (SAP code 101835)
Movement: automatic, Daniel Roth Caliber DR 1307; ø 25.6 mm, height 6.78 mm; 26 jewels; 28,800 vph; 45-hour power reserve
Functions: hours (digital, jumping), minutes (retrograde), subsidiary seconds (segment display with double hand); additional 24-hour display (2nd time zone)
Case: rose gold, 43 × 46 mm, height 15.2 mm; sapphire crystal; transparent case back; pusher to advance 24-hour display; water-resistant to 3 atm
Band: reptile skin, double folding clasp
Price: $51,000

Ammiraglio del Tempo
Reference number: BRRP50BGLDEMR
Movement: manually wound, Daniel Roth Caliber 7301; ø 38 mm, height 9.38 mm; 56 jewels; 14,400 vph; chronometer escapement; Westminster chimes, 4 hammers and gongs, constant force mechanism, cylindrical balance spring, triple shock-absorbing system; 48-hour power reserve
Functions: hours, minutes; minute repeater
Case: rose gold, 45.75 × 50 mm, height 14.9 mm; sapphire crystal; transparent case back
Remarks: mobile lug at 7 o'clock activates chimes
Band: reptile skin, buckle
Price: $374,000

Caliber BVL 138 Finissimo
Automatic; flying platinum microrotor; flying single spring barrel, 60-hour power reserve
Functions: hours, minutes, subsidiary seconds; date
Diameter: 36 mm
Height: 2.23 mm
Jewels: 23
Balance: glucydur
Frequency: 21,600 vph
Balance spring: flat hairspring index for fine adjustment
Shock protection: Incabloc
Remarks: finely finished with côtes de Genève

Caliber BVL 206
Manually wound; flying 1-minute tourbillon; skeletonized mainplate; single spring barrel, 64-hour power reserve
Functions: hours, minutes
Diameter: 34 mm
Height: 5 mm
Frequency: 21,600 vph
Remarks: bridges with black DLC coating and green luminescent bar for hour markers

Caliber BVL 288
Automatic; flying 1-minute tourbillon; hubless peripheral rotor; single spring barrel, 55-hour power reserve
Functions: hours, minutes
Diameter: 36 mm
Height: 1.95 mm
Jewels: 24
Frequency: 21,600 vph
Remarks: engine of thinnest automatic wristwatch—including a tourbillon

CARL F. BUCHERER

While luxury watch brand Carl F. Bucherer is still rather young, the Lucerne-based Bucherer jewelry dynasty behind it draws its vast know-how from more than ninety years of experience in the conception and design of fine wristwatches.

In 2005, Bucherer joined the Sainte-Croix-headquartered partner, Techniques Horlogères Appliquées SA (THA), to manufacture its own movement. THA was integrated into the Bucherer Group and the watch company renamed Carl F. Bucherer Technologies SA (CFBT). The Sainte-Croix operation is led by technical director Dr. Albrecht Haake, who oversees a staff of about twenty. Dr. Haake is currently focusing much of his energy on furthering the capacities at the workshop. "Industrialization is not a question of cost, but rather a question of quality," says Haake.

The Swiss company also expanded its Lengnau location to create a competence center that can focus on manufacturing its own movements as well as in-house watches. The automatic caliber with the peripheral rotor went through a thorough revamping process with the idea of industrializing it. The general goal is to use it for new models that will consolidate the young brand's status as a manufacture. For its 130th birthday, the company decided to create a special "floating" tourbillon for its Manero collection. Not only does it feature a hubless ring rotor, but the power transmission to the tourbillon is done from the side, making it invisible to the observer.

Bucherer AG
Carl F. Bucherer
Langensandstrasse 27
CH-6002 Lucerne
Switzerland

Tel.:
+41-41-369-7070

Fax:
+41-41-369-7072

E-mail:
info@carl-f-bucherer.com

Website:
www.carl-f-bucherer.com

Founded:
1919, repositioned under the name Carl F. Bucherer in 2001

Number of employees:
approx. 200

Annual production:
approx. 30,000 watches

U.S. distributor:
Carl F. Bucherer North America
1805 South Metro Parkway
Dayton, OH 45459
937-291-4366
info@cfbna.com; www.carl-f-bucherer.com

Most important collections/price range:
Patravi, Manero, Alacria and Pathos / core price segment $5,000 to $30,000

Manero Flyback

Reference number: 0010919.03.43.01
Movement: automatic, Caliber CFB 1970; ø 30.4 mm, height 7.9 mm; 25 jewels; 28,800 vph; 42-hour power reserve
Functions: hours, minutes, subsidiary seconds; flyback chronograph; date
Case: rose gold, ø 43 mm, height 14.45 mm; sapphire crystal; transparent case back; water-resistant to 3 atm
Band: reptile skin, buckle
Price: $16,900

Manero Flyback

Reference number: 00.10919.08.33.01
Movement: automatic, Caliber CFB 1970; ø 30.4 mm, height 7.9 mm; 25 jewels; 28,800 vph; 42-hour power reserve
Functions: hours, minutes, subsidiary seconds; flyback chronograph; date
Case: stainless steel, ø 43 mm, height 14.45 mm; sapphire crystal; transparent case back; water-resistant to 3 atm
Band: reptile skin, buckle
Price: $6,200

Manero Flyback

Reference number: 00.10919.03.33.01
Movement: automatic, Caliber CFB 1970; ø 30.4 mm, height 7.9 mm; 25 jewels; 28,800 vph; 42-hour power reserve
Functions: hours, minutes, subsidiary seconds; flyback chronograph; date
Case: rose gold, ø 43 mm, height 14.45 mm; sapphire crystal; transparent case back; water-resistant to 3 atm
Band: reptile skin, buckle
Price: $16,900

CARL F. BUCHERER

Manero Peripheral 43mm
Reference number: 00.10921.08.33.01
Movement: automatic, Caliber CFB A2050; ø 30.6 mm, height 5.28 mm; 33 jewels; 28,800 vph; hubless peripheral rotor with tungsten oscillating mass; 55-hour power reserve; COSC-certified chronometer
Functions: hours, minutes, subsidiary seconds; date
Case: stainless steel, ø 43.1 mm, height 11.2 mm; sapphire crystal; transparent case back; water-resistant to 3 atm
Band: reptile skin, buckle
Price: $6,800
Variations: white dial ($6,800); pink gold ($17,600)

Manero Peripheral 43mm
Reference number: 00.10921.08.23.21
Movement: automatic, Caliber CFB A2050; ø 30.6 mm, height 5.28 mm; 33 jewels; 28,800 vph; hubless peripheral rotor with tungsten oscillating mass; 55-hour power reserve; COSC-certified chronometer
Functions: hours, minutes, subsidiary seconds; date
Case: stainless steel, ø 43.1 mm, height 11.2 mm; sapphire crystal; transparent case back; water-resistant to 3 atm
Band: stainless steel, folding clasp
Price: $7,200
Variations: reptile skin strap ($6,800)

Manero PowerReserve
Reference number: 00.10912.03.13.01
Movement: automatic, Caliber CFB A1011; ø 32 mm, height 6.3 mm; 33 jewels; 21,600 vph; 55-hour power reserve
Functions: hours, minutes, subsidiary seconds; power reserve indicator; large date, weekday
Case: rose gold, ø 42.5 mm, height 12.54 mm; sapphire crystal; transparent case back; water-resistant to 3 atm
Band: reptile skin, buckle
Price: $21,000

Manero Tourbillon Double Peripheral
Reference number: 00.10920.03.13.01
Movement: automatic, Caliber CFB T3000; ø 36.5 mm, height 6.66 mm; 32 jewels; 21,600 vph; silicon escapement; floating 1-minute tourbillon with invisible peripheral drive; hubless peripheral rotor with tungsten oscillating mass; 65-hour power reserve; COSC-certified chronometer
Functions: hours, minutes, subsidiary seconds (on tourbillon cage)
Case: rose gold, ø 43.1 mm, height 11.57 mm; sapphire crystal; transparent case back; water-resistant to 3 atm
Band: reptile skin, folding clasp
Price: $68,000

Patravi ScubaTec
Reference number: 00.10632.22.53.01
Movement: automatic, Caliber CFB 1950.1; ø 26.2 mm, height 4.6 mm; 25 jewels; 28,800 vph; 38-hour power reserve; COSC-certified chronometer
Functions: hours, minutes, sweep seconds; date
Case: rose gold, ø 44.6 mm, height 13.45 mm; unidirectional bezel with ceramic insert, 0-60 scale; sapphire crystal; screw-in crown; helium valve and crown protection in blackened titanium; water-resistant to 50 atm
Band: rubber, folding clasp with extension link
Price: $23,600
Variations: rose gold and stainless steel ($9,600)

Patravi ScubaTec
Reference number: 00.10632.23.33.22
Movement: automatic, Caliber CFB 1950.1; ø 26.2 mm, height 4.6 mm; 25 jewels; 28,800 vph; 38-hour power reserve; COSC-certified chronometer
Functions: hours, minutes, sweep seconds; date
Case: stainless steel, ø 44.6 mm, height 13.45 mm; unidirectional bezel with ceramic insert, 0-60 scale; sapphire crystal; screw-in crown; helium valve; water-resistant to 50 atm
Band: stainless steel, folding clasp with extension link
Price: $6,700
Variations: rubber strap ($6,200)

CARL F. BUCHERER

Patravi TravelTec
Reference number: 00.10620.08.33.02
Movement: automatic, Caliber CFB 1901.1; ø 28.6 mm, height 7.3 mm; 39 jewels; 28,800 vph; 42-hour power reserve; COSC-certified chronometer
Functions: hours, minutes, subsidiary seconds; additional 24-hour display (2nd time zone); chronograph; date
Case: stainless steel, ø 46.6 mm, height 15.5 mm; pusher-activated, bidirectional inner bezel with 24-hour division for a 3rd time zone; sapphire crystal; screw-in crown; water-resistant to 5 atm
Band: rubber, folding clasp
Price: $10,900

Adamavi
Reference number: 00.10314.08.13.21
Movement: automatic, Caliber CFB 1950 (base ETA 2824-2); ø 25.6 mm, height 4.6 mm; 25 jewels; 28,800 vph; 38-hour power reserve
Functions: hours, minutes, sweep seconds; date
Case: stainless steel, ø 39 mm, height 8.77 mm; sapphire crystal; water-resistant to 3 atm
Band: stainless steel Milanese mesh bracelet, folding clasp
Price: $2,600

Adamavi
Reference number: 00.10316.07.36.01
Movement: automatic, Caliber CFB 1968 (base ETA 2895-2); ø 25.6 mm, height 4.35 mm; 27 jewels; 21,600 vph; 42-hour power reserve
Functions: hours, minutes, subsidiary seconds
Case: stainless steel, ø 39 mm, height 7.1 mm; bezel and crown in rose gold; sapphire crystal; water-resistant to 3 atm
Band: reptile skin, buckle
Price: $3,200

Adamavi
Reference number: 00.10320.08.15.22
Movement: automatic, Caliber CFB 1963 (base ETA 2681); ø 20 mm, height 4.8 mm; 25 jewels; 28,800 vph; 38-hour power reserve
Functions: hours, minutes, sweep seconds; date
Case: stainless steel, ø 31 mm, height 8.3 mm; sapphire crystal; transparent case back; water-resistant to 3 atm
Band: stainless steel, folding clasp
Price: $2,700

CFB T3000
Automatic; floating 1-minute tourbillon; silicon escapement, bidirectional peripheral tungsten rotor turning on edge of movement, on spring-based bearings; precision adjustment mechanism; single spring barrel, 65-hour power reserve; COSC-certified chronometer
Functions: hours, minutes, subsidiary seconds; date
Diameter: 36.5 mm
Height: 6.6 mm
Jewels: 32
Balance: glucydur
Frequency: 21,600 vph
Balance spring: flat hairspring
Shock protection: Incabloc

CFB A2050
Automatic; bidirectional peripheral tungsten rotor turning on edge of movement, on spring-based bearings; precision adjustment mechanism; single spring barrel, 55-hour power reserve
Base caliber: CFB A2000
Functions: hours, minutes, subsidiary seconds; date
Diameter: 30.6 mm
Height: 5.28 mm
Jewels: 33
Balance: glucydur
Frequency: 28,800 vph
Balance spring: flat hairspring
Shock protection: Incabloc

Our timeless classic, rewritten in bronze. With an ever-changing patina that reflects the passing years. A single red hand that marks the present day. And a future that's yet to be written.

The story continues.

#GoYourOwnWay

**Big Crown Pointer Date
80th Anniversary Edition**

HÖLSTEIN 1904

CARTIER

Since the Richemont Group's founding, Cartier has played an important role in the luxury concern as its premier brand and instigator of turnover. Although it took a while for Cartier to find its footing and convince the male market of its masculinity, any concerns about Cartier's seriousness and potential are being dispelled by facts. "We aimed to become a key player in *haute horlogerie*, and we succeeded," said CEO Bernard Fornas at a July 2012 press conference at the company's main manufacturing site in La Chaux-de-Fonds. The company is growing by leaps and bounds—a components manufacturing site employing 400 people is being built at the growing Richemont campus in Meyrin (Geneva).

It was Richemont Group's purchase of the Roger Dubuis *manufacture* in Geneva a few years ago that paved the way to the brand's independence and vertical integration. Under its brilliant head of fine watchmaking, Carole Forestier-Kasapi, Cartier has become a serious producer of movements, among them the 1904, which made its debut in the Calibre model. With a diameter of 42 mm, this strikingly designed men's watch is also well positioned in the segment. The designation 1904 MC is a reference to the year in which Louis Cartier developed the first wristwatch made for men—a pilot's watch custom designed for his friend and early pioneer of aviation, Alberto Santos-Dumont.

The automatic movement is a largely unadorned, yet efficient, machine, powered by twin barrels. The central rotor sits on ceramic ball bearings, and the adjustment of the conventional escapement is by excenter screw. It is available for chronographs or diver's watches. But mainly, it has positioned Cartier as one of the most serious and effective makers of high-end watches in a very competitive industry.

Cartier Joaillerie
Branch of Richemont Intl. SA
Rue André-De-Garrini 3
CH-1217 Meyrin
Switzerland

Tel:
+41-22-808-2500

Fax:
+41-22-808-2502

E-mail:
info@cartier.ch

Website:
www.cartier.ch

Founded:
1847

Number of employees:
approx. 1,300 (watch manufacturing)

U.S. distributor:
Cartier North America
645 Fifth Avenue
New York, NY 10022
1-800-CARTIER
www.cartier.us

Most important collections:
Ballon Bleu, Calibre, Clé, Drive, Pasha, Rotonde de Cartier, Santos, Tank

Santos de Cartier
Reference number: WSSA0009
Movement: automatic, Cartier Caliber 1847 MC; ø 25.6 mm; 23 jewels; 28,800 vph; 40-hour power reserve
Functions: hours, minutes, sweep seconds; date
Case: stainless steel, ø 39.8 mm, height 9.08 mm; sapphire crystal; crown with spinel cabochon; water-resistant to 10 atm
Band: stainless steel, double folding clasp
Remarks: comes with additional calfskin strap
Price: $6,850
Variations: 35-mm case ($6,250)

Santos de Cartier
Reference number: W2SA0006
Movement: automatic, Cartier Caliber 1847 MC; ø 25.6 mm; 23 jewels; 28,800 vph; 40-hour power reserve
Functions: hours, minutes, sweep seconds; date
Case: stainless steel, ø 39.8 mm, height 9.08 mm; bezel yellow gold; sapphire crystal; crown with spinel cabochon; water-resistant to 10 atm
Band: calfskin, double folding clasp
Remarks: comes with additional stainless steel bracelet
Price: $10,400
Variations: 35-mm case ($9,100)

Santos de Cartier
Reference number: WGSA0011
Movement: automatic, Cartier Caliber 1847 MC; ø 25.6 mm; 23 jewels; 28,800 vph; 40-hour power reserve
Functions: hours, minutes, sweep seconds; date
Case: pink gold, ø 39.8 mm, height 9.08 mm; sapphire crystal; crown with spinel cabochon; water-resistant to 10 atm
Band: calfskin, double folding clasp
Remarks: comes with additional reptile skin strap
Price: $20,400
Variations: 35-mm case ($17,900)

CARTIER

Santos de Cartier
Reference number: WGSA0009
Movement: automatic, Cartier Caliber 1847 MC; ø 25.6 mm; 23 jewels; 28,800 vph; 40-hour power reserve
Functions: hours, minutes, sweep seconds; date
Case: yellow gold, ø 39.8 mm, height 9.08 mm; sapphire crystal; crown with spinel cabochon; water-resistant to 10 atm
Band: yellow gold, double folding clasp
Remarks: comes with additional reptile skin strap
Price: $37,000
Variations: 35-mm case ($33,300)

Santos de Cartier
Reference number: WGSA0008
Movement: automatic, Cartier Caliber 1847 MC; ø 25.6 mm; 23 jewels; 28,800 vph; 40-hour power reserve
Functions: hours, minutes, sweep seconds; date
Case: pink gold, ø 35.1 mm, height 8.83 mm; sapphire crystal; crown with spinel cabochon; water-resistant to 10 atm
Band: reptile skin, double folding clasp
Remarks: comes with additional pink gold bracelet
Price: $33,300
Variations: 39-mm case ($37,000)

Santos de Cartier Skeleton
Reference number: WHSA0007
Movement: manually wound, Cartier Caliber 9611 MC; 28.6 × 28.6 mm, height 3.97 mm; 20 jewels; 21,600 vph; skeleton movement integrating Roman hour numeral; 2 spring barrels, 72-hour power reserve
Functions: hours, minutes
Case: stainless steel, ø 39.8 mm, height 9.08 mm; bezel screwed to case back with 8 screws; sapphire crystal; transparent case back; crown with sapphire cabochon; water-resistant to 10 atm
Band: stainless steel, double folding clasp
Price: $26,800
Variations: rose gold ($63,500)

Drive de Cartier Extra-Flat
Reference number: WSNM0011
Movement: manually wound, Cartier Caliber 430 MC; ø 20 mm, height 2.15 mm; 18 jewels; 21,600 vph; 43-hour power reserve
Functions: hours, minutes
Case: stainless steel, 38 × 39 mm, height 6.6 mm; sapphire crystal; water-resistant to 3 atm
Band: reptile skin, buckle
Price: $5,600
Variations: yellow gold ($15,400)

Drive de Cartier Moon Phases
Reference number: WSNM0008
Movement: automatic, Cartier Caliber 1904-LU MC; ø 25 mm, height 5.2 mm; 25 jewels; 28,800 vph; 48-hour power reserve
Functions: hours, minutes; moon phase
Case: stainless steel, 40 × 41 mm, height 12 mm; sapphire crystal
Band: reptile skin, double folding clasp
Price: $7,850
Variations: rose gold ($21,100)

Drive de Cartier
Reference number: WSNM0004
Movement: automatic, Cartier Caliber 1904-PS MC; ø 24.9 mm, height 4.5 mm; 27 jewels; 28,800 vph; 48-hour power reserve
Functions: hours, minutes, subsidiary seconds; date
Case: stainless steel, 40 × 41 mm, height 11.3 mm; sapphire crystal; water-resistant to 3 atm
Band: reptile skin, double folding clasp
Price: $6,250
Variations: black dial; pink gold (price on request)

CARTIER

Drive de Cartier Extra-Flat
Reference number: WGNM0006
Movement: manually wound, Cartier Caliber 430 MC; ø 20 mm, height 2.15 mm; 18 jewels; 21,600 vph; 43-hour power reserve
Functions: hours, minutes
Case: pink gold, 38 × 39 mm, height 6.6 mm; sapphire crystal; water-resistant to 3 atm
Band: reptile skin, buckle
Price: $15,400
Variations: white gold ($16,600)

Drive de Cartier Second Time Zone
Reference number: WSNM0005
Movement: automatic, Cartier Caliber 1904-FU MC; ø 25 mm, height 5.2 mm; 28 jewels; 28,800 vph; finely finished with côtes de Genève; 48-hour power reserve
Functions: hours, minutes, subsidiary seconds; additional 12-hour display (2nd time zone, retrograde); day/night indicator; large date
Case: stainless steel, 40 × 41 mm, height 12.63 mm; sapphire crystal; water-resistant to 3 atm
Band: reptile skin, double folding clasp
Price: $8,750
Variations: rose gold ($22,500)

Rotonde de Cartier Skeleton Mysterious Double Tourbillon
Reference number: WHRO0039
Movement: manually wound, Cartier Caliber 9465 MC; ø 35 mm, height 5 mm; 26 jewels; 21,600 vph; skeleton movement integrating Roman hour numeral; double tourbillon located between two sapphire disks; 52-hour power reserve; Geneva Seal
Functions: hours, minutes (off-center)
Case: platinum, ø 45 mm, height 12.4 mm; sapphire crystal; transparent case back; crown with sapphire cabochon; water-resistant to 3 atm
Band: reptile skin, double folding clasp
Price: $207,000; limited to 30 pieces

Rotonde de Cartier Mysterious Day & Night
Reference number: WHRO0043
Movement: manually wound, Cartier Caliber 9982 MC; ø 32.58 mm, height 4.85 mm; 28,800 vph; 48-hour power reserve
Functions: hours, minutes (retrograde)
Case: white gold, ø 40 mm, height 11.7 mm; sapphire crystal; transparent case back; crown with sapphire cabochon; water-resistant to 3 atm
Band: reptile skin, double folding clasp
Price: $71,000
Variations: pink gold ($66,500)

Rotonde de Cartier Second Time Zone
Reference number: W1556368
Movement: automatic, Cartier Caliber 1904-FU MC; ø 25.6 mm; 28,800 vph; 48-hour power reserve
Functions: hours, minutes, subsidiary seconds; additional retrograde 12-hour display (2nd time zone), day/night indicator; large date
Case: stainless steel, ø 42 mm, height 11.96 mm; sapphire crystal; transparent case back; water-resistant to 3 atm
Band: reptile skin, folding clasp
Price: $9,300
Variations: pink gold ($24,900)

Rotonde de Cartier Power Reserve
Reference number: W1556369
Movement: manually wound, Cartier Caliber 9753 MC; ø 20.79 mm; 20 jewels; 21,600 vph; 40-hour power reserve
Functions: hours, minutes; power reserve indicator; date
Case: stainless steel, ø 40 mm, height 8.94 mm; sapphire crystal
Band: reptile skin, folding clasp
Price: $8,350
Variations: pink gold ($21,499)

Rotonde de Cartier Chronograph
Reference number: WSRO0002
Movement: automatic, Cartier Caliber 1904-CH MC; ø 25.6 mm, height 5.72 mm; 35 jewels; 28,800 vph; 2 spring barrels, 48-hour power reserve
Functions: hours, minutes; chronograph; date
Case: stainless steel, ø 40 mm, height 12.15 mm; sapphire crystal; transparent case back; water-resistant to 3 atm
Band: reptile skin, folding clasp
Price: $9,050
Variations: pink gold ($23,500); white gold (price on request)

Calibre de Cartier Diver Blue
Reference number: WGCA0009
Movement: automatic, Cartier Caliber 1904-PS MC; ø 25.6 mm, height 4 mm; 27 jewels; 28,800 vph; 2 spring barrels, 47-hour power reserve
Functions: hours, minutes, subsidiary seconds; date
Case: pink gold, ø 42 mm, height 11 mm; unidirectional bezel with ceramic insert, 0-60 scale; sapphire crystal; screw-in crown; water-resistant to 30 atm
Band: calfskin with rubber overlay, buckle
Price: on request
Variations: stainless steel ($7,900); stainless steel with pink gold bezel ($10,200)

Calibre de Cartier Diver
Reference number: W7100056
Movement: automatic, Cartier Caliber 1904 MC; ø 25.6 mm, height 4 mm; 27 jewels; 28,800 vph; 2 spring barrels, 47-hour power reserve
Functions: hours, minutes, subsidiary seconds; date
Case: stainless steel, ø 42 mm, height 11 mm; unidirectional bezel with black DLC coating and 0-60 scale; sapphire crystal; screw-in crown; water-resistant to 30 atm
Band: rubber, buckle
Price: $7,900
Variations: black DLC treatment ($8,950)

Ballon Bleu de Cartier
Reference number: WSBB0025
Movement: automatic, Cartier Caliber 1847 MC; ø 25.6 mm; 23 jewels; 28,800 vph; 48-hour power reserve
Functions: hours, minutes, sweep seconds; date
Case: stainless steel, ø 42 mm, height 13 mm; sapphire crystal; crown with spinel cabochon; water-resistant to 3 atm
Band: reptile skin, double folding clasp
Price: $6,250

Tank Américaine
Reference number: WSTA0018
Movement: automatic, ETA Caliber 2000-1; ø 20 mm, height 3.6 mm; 20 jewels; 28,800 vph; 36-hour power reserve
Functions: hours, minutes, sweep seconds
Case: stainless steel, 26.6 × 45.1 mm, height 9.65 mm; sapphire crystal; crown with spinel cabochon; water-resistant to 3 atm
Band: reptile skin, double folding clasp
Price: $5,750
Variations: rose gold (price on request)

Tank MC
Reference number: WSTA0010
Movement: automatic, Cartier Caliber 1904-PS MC; ø 25.6 mm, height 4 mm; 27 jewels; 28,800 vph; 2 spring barrels, 48-hour power reserve
Functions: hours, minutes, subsidiary seconds; date
Case: stainless steel, 34.3 × 44 mm, height 9.5 mm; sapphire crystal; transparent case back; water-resistant to 3 atm
Band: reptile skin, folding clasp
Price: $7,000
Variations: white dial; rose gold ($24,800)

CASIO

Well before the electronic wrist-borne airplane dashboards that thrill the neophiles of today, there was Casio. Many who came of age in the 1970s might remember watches with alarms, or calculators, that were very affordable. While you couldn't do too much with them, they did put the fun in function and made you feel like Dick Tracy, or even Mr. Spock. But as an electronics company, while certainly not hurting, Casio was not creating a buzz. Until 1983. And the impetus came from a young man.

Kikuo Ibe, a young man and Casio employee, dropped and shattered his own watch and decided it was time to set a standard. He was determined to build a watch that could withstand a fall from a three-story building as well as survive submersion, dust, dirt, and general abuse that would put G Shock on top of the tough timepiece pyramid.

Launched in 1983, Casio's G Shock has become quite a sensation the world 'round. Built to withstand shocks that would destroy lesser watches, this remarkable timepiece—essentially, the ultimate tool watch—had another major advantage: It was democratically priced.

Since its inception, G Shock has gone through hundreds and even thousands of variations in size, style, and functionality while retaining the common core value of creating the world's toughest, most resilient timepieces possible. During G Shock's evolution, Casio brought forth a multitude of functions and connectivity to the wristwatch and effectively pre-dated the modern tech-wear trend.

Having developed its massive audience over the last three decades, G Shock has recently begun adding higher-end designs. These new models retain the robust nature that makes a G Shock what it is, while incorporating new materials as well as Japanese-inspired decorations like the "hammered" bezel in limited edition releases. These new variations place the popular watch squarely into collector-level price points with special editions from lines like the MTG, MRG, and Master of G watches.

Casio
6-2, Hon-machi 1-chome
Shibuya-ku
Tokyo 151-8543
Japan

Tel.:
973-361-5400

Fax:
973-537-8926

E-mail:
info@casio.com

Website:
http://www.casio.com

Founded:
1957

Number of employees:
12,298

U.S. distributor:
Casio America
570 Mt. Pleasant Ave.
Dover, NJ 07801

Most important collections/price range:
MR-G / $2,800 to $7,400; Baby-G, G-Shock, Edifice, Oceanus, Wave Ceptor / $100 to $5,000

Rangeman
Reference number: GPR B1000
Movement: quartz
Functions: digital hours, minutes, seconds; GPS with backtrack, waypoint memory; digital compass, barometer, altimeter, thermometer, depth gauge; chronograph, countdown; date with sunrise/sunset, tides, moon phase, world time
Case: resin with carbon fiber inserts, 57.7 mm × 20.2 mm, sapphire crystal, water-resistant to 20 atm
Band: rubber, folding clasp
Remarks: multifunctional digital display with solar and magnetic induction charging; Bluetooth connectivity
Price: $800
Variations: different colors

MT-G
Reference number: MTG B100-1A
Movement: quartz, solar power multifunctional analog display
Functions: analog hours, minutes; chronograph, countdown timer; date with calendar, world time, alarm, GPS
Case: stainless steel, 51.7 mm × 14.4 mm, sapphire crystal, water-resistant to 20 atm
Band: stainless steel, folding clasp
Remarks: radio frequency atomic time control, phone finder, Bluetooth connectivity
Price: $800
Variations: different colors

MR-G
Reference number: MRGG2000HB-1A
Movement: quartz regulated analog display, power reserve up to 23 months from full solar charge on low power mode
Functions: analog hours, minutes; chronograph, countdown timer; date, day, world time alarm, perpetual calendar, LED light
Case: stainless steel with hand-decorated bezel in Japanese tsuiki hammered style, 49.8 mm × 14.9 mm, water-resistant to 20 atm
Band: titanium with DLC treatment, folding safety clasp
Remarks: solar powered, GPS hybrid radio-controlled time synchronization; Bluetooth connectivity
Price: $5,000
Variations: Cobarion alloy bezel ($3,700), titanium case with black DLC bezel ($2,800)

Chanel
135, avenue Charles de Gaulle
F-92521 Neuilly-sur-Seine Cedex
France

Tel.:
+33-1-41-92-08-33

Website:
www.chanel.com

Founded:
1914

Distribution:
retail and 200 Chanel boutiques worldwide

U.S. distributor:
Chanel Fine Jewelry and Watches
600 Madison Avenue, 19th Floor
New York, NY 10022
212-715-4741
212-715-4155 (fax)
www.chanel.com

Most important collections:
J12, Première, Boy.Friend, Monsieur de Chanel

CHANEL

After putting the occasional jewelry watch onto the market earlier, the family-owned Chanel opened its own horology division in 1987, a move that gave the brand instant access to the world of watchmaking art. Chanel boasts its own studio and logistics center, both in La Chaux-de-Fonds. While the brand's first collections were directed exclusively at its female clientele, it was actually with the rather simple and masculine J12 that Chanel finally achieved a breakthrough. Designer Jacques Helleu says he mainly designed the unpretentious ceramic watch for himself. "I wanted a timeless watch in glossy black," shares the likable eccentric. Indeed, it's not hard to imagine that the J12 will still look modern a number of years down the road—especially given the fact that the watch now comes in white and shiny polished titanium/ceramic as well.

The J12 collection showpiece, the Rétrograde Mystérieuse, was a stroke of genius—courtesy of the innovative think tank Renaud et Papi. It instantly propelled Chanel into the world of *haute horlogerie*. Lately, it entirely redid the ladies' watch Première, even though the change is not obvious at first glance. The octagonal shape of Place Vendôme in Paris (home of the brand) and the famous Chanel No. 5 bottle stopper are still there, and the simple 1980s style, but with a narrower bezel and adapted hands.

Chanel has started steering toward a younger, dynamic crowd with two collections that suggest a rapprochement between the sexes: The "Vendôme" rectangular Boy.Friend marries a feeling of subdued luxury and asceticism, and is as such so fascinating that its diamond-studded version won the Ladies' Prize at the 2018 edition of the GPHG. The Monsieur de Chanel is a purist, forty-millimeter watch with jumping hour and retrograde minutes driven by the Caliber 1. As for the Première Camélia, with the Caliber 2 inside, it continues in this vein, with a movement skeletonized to form a camellia. The watch comes in three versions bearing various amounts of diamonds.

Première Camélia Skeleton
Reference number: H5252
Movement: manually wound, Chanel Caliber 2; ø 32 mm, height 5.5 mm; 21 jewels; 28,800 vph; 1 barrel spring; skeletonized movement shaped like a camellia; 48-hour power reserve
Functions: hours, minutes
Case: white gold, ø 37 mm × 28.5 mm × 10.6 mm; set with 47 diamonds, bezel set with 94 diamonds; crown set with 11 diamonds; sapphire crystal; transparent back; water-resistant to 3 atm
Band: satin, double folding clasp
Remarks: movement set with 246 diamonds
Price: on request
Variations: bezel, hands and crown set with diamonds and dial only set with diamonds ($183,500); bracelet, bezel movement crown set with diamonds ($444,200)

Monsieur de Chanel
Reference number: H4800
Movement: manually wound, Chanel Caliber I; ø 32 mm, height 5.5 mm; 30 jewels; 28,800 vph; 2 spring barrels; plate/bridges with perlage; 72-hour power reserve
Functions: hours (digital, jumping), minutes (retrograde), subsidiary seconds
Case: beige gold, ø 40 mm, height 10.4 mm; sapphire crystal; transparent back; water-resistant to 3 atm
Band: reptile skin, buckle
Remarks: opaline dial
Price: $34,500
Variations: white gold ($36,000)

Boy.Friend
Reference number: H4862
Movement: manually wound, Chanel Caliber 3; ø 23.3 mm, height 2.5 mm; 17 jewels; skeletonized movement with ADLC on bridges; 55-hour power reserve
Functions: hours, minutes, small seconds
Case: beige gold, ø 37 mm × 28.6 mm × 8.4 mm; sapphire crystal; crown with onyx cabochon; water-resistant to 3 atm
Band: reptile skin, gold buckle
Price: $40,600; limited to 1,200 pieces
Variations: bezel set with diamonds ($51,400)

CHOPARD

The Chopard *manufacture* was founded by Louis-Ulysse Chopard in 1860 in the tiny village of Sonvillier in the Jura mountains of Switzerland. In 1963, it was purchased by Karl Scheufele, a goldsmith from Pforzheim, Germany, and revived as a producer of fine watches and jewelry.

The past eighteen years have seen a breathtaking development, when Karl Scheufele's son, Karl-Friedrich, and his sister, Caroline, decided to create watches with in-house movements, thus restoring the old business launched by Louis-Ulysse back in the nineteenth century.

In the 1990s, literally out of nowhere, Chopard opened up its watchmaking *manufacture* in the sleepy town of Fleurier in the Val-de-Travers, which had not yet experienced the revival of the mechanical watch. Karl-Friedrich Scheufele was convinced that the future of the industry lay in producing high-end timepieces, in spite of what many competitors were saying. The success of the L.U.C models with their own calibers silenced the doubters, even more so when ETA began restricting its sales of base calibers to the industry. Over twenty years later, Chopard's own Fleurier Ebauches SA has restored Fleurier's tradition as a hub of *ébauche* (movement kits) production. Chopard now has a line-up of eleven calibers, ranging from simple three-hander automatics to a tourbillon, a perpetual calendar, chronographs, an ultra-high-frequency chronometer, and a minute repeater.

The company also continues to support the Geneva Watchmaking School with special *ébauches* for the students, a demonstration of its commitment to the industry. With its wide range of *manufacture* watch models and over 160 boutiques worldwide, the brand enjoys firm footing in the rarefied air of *haute horlogerie*.

Chopard & Cie. SA
8, rue de Veyrot
CH-1217 Meyrin (Geneva)
Switzerland

Tel.:
+41-22-719-3131

E-mail:
info@chopard.ch

Website:
www.chopard.ch

Founded:
1860

Distribution:
160 boutiques

U.S. distributor:
Chopard USA
75 Valencia Ave, Suite 1200
Coral Gables, FL 33134
1-800-CHOPARD
www.us.chopard.com

Most important collections/price range:
Superfast / $9,860 to $32,900; L.U.C / beginning at $8,190; Imperiale / beginning at $5,310; Classic Racing / $5,040 to $41,900; Happy Sport / beginning at $5,300

Mille Miglia Chronograph 2018

Reference number: 168589-3006
Movement: automatic, ETA Caliber 2894-2; ø 28.6 mm, height 6.1 mm; 37 jewels; 28,800 vph; 42-hour power reserve; COSC-certified chronometer
Functions: hours, minutes, subsidiary seconds; chronograph; date
Case: stainless steel, ø 42 mm, height 12.67 mm; sapphire crystal; water-resistant to 5 atm
Band: calfskin, buckle
Remarks: "Chopard & Mille Miglia: 30 anni di passione" engraved on case back
Price: $5,620; limited to 1,000 pieces

Mille Miglia GTS Power Control Grigio Speziale

Reference number: 168566-3007
Movement: automatic, Chopard Manufacture Caliber 01.08-C; ø 28.8 mm, height 4.95 mm; 40 jewels; 28,800 vph; 60-hour power reserve; COSC-certified chronometer
Functions: hours, minutes, sweep seconds; power reserve indicator; date
Case: titanium, ø 43 mm, height 11.43 mm; sapphire crystal; transparent case back; screw-in crown, with gray PVD coating; water-resistant to 10 atm
Band: textile, folding clasp
Price: $8,220; limited to 1,000 pieces

Grand Prix de Monaco Historique 2018 Race Edition

Reference number: 168570-3004
Movement: automatic, ETA Caliber 7750; ø 30.4 mm, height 7.9 mm; 25 jewels; 28,800 vph; 46-hour power reserve; COSC-certified chronometer
Functions: hours, minutes, subsidiary seconds; chronograph; date
Case: stainless steel, ø 44.5 mm, height 14.1 mm; sapphire crystal; screw-in crown; water-resistant to 10 atm
Band: textile, folding clasp
Price: $7,390; limited to 250 pieces

Mille Miglia 2017 Race Edition
Reference number: 168571-3002
Movement: automatic, ETA Caliber 7750; ø 30.4 mm, height 7.9 mm; 25 jewels; 28,800 vph; 48-hour power reserve; COSC-certified chronometer
Functions: hours, minutes, subsidiary seconds; chronograph; date
Case: stainless steel, ø 44 mm, height 13.79 mm; sapphire crystal; water-resistant to 10 atm
Band: rubber, folding clasp
Price: $7,180; limited to 1,000 pieces

Mille Miglia Classic Chronograph
Reference number: 168589-3002
Movement: automatic, ETA Caliber 2894-2; ø 28.6 mm, height 6.1 mm; 37 jewels; 28,800 vph; 42-hour power reserve; COSC-certified chronometer
Functions: hours, minutes, subsidiary seconds; chronograph; date
Case: stainless steel, ø 42 mm, height 12.67 mm; sapphire crystal; transparent case back; water-resistant to 5 atm
Band: rubber, buckle
Price: $5,260

Classic Racing Superfast Porsche 919 Edition
Reference number: 168535-3002
Movement: automatic, Chopard Manufacture Caliber 03.05-M; ø 28.8 mm, height 7.6 mm; 45 jewels; 28,800 vph; 60-hour power reserve; COSC-certified chronometer
Functions: hours, minutes, subsidiary seconds; flyback chronograph; date
Case: stainless steel, ø 45 mm, height 15.18 mm; sapphire crystal; transparent case back; screw-in crown; water-resistant to 10 atm
Band: rubber, folding clasp
Price: $12,300; limited to 919 pieces

L.U.C Time Traveler One
Reference number: 161942-9001
Movement: automatic, L.U.C Caliber 01.05-L; ø 35.3 mm, height 6.52 mm; 39 jewels; 28,800 vph; 60-hour power reserve; COSC-certified chronometer
Functions: hours, minutes, sweep seconds; world time display (2nd time zone); date
Case: rose gold, ø 42 mm, height 12.09 mm; crown-activated scale ring, with reference city names; sapphire crystal; transparent case back; water-resistant to 5 atm
Band: reptile skin, buckle
Price: $37,900

L.U.C Quattro
Reference number: 161926-5004
Movement: manually wound, L.U.C Caliber 98.01-L; ø 28.6 mm, height 3.7 mm; 39 jewels; 28,800 vph; 4 spring barrels, swan-neck fine adjustment, gold rotor; 216-hour power reserve; Geneva Seal, COSC-certified chronometer
Functions: hours, minutes, subsidiary seconds; power reserve indicator; date
Case: rose gold, ø 43 mm, height 8.84 mm; sapphire crystal; transparent case back; water-resistant to 5 atm
Band: reptile skin, buckle
Price: $25,800; limited to 50 pieces

L.U.C All in One
Reference number: 161925-9003
Movement: manually wound, L.U.C Caliber 05.01-L; ø 33 mm, height 11.75 mm; 42 jewels; 28,800 vph; 1-minute tourbillon; 170-hour power reserve; Geneva Seal; COSC-certified chronometer
Functions: hours, minutes, subsidiary seconds; day/night indicator, power reserve indicator, time equation, sunrise/sunset (on movement side); perpetual calendar with large date, weekday, month, orbital astronomical moon phase, leap year
Case: white gold, ø 46 mm, height 18.5 mm; sapphire crystal; transparent case back; water-resistant to 3 atm
Band: reptile skin, buckle
Price: on request; limited to 10 pieces

CHOPARD

L.U.C Perpetual Chrono
Reference number: 161973-5002
Movement: manually wound, L.U.C Caliber 03.10-L; ø 33 mm, height 8.32 mm; 42 jewels; 28,800 vph; mainplate and balance cock of German silver; 60-hour power reserve; Geneva Seal, COSC-certified chronometer
Functions: hours, minutes, sweep seconds; day/night indicator; flyback chronograph; perpetual calendar with large date, weekday, month, moon phase, leap year
Case: rose gold, ø 45 mm, height 15.06 mm; sapphire crystal; transparent case back; water-resistant to 3 atm
Band: reptile skin, folding clasp
Price: $91,100; limited to 20 pieces
Variations: platinum

L.U.C Perpetual Chrono
Reference number: 161973-9001
Movement: manually wound, L.U.C Caliber 03.10-L; ø 33 mm, height 8.32 mm; 42 jewels; 28,800 vph; mainplate and balance cock of German silver; 60-hour power reserve; Geneva Seal, COSC-certified chronometer
Functions: hours, minutes, sweep seconds; day/night indicator; flyback chronograph; perpetual calendar with large date, weekday, month, moon phase, leap year
Case: platinum, ø 45 mm, height 15.06 mm; sapphire crystal; transparent case back; water-resistant to 3 atm
Band: reptile skin, folding clasp
Price: on request; limited to 20 pieces

L.U.C Perpetual Twin
Reference number: 168561-3001
Movement: automatic, L.U.C Caliber 96.51-L; ø 33 mm, height 6 mm; 32 jewels; 28,800 vph; 2 spring barrels, heavy metal microrotor; côtes de Genève; 58-hour power reserve; COSC-certified chronometer
Functions: hours, minutes, subsidiary seconds; perpetual calendar with large date, weekday, month, leap year
Case: stainless steel, ø 43 mm, height 11.47 mm; sapphire crystal; transparent case back; water-resistant to 3 atm
Band: reptile skin, buckle
Price: $24,700

L.U.C Tourbillon QF Fairmined
Reference number: 161929-5006
Movement: manually wound, L.U.C Caliber 02.13-L; ø 29.7 mm, height 6.1 mm; 33 jewels; 28,800 vph; 1-minute tourbillon, bridges with côtes de Genève; 216-hour power reserve; COSC-certified chronometer, Qualité Fleurier
Functions: hours, minutes, subsidiary seconds; power reserve indicator
Case: rose gold, ø 43 mm, height 11.15 mm; sapphire crystal; transparent case back; water-resistant to 3 atm
Band: reptile skin, buckle
Remarks: case made of certified fair-mined gold
Price: on request; limited to 25 pieces

L.U.C GMT One
Reference number: 168579-3001
Movement: automatic, L.U.C Caliber 01.10-L; ø 31.9 mm, height 5.95 mm; 31 jewels; 28,800 vph; bridges with côtes de Genève; 60-hour power reserve; COSC-certified chronometer
Functions: hours, minutes, sweep seconds; additional 24-hour display (2nd time zone); date
Case: stainless steel, ø 42 mm, height 11.71 mm; sapphire crystal; transparent case back; water-resistant to 5 atm
Band: reptile skin, buckle
Price: $10,200
Variations: rose gold ($20,600)

L.U.C XPS Twist QF Fairmined
Reference number: 161945-5001
Movement: automatic, L.U.C Caliber 96-09-L; ø 27.4 mm, height 3.3 mm; 29 jewels; 28,800 vph; 2 spring barrels, gold microrotor; 65-hour power reserve; COSC-certified chronometer, Qualité Fleurier
Functions: hours, minutes, subsidiary seconds
Case: rose gold, ø 40 mm, height 7.2 mm; sapphire crystal; transparent case back; screw-in crown; water-resistant to 3 atm
Band: reptile skin, buckle
Remarks: case made of certified fair-traded gold
Price: $19,900; limited to 250 pieces

CHOPARD

L.U.C Lunar One
Reference number: 161927-5001
Movement: automatic, L.U.C Caliber 96.13-L; ø 33 mm, height 6 mm; 32 jewels; 28,800 vph; 65-hour power reserve; Geneva Seal, COSC-certified chronometer
Functions: hours, minutes, subsidiary seconds; additional 24-hour display (2nd time zone); perpetual calendar with large date, weekday, month, orbital moon phase display, leap year
Case: rose gold, ø 43 mm, height 11.47 mm; sapphire crystal; transparent case back; water-resistant to 5 atm
Band: reptile skin, folding clasp
Price: $60,600
Variations: diamond bezel ($94,000); white gold ($60,600)

L.U.C Lunar Big Date
Reference number: 161969-1001
Movement: automatic, L.U.C Caliber 96.20-L; ø 33 mm, height 5.25 mm; 33 jewels; 28,800 vph; côtes de Genève; 65-hour power reserve; COSC-certified chronometer
Functions: hours, minutes, subsidiary seconds; large date, orbital moon phase display
Case: white gold, ø 42 mm, height 11.04 mm; sapphire crystal; transparent case back; water-resistant to 5 atm
Band: reptile skin, buckle
Price: $30,300
Variations: rose gold ($30,300)

L.U.C Regulator
Reference number: 161971-5001
Movement: manually wound, L.U.C Caliber 98.02-L; ø 30.4 mm, height 4.9 mm; 39 jewels; 28,800 vph; 4 spring barrels, bridges with côtes de Genève; 216-hour power reserve; Geneva Seal, COSC-certified chronometer
Functions: hours (off-center), minutes, subsidiary seconds; additional 24-hour display (2nd time zone); power reserve indicator; date
Case: rose gold, ø 43 mm, height 9.78 mm; sapphire crystal; transparent case back; water-resistant to 3 atm
Band: reptile skin, buckle
Price: $32,000

L.U.C XP
Reference number: 168592-3001
Movement: automatic, L.U.C Caliber 96.53; ø 27.4 mm, height 3.3 mm; 27 jewels; 28,800 vph; heavy metal rotor, bridges with côtes de Genève; 58-hour power reserve
Functions: hours, minutes
Case: stainless steel, ø 40 mm, height 7.2 mm; sapphire crystal; transparent case back; water-resistant to 3 atm
Band: textile, buckle
Price: $8,810

L.U.C XPS
Reference number: 161948-5001
Movement: automatic, L.U.C Caliber 96.12-L; ø 27.4 mm, height 3.3 mm; 29 jewels; 28,800 vph; gold rotor, bridges with côtes de Genève; 65-hour power reserve; COSC-certified chronometer
Functions: hours, minutes, subsidiary seconds
Case: rose gold, ø 40 mm, height 7.2 mm; sapphire crystal; transparent case back; water-resistant to 3 atm
Band: reptile skin, buckle
Price: $16,500

L.U.C XPS 1860 Edition
Reference number: 161946-5001
Movement: automatic, L.U.C Caliber 96.01-L; ø 27.4 mm, height 3.3 mm; 29 jewels; 28,800 vph; 2 spring barrels, gold rotor; 65-hour power reserve; Geneva Seal, COSC-certified chronometer
Functions: hours, minutes, subsidiary seconds; date
Case: rose gold, ø 40 mm, height 7.2 mm; sapphire crystal; transparent case back; water-resistant to 3 atm
Band: reptile skin, buckle
Price: $21,700; limited to 250 pieces

CHOPARD

Caliber L.U.C 03.10-L

Manually wound; mainplate and balance cock of German silver; single spring barrel, 60-hour power reserve; Geneva Seal, COSC-certified chronometer
Functions: hours, minutes, sweep seconds; day/night indicator; flyback chronograph; perpetual calendar with large date, weekday, month, moon phase, leap year
Diameter: 33 mm
Height: 8.32 mm
Jewels: 42
Balance: Variner with 4 weighted screws
Frequency: 28,800 vph
Balance spring: flat hairspring

Caliber L.U.C 96.01-L

Automatic; gold microrotor; double spring barrel, 65-hour power reserve; Geneva Seal, COSC-certified chronometer
Functions: hours, minutes, subsidiary seconds; date
Diameter: 27.4 mm
Height: 3.3 mm
Jewels: 29
Balance: glucydur
Frequency: 28,800 vph
Balance spring: flat hairspring, Nivarox 1

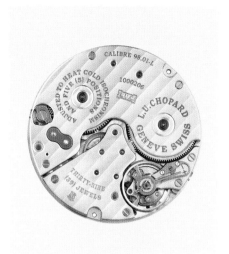

Caliber L.U.C 98.01-L

Manually wound; swan-neck fine regulation; quadruple spring barrel running in twin series barrel springs, 216-hour power reserve; Geneva Seal, COSC-certified chronometer
Functions: hours, minutes, subsidiary seconds; power reserve indicator; date
Diameter: 28.6 mm
Height: 3.7 mm
Jewels: 39
Frequency: 28,800 vph
Balance spring: Breguet hairspring

Caliber L.U.C 02.13-L

Manually wound; 1-minute tourbillon; quadruple spring barrel, 216-hour power reserve; Geneva Seal, COSC-certified chronometer, Qualité Fleurier
Functions: hours, minutes, subsidiary seconds (on tourbillon cage); power reserve indicator
Diameter: 29.7 mm
Height: 6.1 mm
Jewels: 33
Frequency: 28,800 vph
Balance spring: flat hairspring

Caliber L.U.C 96.51-L

Automatic; microrotor; double spring barrel, 58-hour power reserve; COSC-certified chronometer
Functions: hours, minutes, subsidiary seconds; perpetual calendar with large date, weekday, month, leap year
Diameter: 33 mm
Height: 6 mm
Jewels: 32
Balance: glucydur
Frequency: 28,800 vph
Balance spring: flat hairspring, Nivarox 1

Caliber L.U.C 96.20-L

Automatic; gold microrotor; double spring barrel, 65-hour power reserve; Geneva Seal, COSC-certified chronometer
Functions: hours, minutes, subsidiary seconds; large date, orbital moon phase display
Diameter: 33 mm
Height: 5.25 mm
Jewels: 33
Balance: glucydur
Frequency: 28,800 vph
Balance spring: flat hairspring, Nivarox 1
Remarks: 296 parts

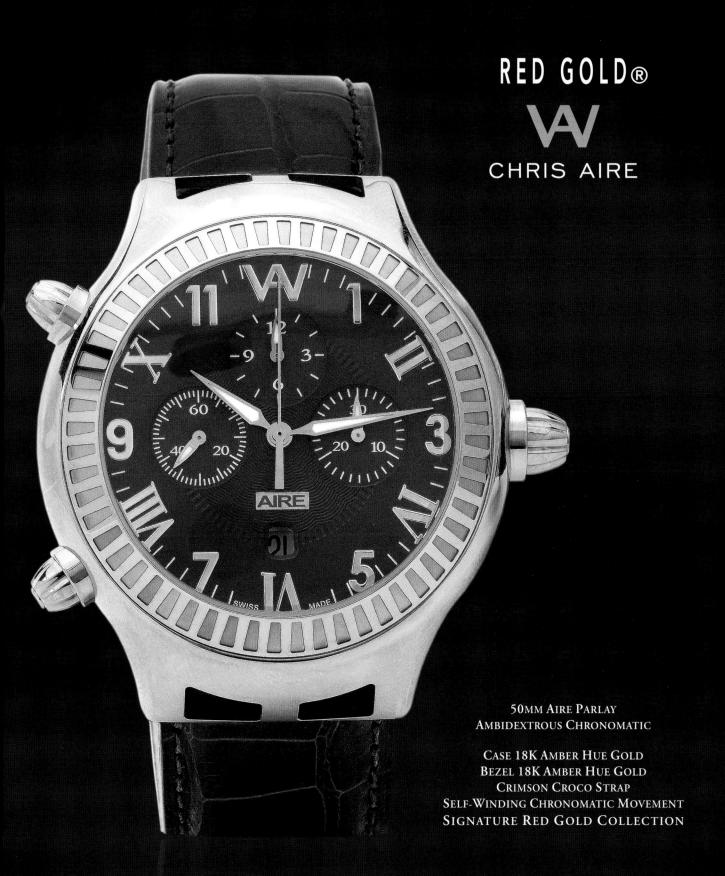

CHRISTOPHE CLARET

Individuals like Christophe Claret are authentic horological engineers who eat, drink, and breathe watchmaking and have developed careers based on pushing the envelope to the very edge of what's possible.

By the age of twenty-three, the Lyon-born Claret was in Basel alongside Journe, Calabrese, and other independents, where he was spotted by the late Rolf Schnyder of Ulysse Nardin and commissioned to make a minute repeater with jacquemarts. In 1989, he opened his *manufacture*, a nineteenth-century mansion tastefully extended with a state-of-the-art machining area. Indeed, Claret embraces wholeheartedly the potential in modern tools to create the precise pieces needed to give physical expression to exceedingly complex ideas.

Twenty years after establishing his business, and in the midst of the massive subprime recession, Claret finally launched his own complex watches. The DualTow has hours and minutes on two tracks plus a minute repeater. Then came the Adagio, again a minute repeater, with a clear dial that manages a second time zone and large date. In 2011, Claret wowed the watch world with a humorous, on-the-wrist gambling machine telling time and playing blackjack, craps, or roulette. It was followed by the stunning X-TREM-1, a turbocharged DualTow with two spheres controlled by magnets hovering along the numeral tracks to tell the time plus a tourbillon. For 2017, the model was further developed in a partnership with the high-end hip jeweler StingHD.

Whatever Claret produces—the Margot, for women, the art-laden Aventicum, or the Adagio, a minute repeater with an almost Dadaist dial—his signature is always present: a total dedication to power mechanics and an infallible sense of style. The winding rotor of the Marguerite, a high-end mechanical ladies' watch, features rubies that will point to "He loves me" or "He loves me not" engraved on the case back and separated by a little heart.

Christophe Claret SA
Route du Soleil d'Or 2
CH-2400 Le Locle
Switzerland

Tel.:
+41-32-933-0000

Fax:
+41-32-933-8081

E-mail:
info@christopheclaret.com

Website:
www.christopheclaret.com

Founded:
manufacture 1989, brand 2009

Number of employees:
70

Distribution:
Contact the *manufacture* directly.

Most important collections:
Traditional complications (Maestro/Mecca/Allegro/Aventicum/Maestoso/Kantharos/Soprano), Extreme line (X-TREM-1), gaming watches (Poker/Baccara/Blackjack), and ladies' complications line (Margot, Layla, Marguerite)

Maestro Mamba
Reference number: MTR.DMC16.230-258 MAMBA
Movement: manually wound, Christophe Claret Caliber DMC16; ø 36.25 mm, height 10.5 mm; 33 jewels; 21,600 vph; inverted movement construction with balance on dial; 168-hour power reserve
Functions: hours, minutes, manually activated memo display; large date (double digit) on pyramid base
Case: black PVD-coated titanium, ø 42 mm, height 16 mm; sapphire crystal; transparent case back
Band: reptile skin, folding clasp
Remarks: comes with extra python leather strap
Price: CHF 96,000; limited to 28 pieces
Variations: orange with pantheropis strap

Adagio
Reference number: MTR.SLB88.902
Movement: manually wound, Christophe Claret Caliber SLB88; ø 34 mm, height 8.4 mm; 46 jewels; 18,000 vph; minute repeater with cathedral gongs; 48-hour power reserve
Functions: hours, minutes, subsidiary seconds (disk display); additional 12-hour display (2nd time zone), day/night indicator; minute repeater; large date
Case: pink gold, ø 44 mm, height 13.9 mm; bezel and lugs set with baguette diamonds; sapphire crystal; transparent case back
Band: reptile skin, folding clasp
Remarks: unique piece
Price: CHF 490,000

Marguerite
Reference number: MTR.MT113.040-070
Movement: automatic, Christophe Claret Caliber MT113; ø 30.1 mm, height 9.05 mm; 36 jewels; 28,800 vph; 2 spring barrels, 72-hour power reserve
Functions: hours, minutes (with butterfly hands); dial-side text can be called up by pusher, alternating with numerals (customizable)
Case: pink gold, ø 36.9.mm, height 11.53 mm; bezel and lugs set with 98 diamonds; sapphire crystal; transparent case back; water-resistant to 3 atm
Remarks: butterflies as hands; "He loves me, he loves me not . . ." engraved on winding rotor to indicate yes/no
Price: CHF 65,000; limited to 30 pieces each style
Variations: rose or gray gold case with different diamond settings and in red, green, or blue

Chronoswiss AG
Löwenstrasse 16a
CH-6004 Lucerne
Switzerland

Tel.:
+41-41-552-2100

Fax:
+41-41-552-2109

E-mail:
mail@chronoswiss.com

Website:
www.chronoswiss.com

Founded:
1983

Number of employees:
approx. 30

Annual production:
up to 5,000 wristwatches

U.S. distributor:
Chronoswiss US Service Office
Shami Fine Watchmaking
Adam Shami
155 Willowbrook Blvd.
Suite 320
Wayne, NJ 07470
Tel.: 973-785-0004
Fax: 973-785-0055

Most important collections/price range:
approx. 30 models including Sirius Regulator, Sirius Triple Date, Sirius Perpetual Calendar, Sirius Artist, Timemaster Big Date, Timemaster Chronograph GMT / approx. $3,550 to $47,000

CHRONOSWISS

Chronoswiss has been assembling its signature watches—which boast such features as coin edge bezels and onion crowns—since 1983. Founder Gerd-Rüdiger Lang loved to joke about having "the only Swiss watch factory in Germany," as the brand has always adhered closely to the qualities of the Swiss watch industry while still contributing a great deal to reviving mechanical watches from its facilities in Karlsfeld near Munich, with concepts and designs "made in Germany."

There was more. Lang also created regulator watches in the 1980s, a pioneering idea that found many fans of new ways to tell the time. This is also part of the brand's history and DNA, a remarkable feat, and somewhat anachronistic back then. Chronoswiss has always been a little on the edge of the industry in terms of style and technical developments. It created the enduring *manufacture* caliber C.122—based on an old Enicar automatic movement with a patented rattrapante mechanism—and its Chronoscope chronograph has earned a solid reputation for technical prowess. The Pacific and Sirius models, additions to the classic collection, point the company in a new stylistic direction designed to help win new buyers and the attention of the international market.

In March 2012, a Swiss couple, Oliver and Eva Ebstein, purchased Chronoswiss and soon afterward moved the company headquarters to Lucerne, Switzerland. The Ebsteins are watch enthusiasts and intended from the start to keep Chronoswiss as an independent family-run company, though now with a direct connection to Switzerland. There were to be no great shake-ups.

Recent models reveal the brand to be faithful to its regulator and coin edges, and the large crown, though the dial has acquired a three-dimensional design. The Flying Regulator and the Regulator Jumping Hour find the minute hand hovering freely over the dial, with the hours and seconds on bridges. The name "Atelier Lucerne" appears on the dial, a hint of the new premises, but the trusty C.122 beats inside many of these new models.

Flying Regulator Open Gear Anniversary Edition

Reference number: CH-8751R-BLSI
Movement: automatic, Chronoswiss Caliber C.299; ø 35.2 mm, height 6.11 mm; 31 jewels; 28,800 vph; hand mechanism (transmission wheel) relocated to dial side; finely finished movement; 42-hour power reserve
Functions: hours (off-center), minutes, subsidiary seconds
Case: pink gold, ø 41 mm, height 13.85 mm; sapphire crystal; transparent case back; water-resistant to 10 atm
Band: reptile skin, buckle
Remarks: special edition for 35th anniversary of brand with hand-guillochéed dials, 35 pieces each
Price: $16,840
Variations: stainless steel ($8,110)

Flying Regulator Open Gear

Reference number: CH-8753-SISI
Movement: automatic, Chronoswiss Caliber C.299; ø 35.2 mm, height 6.11 mm; 31 jewels; 28,800 vph; hand mechanism (transmission wheel) relocated to dial side; finely finished movement; 42-hour power reserve
Functions: hours (off-center), minutes, subsidiary seconds
Case: stainless steel, ø 41 mm, height 13.85 mm; sapphire crystal; transparent case back; water-resistant to 10 atm
Band: reptile skin, buckle
Price: $6,680
Variations: pink gold ($15,910)

Flying Regulator Open Gear

Reference number: CH-8753-BKBK
Movement: automatic, Chronoswiss Caliber C.299; ø 35.2 mm, height 6.11 mm; 31 jewels; 28,800 vph; hand mechanism (transmission wheel) relocated to dial side; finely finished movement; 42-hour power reserve
Functions: hours (off-center), minutes, subsidiary seconds
Case: stainless steel, ø 41 mm, height 13.85 mm; sapphire crystal; transparent case back; water-resistant to 10 atm
Band: reptile skin, buckle
Price: $6,680

CHRONOSWISS

Flying Regulator Night and Day Limited Edition
Reference number: CH-8763-BKRE
Movement: automatic, Chronoswiss Caliber C.296; ø 25.2 mm, height 4.35 mm; 27 jewels; 28,800 vph; skeletonized rotor; finely finished movement; 42-hour power reserve
Functions: hours (off-center), minutes, subsidiary seconds; day/night indicator; date
Case: stainless steel, ø 41 mm, height 13.85 mm; sapphire crystal; transparent case back; water-resistant to 10 atm
Band: reptile skin, buckle
Remarks: special edition for 30th anniversary of Regulator
Price: $7,190

Flying Regulator Night and Day
Reference number: CH-8763-BLBL
Movement: automatic, Chronoswiss Caliber C.296; ø 25.2 mm, height 4.35 mm; 27 jewels; 28,800 vph; skeletonized rotor; finely finished movement; 42-hour power reserve
Functions: hours (off-center), minutes, subsidiary seconds; day/night indicator; date
Case: stainless steel, ø 41 mm, height 13.85 mm; sapphire crystal; transparent case back; water-resistant to 10 atm
Band: reptile skin, buckle
Price: $6,790

Flying Grand Regulator Skeleton
Reference number: CH-6723S-SISI
Movement: manually wound, Chronoswiss Caliber C.677S; ø 37.2 mm, height 4.5 mm; 17 jewels; 18,000 vph; screw balance, swan-neck fine adjustment, skeletonized mainplate, bridges, and gearwheels; finely finished movement; 46-hour power reserve
Functions: hours (off-center), minutes, subsidiary seconds
Case: stainless steel, ø 44 mm, height 12.48 mm; sapphire crystal; transparent case back; water-resistant to 3 atm
Band: reptile skin, folding clasp
Remarks: special edition for 30th anniversary of Regulator
Price: $8,930

Flying Grand Regulator
Reference number: CH-6725-REBK
Movement: manually wound, Chronoswiss Caliber C.678; ø 37.2 mm, height 4.5 mm; 17 jewels; 18,000 vph; screw balance, swan-neck fine adjustment; finely finished movement; 46-hour power reserve
Functions: hours (off-center), minutes, subsidiary seconds
Case: stainless steel with black DLC treatment, ø 44 mm, height 12.48 mm; sapphire crystal; transparent case back; water-resistant to 3 atm
Band: reptile skin, folding clasp
Price: $9,660; limited to 30 pieces
Variations: black dial (limited to 15 pieces, $9,150)

Flying Regulator Manufacture
Reference number: CH-1243.3-BLBL
Movement: automatic, Chronoswiss Caliber C.122; ø 26.8 mm, height 5.3 mm; 30 jewels; 21,600 vph; skeletonized rotor; finely finished movement; 40-hour power reserve
Functions: hours (off-center), minutes, subsidiary seconds
Case: stainless steel, ø 40 mm, height 12 mm; sapphire crystal; transparent case back; water-resistant to 3 atm
Band: reptile skin, buckle
Price: $7,000
Variations: black DLC treatment ($7,570)

Flying Regulator Manufacture
Reference number: CH-1241.3R-SISI
Movement: automatic, Chronoswiss Caliber C.122; ø 26.8 mm, height 5.3 mm; 30 jewels; 21,600 vph; skeletonized rotor; finely finished movement; 40-hour power reserve
Functions: hours (off-center), minutes, subsidiary seconds
Case: pink gold, ø 40 mm, height 12 mm; sapphire crystal; transparent case back; water-resistant to 3 atm
Band: reptile skin, buckle
Price: $16,550

CHRONOSWISS

Flying Regulator Jumping Hour
Reference number: CH-8321R-BKBK
Movement: automatic, Chronoswiss Caliber C.283; ø 30 mm, height 5.35 mm; 27 jewels; 28,800 vph; skeletonized rotor; finely finished movement; 42-hour power reserve
Functions: hours (digital, jumping), minutes (off-center), subsidiary seconds
Case: pink gold, ø 40 mm, height 12 mm; sapphire crystal; transparent case back; water-resistant to 3 atm
Band: reptile skin, buckle
Price: $17,060

Flying Regulator Jumping Hour
Reference number: CH-8323-GRGR
Movement: automatic, Chronoswiss Caliber C.283; ø 30 mm, height 5.35 mm; 27 jewels; 28,800 vph; skeletonized rotor; finely finished movement; 42-hour power reserve
Functions: hours (digital, jumping), minutes (off-center), subsidiary seconds
Case: stainless steel, ø 40 mm, height 12 mm; sapphire crystal; transparent case back; water-resistant to 3 atm
Band: reptile skin, buckle
Price: $7,530
Variations: black DLC treatment ($8,040)

Regulator Classic Date
Reference number: CH-8733-BLBK
Movement: automatic, Chronoswiss Caliber C.292; ø 30 mm, height 6.35 mm; 30 jewels; 28,800 vph; finely finished movement; 38-hour power reserve
Functions: hours (off-center), minutes, subsidiary seconds; date
Case: stainless steel, ø 41 mm, height 12.93 mm; sapphire crystal; transparent case back; water-resistant to 10 atm
Band: reptile skin, folding clasp
Price: $3,990
Variations: silver and black dial

Artist Regulator Jumping Hour
Reference number: CH-8323E-BL
Movement: automatic, Chronoswiss Caliber C.283; ø 29.4 mm, height 5.35 mm; 27 jewels; 28,800 vph; skeletonized, hand-guilloché on rotor; finely finished movement, partially guillochéed by hand; 42-hour power reserve
Functions: hours (digital, jumping), minutes (off-center), subsidiary seconds
Case: stainless steel, ø 40 mm, height 9.75 mm; sapphire crystal; transparent case back; water-resistant to 3 atm
Band: reptile skin, buckle
Remarks: dial with hand-guilloché and enamel
Price: $10,080

Sirius Chronograph Moonphase
Reference number: CH-7541-LR
Movement: automatic, Chronoswiss Caliber C.755 (base ETA 7750); ø 30 mm, height 7.9 mm; 25 jewels; 28,800 vph; perlage on movement, côtes de Genève, skeletonized rotor; finely finished movement; 46-hour power reserve
Functions: hours, minutes, subsidiary seconds; chronograph; date, moon phase
Case: pink gold, ø 41 mm, height 15.45 mm; sapphire crystal; transparent case back; water-resistant to 3 atm
Band: reptile skin, buckle
Price: $18,730
Variations: stainless steel ($6,860)

Sirius Chronograph Skeleton
Reference number: CH-7543S
Movement: automatic, Chronoswiss Caliber C.741 S (base ETA 7750); ø 30 mm, height 7.9 mm; 25 jewels; 28,800 vph; entirely skeletonized movement with ribbing; 46-hour power reserve
Functions: hours, minutes, subsidiary seconds; chronograph; date
Case: stainless steel, ø 41 mm, height 15.45 mm; sapphire crystal; transparent case back; water-resistant to 3 atm
Band: reptile skin, buckle
Remarks: skeletonized dial
Price: $10,190
Variations: pink gold ($22,480)

CLAUDE MEYLAN

Claude Meylan
Route de l'Hôtel de Ville 2
CH-1344 L'Abbaye
Switzerland

Tel.:
+41-21 841 14 57

E-mail:
info@claudemeylan.com

Website:
www.claudemeylan.com

Founded:
originally mid-18th century; revived in the mid-20th century, and purchased in 2011

Number of employees:
7

Annual production:
approx. 1,000 pieces

U.S. distributor:
Please contact brand for information.

Most important collections/price range:
Tortue, Ligne Lac / $4,500 to $6,850; Légendes series / up to $33,000

In the quest for recognition, many companies, especially the smaller ones, look for a niche in which they can excel. The Swiss brand Claude Meylan, located in L'Abbaye near Joux Lake in the heart of watch country, specializes in skeletonization, which is the art of removing as much material as possible from bridges, plates, the dial, even the hands. The exercise is not just for fun. First, it transforms a watch, making it transparent and allowing a view of the mechanical innards. Second, it allows for imaginative designs using what's left of the material, notably the bridges. These can be either abstract or representative.

Skeletonization has become fairly popular in recent years, but it's not as simple as it might sound. As the various metal components are hollowed out and properly finished with chamfering and sanding, the tensions within the material change. This can then have a deleterious effect on the functioning of the mechanism, since the bridges and plates are in fact used to hold and stabilize the movement.

Claude Meylan, who worked at a local watch company, actually had watchmaking quite literally in his blood. One of his ancestors, a fellow named Samuel-Olivier Meylan, brought the craft to the Joux Valley in 1748. Over a century later, his great-grandfather and grandfather were active as watchmakers with a reputation for skeletonizing watches.

In 1988, Claude Meylan founded his company. It was taken over soon after by another watchmaker, Henri Berney, who kept up the old tradition. In 2011, the next CEO took charge, Philippe Belais, who also heads Vaudaux, a maker of high-end boxes and cases in Geneva.

The company has five main collections, all relating in some way to the region: Lac, for Joux Lake; l'Abbaye; Légendes (exploring local tales); Lionne, the river that flows by the workshops; and, finally, Tortue, whose tonneau case is reminiscent of a turtle. The brand's products, which show many different aspects of the art of skeletonization, do live up to its tagline, "sculptors of time." The Tortue line, for example, features delicate vine-like elements spreading across the movement, while one skeleton in the Lionne collection looks like a snowflake.

Lionne 6040-O

Reference number: 6040 O
Movement: automatic, ETA Caliber 2892-2; ø 25.6 mm, height 3.6 mm; 21 jewels; 28,800 vph; 42-hour power reserve
Functions: hours, minutes, sweep seconds
Case: stainless steel, ø 35 mm, height 10.9 mm; sapphire crystal; transparent case back; water-resistant to 3 atm
Band: calfskin, buckle
Price: $4,750

Skeleton Tortue

Reference number: 6047
Movement: manually wound, Claude Meylan Caliber 165CM16; ø 40 mm wide, height 4.5 mm; 17 jewels; 18,000 vph; openworked dial, bi-colored black/rose movement; 38-hour power reserve
Functions: hours, minutes, subsidiary seconds
Case: stainless steel, ø 40 × 43 mm, height 12 mm; sapphire crystal; transparent case back; water-resistant to 3 atm
Band: leather, buckle
Price: $4,500
Variations: comes in a variety of color schemes

Eclipse

Reference number: 6044 N
Movement: manually wound, Unitas Caliber 6497; ø 36.6 mm, height 4.5 mm; 17 jewels; 18,000 vph; openworked dial, decorated rhodium-plated bridges; 46-hour power reserve
Functions: hours, minutes, subsidiary seconds
Case: stainless steel with black PVD coating, ø 42 mm, height 11 mm; sapphire crystal; transparent case back; water-resistant to 3 atm
Band: leather, buckle
Price: $3,400

Montres Corum Sàrl
Rue du Petit-Château 1
Case postale 374
CH-2301 La Chaux-de-Fonds
Switzerland

Tel.:
+41-32-967-0670

Fax:
+41-32-967-0800

E-mail:
info@corum.ch

Website:
www.corum.ch

Founded:
1955

Number of employees:
160 worldwide

Annual production:
16,000 watches

U.S. distributor:
Montres Corum USA
CWJ BRANDS
1551 Sawgrass Corporate Parkway
Suite 109
Sunrise, FL 33323
954-279-1220; 954-279-1780 (fax)
www.corum.ch

Most important collections/price range:
Admiral's Cup, Golden Bridge, Bubble, and Heritage, Romvlvs and Artisan, 150 models in total /approx. $1,500 to over $1,000,000

CORUM

Founded in 1955, Switzerland's youngest luxury watch brand, Corum, celebrated sixty years of unusual—and sometimes outlandish—case and dial designs in 2015. The brand has had quite a busy history, but still by and large remains true to the collections launched by founders Gaston Ries and his nephew René Bannwart: the Admiral's Cup, Bridges, and Heritage. Among Corum's most iconic pieces is the legendary Golden Bridge baguette movement, which has received a complete makeover in recent years with the use of modern materials and complicated mechanisms. The development of these extraordinary movements required great watchmaking craftsmanship and expansion and modernization of the product development department.

The Bridges collection has always been an eye-catcher with its unusual movement, originally the brainchild of the great watchmaker Vincent Calabrese. Its introduction was a milestone in watchmaking history. And the Golden Bridge recently acquired a new highlight: in the Golden Bridge Tourbillon Panoramique with all components appearing to float in thin air.

To secure its financial future, Corum was sold to China Haidian Group (now Citychamp Watch & Jewellery Group) for over $90 million in April 2013. CEO Antonio Calce welcomed the move not only for the financial independence it was to bring, but also for the access it allows to the crucial Chinese market.

Today, the sporty Admiral's Cup collection is divided into two families: the classical Legend and the more athletic AC-One 45. The colorful nautical number flags have returned to the Admiral's Cup dials. For the brand's sixtieth birthday in 2015, it produced a Legend with a flying tourbillon and revived the remarkable Bubble in a limited series. It earned its moniker from the domed shape of the crystal. Success breeds iterations. Davide Traxler, the new head of Corum, decided on a bold strategy to refresh the brand by enlarging the number of culty Bubbles, which now come in many different shapes and sizes. In September 2017, Jérôme Briard took over for Traxler, but this has not slowed the rise of the successful and hip Bubbles.

Golden Bridge Rectangle

Reference number: B113/03044
Movement: manually wound, Caliber CO 113; 4.9 × 34 mm, height 3 mm; 19 jewels; 28,800 vph; baguette movement, bridges and mainplate, hand-engraved
Functions: hours, minutes
Case: pink gold, 29.5 × 42.2 mm, height 9.3 mm; sapphire crystal; transparent case back; water-resistant to 3 atm
Band: reptile skin, triple folding clasp
Remarks: baguette movement surrounded by 3D structures shaped like Roman numerals
Price: $36,900

Golden Bridge Stream

Reference number: B313/03371
Movement: automatic, Caliber CO 313; 11.25 × 33.18 mm; 26 jewels; 28,800 vph; variable inertia balance, baguette movement with gold bridges and plates, linear winding with sliding platinum weight; 40-hour power reserve
Functions: hours, minutes
Case: pink gold, 31 × 42.2 mm, height 14.7 mm; sapphire crystal; transparent case back; water-resistant to 3 atm
Band: reptile skin, triple folding clasp
Remarks: baguette movement with linear winding with sliding platinum weight flanked by 3D microstructures reminiscent of Golden Gate Bridge
Price: $60,000; limited to 88 pieces

Golden Bridge Round

Reference number: B113/03010
Movement: manually wound, Caliber CO 113; 4.9 × 34 mm, height 3 mm; 19 jewels; 28,800 vph; baguette movement, bridges and mainplate, hand-engraved
Functions: hours, minutes
Case: pink gold, ø 43 mm, height 8.8 mm; sapphire crystal; transparent case back; water-resistant to 3 atm
Band: reptile skin, triple folding clasp
Remarks: baguette movement flanked by 3D microstructures
Price: $41,700
Variations: diamond bezel ($48,700)

CORUM

Admiral Legend 42
Reference number: A395/03595
Movement: automatic, Caliber CO 395 (base ETA 2892-A2); ø 25.9 mm; 27 jewels; 28,800 vph; 42-hour power reserve
Functions: hours, minutes, subsidiary seconds; date
Case: stainless steel, ø 42 mm, height 9.5 mm; bezel with blue rubber coating; sapphire crystal; transparent case back; water-resistant to 5 atm
Band: rubber, triple folding clasp
Price: $4,100

Admiral Legend 42
Reference number: A395/03596
Movement: automatic, Caliber CO 395 (base ETA 2892-A2); ø 25.9 mm; 27 jewels; 28,800 vph; 42-hour power reserve
Functions: hours, minutes, subsidiary seconds; date
Case: stainless steel, ø 42 mm, height 9.5 mm; bezel in rose gold; sapphire crystal; transparent case back; crown in rose gold; water-resistant to 5 atm
Band: calfskin, triple folding clasp
Price: $8,300

Admiral Legend 42 Chronograph
Reference number: A984/03597
Movement: automatic, Caliber CO 984 (base ETA 2894-2); ø 32 mm, height 6.1 mm; 37 jewels; 28,800 vph; 42-hour power reserve
Functions: hours, minutes, subsidiary seconds; chronograph; date
Case: stainless steel, ø 42 mm, height 12.3 mm; bezel with blue rubber coating; sapphire crystal; transparent case back; water-resistant to 3 atm
Band: rubber, triple folding clasp
Price: $6,100

Admiral 45
Reference number: A082/02887
Movement: automatic, Caliber CO 082; ø 25.6 mm; 21 jewels; 28,800 vph; mainplate and skeletonized rotor; 42-hour power reserve
Functions: hours, minutes, sweep seconds; date
Case: bronze, ø 45 mm, height 13.3 mm; sapphire crystal; transparent case back; water-resistant to 30 atm
Band: calfskin, triple folding clasp
Remarks: wooden dial
Price: $7,800

Admiral 45 Chronograph
Reference number: A116/03574
Movement: automatic, Caliber CO 116 (base ETA 2892-A2 with Dubois Dépraz module); ø 28.6 mm, height 6.1 mm; 39 jewels; 28,800 vph; rotor with black PVD coating; 42-hour power reserve
Functions: hours, minutes, subsidiary seconds; chronograph; date
Case: stainless steel, ø 45 mm, height 14.3 mm; sapphire crystal; transparent case back; water-resistant to 30 atm
Band: calfskin, triple folding clasp
Remarks: wooden dial
Price: $7,500

Admiral 45 Skeleton
Reference number: A082/03685
Movement: automatic, Caliber CO 082; ø 25.6 mm; 21 jewels; 28,800 vph; skeletonized mainplate, bridges, and rotor; 42-hour power reserve
Functions: hours, minutes, sweep seconds; date
Case: titanium with black PVD coating, ø 45 mm, height 13.3 mm; sapphire crystal; transparent case back; water-resistant to 30 atm
Band: rubber, triple folding clasp
Price: $9,800

CORUM

Heritage La Grande Vie
Reference number: Z082/03590
Movement: automatic, Caliber CO 082; ø 25.6 mm; 21 jewels; 28,800 vph; 42-hour power reserve
Functions: hours, minutes
Case: titanium, ø 42 mm, height 9.05 mm; sapphire crystal; transparent case back; water-resistant to 5 atm
Band: reptile skin, buckle
Price: $3,500
Variations: green dial and strap

Heritage La Grande Vie
Reference number: Z082/03588
Movement: automatic, Caliber CO 082; ø 25.6 mm; 21 jewels; 28,800 vph; 42-hour power reserve
Functions: hours, minutes
Case: titanium, ø 42 mm, height 9.05 mm; sapphire crystal; transparent case back; water-resistant to 5 atm
Band: reptile skin, buckle
Price: $3,500
Variations: green dial and strap

Heritage Coin Watch
Reference number: C082/03167
Movement: automatic, Caliber CO 082; ø 25.6 mm; 21 jewels; 28,800 vph; 42-hour power reserve
Functions: hours, minutes
Case: yellow gold, ø 43 mm, height 7.6 mm; sapphire crystal
Band: reptile skin, buckle
Remarks: dial and case back manufactured of a "Double Eagle" gold dollar coin
Price: $23,000

Bubble 47 Skeleton
Reference number: L082/03162
Movement: automatic, Caliber CO 082; ø 25.94 mm; 26 jewels; 28,800 vph; 42-hour power reserve
Functions: hours, minutes, sweep seconds
Case: stainless steel, ø 47 mm, height 18.8 mm; sapphire crystal; transparent case back; water-resistant to 10 atm
Band: rubber, buckle
Remarks: vaulted sapphire crystal; skeletonized dial
Price: $6,000

Big Bubble Magical 52
Reference number: L390/03640
Movement: automatic, Caliber CO 390; ø 29.9 mm; 30 jewels; 28,800 vph; 65-hour power reserve
Functions: hours, minutes (mysterious time display without visible hands)
Case: titanium, ø 52 mm, height 20.2 mm; sapphire crystal; transparent case back; water-resistant to 10 atm
Band: rubber, buckle
Price: $5,500; limited to 88 pieces

Bubble 47 Central Tourbillon
Reference number: L406/03664
Movement: automatic, Caliber CO 406; ø 29.9 mm; 38 jewels; 28,800 vph; central, flying 1-minute tourbillon; 65-hour power reserve
Functions: hours, minutes, (mysterious time display without visible hands)
Case: titanium, ø 47 mm, height 19.6 mm; sapphire crystal; transparent case back; water-resistant to 10 atm
Band: rubber with textile overlay, buckle
Price: $76,500

CUERVO Y SOBRINOS

Cuba means a lot of things to different people. Today it seems to be the last bastion of genuine retro in an age of frenzied technology. However, turn the clock back to the early twentieth century and you find that Ramón Rio y Cuervo and his sister's sons (his nephews, the "sobrinos" of the brand name) kept a watchmaking workshop and an elegant store on Quinta Avenida, where they sold fine Swiss pocket watches—and more modest American models as well. With the advent of tourism from the coast of Florida, their business developed with wristwatches, whose dials Don Ramón soon had printed with *Cuervo y Sobrinos*—"Cuervo and Nephews."

An Italian watch enthusiast, Marzio Villa, resuscitated Cuervo y Sobrinos in 2002 and started manufacturing in the Italian-speaking region of Switzerland and in cooperation with various Swiss watchmakers. The tagline "Latin heritage, Swiss manufacture" says it all. These timepieces epitomize—or even romanticize—the island's heyday. The colors hint at cigar leaves and sepia photos in frames of old gold. The lines are at times elegant and sober, like the Esplendidos, or radiate the ease of those who still have time on their hands, like the Prominente. Playfulness is also a Cuervo y Sobrinos quality: The Piratas have buttons shaped like the muzzle of a blunderbuss, a cannonball crown, and a porthole flange. At the time of writing, the brand has once again changed hands . . .

CyS SA
Rue du Doubs, 6
CH-2340
Switzerland

Tel.:
+41 21 552 18 82

E-mail:
info@cuervoysobrinos.com

Website:
www.cuervoysobrinos.com

Founded:
1882

Annual production:
3,500 watches

U.S. distributor:
Provenance Gems
500 East Broward Blvd., Suite 1710
Fort Lauderdale, FL 33394
ines@provenancegems.com
800-305-3869

Most important collections/price range:
Historiador, Prominente, Torpedo, Robusto / $2,000 to $20,000; higher for perpetual calendars and tourbillon models

Historiador Vuelo

Reference number: 3201.1I
Movement: automatic, CYS 8120 and Dubois Dépraz 30342; ø 26.20 mm, height 6.8 mm; 51 jewels; 28,800 vph; 40-hour power reserve; rotor with fan decoration and CyS engraving
Functions: hours, minutes, subsidiary seconds; sweep seconds; time zone and 24-hour indication; date at 6 o'clock
Case: stainless steel, 44 mm, height 13.65 mm; sapphire crystal; tachymeter on bezel; transparent case back; water-resistant to 3 atm
Band: reptile skin, folding clasp
Price: $7,600

Torpedo Pirata Chrono Day-Date

Reference number: 3051.1NDD
Movement: automatic, Caliber Valjoux 7750; ø 30 mm, height 7.9 mm; 25 jewels; 28,800 vph, 48-hour power reserve
Functions: hours, minutes, subsidiary seconds; chronograph; day, date
Case: stainless steel and titanium, black DLC coating, ø 45 mm, height 15.5 mm; sapphire crystal; screwed see-through case back with unique "Pirata" logo engraving; water-resistant to 3 atm
Band: reptile skin, buckle
Remarks: "Pirata de el Tiempo de la vida" on bezel
Price: $7,400

Historiador GMT

Reference number: 3196.IC
Movement: automatic, ETA Caliber 2893-1; ø 25.60 mm, height 4.10 mm; 21 jewels; 28,800 vph; 42-hour power reserve
Functions: hours, minutes, sweep seconds; GMT (on dial); date
Case: stainless steel, ø 40 mm, height 9.9 mm; sapphire crystal; integrated and protected key winding crown; water-resistant to 3 atm
Band: reptile skin, folding clasp
Remarks: vintage look inspired by 1950s CyS timepieces
Price: $3,850

STYLE
SOPHISTICATION
SECURITY

High-security luxury safes and watch winders expertly crafted to keep your collection securely organized and running in top form.

View the complete collection of fine watchwinders at *Orbita.com* & visit *BrownSafe.com* for a full lineup of high-security luxury safes featuring Orbita watchwinders.

BROWN SAFE
Ph.(760) 233-2293
www.BrownSafe.com

ORBITA WATCHWINDERS
Ph.(800) 800-4436
www.Orbita.com

CVSTOS

Cvstos
2, rue Albert Richard
CH-1201 Geneva
Switzerland

Tel.:
+41-22-989-1010

Fax:
+41-22-989-1019

E-mail:
info@cvstos.com

Website:
www.cvstos.com

Founded:
2005

U.S. distributor:
Cvstos USA, Inc.
207 W. 25th Street, 8th Floor
New York, NY 10001
212-463-8898

Most important collections/price range:
Challenge, Challenge-R, Concept-S, Sea Liner, High Fidelity / $10,000 to $315,000

Today's big buzzword is "disruption," which usually has nasty connotations for legacy industries and esthetics. In the microworld of watches, though, disruption is often attractive to watch lovers seeking something out of the ordinary. The target of disruption at Cvstos (pronounced coo-stos, and meaning "guardian" in Latin) is the dial. Technology rules the roost at the brand, and thus these extroverted, stately timepieces show, quite literally, what they're made of thanks to extensive skeletonization. The dials are nevertheless busy, maximalistic, with a complex look cultivated throughout the collection. It's definitely *haute horlogerie*, but targets a clientele that doesn't necessarily need elements such as *côtes de Genève*, gold, and guilloché to fulfill their watchmaking ideal.

Although it may not seem so at first sight, a great deal of watchmaking know-how goes into the making of a Cvstos, and that is no surprise for a brand that is the spiritual child of Sassoun Sirmakes, son of Vartan Sirmakes, the man who led Genevan watchmaker Franck Muller to world fame. Under the tutelage of his father, the cofounder of the Watchland *manufacture* in Genthod, young Sassoun was introduced to the hands-on side of watchmaking. In 2005, fate brought him together with designer and watchmaker Antonio Terranova, who had made a name for himself in the Swiss watch industry with the timepieces he designed and produced for leading brands. Even though he had freelanced for some of the more staid brands, Terranova had the heart of an avant-gardist willing to break free of the constraints of traditional forms—he collaborated on some of the early Richard Mille pieces, for example. This year, the brand collaborated with Dutch interior designer Eric Kuster, who has an affinity for camouflage colors, green and beige, but also for the color of his own country, orange. So a Cvstos has all the thrilling complications, GMT, tourbillons, perpetual calendars, and more, but expect some technoid materials, high engineering art, and now some unusual color schemes.

Challenge Chrono II Eric Kuster Limited Edition

Movement: automatic, Caliber CVS-577; ø 30.4 mm, height 7.9 mm; 25 jewels; 28,800 vph; special black finish on movement; skeletonized; 42-hour power reserve
Functions: hours, minutes, subsidiary seconds; power reserve indicator; chronograph; date
Case: titanium and aluminum, 41 × 53.7 mm, height 13.35 mm; sapphire crystal; transparent case back; water-resistant to 10 atm
Band: rubber, folding clasp
Remarks: in cooperation with Dutch interior designer Eric Kuster
Price: $17,500; limited to 14 pieces

Challenge Chrono II Brancard

Movement: automatic, Caliber CVS-577; ø 30.4 mm, height 7.9 mm; 25 jewels; 28,800 vph; special black finish on movement; skeletonized; 42-hour power reserve
Functions: hours, minutes, subsidiary seconds; power reserve indicator; chronograph; date
Case: titanium and pink gold, 41 × 53.7 mm, height 13.35 mm; sapphire crystal; transparent case back; water-resistant to 10 atm
Band: rubber, folding clasp
Price: $26,000

Challenge Jetliner II Brancard

Movement: automatic, Caliber CVS-350; ø 25.6 mm, height 4.6 mm; 21 jewels; 28,800 vph; special movement finish; partially skeletonized; titanium rotor with tungsten oscillating mass; 42-hour power reserve
Functions: hours, minutes, sweep seconds; date
Case: titanium and pink gold, 41 × 53.7 mm, height 16 mm; sapphire crystal; transparent case back; water-resistant to 10 atm
Band: reptile skin, folding clasp
Price: $17,000

Czapek & Cie.
Rue Saint-Léger 2
CH-1205 Genève
Switzerland

Tel.:
+41-76-815-1845

E-mail:
info@czapek.com

Website:
www.czapek.com

Founded:
2015

U.S. distributor:
Horology Works
11 Flagg Road
West Hartford, CT 06117
860-986-9676
info@horologyworks.com

Most important collections:
Quai des Bergues men's and ladies' watches, up to $28,200; Place Vendôme up to $110,500

CZAPEK & CIE.

Until recently, the name Czapek was literally unknown. Born in Bohemia (Czech Republic today) in 1811, Frantiszek Czapek fought in the failed Polish insurrection of 1832 against Russia and then fled to Geneva. In 1839, he joined fellow Pole Jean de Patek. When the contract expired in 1845, Patek decided on a partnership with Mr. Philippe, inventor of the keyless watch. Czapek went on to become purveyor of watches to Emperor Napoleon III and author of a book on watches. Then he vanished without a trace sometime in the 1860s.

Entrepreneur, art specialist, and occasional watch collector Harry Guhl bought the name and brought together a management team. They chose Czapek's model No. 3430 as a base upon which to build up a new brand. It's an intriguing piece, with elongated Roman numerals and superbly cut fleur-de-lys hands. The piece also has two oddly placed subdials at 7:30 and 4:30, one for small seconds, the other featuring a clever double hand for the seven-day power reserve and days of the week. It would serve as a model for the new brand's portfolio.

The team sought out some of the best suppliers in Switzerland, including Jean-François Mojon for the in-house calibers (double barrel spring, open ratchets), Donzé for the grand-feu dials with the secret signature, and Aurélien Bouchet for the fine fleur-de-lys hands. Funding came from subscribers over the crowd equity sites Raizers and Crowd for Angels.

The first collection, the Quai des Bergues, was announced in November 2015. A year later, the model with the name No. 33bis picked up the Public Prize of the Grand Prix d'Horlogerie de Genève. For their next evolutions, Czapek & Cie. decided on a ladies' version, with diamonds. Next came the Place Vendôme in homage to the great square in Paris where Czapek actually had a boutique. The two famous subdials contain an unusual suspended tourbillon and a second time zone, respectively.

The Faubourg de Cracovie, 2018, is a classic chronograph, but with a grand-feu dial, and, once again, the fine hands that reach all the way back to Czapek's day.

Quai des Bergues No. 25 Aqua Blue

Movement: manually wound, Czapek Caliber SXH1; ø 32 mm, height 4.75 mm; 21,600 vph; 2 barrel springs, double open ratchets; 168-hour power reserve
Functions: hours, minutes, subsidiary seconds; power reserve indicator; weekdays
Case: XO steel, ø 42.5 mm, height 11.9 mm; sapphire crystal; transparent case back; grand-feu enamel dial; water-resistant to 5 atm
Band: reptile skin, buckle
Remarks: special hand-made ricochet guilloché on dial
Price: $16,200; limited to 15 pieces per year
Variations: various case materials and dial colors

Faubourg de Cracovie

Movement: manually wound, Czapek Caliber SXH3; ø 30 mm, height 6.95 mm; 36,000 vph; 2 barrel springs, double open ratchets; 65-hour power reserve; diamond polished bridges, snailed "trottoirs"
Functions: hours, minutes, subsidiary seconds; chronograph; date
Case: stainless steel, ø 41.5 mm, height 10.8 mm; sapphire crystal; transparent case back; grand-feu enamel dial; water-resistant to 3 atm
Band: reptile skin, folding clasp
Remarks: blued-steel fleur-de-lys hands; secret Czapek signature on dial
Price: $24,800
Variations: modern rhodium-plated hands; hand-crafted guilloché dial in blue or black

Place Vendôme

Movement: manually wound, Czapek Caliber SXH2; ø 34.8 mm, height 9.8 mm; 21,600 vph; 1-minute tourbillon (off-center); open ratchet; finely decorated bridges and mainplate; 60-hour power reserve
Functions: hours, minutes, subsidiary seconds (on tourbillon cage); 2nd time zone; day/night indicator; power reserve indicator
Case: rose gold, ø 43.5 mm, height 14.6 mm; sapphire crystal; transparent case back; grand-feu enamel dial; water-resistant to 3 atm
Band: reptile skin, rose gold buckle
Remarks: fleur-de-lys hands
Price: $97,400
Variations: platinum ($110,500)

DAVOSA

One of the more important brands occupying the lower segment of the market is Davosa, which offers pilot watches, quality divers (with helium valve), dress watches, and ladies' watches, all at very affordable prices. The brand has even come out with an apnea training watch that cleverly comes out of its case. The company uses solid Swiss movements, which it occasionally modifies for its own designs, or it experiments with special coatings like the "gun" PVD coating on the latest Argonautics, which is dark green.

To create a broad portfolio requires experience, and that is something Davosa has in spades. The company was founded in 1891. Back then, farmer Abel Frédéric Hasler spent the long winter months in Tramelan, in Switzerland's Jura mountains, making silver pocket watch cases. Later, two of his brothers ventured out to the city of Geneva and opened a watch factory. The third brother also opted to engage with the watch industry and moved to Biel. The entire next generation of Haslers went into watchmaking as well.

The name Hasler & Co. appeared on the occasional package mailed in Switzerland or overseas. Playing the role of unassuming private-label watchmakers, the Haslers remained in the background and let their customers in Europe and the United States run away with the show. It wasn't until after World War II that brothers Paul and David Hasler dared produce their own timepieces.

The long experience with watchmaking and watches culminated in 1987 with the brothers developing their own line of watches under the brand name Davosa. The Haslers then signed a partnership with the German distributor Bohle. In Germany, mechanical watches were experiencing a boom, so the brand was able to evolve quickly. In 2000, Corinna Bohle took over as manager of strategic development. Davosa now reaches well beyond Switzerland's borders and has become an integral part of the world of mechanical watches.

Hasler & Co. SA
CH-2543 Lengnau
Switzerland

E-mail:
info@davosa.com

Website:
www.davosa.com

Founded:
1881

U.S. distributor:
Davosa U.S.A
4446 Carver St.
Lake Worth, FL 33461
877-DAVOSA1
info@davosa-usa.com
www.davosa-usa.com

Most important collections/price range:
Apnea Diver, Argonautic, Classic, Military, Titanium, Gentleman, Pilot, Ternos / $500 to $2,000

Argonautic BG Automatic
Reference number: 161.522.90
Movement: automatic, Sellita Caliber SW200-1; ø 25.6 mm, height 4.6 mm; 26 jewels; 28,800 vph; 38-hour power reserve
Functions: hours, minutes, sweep seconds; date
Case: stainless steel, ø 42.5 mm, height 13.5 mm; unidirectional bezel with ceramic insert, 0-60 scale; sapphire crystal; screw-in crown; helium valve; water-resistant to 30 atm
Band: stainless steel, folding clasp, with safety lock and extension link
Price: $799
Variations: various colors

Argonautic Bronze Automatic Limited Edition
Reference number: 161.581.55
Movement: automatic, Sellita Caliber SW200-1; ø 25.6 mm, height 4.6 mm; 26 jewels; 28,800 vph; 38-hour power reserve
Functions: hours, minutes, sweep seconds; date
Case: bronze, ø 42 mm, height 13.5 mm; unidirectional bezel in stainless steel with black PVD coating and ceramic inlay, 0-60 scale; sapphire crystal; screw-in crown; helium valve; water-resistant to 30 atm
Band: calfskin, buckle
Price: $1,199
Variations: various colors

Ternos Ceramic Automatic
Reference number: 161.555.62
Movement: automatic, ETA Caliber 2824-2; ø 25.6 mm, height 4.6 mm; 25 jewels; 28,800 vph; 38-hour power reserve
Functions: hours, minutes, sweep seconds; date
Case: stainless steel, ø 40 mm, height 12.5 mm; unidirectional bezel with rose gold PVD coating and ceramic inlay, 0-60 scale; sapphire crystal; screw-in crown; water-resistant to 20 atm
Band: stainless steel, folding clasp, with safety lock and extension link
Price: $859
Variations: various colors

DAVOSA

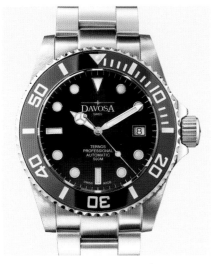

Ternos Professional TT Automatic
Reference number: 161.559.45
Movement: automatic, Sellita Caliber SW200-1; ø 25.6 mm, height 4.6 mm; 25 jewels; 28,800 vph; 38-hour power reserve
Functions: hours, minutes, sweep seconds; date
Case: stainless steel, ø 42 mm, height 15.5 mm; unidirectional bezel with ceramic insert, 0-60 scale; sapphire crystal; screw-in crown; helium valve; water-resistant to 50 atm
Band: stainless steel, folding clasp, with safety lock and extension link
Price: $849
Variations: various colors

Argonautic Lumis Colour
Reference number: 161.520.40
Movement: automatic, ETA Caliber 2824-2; ø 25.6 mm, height 4.6 mm; 25 jewels; 28,800 vph; 38-hour power reserve
Functions: hours, minutes, sweep seconds; date
Case: stainless steel, ø 42 mm, height 14 mm; unidirectional bezel with ceramic insert, 0-60 scale; sapphire crystal; screw-in crown; helium valve; water-resistant to 30 atm
Band: stainless steel Milanese mesh bracelet, folding clasp with safety lock
Remarks: hands and indices illuminated with tritium gas tubes
Price: $869

Apnea Diver Automatic
Reference number: 161.568.55
Movement: automatic, Sellita Caliber SW200-1; ø 25.6 mm, height 4.6 mm; 25 jewels; 28,800 vph; 38-hour power reserve
Functions: hours, minutes, sweep seconds
Case: stainless steel, ø 46 mm, height 12.5 mm; unidirectional bezel with ceramic insert, 0-60 scale; sapphire crystal; screw-in crown; water-resistant to 20 atm
Band: rubber, buckle
Remarks: case can be removed and set upright for breathing training
Price: $999
Variations: partial or total black PVD coating ($1,049/$1,099)

Evo 1908 Automatic
Reference number: 161.575.36
Movement: automatic, Sellita Caliber SW260; ø 25.6 mm, height 5.6 mm; 31 jewels; 28,800 vph; 38-hour power reserve
Functions: hours, minutes, subsidiary seconds; date
Case: stainless steel, 36 × 39.5 mm, height 12 mm; sapphire crystal; transparent case back; water-resistant to 5 atm
Band: calfskin, buckle
Price: $999
Variations: various straps and dials

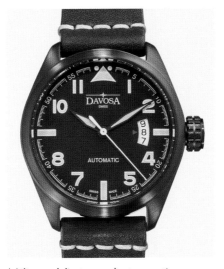

Military Vintage Automatic
Reference number: 161.511.84
Movement: automatic, ETA Caliber 2824-2; ø 25.6 mm, height 4.6 mm; 25 jewels; 28,800 vph; 38-hour power reserve
Functions: hours, minutes, sweep seconds; date
Case: stainless steel with black PVD coating, ø 42 mm, height 12.5 mm; sapphire crystal; screw-in crown; water-resistant to 20 atm
Band: calfskin, buckle
Remarks: comes with additional textile strap
Price: $749
Variations: various dials; without PVD coating ($699)

Pilot Automatic Chronograph
Reference number: 161.004.56
Movement: automatic, ETA Caliber 7750; ø 30 mm, height 7.9 mm; 25 jewels; 28,800 vph; finely finished with côtes de Genève, blued screws, perlage on bridges; 42-hour power reserve
Functions: hours, minutes, subsidiary seconds; chronograph; date, weekday
Case: stainless steel, ø 42 mm, height 15.6 mm; sapphire crystal; transparent case back; water-resistant to 10 atm
Band: calfskin, buckle
Price: $1,799
Variations: stainless steel bracelet ($1,899)

DETROIT WATCH COMPANY

Founders Patrick Ayoub and Amy Ayoub launched Detroit Watch Company in 2013 with the first and only mechanical timepieces designed and assembled in Detroit, Michigan. Patrick, a car designer, and Amy, an interior designer, share a passion for original design and timepieces and have worked hard to develop their brand, which draws inspiration from, and celebrates, the city of "Détroit."

Detroit means a lot of things to different people, and because the history of the people and places has shaped the city, Detroit's stories are also part of the Detroit Watch Company's collective story, which deserves to be told. Their introductory timepiece, the 1701, for instance, commemorates Antoine de la Mothe Cadillac, Knight of St. Louis, who, with his company of colonists, arrived at Détroit on July 24, 1701, and on that day, under the patronage of Louis XIV and protected by the flag of France, the city of Détroit, then called Fort Pontchartrain, was founded.

People phoning Detroit will understand why the company came out with a watch named 313—it's the area code of the city that brought cars and Motown (*motor + town*) music to the world. Needless to say, the dial looks like an old-fashioned phone dial. And where did those cars ride and race informally? On Woodward Avenue, the first mile of concrete highway in the USA, where carriages once rolled. It's the name for a collection of sporty chronographs.

The Detroit Watch Company timepieces are beautifully designed and hand-assembled in-house, and may be purchased directly through the Detroit Watch Company website.

Detroit Watch Company, LLC
P.O. Box 60
Birmingham, MI 48012

Tel:
248-321-5601

E-mail:
info@detroitwatchco.com

Founded:
2013

Number of employees:
3

Annual production:
500 watches

Distribution:
direct sales only

Most important collections/price range:
M1 Woodward Moonphase, 1701 Pontchartrain GMT, 1701 L'Horloge; B24 Liberator / $998 to $2,650

M1 Woodward Moonphase

Reference number: DWC M1W-Moonphase
Movement: automatic, ETA Caliber 7751; ø 30 mm, height 7.9 mm; 25 jewels; 28,800 vph; 48-hour power reserve
Functions: hours, minutes, subsidiary seconds; chronograph; date, day, month, moon phase
Case: stainless steel, ø 42 mm, height 14.5 mm, sapphire crystal with antireflective coating, screw-down exhibition case back with engraving; water-resistant to 5 atm
Band: calfskin, folding clasp
Price: $2,550

City Collection 313

Reference number: DWC-CITY-313
Movement: automatic, ETA Caliber 2824-2, ø 26.6 mm, height 4.6 mm; 26 jewels; 28,800 vph; 38-hour power reserve
Functions: hours, minutes, subsidiary seconds
Case: stainless steel, ø 42 mm, height 9.4 mm, sapphire crystal with antireflective coating, screw-down case back with engraving; water-resistant to 5 atm
Band: calfskin, buckle
Price: $1,100
Variations: black on black dial, transparent case back ($1,250)

1701 Pontchartrain GMT

Reference number: DWC-1701GMT-S1
Movement: automatic, ETA Caliber 2893-2, ø 26.6 mm, height 4.1 mm; 21 jewels; 28,800 vph; 42-hour power reserve
Functions: hours, minutes, subsidiary seconds; 2nd time zone, date
Case: stainless steel, ø 42 mm, height 9.4 mm; sapphire crystal with antireflective coating, screw-down case back with engraving; water-resistant to 5 atm
Band: calfskin, buckle
Price: $1,395
Variations: transparent case back ($1,595)

DUMANÈGE

Montres duManège Sàrl
Rue du manège 16
CH-2300 La Chaux-de-Fonds

Tel.:
+41-32-913-32-33

E-mail:
info@dumanege.com

Website:
www.dumanege.com

Founded:
2011

Number of employees:
4

Annual production:
approx. 300 watches

U.S. distributors:
Brands Consulting, LLC
Thierry Chaunu
Wells Fargo Bank, N.A.
420 Montgomery
San Francisco, CA 94104
646-732-1822
thierrychaunu@gmail.com

Most important collections/price range:
DM Exploration / $7,500 to $18,900; Heritage / $3,175 to 10,000; Heritage Art / $12,500 and higher

Strictly speaking, the term "Swiss-made" is something of a misnomer when applied to watches. With a few exceptions, the horological hotbed of the country is, in fact, the narrow French-speaking region in the west of the country. Geneva is only one of the hubs here, of course, the real forge for Swiss-watches being located in the stark Jura mountains, especially La-Chaux-de-Fonds and neighboring Le Locle, joint members of the UNESCO World Heritage club.

But the names of these towns never appear on a watch, oddly. Until 2014, when a young *chaudefonnier*, Julien Fleury, decided to put the name of his native city on a dial. As a graphic designer with training in jewelry, he had a clear idea of what his watches were meant to look like. And he had a vision for developing his brand.

He called upon family and friends, pulled all stops with contacts, and managed to get a collection of large, sportive watches onto the market, which he sold by subscription. He called it DM Exploration, the initials standing for his brand's name, duManège, a reference to the magnificent nineteenth-century riding school in La-Chaux-de-Fonds that was later used for low-cost cooperative living.

Inside the watch, a trusty Technotime movement. Today, the Exploration comes as a monopusher chronograph with a Valjoux engine. On the dial, "La-Chaux-de-Fonds."

Would he make it? Fleury is a marathon runner. He paced himself and looked to the future.

After Exploration came the Heritage line, a three-hander with date exuding the charm of a classic Swiss watch. Its main line features an ivory-white dial. The Heritage's sparsely populated dial allowed Fleury to not only exercise his talents as a jeweler, but also to explore a number of decorative crafts, from miniature painting (for bespoke commissioned pieces) to grand-feu enameling. He has used it for minimalist pieces as well, like the Meteorite, with a slab of Muonionalusta from a landing site in northern Sweden, or for a deep blue aventurine dial recalling a night sky.

DM Exploration

Reference number: DM.E.CMP-A.TI.45.100-1
Movement: automatic, Valjoux 7750; ø 30 mm, height 7.9 mm; 29 jewels; 28,800 vph; special black duManège treatment on bridges; 48-hour power reserve
Functions: hours, minutes, subsidiary seconds; monopusher chronograph
Case: titanium, ø 45.5 mm, height 14 mm; sapphire crystal; bezel with black DLC treatment; transparent case back; water-resistant to 5 atm
Remarks: black opaline, with blue touches
Band: rubber, buckle
Price: $5,650
Variations: steel ($5,200); titanium with black DLC ($6,850); pink gold ($14,700)

Heritage

Reference number: DM.H.Q-A.WG.42.132-1
Movement: manually wound, ETA Caliber 2892-2; ø 25.6 mm, height 3.6 mm; 21 jewels; 28,800 vph; 42-hour power reserve
Functions: hours, minutes, sweep seconds
Case: white gold, ø 42 mm, height 9.5 mm; sapphire crystal; water-resistant to 3 atm
Band: reptile skin, folding clasp
Remarks: blue aventurine dial
Price: $10,350; limited edition of 26 pieces
Variations: red gold ($10,350); stainless steel ($2,950)

Heritage Meteorite

Reference number: DM.H.Q-A.SS.42.132-2
Movement: manually wound, ETA Caliber 2892-2; ø 25.6 mm, height 3.6 mm; 21 jewels; 28,800 vph; 42-hour power reserve
Functions: hours, minutes, sweep seconds
Case: stainless steel, ø 42 mm, height 9.5 mm; sapphire crystal; water-resistant to 3 atm
Band: reptile skin, folding clasp
Remarks: dial of Muonionalusta meteorite from northern Sweden
Price: $2,950

EBERHARD & CO.

Chronographs weren't always the main focus of the Eberhard & Co. brand. In 1887, Georges-Emile Eberhard rented a workshop in La Chaux-de-Fonds to produce a small series of pocket watches, but it was the unstoppable advancement of the automotive industry that gave the young company its inevitable direction. By the 1920s, Eberhard was producing timekeepers for the first auto races. In Italy, Eberhard & Co. functioned well into the 1930s as the official timekeeper for all important events relating to motor sports. And the Italian air force later commissioned some split-second chronographs from the company, one of which went for 56,000 euros at auction.

Eberhard & Co. is still doing well, thanks to the late Massimo Monti. In the 1990s, he associated the brand with legendary racer Tazio Nuvolari. The company dedicated a chronograph collection to Nuvolari and sponsored the annual Gran Premio Nuvolari vintage car rally in his hometown of Mantua.

With the launch of its four-counter chronograph, this most Italian of Swiss watchmakers underscored its expertise and ambitions where short time/sports time measurement is concerned. Indeed, Eberhard & Co.'s Chrono 4 chronograph, featuring four little counters all in a row, has brought new life to the chronograph in general. CEO Mario Peserico has continued to develop it, putting out versions with new colors and slightly altered looks.

The brand is pure vintage, so it will come as no surprise that it regularly reissues and updates some of its older, popular models, like the two-totalizer Contograph chrono from the 1960s, which originally allowed the user to calculate phone units exactly (*conto* = bill). In 2016, it relaunched the venerable 1950s Scafograph, a very clean diver with streamlined hands and a ceramic bezel. And 2017 brought a crop of special editions of the brand's most iconic model, the Chrono 4.

Eberhard & Co.
5, rue du Manège
CH-2502 Biel/Bienne
Switzerland

Tel.:
+41-32-342-5141

Fax:
+41-32-341-0294

E-mail:
info@eberhard-co-watches.ch

Website:
www.eberhard-co-watches.ch

Founded:
1887

Distribution:
Contact main office for information on U.S. distribution

Most important collections/price range:
Chrono 4; 8 Jours, Tazio Nuvolari; Extra-fort; Gilda; Contograf; Scafograf

Scafograf GMT
Reference number: 41038
Movement: automatic, ETA Caliber 2893-2; ø 25.6 mm, height 4.1 mm; 21 jewels; 28,800 vph; 42-hour power reserve
Functions: hours, minutes, sweep seconds; additional 24-hour display (2nd time zone); date
Case: stainless steel, ø 43 mm, height 11.8 mm; bidirectional bezel with ceramic insert, with 0-24 scale; sapphire crystal; screw-in crown; water-resistant to 10 atm
Band: stainless steel, folding clasp
Price: $4,700
Variations: blue dial; rubber strap ($3,880)

Chrono 4 130
Reference number: 31129
Movement: automatic, Eberhard Caliber EB 251-12 1/2 (base ETA 2894-2); ø 33 mm, height 7.5 mm; 53 jewels; 28,800 vph; 4 counters in a row
Functions: hours, minutes, subsidiary seconds; additional 24-hour display; chronograph; date
Case: stainless steel, ø 42 mm, height 13.3 mm; sapphire crystal; screw-in crown; water-resistant to 5 atm
Band: calfskin, buckle
Price: $6,480
Variations: various dials

8 Jours Grande Taille
Reference number: 21027
Movement: manually wound, Eberhard Caliber EB 896 (base ETA 7001); ø 34 mm, height 5 mm; 25 jewels; 21,600 vph; 2 spring barrels, 192-hour power reserve
Functions: hours, minutes, subsidiary seconds; power reserve indicator
Case: stainless steel, ø 41 mm, height 10.85 mm; sapphire crystal; transparent case back; water-resistant to 3 atm
Band: reptile skin, buckle
Price: $5,350
Variations: black dial

EBERHARD & CO.

Scafograf 300
Reference number: 41034
Movement: automatic, ETA Caliber 2824-2; ø 25.6 mm, height 4.6 mm; 25 jewels; 28,800 vph; 42-hour power reserve
Functions: hours, minutes, sweep seconds; date
Case: stainless steel, ø 43 mm, height 12.6 mm; unidirectional bezel with ceramic insert, 0-60 scale; sapphire crystal; screw-in crown; helium valve; water-resistant to 30 atm
Band: rubber, buckle
Price: $3,260
Variations: blue indices; stainless steel bracelet ($4,070)

Scafograf GMT "The Black Sheep" Limited Edition
Reference number: 41040
Movement: automatic, ETA Caliber 2893-2; ø 25.6 mm, height 4.1 mm; 21 jewels; 28,800 vph; 42-hour power reserve
Functions: hours, minutes, sweep seconds; additional 24-hour display (2nd time zone); date
Case: stainless steel with black DLC coating, ø 43 mm, height 11.8 mm; bidirectional bezel with ceramic inlays with 0-24 scale; sapphire crystal; screw-in crown; water-resistant to 10 atm
Band: rubber, buckle
Price: $5,130; limited to 500 pieces

Extra-Fort Automatic
Reference number: 41029
Movement: automatic, Sellita Caliber SW200-1; ø 35.6 mm, height 4.6 mm; 26 jewels; 28,800 vph; 42-hour power reserve
Functions: hours, minutes, sweep seconds; date
Case: stainless steel, ø 40 mm, height 10.07 mm; sapphire crystal; screw-in crown; water-resistant to 5 atm
Band: reptile skin, buckle
Price: $3,300
Variations: silver white or black dial

Nuvolari Legend
Reference number: 31138
Movement: automatic, ETA Caliber 7750; ø 30 mm, height 7.9 mm; 25 jewels; 28,800 vph; 42-hour power reserve
Functions: hours, minutes; chronograph
Case: stainless steel, ø 43 mm, height 13.5 mm; sapphire crystal; transparent case back; screw-in crown; water-resistant to 3 atm
Band: calfskin, buckle
Remarks: gold, stylized Alfa Romeo on rotor
Price: $5,790

Tazio Nuvolari Gold Car
Reference number: 31038.5
Movement: automatic, ETA Caliber 7750; ø 30 mm, height 7.9 mm; 25 jewels; 28,800 vph; 42-hour power reserve
Functions: hours, minutes; chronograph
Case: stainless steel, ø 43 mm, height 13 mm; sapphire crystal; transparent case back; screw-in crown; water-resistant to 3 atm
Band: reptile skin, buckle
Remarks: gold, stylized Alfa Romeo on rotor
Price: $5,530

Tazio Nuvolari Data
Reference number: 31066
Movement: automatic, ETA Caliber 7750; ø 30 mm, height 7.9 mm; 25 jewels; 28,800 vph; 42-hour power reserve
Functions: hours, minutes; chronograph; date
Case: stainless steel, ø 43 mm, height 13 mm; sapphire crystal; screw-in crown; water-resistant to 3 atm
Band: reptile skin, buckle
Price: $5,170

ETERNA

Eterna is a milestone in watchmaking. Founded in 1856 in what was then a village, Grenchen, the company, named Dr. Girard & Schild, became a *manufacture*, producing pocket watches under Urs Schild in 1870. Among its earliest claims to fame was the first wristwatch with an alarm, released in 1908, by which time the company had taken on the name Eterna. Forty years later came the legendary Eterna-matic, featuring micro ball bearings for an automatic winding rotor. At the slightest movement of the watch, the rotor began to turn and set in motion what was another newly developed system of two ratchet wheels, which, independent of the rotational direction, lifted the mainspring over the automatic gears. Today, the five micro ball bearings used to cushion that rotor are the inspiration for Eterna's stylized pentagon-shaped logo.

In 2007, the company launched its Caliber 39, an automatic chronograph with three totalizers. This led to an entire family of eighty-eight versions. Modules for additional indicators or functions could be affixed with just a few screws or connected by way of a bridge. Thanks to the famous ball bearing–mounted Spherodrive winding mechanism, the movement has a power reserve of sixty-eight hours.

So, ironically perhaps, Eterna, whose original movement division became a separate company called ETA (now with the Swatch Group), has returned to building its own movements. It was acquired by F.A. Porsche Beteiligungen GmbH and started manufacturing for Porsche Design. But in 2011, International Volant Ltd., a wholly owned subsidiary of Citychamp, bought up the Porsche-owned shares in Eterna, opening many opportunities in Asia through its chain of retailers. In March 2014, Eterna and Porsche finally separated, freeing up Eterna's technical and financial resources to focus on growth. The company released new iterations of older models, notably the Super KonTiki Chronograph, a flyback equipped with the 31916A in-house caliber, and a new interpretation of the original Super KonTiki (no chronograph) that accompanied Thor Heyerdahl on his legendary expedition. But if the production of wristwatches seems to have slowed a bit, the company is drawing the attention of other, smaller brands seeking a flexible movement for their production.

Eterna SA
Schützenstrasse 40
CH-2540 Grenchen
Switzerland

Tel.:
+41-32-654-7211

Website:
www.eterna.com

Founded:
1856

Number of employees:
approx. 80

U.S. distributor:
CWJ Brands
1551 Sawgrass Blvd. Unit 109
Sunrise, FL 33323
954-279-1220

Most important collections:
KonTiki, Eternity, 1948

KonTiki Bronze Manufacture

Reference number: 1291.78.49.1422
Movement: automatic, Eterna Caliber 3902A; ø 30.4 mm, height 5.6 mm; 30 jewels; 28,800 vph; 65-hour power reserve
Functions: hours, minutes, sweep seconds
Case: bronze, ø 44 mm, height 14.05 mm; unidirectional bezel with ceramic insert, 0-60 scale; sapphire crystal; transparent case back; water-resistant to 20 atm
Band: calfskin, buckle
Price: $2,950; limited to 300 pieces

1940 Telemeter Chronograph Flyback Bronze Manufacture

Reference number: 7950.78.54.1416
Movement: automatic, Eterna Caliber 3916A; ø 30.4 mm, height 7.9 mm; 35 jewels; 28,800 vph; 60-hour power reserve
Functions: hours, minutes, subsidiary seconds; flyback chronograph; date
Case: bronze, ø 42 mm, height 14.1 mm; sapphire crystal; transparent case back; water-resistant to 5 atm
Band: calfskin, buckle
Price: $5,100; limited to 100 piece

KonTiki Diver Gent

Reference number: 1290.41.89.1418
Movement: automatic, Sellita Caliber SW200-1; ø 25.6 mm, height 4.6 mm; 26 jewels; 28,800 vph; 38-hour power reserve
Functions: hours, minutes, sweep seconds; date
Case: stainless steel, ø 44 mm, height 12.2 mm; unidirectional bezel with ceramic insert, 0-60 scale; sapphire crystal; water-resistant to 20 atm
Band: rubber, folding clasp
Price: $1,800

ETERNA

KonTiki Adventure

Reference number: 1910.79.50.1428
Movement: automatic, Soprod Caliber A10; ø 25.6 mm, height 3.6 mm; 25 jewels; 28,800 vph; 42-hour power reserve
Functions: hours, minutes, sweep seconds
Case: stainless steel with brown PVD coating, ø 44 mm, height 11 mm; unidirectional bezel, 0-60 scale; sapphire crystal; transparent case back; water-resistant to 20 atm
Band: calfskin, buckle
Price: $2,750

Lady Diver Gent

Reference number: 1282.64.69.1420
Movement: quartz
Functions: hours, minutes, sweep seconds; date
Case: stainless steel with rose gold PVD coating, ø 36 mm, height 11.35 mm; unidirectional bezel with ceramic insert, 0-60 scale; sapphire crystal; water-resistant to 20 atm
Band: calfskin, buckle
Price: $1,200

Lady Diver Gent

Reference number: 1282.41.66.1419
Movement: quartz
Functions: hours, minutes, sweep seconds; date
Case: stainless steel, ø 36 mm, height 11.35 mm; unidirectional bezel with ceramic insert, 0-60 scale; sapphire crystal; water-resistant to 20 atm
Band: calfskin, buckle
Price: $1,100

Caliber 3916A

Automatic; rotor on ball bearings; single spring barrel, 65-hour power reserve
Functions: hours, minutes, subsidiary seconds, flyback chronograph, date
Diameter: 30.4 mm
Height: 7.9 mm
Jewels: 35
Balance: glucydur
Frequency: 28,800 vph
Balance spring: flat hairspring
Shock protection: Incabloc

Caliber 3902M

Manually wound; completely skeletonized with matte finish; single spring barrel, 65-hour power reserve
Functions: hours, minute, sweep seconds
Diameter: 30.4 mm
Height: 4.95 mm
Jewels: 20
Balance: glucydur
Frequency: 28,800 vph
Balance spring: flat hairspring
Shock protection: Incabloc

Caliber 3902A

Automatic; rotor on ball bearings; single spring barrel, 65-hour power reserve
Functions: hours, minutes, sweep seconds
Diameter: 30.4 mm
Height: 5.6 mm
Jewels: 30
Balance: glucydur
Frequency: 28,800 vph
Balance spring: flat hairspring
Shock protection: Incabloc

FABERGÉ

Peter Carl Fabergé (1846–1920), son of a St. Petersburg jeweler of French Protestant stock and supplier to the Romanovs, is a legend. In 1885, he was commissioned by Tsar Alexander III to produce a special Easter egg for the tsarina. He did so, employing the best craftspeople of the time, and in the process catapulted himself into the good graces of the Romanovs. This also meant exile when the Bolsheviks took over in 1918. His sons set up a jewelry and restoration business in Paris.

After being sold several times during the twentieth century, the name Fabergé finally ended up being owned by Pallinghurst, a holding company with investments in mining that include the famous Gemfields, a specialist in colored stones.

In 2013, Fabergé decided to launch a new portfolio of watches. Utilizing colored stones and platinum was a foregone conclusion. Victor Mayer, a former licensee, took on the enamel guilloché dials of the Fabergé Flirt core collection, which received a Vaucher movement. For the men's watch, Renaud & Papi produced a subtly modern flying tourbillon with a geometrically openworked dial. But the pièce de résistance, the Lady Compliquée, was assigned to Jean-Marc Wiederrecht of Agenhor, who created a movement driving a retrograde peacock's tail (Peacock) or a wave of frost (Winter) to display the minutes, while the hours circle the dial in the opposite direction. It won a prize at the prestigious Grand Prix d'Horlogerie de Genève.

Sourcing quality continues to pay up for Fabergé with a new crop of outstanding watches, like the iteration of the Visionnaire DTZ, which features an almost invisible rotor oscillating just on the edge of the dial and a second time zone under a magnifying glass in the middle. The Visionnaire Chronograph has the three chrono hands stacked atop each other in the middle of the dial and required an entirely new module from Agenhor. The new model, Dynamique, has been primped with restive orange touches. Meanwhile, the Lady Libertine collection continues to grow, with fascinating gem layouts.

Fabergé
1 New Burlington Place, 4th level
London W1S 2HR
Great Britain

Tel.:
+44-20-7518-3400

E-mail:
information@faberge.com

Website:
www.faberge.com

Founded:
1842, current watch department relaunched 2013

Annual production:
approx. 350 watches

U.S. distributor:
Contact: sales@faberge.com

Most important collections:
Fabergé Flirt; Summer in Provence; Fabergé Lady Compliquée; Fabergé Visionnaire (DTZ & Chronograph); Fabergé Altruist, Fabergé Dalliance

Lady Compliquée Peacock Ruby

Movement: manually wound, Caliber AGH6901 exclusive for Fabergé; ø 32.7 mm, height 3.58 mm; 38 jewels; 21,600 vph; 50-hour power reserve
Functions: hours (on disk at crown), minutes (retrograde)
Case: platinum, 38 mm, height 12.90 mm; 54 diamonds on bezel; transparent back; sapphire crystal; water-resistant to 3 atm
Band: reptile skin, platinum buckle
Remarks: after the 1908 Fabergé Peacock Egg; dial set with rubies and diamonds, hand-engraved peacock on dial
Price: $89,000
Variations: Lady Compliquée Peacock Emerald, Lady Compliquée Peacock Black Sapphire

Fabergé Flirt 39 mm

Movement: automatic, Vaucher Caliber 3000; ø 23.3 mm, height 3.9 mm; 28 jewels; 28,800 vph; white gold rotor; 50-hour power reserve
Functions: hours, minutes
Case: white gold, ø 39 mm; bezel set with 51 diamonds, crown with moonstone; transparent case back; sapphire crystal; water resistant to 3 atm
Remarks: green enamel guilloché dial
Band: reptile skin, buckle
Price: $34,500
Variations: different dial colors; 36-mm case

Lady Compliquée Peacock Black

Movement: manually wound, Caliber AGH 6901 exclusive for Fabergé; ø 32.7 mm, height 3.58 mm; 38 jewels; 21,600 vph; 50-hour power reserve
Functions: hours (on disk at crown), minutes (retrograde)
Case: white gold, 38 mm; transparent back; sapphire crystal; water-resistant to 3 atm
Band: reptile skin, white gold buckle
Remarks: black lacquer dial, black-painted mother-of-pearl hour ring
Variations: rose gold
Price: $34,500

FABERGÉ

Fabergé Visionnaire DTZ

Movement: automatic, Caliber AGH 6924 exclusive for Fabergé; ø 34.8, height 8.3 mm; 30 jewels; 21,600 vph; 50-hour power reserve; mainplate, bridges with côtes de Genève
Functions: hours, minutes, central dual time zone (24-hour indication)
Case: yellow gold and titanium, ø 43 mm; sapphire crystal, transparent case back; water-resistant to 5 atm
Remarks: mysterious rotor with blue coloring on dial side and TC3 luminescent coating
Band: reptile skin, rose gold and titanium folding clasp
Price: $29,500
Variations: white gold and titanium ($29,500)

Fabergé Visionnaire Chronograph Dynamique

Movement: automatic, Caliber AGH 6361; ø 34.40, height 7.17 mm; 67 jewels; 21,600 vph; 60-hour power reserve; mainplate, bridges with côtes de Genève
Functions: hours, minutes; central chronograph
Case: black ceramic and dark gray DLC treated titanium, ø 43 mm, height 14.34 mm; sapphire crystal, transparent case back; water-resistant to 5 atm
Remarks: black dial with mysterious rotor on dial side; TC1 luminescent coating
Band: reptile skin, folding clasp
Price: $34,500
Variations: rose gold and titanium with opaline dial ($39,500)

Fabergé Altruist

Movement: automatic, Vaucher Caliber 3000; ø 23.3 mm, height 3.9 mm; 28 jewels; 28,800 vph; white gold rotor; 50-hour power reserve
Functions: hours, minutes, sweep seconds
Case: white gold, ø 41 mm; transparent case back; sapphire crystal
Band: reptile skin, white gold buckle
Remarks: blue enamel and guilloché dial
Price: $21,000
Variations: rose gold without enamel dial ($17,500)

Fabergé Lady Libertine I

Movement: manually wound, Caliber AGH 6911 exclusive for Fabergé, ø 30 mm, height 3.2 mm; 15 jewels; 21,600 vph; 50-hour power reserve; bridges with côtes de Genève
Functions: hours, minutes
Case: rose gold, 36 mm; sapphire crystal; transparent case back; bezel set with diamonds; water-resistant to 1 atm
Remarks: circular central space with representation of terrain in Zambia where emeralds for dial are mined; stylized arrowheads to point to hours and minutes
Band: reptile skin, rose gold buckle
Price: on request

Fabergé Dalliance Ruby

Movement: manually wound, Caliber AGH 6911 exclusive for Fabergé, ø 30 mm, height 3.2 mm; 15 jewels; 21,600 vph; 50-hour power reserve; mainplate circular-grained, bridges with côtes de Genève
Functions: hours, minutes
Case: white gold, ø 39.5 mm; sapphire crystal; transparent case back; bezel set with rubies; water-resistant to 1 atm
Remarks: dial set with diamonds; central section of dial can be personalized
Band: reptile skin, rose gold buckle
Price: on request

Fabergé Lady Libertine II

Movement: manually wound, Caliber AGH 6911 exclusive for Fabergé; ø 30 mm, height 3.2 mm; 15 jewels; 21,600 vph; 50-hour power reserve; bridges with côtes de Genève
Functions: hours, minutes
Case: white gold, ø 36 mm; sapphire crystal; transparent case back; bezel set with diamonds; water-resistant to 1 atm
Remarks: circular central space with representation of terrain in Zambia where emeralds for dial are mined; stylized arrowheads to point to hours and minutes
Band: reptile skin, white gold buckle
Price: on request

F.P. JOURNE

Born in Marseilles in 1957, François-Paul Journe might have become something else had he concentrated in school. He was kicked out and went to Paris, where he completed watchmaking school before going to work for his watchmaking uncle. And he has never looked back. By the age of twenty he had made his first tourbillon and soon was producing watches for connoisseurs.

He then moved to Switzerland, where he started out with handmade creations for a limited clientele and developing the most creative and complicated timekeepers for other brands before taking the plunge and founding his own in the heart of Geneva. The timepieces he basically single-handedly and certainly single-mindedly—hence his tagline *invenit et fecit*—conceives and produces are of such extreme complexity that it is no wonder they leave his workshop in relatively small quantities. Journe has won numerous top awards, some several times over. He particularly values the Prix de la Fondation de la Vocation Bleustein-Blanchet, since it came from his peers.

His collection is divided into two pillars: the automatic Octa line with its more readily understandable complications and the manually wound Souveraine line, containing horological treasures that can't be found anywhere else. The latter includes a *grande sonnerie*, a minute repeater, a constant force tourbillon with deadbeat seconds, and even a timepiece with two escapements beating in resonance—and providing chronometer-precise timekeeping. Journe, calmly, never stops surprising the watch world. The Centigraphe has a one-hundredth of a second chronograph with a lever rather than a pusher. His Elégante models, for women and men, feature a microprocessor that electronically notes when the watch is immobile, stops the mechanical movement to preserve energy, and restarts it when the watch is once again in motion. And rumor has it, he's working on an astronomical watch . . .

Montres Journe SA
17 rue de l'Arquebuse
CH-1204 Geneva
Switzerland

Tel.:
+41-22-322-09-09

Fax:
+41-22-322-09-19

E-mail:
info@fpjourne.com

Website:
www.fpjourne.com

Founded:
1999

Number of employees:
135

Annual production:
850–900 watches

U.S. distributor:
Montres Journe America
4330 NE 2nd Avenue
Miami, FL 33137
305-572-9802
phalimi@fpjourne.com

Most important collections:
Souveraine, Octa, Vagabondage, Elegante (Prices are in Swiss francs. Use daily exchange rate for calculations.)

Sonnerie Souveraine

Movement: manually wound, F.P.Journe Caliber 1505; ø 35.8 mm, height 7.8 mm; 42 jewels; 21,600 vph; rose gold plate and bridges; repeater chimes hours/quarter hours automatically, minute repeater on demand; on/off function; 422 components; 10 patents
Functions: hours, minutes (off-center), subsidiary seconds; grande sonnerie; power reserve indicator; chime indicator
Case: stainless steel, ø 42 mm, height 12.25 mm; sapphire crystal; screw-in crown and pusher; transparent case back
Band: 2 reptile skin straps, double folding clasp and 1 stainless steel bracelet
Price: CHF 771,200

Centigraphe Souverain Sport

Movement: manually wound, F.P.Journe Caliber 1506 in rose gold; ø 34.4 mm, height 5.6 mm; 26 jewels; 21,600 vph; aluminum movement; 80-hour power reserve (with chronograph off)
Functions: hours, minutes; 1-second, 20-second, and 10-minute chronograph subdials at 10, 2, and 6 o'clock
Case: titanium, ø 42 mm, height 11.6 mm; sapphire crystal; transparent case back
Band: rubber strap or titanium bracelet, folding clasp
Price: CHF 45,400
Variations: gold (CHF 60,700) or platinum (CHF 64,600)

Octa Divine

Movement: automatic, F.P.Journe Caliber 1300.3 in rose gold; ø 30.8 mm, height 5.7 mm; 39 jewels; 21,600 vph; up to 120–chronometric hour power reserve; pink gold guillochéed rotor, plate, and bridges
Functions: hours, minutes, subsidiary seconds; large date; moon phase; power reserve indicator
Case: red gold, ø 42 mm, height 10.6 mm; sapphire crystal; transparent case back
Band: reptile skin, platinum buckle
Remarks: white gold dial and silver hour circle silver guilloché dial with clous de Paris
Price: CHF 49,900
Variations: platinum or red gold with 40- or 42-mm case (gold 40-mm case, CHF 46,100)

F.P. JOURNE

Chronomètre Optimum
Movement: manually wound, F.P.Journe Caliber 1501 in rose gold; ø 34.4 mm, height 3.75 mm; 44 jewels; 21,600 vph; pink gold plate and bridges; double barrel, constant force remontoire, EPHB high-performance biaxial escapement, balance spiral with Phillips curve, deadbeat seconds on back
Functions: hours, minutes, subsidiary seconds; power reserve indicator
Case: platinum, ø 40 mm, height 10.1 mm; sapphire crystal; transparent case back
Band: reptile skin, buckle
Price: CHF 92,100
Variations: red gold (CHF 88,200)

Chronomètre à Résonance
Movement: manually wound, F.P.Journe Caliber 1499.3 in rose gold; ø 32.6 mm, height 4.2 mm; 36 jewels; 21,600 vph; unique concept of 2 escapements mutually influencing and stabilizing each other through resonance; pink gold plate and bridges
Functions: hours, minutes, subsidiary seconds; second time zone; power reserve indicator
Case: platinum, ø 40 mm, height 9 mm; sapphire crystal; transparent case back
Band: reptile skin, platinum buckle
Price: CHF 84,300
Variations: red gold (CHF 80,400)

Chronomètre Bleu
Movement: manually wound, F.P.Journe Caliber 1304 in rose gold; ø 30.4 mm, height 3.75 mm; 22 jewels; 21,600 vph; pink gold plate and bridges; chronometer balance with "invisible" connection to gear train; 2 spring barrels
Functions: hours, minutes, subsidiary seconds
Case: tantalum, ø 39 mm, height 8.6 mm; sapphire crystal; transparent case back
Band: reptile skin, tantalum buckle
Price: CHF 23,400

Tourbillon Souverain
Movement: manually wound, F.P.Journe Caliber 1403 in rose gold; ø 32.4 mm, height 7.15 mm; 26 jewels; 21,600 vph; tourbillon with constant force; balance with variable inertia; mainplate with côtes de Genève; 42-hour power reserve
Functions: hours, minutes, subsidiary deadbeat seconds; power reserve indicator
Case: platinum, ø 40 mm, height 9.9 mm; sapphire crystal; transparent case back
Band: calfskin, platinum buckle
Price: CHF 167,700
Variations: red gold case (on request)

Elégante 48 mm
Movement: electromechanical, F.P.Journe Caliber 1210; 28.5 x 28.3 mm, height 3.13 mm; 18 jewels; quartz frequency 32,000 Hz; autonomy: daily use up to 10 years, 18 years in standby mode
Functions: hours, minutes, subsidiary seconds; motion detector with inertia weight at 4:30
Case: titanium, 48 × 40 mm, height 7.35 mm; sapphire crystal; transparent case back
Band: navy blue rubber strap, titanium buckle
Remarks: standby mode after 30 minutes motionless, microprocessor keeps time, restarts automatically, sets time when watch put back on; luminescent dial
Price: CHF 11,500
Variations: in 48 mm: titanium with diamonds (CHF 24,500); various strap colors

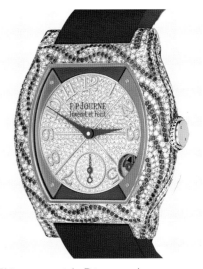

Elégante with Diamonds
Movement: electromechanical, F.P.Journe Caliber 1210; 28.5 × 28.3 mm, height 3.13 mm; 18 jewels; quartz frequency 32,000 Hz; autonomy: 100-year/ 18 years in standby mode; pink gold
Functions: hours, minutes, subsidiary seconds; motion detector with inertia weight at 4:30
Case: platinum, 40 × 35 mm, height 7.35 mm; pave-set with diamonds and sapphires; sapphire crystal; transparent case back
Band: navy blue rubber strap, pink gold buckle
Remarks: standby mode after 30 minutes motionless, microprocessor keeps time, restarts automatically, sets time when watch put back on; black sapphire dial
Price: CHF 119,000; **Variations:** titanium with rubber strap (CHF 15,200); platinum with rubber or snake strap (CHF 29,200)

FRANCK MULLER

Francesco "Franck" Muller has been considered one of the great creative minds in the industry ever since he designed and built his first tourbillon watch back in 1986. In fact, he never ceased amazing his colleagues and competition ever since, with his astounding timepieces that combined complications in a new and imaginative manner.

But a while ago, the "master of complications" stepped away from the daily business of the brand, leaving space for the person who had paved young Muller's way to fame, Vartan Sirmakes. It was Sirmakes, previously a specialist in watch cases, who had contributed to the development of the double-domed, tonneau-shaped Cintrée Curvex case, with its elegant, 1920s retro look. The complications never stop, either. Franck Muller created the Gigatourbillons, which are 20 millimeters across, and the Revolution series has a tourbillon that rises toward the crystal.

Even a brand that prides itself on a traditional look must make some concessions to modern esthetics. The more recent Vanguards and pieces like the newly released Skafander reveal an edginess that generates attractive tensions in that traditional Art Deco tonneau-shaped case that is so typical of the brand.

Muller and Sirmakes founded the Franck Muller Group Watchland in 1997. The Group now holds the majority interest in thirteen other companies, eight of which are watch brands. During the 2009 economic crisis, the company downsized somewhat, but it was only a glitch in an otherwise well-planned-out strategy to focus on developing complicated watches, like the Vanguard series, which has gone through numerous iterations, including being skeletonized. The far-reaching synergies within the Group mean that the success of the leader is indeed trickling laterally to the other participants, like Barthelay, Backes & Strauss, ECW, Martin Braun, Pierre Kunz, Rodolphe, Smalto Timepieces, and Roberto Cavalli by Franck Muller.

Groupe Franck Muller Watchland SA
22, route de Malagny
CH-1294 Genthod
Switzerland

Tel.:
+41-22-959-8888

Fax:
+41-22-959-8882

E-mail:
info@franckmuller.ch

Website:
www.franckmuller.com

Founded:
1997

Number of employees:
approx. 500 (estimated)

U.S. distributor:
Franck Muller USA, Inc.
207 W. 25th Street, 8th Floor
New York, NY 10001
212-463-8898
www.franckmuller.com

Most important collections:
Giga, Aeternitas, Revolution, Evolution 3-1, Vanguard, Cintrex

Vanguard Gravity Yachting Skeleton
Reference number: V 45 T GRAVITY CS YACHT SQT
Movement: manually wound, FM Caliber CS-03; 38.4 × 39.6 mm, height 9.1 mm; 24 jewels; 18,000 vph; 1-minute tourbillon, skeletonized movement; 120-hour power reserve
Functions: hours, minutes
Case: white gold, 44 × 53.7 mm, height 12.65 mm; sapphire crystal; transparent case back; water-resistant to 3 atm
Band: rubber with textile overlay, buckle
Price: $141,600

Vanguard S6 Yachting
Reference number: V 45 S6 SQT YACHTING
Movement: manually wound, FM Caliber 1740VS; 37.05 × 40.2 mm, height 6 mm; 21 jewels; 18,000 vph; inverted, skeletonized movement 168-hour power reserve
Functions: hours, minutes, subsidiary seconds; fine adjustment
Case: rose gold, 44 × 53.7 mm, height 12.65 mm; sapphire crystal; transparent case back; water-resistant to 3 atm
Band: rubber with textile overlay, buckle
Price: $45,400

Vanguard Yachting Chronograph
Reference number: V 45 CC DT YACHT
Movement: automatic, FM Caliber 7000; ø 30 mm, height 7.9 mm; 27 jewels; 28,800 vph; 48-hour power reserve
Functions: hours, minutes, subsidiary seconds; chronograph; date
Case: rose gold, 44 × 53.7 mm, height 15.8 mm; sapphire crystal; transparent case back
Band: rubber with textile overlay, buckle
Price: $25,800

FRANCK MULLER

Vanguard Tourbillon Minute Repeater
Reference number: V 50 L RMT SQT
Movement: automatic; ø 36.5 mm, height 6.91 mm; 33 jewels; 18,000 vph; 1-minute tourbillon, skeletonized movement; 60-hour power reserve
Functions: hours, minutes, subsidiary seconds (on tourbillon cage); minute repeater
Case: white gold, 46 × 55.9 mm, height 13.7 mm; sapphire crystal; transparent case back
Band: reptile skin, buckle
Price: $340,000

Vanguard Gravity Skeleton
Reference number: V 45 T GR CS SQT
Movement: manually wound, FM Caliber CS-03; 38.4 × 39.6 mm, height 8.7 mm; 25 jewels; 18,000 vph; 1-minute tourbillon, skeletonized movement; 120-hour power reserve
Functions: hours, minutes
Case: titanium with black PVD coating, 44 × 53.7 mm, height 15.1 mm; sapphire crystal; transparent case back; water-resistant to 3 atm
Band: reptile skin, buckle
Price: $130,000

Vanguard World Timer GMT
Reference number: V 45 HU AC BR 5N (5N)
Movement: automatic, FM Caliber 2802-GMT24HM; 26 × 31 mm, height 6 mm; 23 jewels; 28,800 vph; 42-hour power reserve
Functions: hours, minutes, sweep seconds; world time display (2nd time zone)
Case: stainless steel, 44 × 53.7 mm, height 14.5 mm; bezel in rose gold; sapphire crystal; crown in rose gold; water-resistant to 3 atm
Band: rubber with reptile skin overlay, folding clasp
Price: $24,700

Vanguard Golf
Reference number: V 45 C GOLF TT BR.NR
Movement: automatic, FM Caliber 800CGS; ø 26.2 mm, height 5.6 mm; 21 jewels; 28,800 vph; 42-hour power reserve
Functions: hours, minutes, sweep seconds, manual stroke counter
Case: titanium, 44 × 53.7 mm, height 13.7 mm; sapphire crystal; transparent case back; water-resistant to 3 atm
Band: rubber, buckle
Price: $11,000

Vanguard Master Banker
Reference number: V 45 MB SC DT SQT
Movement: automatic, FM Caliber 2800-MBSCDT SQT; 26.2 × 29.9 mm, height 5.75 mm; 23 jewels; 28,800 vph; skeletonized movement; 42-hour power reserve
Functions: hours, minutes, sweep seconds; 2 additional 12-hour displays (2nd and 3rd time zones); date
Case: rose gold, 44 × 53.7 mm, height 13.65 mm; sapphire crystal; transparent case back; water-resistant to 3 atm
Band: rubber with reptile skin overlay, buckle
Price: $38,800
Variations: stainless steel ($28,800)

Vanguard Slim Mécanique
Reference number: V 45 S (NR)
Movement: manually wound, FM Caliber 2250; ø 23.7 mm, height 2.6 mm; 21,600 vph; 45-hour power reserve
Functions: hours, minutes, subsidiary seconds
Case: stainless steel, 44 × 53.7 mm, height 9.5 mm; sapphire crystal
Band: rubber with reptile skin overlay, buckle
Price: $7,700
Variations: pink gold ($16,700)

FRANCK MULLER

Master Diving
Reference number: 2083 CC
Movement: automatic, FM Caliber 7750 (base ETA Caliber 7753); ø 30 mm, height 7.9 mm; 25 jewels; 28,800 vph; 42-hour power reserve
Functions: hours, minutes, subsidiary seconds; chronograph
Case: stainless steel, ø 46.3 mm, height 13.8 mm; unidirectional bezel, with 0-60 scale; sapphire crystal; water-resistant to 10 atm
Band: rubber, folding clasp
Price: $15,800
Variations: various dial colors

Skafander
Reference number: SKF 46 DV SC DT AC BR (AC)
Movement: automatic, FM Caliber 0800-SK; ø 31.5 mm, height 5.47 mm; 24 jewels; 28,800 vph; 42-hour power reserve
Functions: hours, minutes, sweep seconds
Case: stainless steel, 46 × 57 mm, height 15.6 mm; crown-activated scale ring, with 0-60 scale; sapphire crystal; screw-in crown; pusher with blocking mechanism to protect from accidental activation; water-resistant to 10 atm
Band: rubber, buckle
Price: $13,800

Endurance Tourbillon Gravity
Reference number: END 47.5 T GR CS SQT
Movement: manually wound, FM Caliber CS-03.R.SQ; ø 42.2 mm, height 9.1 mm; 24 jewels; 18,000 vph; off-center 1-minute tourbillon with cage made of convex structures, skeletonized movement; 120-hour power reserve
Functions: hours, minutes
Case: white gold, ø 47.7 mm, height 13.4 mm; sapphire crystal; transparent case back; water-resistant to 3 atm
Band: rubber, buckle
Price: $115,800

Vanguard Lady
Reference number: V 32 SC AT FO (NR)
Movement: automatic, FM Caliber 2671 (base ETA 2671); ø 17.5 mm, height 4.8 mm; 25 jewels; 28,800 vph; 38-hour power reserve
Functions: hours, minutes, sweep seconds
Case: stainless steel, 32 × 42.3 mm, height 9.9 mm; sapphire crystal
Band: reptile skin, buckle
Price: $7,800

Vanguard Lady Gravity
Reference number: V 35 T GR CS
Movement: manually wound, FM Caliber L03; 29.5 × 34.9 mm, height 5.7 mm; 23 jewels; 21,600 vph; off-center 1-minute tourbillon with cage made of convex structures; 120-hour power reserve
Functions: hours, minutes
Case: white gold, 35 × 46.3 mm, height 11.5 mm; sapphire crystal; transparent case back; water-resistant to 3 atm
Band: rubber with reptile skin overlay, buckle
Price: $100,600

Vanguard Heart Skeleton
Reference number: V 35 S6 SQT HEART
Movement: manually wound, FM Caliber 1540V-S; 29.5 × 34.9 mm, height 5.9 mm; 21 jewels; 18,000 vph; skeletonized movement with heart-shaped, color-contrasted bridges; 96-hour power reserve
Functions: hours, minutes, subsidiary seconds
Case: rose gold, 35 × 46.3 mm, height 10.4 mm; sapphire crystal; transparent case back; water-resistant to 3 atm
Band: rubber with reptile skin overlay, buckle
Price: $28,800

Frédérique Constant SA
Chemin du Champ des Filles 32
CH-1228 Plan-les-Ouates (Geneva)
Switzerland

Tel.:
+41-22-860-0440

Fax:
+41-22-860-0464

E-mail:
info@frederique-constant.com

Website:
www.frederique-constant.com

Founded:
1988

Number of employees:
100

Annual production:
approx. 146,000 watches

U.S. distributor:
Alpina Frederique Constant USA
350 5th Avenue, 29th Fl.
New York, NY 10118
646-438-8124
lmellor@usa.frederique-constant.com

Most important collections/price range:
Tourbillon Perpetual Calendar Manufacture / from approx. $21,995 to $32,995; Hybrid Manufacture / from approx. $3,695 to $3,895; Worldtimer Manufacture / from approx. $4,195; Vintage Rally Healey / from approx. $2,795 to $3,095; Ladies' Double Heart Beat Automatic / from approx. $1,695 to $1,995

FRÉDÉRIQUE CONSTANT

Peter and Aletta Stas, the Dutch couple who founded Frédérique Constant, have always sought to make high-end watches for consumers without deep pockets. So high-end, in fact, that in 2004 they went public with the brand's first movement produced entirely in-house and equipped with innovative silicon components. The move was in line with the other strategy of staying independent, and it was crowned a success. The Heart Beat calibers proved to be reliable, popular, and affordable.

Since 1991, the Dutch couple have genuinely lived up to the tagline they use for their Swiss brand: "live your passion." The watch brand, named for Aletta's great-grandmother Frédérique Schreiner and Peter's great-grandfather Constant Stas, was conceived in the late 1980s. The new company had its work cut out for it: Frédérique Constant had to compete in a watch market truly saturated with brands.

After the success of the Heart Beat *manufacture* mode, the Stases decided to invest in their own watch factory, an impressive, four-floor facility with ample room for a spacious atelier, administrative offices, conference rooms, a fitness area, and a cafeteria, in Geneva's industrial Plan-les-Ouates. Frédérique Constant moved into its new home in 2006, joined shortly after by sister brand Alpina. The Heart Beat collection continues to make waves, but the brand is growing in other directions as well, seeking to bridge the gap between fans of fine watchmaking and users of electronic nannies. The Horological Smartwatch, equipped with a quartz movement, connects with mobile phones and other electronic devices and can display data on its analog dial.

In 2016, much to the surprise of the watch world, the Stases decided it was time to find an investor to carry on their work. Frédérique Constant and Alpina were sold to Citizen. The two founders agreed to stay on as CEOs for four years.

Tourbillon Perpetual Calendar Manufacture

Reference number: FC-975S4H6
Movement: automatic, Caliber FC-975; ø 30 mm, height 6.67 mm; 33 jewels; 28,800 vph; 1-minute tourbillon; silicon escapement; 38-hour power reserve
Functions: hours, minutes, subsidiary seconds (on tourbillon cage); perpetual calendar with date, weekday, month, leap year
Case: stainless steel, ø 42 mm, height 12.5 mm; sapphire crystal; transparent case back; water-resistant to 5 atm
Band: reptile skin, folding clasp
Price: $22,995; limited to 88 pieces

Hybrid Manufacture

Reference number: FC-750DG4H6
Movement: automatic, Caliber FC-750; ø 30 mm, height 6.28 mm; 33 jewels; 28,800 vph; additional quartz movement for smart functions; 42-hour power reserve
Functions: hours, minutes, sweep seconds; additional 24-hour display (2nd time zone); smart functions like activity and sleep, alarm, rate analysis of mechanical movement; date
Case: stainless steel, ø 42 mm, height 12.84 mm; sapphire crystal; transparent case back; water-resistant to 5 atm
Band: reptile skin, folding clasp
Price: $3,795; limited to 888 pieces

Worldtimer Manufacture

Reference number: FC-718GRWM4H6
Movement: automatic, Caliber FC-718; ø 30 mm, height 6.2 mm; 26 jewels; 28,800 vph; 42-hour power reserve
Functions: hours, minutes, sweep seconds; world time indicator (2nd time zone); date
Case: stainless steel, ø 42 mm, height 12.4 mm; crown-activated scale ring, with 0-24 scale and city reference names; sapphire crystal; transparent case back; water-resistant to 3 atm
Band: reptile skin, folding clasp
Price: $4,195

FRÉDÉRIQUE CONSTANT

Slimline Moonphase Manufacture
Reference number: FC-705GR4S6
Movement: automatic, Caliber FC-705, ø 27.5 mm, height 6.2 mm; 26 jewels; 28,800 vph; finely finished with côtes de Genève; 42-hour power reserve
Functions: hours, minutes; date, moon phase
Case: stainless steel, ø 42 mm, height 11.3 mm; sapphire crystal; transparent case back; water-resistant to 3 atm
Band: reptile skin, folding clasp
Price: $2,995

Vintage Rally Chronograph
Reference number: FC-397HGR5B6
Movement: automatic, Caliber FC-397 (base Sellita SW500); ø 30 mm, height 7.9 mm; 25 jewels; 28,800 vph; 46-hour power reserve
Functions: hours, minutes, subsidiary seconds; chronograph
Case: stainless steel, ø 42 mm, height 14.45 mm; sapphire crystal; transparent case back; water-resistant to 5 atm
Band: calfskin, buckle
Price: $2,795

Ladies' Automatic Double Heart Beat
Reference number: FC-310LGDHB3B4
Movement: automatic, Caliber FC-310 (base Sellita SW200-1); ø 25.6 mm, height 4.6 mm; 26 jewels; 28,800 vph; partially skeletonized mainplate under escapement; 38-hour power reserve
Functions: hours, minutes, sweep seconds
Case: rose gold–plated stainless steel, ø 36 mm, height 9.85 mm; sapphire crystal; transparent case back; water-resistant to 5 atm
Band: reptile skin, folding clasp
Remarks: openworked dial
Price: $1,695

Classics Automatic Heart Beat
Reference number: FC-310MV5B4
Movement: automatic, Caliber FC-310 (base Sellita SW200-1); ø 25.6 mm, height 4.6 mm; 26 jewels; 28,800 vph; partially skeletonized mainplate under escapement; 38-hour power reserve
Functions: hours, minutes, sweep seconds
Case: rose gold–plated stainless steel, ø 40 mm, height 10.5 mm; sapphire crystal; water-resistant to 5 atm
Band: calfskin, buckle
Remarks: openworked dial
Price: $1,550

Ladies' Automatic
Reference number: FC-303LGD3B6
Movement: automatic, Caliber FC-303 (base Sellita SW200-1); ø 25.6 mm, height 4.6 mm; 26 jewels; 28,800 vph; 38-hour power reserve
Functions: hours, minutes, sweep seconds; date
Case: stainless steel, ø 36 mm, height 9.85 mm; sapphire crystal; transparent case back; water-resistant to 5 atm
Band: reptile skin, folding clasp
Price: $1,595

Classics Index Automatic
Reference number: FC-303MS5B6
Movement: automatic, Caliber FC-303 (base Sellita SW200-1); ø 25.6 mm, height 4.6 mm; 26 jewels; 28,800 vph; 38-hour power reserve
Functions: hours, minutes, sweep seconds; date
Case: stainless steel, ø 40 mm, height 10.3 mm; sapphire crystal; transparent case back; water-resistant to 5 atm
Band: stainless steel, folding clasp
Price: $950

FRÉDÉRIQUE CONSTANT

Runabout
Reference number: FC-303RMN5B4
Movement: automatic, Caliber FC-303 (base Sellita SW200-1); ø 25.6 mm, height 4.6 mm; 26 jewels; 28,800 vph; 38-hour power reserve
Functions: hours, minutes, sweep seconds; date
Case: rose gold–plated stainless steel, ø 42 mm, height 11.5 mm; sapphire crystal; transparent case back; water-resistant to 5 atm
Band: calfskin, folding clasp
Price: $1,895; limited to 2,888 pieces

Horological Smartwatch
Reference number: FC-282ABS5B6
Movement: quartz
Functions: hours, minutes; chronograph, activity and sleep measurement, alarm function, date, SMS messaging
Case: stainless steel, ø 42 mm, height 13.35 mm; sapphire crystal
Band: calfskin, buckle
Price: $795

Classics Chronograph Quartz
Reference number: FC-292MS5B6
Movement: quartz
Functions: hours, minutes, subsidiary seconds; chronograph; date
Case: rose gold–plated stainless steel, ø 40 mm, height 10.26 mm; sapphire crystal; water-resistant to 5 atm
Band: calfskin, buckle
Price: $995

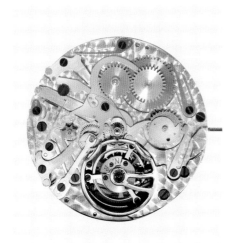

Caliber FC-975
Automatic; 1-minute tourbillon; single spring barrel, 38-hour power reserve
Functions: hours, minutes, subsidiary seconds; perpetual calendar with date, weekday, month, leap year
Diameter: 30 mm
Height: 6.67 mm
Jewels: 33
Balance: silicon
Frequency: 28,800 vph
Balance spring: flat hairspring
Shock protection: Incabloc
Remarks: 250 parts

Caliber FC-750
Automatic; mechanical and electronic movement hybrid; single spring barrel, 42-hour power reserve
Functions: hours, minutes, sweep seconds; smart functions like activity and sleep, alarm, rate analysis of mechanical movement; date
Diameter: 30 mm
Height: 6.28 mm
Jewels: 33
Frequency: 28,800 vph
Balance spring: flat hairspring
Shock protection: Incabloc
Remarks: 180 parts

Caliber FC-718
Automatic; single spring barrel, 42-hour power reserve
Functions: hours, minutes, sweep seconds; world time indicator (2nd time zone); date
Diameter: 30 mm
Height: 6.2 mm
Jewels: 26
Frequency: 28,800 vph
Balance spring: flat hairspring with fine adjustment
Shock protection: Incabloc
Remarks: crown control for all functions

GIRARD-PERREGAUX

When Girard-Perregaux CEO Luigi ("Gino") Macaluso died in 2010, the former minority partner of Sowind Group, PPR (Pinault, Printemps, Redoute), increased its equity stake to 51 percent. Under the leadership of Michele Sofisti since 2011, the brand has been charting a rather bold course that includes some technically sharp developments with the support of a strong development team and an excellently equipped production department. Under his guidance, the company has reduced its multitude of references but continues treading the fine line between fashionable watches and technical miracles. The various combinations of tourbillons and the gold bridges remain the company specialty. The most dazzling talking piece lately has undoubtedly been the Constant Escapement, a new concept that stores energy by buckling an ultrathin silicon blade and then releasing it to the balance wheel. Like many sophisticated systems, it was born of the banal: Inventor Nicolas Déhon was absentmindedly bending a train ticket one day when he was suddenly struck by a simple thought. As the ticket bent, it collected energy that was released in even bursts when it straightened out.

In 2015, Antonio Calce (of Eterna and Corum fame) became the company CEO and launched a freshening-up program, which included its marketing strategies. The Vintage 1945 and the elegant GP 1966 are still the mainstays of the brand, together with the Competizione chronographs. But Calce has also decided to put the ladies' Cat's Eye collection in the limelight and given a serious facelift to the Three Bridges tourbillon.

Like many brands in the Era of Vintage, Girard-Perregaux has been revisiting its past milestones for inspiration, like the Gyromatic HF from 1966. The company also dug up the much coveted Laureato from 1975 and turned it into a flagship of sorts. Something in the octagonal bezel on the round base epitomizes our contemporary sportive-elegant style. The collection now boasts a skeleton version, a chronograph, and even a flying tourbillon model.

Girard-Perregaux
1, Place Girardet
CH-2300 La Chaux-de-Fonds
Switzerland

Tel.:
+41-32-911-3333

Fax:
+41-32-913-0480

Website:
www.girard-perregaux.com

Founded:
1791

Number of employees:
280

Annual production:
approx. 12,000 watches

U.S. distributor:
Girard-Perregaux
Tradema of America, Inc.
7900 Glades Road, Suite 200
Boca Raton, FL 33434
833-GPWATCH
www.girard-perregaux.com

Most important collections/price range:
Laureato / Vintage 1945 / approx. $7,500 to $625,000; ww.tc / $12,300 to $23,800; GP 1966 / $7,500 to $291,000

Laureato 42mm Automatic

Reference number: 81010-11-431-11A
Movement: automatic, GP Caliber 01800-0013; ø 30 mm, height 3.97 mm; 28 jewels; 28,800 vph; 54-hour power reserve
Functions: hours, minutes, sweep seconds; date
Case: stainless steel, ø 42 mm, height 10.88 mm; sapphire crystal; transparent case back; water-resistant to 10 atm
Band: stainless steel, triple folding clasp
Price: $11,600
Variations: silver or gray dial; reptile skin strap ($10,800); titanium/rose gold with reptile skin strap ($16,000), link bracelet ($23,700)

Laureato 42mm Automatic

Reference number: 81010-11-634-11A
Movement: automatic, GP Caliber 01800-0013; ø 30 mm, height 3.97 mm; 28 jewels; 28,800 vph; 54-hour power reserve
Functions: hours, minutes, sweep seconds; date
Case: stainless steel, ø 42 mm, height 10.88 mm; sapphire crystal; transparent case back; water-resistant to 10 atm
Band: stainless steel, triple folding clasp
Price: $11,600
Variations: silver or blue dial; reptile skin strap ($10,800); titanium/rose gold with reptile skin strap ($16,000), link bracelet ($23,700)

Laureato 42mm Automatic

Reference number: 81010-32-631-32A
Movement: automatic, GP Caliber 01800-0013; ø 30 mm, height 3.97 mm; 28 jewels; 28,800 vph; 54-hour power reserve
Functions: hours, minutes, sweep seconds; date
Case: ceramic, ø 42 mm, height 10.88 mm; sapphire crystal; transparent case back; water-resistant to 10 atm
Band: ceramic, double folding clasp
Price: $16,500

GIRARD-PERREGAUX

Laureato Chronograph 42mm
Reference number: 81020-11-131-11A
Movement: automatic, GP Caliber 03300-0137/0138/0141; ø 25.95 mm, height 6.5 mm; 63 jewels; 28,800 vph; 46-hour power reserve
Functions: hours, minutes, subsidiary seconds; chronograph; date
Case: stainless steel, ø 42 mm, height 12.01 mm; sapphire crystal; water-resistant to 10 atm
Band: stainless steel, triple folding clasp
Price: $15,000
Variations: reptile skin strap ($14,200)

Laureato Chronograph 42mm
Reference number: 81020-11-431-11A
Movement: automatic, GP Caliber 03300-0137/0138/0141; ø 25.95 mm, height 6.5 mm; 63 jewels; 28,800 vph; 46-hour power reserve
Functions: hours, minutes, subsidiary seconds; chronograph; date
Case: stainless steel, ø 42 mm, height 11.9 mm; sapphire crystal; water-resistant to 10 atm
Band: stainless steel, triple folding clasp
Price: $15,000
Variations: reptile skin strap ($14,200)

Laureato Chronograph 42mm
Reference number: 81020-52-432-BB4A
Movement: automatic, GP Caliber 03300-0137/0138/0141; ø 25.95 mm, height 6.5 mm; 63 jewels; 28,800 vph; 46-hour power reserve
Functions: hours, minutes, subsidiary seconds; chronograph; date
Case: rose gold, ø 42 mm, height 12.01 mm; sapphire crystal; water-resistant to 10 atm
Band: reptile skin, triple folding clasp
Remarks: comes with additional rubber strap
Price: $33,700

Laureato Skeleton
Reference number: 81015-11-001-11A
Movement: automatic, GP Caliber 01800-0006; ø 30.6 mm, height 4.16 mm; 25 jewels; 28,800 vph; skeletonized movement; 54-hour power reserve
Functions: hours, minutes, subsidiary seconds
Case: stainless steel, ø 42 mm, height 11.13 mm; sapphire crystal; transparent case back; water-resistant to 3 atm
Band: stainless steel, triple folding clasp
Price: $33,600
Variations: rose gold ($63,700)

Laureato Flying Tourbillon Skeleton
Reference number: 99110-52-000-52A
Movement: automatic, GP Caliber 09520-0001; ø 32.5 mm, height 6.2 mm; 28 jewels; 21,600 vph; flying 1-minute tourbillon; gold microrotor; skeletonized movement; 50-hour power reserve
Functions: hours, minutes
Case: rose gold, ø 42 mm, height 10.76 mm; sapphire crystal; transparent case back; crown in rose gold; water-resistant to 3 atm
Band: rose gold, double folding clasp
Price: $129,000
Variations: white gold ($136,000)

Neo-Tourbillon with Three Bridges Skeleton
Reference number: 99295-21-000-BA6A
Movement: automatic, GP Caliber 09400-0011; ø 36 mm, height 9.54 mm; 27 jewels; 21,600 vph; 1-minute tourbillon under 3 bridges, skeletonized movement, titanium bridges with black PVD coating, platinum microrotor; 60-hour power reserve
Functions: hours, minutes
Case: titanium, ø 45 mm, height 15.85 mm; sapphire crystal; transparent case back; water-resistant to 3 atm
Band: reptile skin, triple folding clasp
Price: $145,000

GIRARD-PERREGAUX

Neo Bridges
Reference number: 84000-21-001-BB6A
Movement: automatic, GP Caliber 08400-0001; ø 32 mm, height 5.45 mm; 29 jewels; 21,600 vph; symmetrical skeleton construction; NAC-coated mainplate, PVD-coated bridges; microrotor; 54-hour power reserve
Functions: hours, minutes
Case: titanium, ø 45 mm, height 12.18 mm; sapphire crystal; transparent case back; water-resistant to 3 atm
Band: reptile skin, triple folding clasp
Price: $25,200

Neo-Tourbillon with Three Bridges
Reference number: 99270-21-000-BA6E
Movement: automatic, GP Caliber 09400-0001; ø 36.6 mm, height 8.21 mm; 27 jewels; 21,600 vph; 1-minute tourbillon under 3 bridges, titanium bridges with black PVD coating; platinum microrotor; 60-hour power reserve
Functions: hours, minutes, subsidiary seconds (on tourbillon cage)
Case: rose gold with black PVD coating, ø 45 mm, height 14.45 mm; sapphire crystal; transparent case back; water-resistant to 3 atm
Band: reptile skin, triple folding clasp
Price: $128,000

Constant Escapement L.M.
Reference number: 93505-39-633-BA6J
Movement: manually wound, GP Caliber 09100-0004; ø 39.2 mm, height 8.05 mm; 28 jewels; 21,600 vph; escapement with constant force mechanism with 2 pallet forks and flat, impulse-giving silicon blade spring; 2 spring barrels, 155-hour power reserve
Functions: hours, minutes (off-center), sweep seconds; linear power reserve indicator
Case: composite material (titanium and carbon fiber), ø 46 mm, height 14.84 mm; sapphire crystal; transparent case back; water-resistant to 3 atm
Band: reptile skin, triple folding clasp
Remarks: dedicated to the late head of the brand Luigi "Gino" Macaluso
Price: $103,000

GP 1966 40mm
Reference number: 49555-11-231-BB60
Movement: automatic, GP Caliber 03300-0130; ø 25.6 mm, height 3.36 mm; 27 jewels; 28,800 vph; 46-hour power reserve
Functions: hours, minutes, sweep seconds; date
Case: stainless steel, ø 40 mm, height 8.9 mm; sapphire crystal; transparent case back; water-resistant to 3 atm
Band: reptile skin, buckle
Price: $7,900

GP 1966 ww.tc
Reference number: 49557-11-132-BB6C
Movement: automatic, GP Caliber 03300-0022; ø 25.6 mm, height 5.71 mm; 32 jewels; 28,800 vph; 46-hour power reserve
Functions: hours, minutes, subsidiary seconds; world time indicator, day/night indicator (2nd time zone)
Case: stainless steel, ø 40 mm, height 12 mm; crown-activated scale ring, with reference city names; sapphire crystal; transparent case back; water-resistant to 3 atm
Band: reptile skin, folding clasp
Price: $12,900
Variations: stainless steel bracelet ($13,700); rose gold ($25,000)

GP 1966 38mm
Reference number: 49525-52-131-BK6A
Movement: automatic, GP Caliber 03300-00030; ø 25.6 mm, height 3.36 mm; 27 jewels; 28,800 vph; 46-hour power reserve
Functions: hours, minutes, sweep seconds; date
Case: rose gold, ø 38 mm, height 8.62 mm; sapphire crystal; transparent case back; water-resistant to 3 atm
Band: reptile skin, buckle
Price: $16,900

GIRARD-PERREGAUX

Caliber GP01800-0008
Automatic; single spring barrel, 54-hour power reserve
Functions: hours, minutes, sweep seconds; date
Diameter: 30 mm
Height: 3.97 mm
Jewels: 28
Frequency: 28,800 vph

Caliber GP3300
Automatic; rotor with ceramic ball bearings, stop-seconds mechanism; single spring barrel, 46-hour power reserve
Functions: hours, minutes, sweep seconds or subsidiary seconds at 9 o'clock; date
Diameter: 25.6 mm
Height: 3.2 mm
Jewels: 27
Balance: glucydur
Frequency: 28,800 vph
Balance spring: flat hairspring, fine adjustment
Shock protection: Kif
Remarks: 191 parts

Caliber GP08400
Automatic; symmetrical skeletonized construction; microrotor; single spring barrel, 54-hour power reserve
Functions: hours, minutes
Diameter: 32 mm
Height: 5.45 mm
Jewels: 29
Frequency: 21,600 vph
Remarks: NAC-coated mainplate, PVD-coated bridges

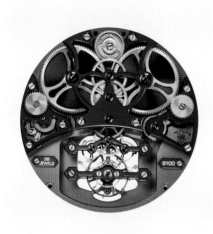

Caliber GP09100
Manually wound; escapement with constant force mechanism with 2 pallet forks and flat, impulse-giving silicon blade spring; double spring barrel, 144-hour power reserve
Functions: hours, minutes (off-center), sweep seconds; power reserve indicator (linear)
Diameter: 39.2 mm
Height: 8.05 mm
Jewels: 28
Frequency: 21,600 vph

Caliber GP09400
Automatic; 1-minute tourbillon, bidirectional winding microrotor; PVD-coated titanium; single spring barrel, 60-hour power reserve
Functions: hours, minutes, subsidiary seconds (on tourbillon cage)
Diameter: 36.6 mm
Height: 8.21 mm
Jewels: 27
Balance: screw balance
Frequency: 21,600 vph
Remarks: modern variation of classic tourbillon under 3 gold bridges

Caliber GP09520
Automatic; flying 1-minute tourbillon; skeletonized movement; single spring barrel, 50-hour power reserve
Functions: hours, minutes
Diameter: 32.5 mm
Height: 6.2 mm
Jewels: 28
Frequency: 21,600 vph

GLASHÜTTE ORIGINAL

Is there a little nostalgia creeping into the designers at Glashütte Original? Or is it just understated ecstasy for older looks? The retro touches that started appearing again a few years ago with the Sixties Square Tourbillon are still in vogue as the company delves into its own past for inspiration, such as the use of a special silver treatment on dials.

Glashütte Original *manufacture* roots go back to the mid-nineteenth century, though the name itself came later. The company, which had a sterling reputation for precision watches, became subsumed in the VEB Glashütter Uhrenbetriebe, a group of Glashütte watchmakers and suppliers who were collectivized as part of the former East German system. After reunification, the company took up its old moniker of Glashütte Original, and in 1995, the *manufacture* released an entirely new collection. Later, it purchased Union Glashütte. In 2000, the Swiss Swatch Group acquired the whole company and invested in expanding the production space at Glashütte Original headquarters.

Manufacturing depth has reached 95 percent. All movements are designed by a team of experienced in-house engineers, while the components comprising them such as plates, screws, pinions, wheels, levers, spring barrels, balance wheels, and tourbillon cages are manufactured in the upgraded production areas. These parts are lavishly finished by hand before assembly by a group of talented watchmakers. Even dials are in-house, ever since the purchase of a dial maker in Pforzheim, Germany, in 2012.

The large and elegant Senator Chronometer is a highlight of recent years with its classic design. It also boasts second and minute hands that automatically jump to zero when the crown is pulled, allowing for extremely accurate time setting. And to prove that the company is not just about tradition, it has even created a Senator-based app.

Glashütter Uhrenbetrieb GmbH
Altenberger Strasse 1
D-01768 Glashütte
Germany

Tel.:
01149-350-53-460

Fax:
01149-350-53-46-10999

E-mail:
info@glashuette-original.com

Website:
www.glashuette-original.com

Founded:
1951 (as VEB Glashütter Uhrenbetriebe)

Annual production:
N/A

U.S. distributor:
Glashütte Original
The Swatch Group (U.S.), Inc.
1200 Harbor Boulevard
Weehawken, NJ 07087
201-271-1400

Most important collections/price range:
Senator, Pano, Vintage, Ladies / $4,900 to $118,600

Grande Cosmopolite Tourbillon

Reference number: 1-89-01-03-03-04
Movement: manually wound, Glashütte Original Caliber 89-01; ø 39.2 mm, height 7.5 mm; 70 jewels; 21,600 vph; flying 1-minute tourbillon; 72-hour power reserve
Functions: hours, minutes, subsidiary seconds (on tourbillon cage); world time indicator with 35 time zones, day/night indicator, power reserve indicator (movement side); perpetual calendar with panorama date, weekday, month, leap year
Case: platinum, ø 48 mm, height 16 mm; sapphire crystal; transparent case back; water-resistant to 5 atm
Band: reptile skin, folding clasp
Price: on request; limited to 25 pieces

PanoLunar Tourbillon

Reference number: 1-93-02-05-05-05
Movement: automatic, Glashütte Original Caliber 93-02; ø 32.2 mm, height 7.65 mm; 48 jewels; 21,600 vph; flying 1-minute tourbillon, screw balance with 18 weighted screws, 2 diamond endstones, blued screws, skeletonized rotor with gold oscillating mass; 48-hour power reserve
Functions: hours, minutes (off-center), subsidiary seconds (on tourbillon cage); panorama date, moon phase
Case: pink gold, ø 40 mm, height 13.1 mm; sapphire crystal; transparent case back; water-resistant to 5 atm
Band: reptile skin, folding clasp
Price: $117,400
Variations: buckle ($115,400)

Senator Cosmopolite

Reference number: 1-89-02-03-02-30
Movement: automatic, Glashütte Original Caliber 89-02; ø 39.2 mm, height 8 mm; 63 jewels; 28,800 vph; swan neck spring to regulate rate; 72-hour power reserve
Functions: hours, minutes, subsidiary seconds; additional 12-hour display (2nd time zone), world time indicator with 35 time zones, day/night indicator, power reserve indicator; panorama date
Case: stainless steel, ø 44 mm, height 14 mm; sapphire crystal; transparent case back; water-resistant to 5 atm
Band: reptile skin, folding clasp
Price: $21,500
Variations: buckle ($21,200); pink gold ($43,500); white gold ($45,300)

Senator Excellence Perpetual Calendar

Reference number: 1-36-02-03-04-30
Movement: automatic, Glashütte Original Caliber 36-02; ø 32.2 mm, height 7.35 mm; 49 jewels; 28,800 vph; silicon hairspring, swan-neck fine adjustment; 100-hour power reserve
Functions: hours, minutes, sweep seconds; perpetual calendar with panorama date, weekday, month, moon phase, leap year
Case: white gold, ø 42 mm, height 12.8 mm; sapphire crystal; transparent case back; water-resistant to 5 atm
Band: reptile skin, folding clasp
Remarks: opening in dial, guillochéed mainplate
Price: $39,900; limited to 100 pieces
Variations: buckle ($37,500)

Senator Excellence Perpetual Calendar

Reference number: 1-36-02-02-05-30
Movement: automatic, Glashütte Original Caliber 36-02; ø 32.2 mm, height 7.35 mm; 49 jewels; 28,800 vph; silicon hairspring, screw balance with 4 regulating screws, swan-neck fine adjustment; 100-hour power reserve
Functions: hours, minutes, sweep seconds; perpetual calendar with panorama date, weekday, month, moon phase, leap year
Case: pink gold, ø 42 mm, height 12.8 mm; sapphire crystal; transparent case back; water-resistant to 5 atm
Band: reptile skin, folding clasp
Price: $37,100
Variations: buckle ($35,100); stainless steel ($22,300)

Senator Excellence

Reference number: 1-36-01-03-02-01
Movement: automatic, Glashütte Original Caliber 36-01; ø 32.2 mm, height 4.45 mm; 27 jewels; 28,800 vph; silicon hairspring, screw balance with 4 regulating screws, swan-neck fine adjustment, Glashütte three-quarter plate with ribbing, blued screws, skeletonized rotor with gold oscillating mass; 100-hour power reserve
Functions: hours, minutes, sweep seconds
Case: stainless steel, ø 40 mm, height 10 mm; sapphire crystal; transparent case back; water-resistant to 5 atm
Band: calfskin, buckle
Price: $9,700
Variations: folding clasp ($10,000)

Senator Excellence Panorama Date

Reference number: 1-36-03-04-02-30
Movement: automatic, Glashütte Original Caliber 36-03; ø 32.2 mm, height 6.7 mm; 41 jewels; 28,800 vph; silicon hairspring, screw balance with 4 regulating screws, swan-neck fine adjustment; 100-hour power reserve
Functions: hours, minutes, sweep seconds; panorama date
Case: stainless steel, ø 42 mm, height 12.2 mm; sapphire crystal; transparent case back; water-resistant to 5 atm
Band: reptile skin, folding clasp
Price: $10,700
Variations: buckle ($10,400); white or silver gray dial ($10,700)

Senator Excellence Panorama Date

Reference number: 1-36-03-03-02-31
Movement: automatic, Glashütte Original Caliber 36-03; ø 32.2 mm, height 6.7 mm; 41 jewels; 28,800 vph; silicon hairspring, screw balance with 4 regulating screws, swan-neck fine adjustment; 100-hour power reserve
Functions: hours, minutes, sweep seconds; panorama date
Case: stainless steel, ø 42 mm, height 12.2 mm; sapphire crystal; transparent case back; water-resistant to 5 atm
Band: reptile skin, folding clasp
Price: $10,700
Variations: buckle ($10,400); blue or white dial ($10,700)

Senator Excellence Panorama Date Moonphase

Reference number: 1-36-04-05-02-31
Movement: automatic, Glashütte Original Caliber 36-04; ø 32.2 mm, height 6.7 mm; 43 jewels; 28,800 vph; silicon hairspring, screw balance with 4 regulating screws, swan-neck fine adjustment; 100-hour power reserve
Functions: hours, minutes, sweep seconds; panorama date, moon phase
Case: stainless steel, ø 42 mm, height 12.2 mm; sapphire crystal; transparent case back; water-resistant to 5 atm
Band: reptile skin, folding clasp
Price: $11,700
Variations: buckle ($11,400); silver gray or blue dial ($11,700)

GLASHÜTTE ORIGINAL

Senator Chronograph Panorama Date
Reference number: 1-37-01-05-02-35
Movement: automatic, Glashütte Original Caliber 37-01; ø 31.6 mm, height 8 mm; 65 jewels; 28,800 vph; swan neck spring to regulate rate; 70-hour power reserve
Functions: hours, minutes, subsidiary seconds; 70-hour power reserve display; flyback chronograph; panorama date
Case: stainless steel, ø 42 mm, height 14.6 mm; sapphire crystal; transparent case back; screw-in crown; water-resistant to 10 atm
Band: calfskin, folding clasp
Price: $14,900
Variations: stainless steel bracelet ($16,400); pink gold ($31,500)

Senator Chronometer
Reference number: 1-58-01-05-34-30
Movement: manually wound, Glashütte Original Caliber 58-01; ø 35 mm, height 6.47 mm; 58 jewels; 28,800 vph; Glashütte three-quarter plate, second reset when crown is pulled allowing precise setting of minutes hand; DIN certified chronometer
Functions: hours, minutes, subsidiary seconds; day/night indicator, power reserve indicator; panorama date
Case: white gold, ø 42 mm, height 12.47 mm; sapphire crystal; transparent case back; water-resistant to 5 atm
Band: reptile skin, folding clasp
Price: $32,200
Variations: buckle ($29,800); rose gold ($30,300)

Senator Observer
Reference number: 100-14-05-02-05
Movement: automatic, Glashütte Original Caliber 100-14; ø 31.15 mm, height 6.5 mm; 60 jewels; 28,800 vph; screw balance, swan-neck fine adjustment; divided three-quarter plate with stripe finish; 55-hour power reserve
Functions: hours, minutes, subsidiary seconds; power reserve indicator; panorama date
Case: stainless steel, ø 44 mm, height 12 mm; sapphire crystal; transparent case back; water-resistant to 5 atm
Band: calfskin, folding clasp
Price: $11,800
Variations: black or gray dial; stainless steel bracelet ($11,800); reptile skin strap ($11,800)

PanoMaticLunar
Reference number: 1-90-02-42-32-05
Movement: automatic, Glashütte Original Caliber 90-02; ø 32.6 mm, height 7 mm; 47 jewels; 28,800 vph; screw balance with 18 weighted screws, duplex swan-neck fine regulation; 42-hour power reserve
Functions: hours, minutes (off-center), subsidiary seconds; panorama date, moon phase
Case: stainless steel, ø 40 mm, height 12.7 mm; sapphire crystal; transparent case back; water-resistant to 5 atm
Band: reptile skin, folding clasp
Price: $11,500
Variations: buckle ($11,200); stainless steel bracelet ($12,800); blue or gray dial ($11,500); pink gold ($23,900)

PanoMaticInverse
Reference number: 1-91-02-01-05-30
Movement: automatic, Glashütte Original Caliber 91-02; ø 38.2 mm, height 7.1 mm; 49 jewels; 28,800 vph; screw balance with 18 weighted screws, duplex swan-neck fine regulation; inverted movement construction, hand-engraved balance cock; 42-hour power reserve
Functions: hours, minutes (off-center), subsidiary seconds; panorama date
Case: pink gold, ø 42 mm, height 12.3 mm; sapphire crystal; transparent case back; water-resistant to 5 atm
Band: reptile skin, folding clasp
Price: $29,700
Variations: buckle ($23,200); stainless steel ($14,900)

PanoReserve
Reference number: 1-65-01-26-12-35
Movement: manually wound, Glashütte Original Caliber 65-01; ø 32.2 mm, height 6.1 mm; 48 jewels; 28,800 vph; duplex swan-neck fine regulation; hand-engraved balance and second cock; 42-hour power reserve
Functions: hours, minutes (off-center), subsidiary seconds; power reserve indicator; panorama date
Case: stainless steel, ø 40 mm, height 11.7 mm; sapphire crystal; transparent case back; water-resistant to 5 atm
Band: reptile skin, folding clasp
Price: $9,700
Variations: buckle ($9,400); stainless steel bracelet ($10,800); pink gold ($20,200)

GLASHÜTTE ORIGINAL

Seventies Chronograph Panorama Date
Reference number: 1-37-02-01-02-70
Movement: automatic, Glashütte Original Caliber 37-02; ø 31.6 mm, height 8 mm; 65 jewels; 28,800 vph; 70-hour power reserve
Functions: hours, minutes, subsidiary seconds; power reserve indicator; flyback chronograph; panorama date
Case: stainless steel, 40 × 40 mm, height 13.5 mm; sapphire crystal; transparent case back; screw-in crown; water-resistant to 10 atm
Band: stainless steel, folding clasp
Price: $16,400
Variations: rubber strap ($14,900); reptile skin strap ($14,900)

Seventies Panorama Date
Reference number: 2-39-47-13-12-04
Movement: automatic, Glashütte Original Caliber 39-47; ø 30.95 mm, height 5.9 mm; 39 jewels; 28,800 vph; swan-neck fine adjustment; 40-hour power reserve
Functions: hours, minutes, sweep seconds; panorama date
Case: stainless steel, 40 × 40 mm, height 11.5 mm; sapphire crystal; transparent case back; screw-in crown; water-resistant to 10 atm
Band: reptile skin, folding clasp
Price: $10,100
Variations: gray or silver dial ($11,600); rubber strap ($10,100); stainless steel bracelet ($10,100)

Sixties Panorama Date
Reference number: 2-39-47-04-02-04
Movement: automatic, Glashütte Original Caliber 39-47; ø 30.95 mm, height 5.9 mm; 39 jewels; 28,800 vph; swan-neck fine adjustment, Glashütte three-quarter plate with strip finish, skeletonized rotor with gold oscillating mass; finely finished movement; 40-hour power reserve
Functions: hours, minutes, sweep seconds; panorama date
Case: stainless steel, ø 42 mm, height 12.4 mm; sapphire crystal; transparent case back; water-resistant to 3 atm
Band: calfskin, buckle
Price: $9,300
Variations: silver, black, or blue dial ($9,300)

Lady Serenade
Reference number: 1-39-22-08-22-04
Movement: automatic, Glashütte Original Caliber 39-22; ø 26 mm, height 4.3 mm; 25 jewels; 28,800 vph; swan-neck fine adjustment, Glashütte three-quarter plate; 40-hour power reserve
Functions: hours, minutes, sweep seconds; date
Case: stainless steel, ø 36 mm, height 10.2 mm; bezel set with 52 diamonds; sapphire crystal; transparent case back; crown with diamond; water-resistant to 5 atm
Band: satin, folding clasp
Remarks: mother-of-pearl dial
Price: $10,000
Variations: buckle ($10,600); various cases, straps, and dials

PanoMatic Luna
Reference number: 1-90-12-03-12-02
Movement: automatic, Glashütte Original Caliber 90-12; ø 32.6 mm, height 7 mm; 47 jewels; 28,800 vph; duplex swan-neck fine adjustment, hand-engraved balance cock; 42-hour power reserve
Functions: hours, minutes, subsidiary seconds; panorama date, moon phase
Case: stainless steel, ø 39.4 mm, height 12 mm; bezel with 64 diamonds; sapphire crystal; transparent case back; crown with diamond; water-resistant to 3 atm
Band: reptile skin, buckle
Remarks: mother-of-pearl dial
Price: $20,400
Variations: white or dark mother-of-pearl dial ($20,400)

Pavonina
Reference number: 1-03-02-05-05-30
Movement: quartz
Functions: hours, minutes
Case: pink gold, 31 × 31 mm, height 7.5 mm; sapphire crystal; crown with diamond; water-resistant to 5 atm
Band: reptile skin, folding clasp
Remarks: mother-of-pearl dial, lugs set with diamonds
Price: $16,000
Variations: various cases, straps, and dials

GLASHÜTTE ORIGINAL

Caliber 36

Automatic; single spring barrel, 100-hour power reserve
Functions: hours, minutes, sweep seconds
Diameter: 32.2 mm
Height: 4.45 mm
Jewels: 27
Balance: screw balance with 4 gold weighted screws
Frequency: 28,800 vph
Balance spring: silicon
Shock protection: Incabloc
Remarks: very finely finished movement, three-quarter plate with Glashütte stripe finish, skeletonized rotor with gold oscillating mass
Related calibers: 36-02 (perpetual calendar), 36-03 (panorama date), 36-04 (panorama date and moon phase)

Caliber 37

Automatic; single spring barrel, 70-hour power reserve
Functions: hours, minutes, subsidiary seconds; power reserve indicator; flyback chronograph; panorama date
Diameter: 31.6 mm
Height: 8 mm
Jewels: 65
Balance: screw balance with 4 gold regulating screws
Frequency: 28,800 vph
Balance spring: flat hairspring, swan-neck fine regulation for the rate
Remarks: finely finished movement, beveled edges, polished steel parts, blued screws, three-quarter plate with Glashütte ribbing, skeletonized rotor with 21-kt gold oscillating mass

Caliber 39

Automatic; single spring barrel, 40-hour power reserve
Functions: hours, minutes, sweep seconds (base caliber)
Diameter: 26.2 mm
Height: 4.3 mm
Jewels: 25
Balance: glucydur
Frequency: 28,800 vph
Balance spring: flat hairspring, swan-neck fine adjustment
Shock protection: Incabloc
Related calibers: 39-55 (GMT, 40 jewels), 38-52 (automatic, 25 jewels), 39-50 (perpetual calendar, 48 jewels), 38-41/39-42 (panorama date, 44 jewels), 39-31 (chronograph, 51 jewels), 39-21/39-22 (date, 25 jewels)

Caliber 58-01

Manually wound, second reset when crown is pulled allowing precise setting of minutes hand; single spring barrel, 44-hour power reserve
Functions: hours, minutes, subsidiary seconds; day/night indicator, power reserve indicator with planetary drive; panorama date
Diameter: 35 mm
Height: 6.5 mm
Jewels: 58
Balance: screw balance with 18 weighted screws
Frequency: 28,800 vph
Balance spring: flat hairspring, swan-neck fine adjustment
Remarks: three-quarter plate with Glashütte ribbing, hand-engraved balance cock

Caliber 61

Manually wound; single spring barrel, 42-hour power reserve
Functions: hours, minutes (off-center), subsidiary seconds; flyback chronograph; panorama date
Diameter: 32.2 mm
Height: 7.2 mm
Jewels: 41
Balance: screw balance with 18 weighted screws
Frequency: 28,800 vph
Balance spring: flat hairspring, swan-neck fine adjustment
Remarks: finely finished movement, beveled edges, polished steel parts, screw-mounted gold chatons, blued screws, bridges, and balance cock with côtes de Genève, hand-engraved balance cock

Caliber 65

Manually wound; single spring barrel, 42-hour power reserve
Functions: hours, minutes (off-center), subsidiary seconds; power reserve indicator; panorama date
Diameter: 32.2 mm
Height: 6.1 mm
Jewels: 48
Balance: screw balance with 18 weighted screws
Frequency: 28,800 vph
Balance spring: flat hairspring, duplex swan-neck fine regulation for rate and beat
Shock protection: Incabloc
Remarks: finely finished movement, three-quarter plate with Glashütte ribbing, hand-engraved balance bridge

GLASHÜTTE ORIGINAL

Caliber 89-02
Automatic; single spring barrel, 72-hour power reserve
Functions: hours, minutes, subsidiary seconds; 2nd time zone, world time with 37 time zones, day/night indicator, power reserve indicator; panorama date
Diameter: 39.2 mm
Height: 8 mm
Jewels: 63
Balance: screw balance with 4 gold regulating screws
Frequency: 28,800 vph
Balance spring: flat hairspring, duplex swan-neck fine regulation for rate and beat
Shock protection: Incabloc
Remarks: winding gears with double sun brushing, three-quarter plate with Glashütte ribbing, hand-engraved balance bridge

Caliber 90
Automatic; single spring barrel, 42-hour power reserve
Functions: hours, minutes (off-center), subsidiary seconds; panorama date, moon phase
Diameter: 32.6 mm
Height: 5.4 mm
Jewels: 28
Balance: screw balance with 18 weighted screws
Frequency: 28,800 vph
Balance spring: flat hairspring, duplex swan-neck fine regulation for rate and beat
Shock protection: Incabloc
Remarks: eccentric, skeletonized, 21-kt gold oscillating weight, hand-engraved balance bridge

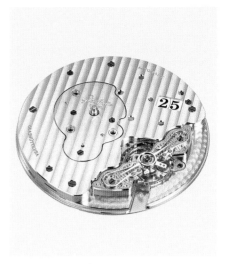

Caliber 91-02
Automatic; inverted movement with rate regulator on dial side; single spring barrel, 42-hour power reserve
Functions: hours, minutes (off-center), subsidiary seconds; panorama date
Diameter: 38.2 mm
Height: 7.1 mm
Jewels: 49
Balance: screw balance with 18 weighted screws
Frequency: 28,800 vph
Balance spring: flat hairspring, duplex swan-neck fine regulation for rate and beat
Shock protection: Incabloc
Remarks: finely finished movement, three-quarter plate with Glashütte ribbing

Caliber 93-03
Automatic; flying tourbillon; single spring barrel, 48-hour power reserve
Functions: hours, minutes (off-center), subsidiary seconds (on tourbillon cage); panorama date; moon phase
Diameter: 32.2 mm
Height: 7.65 mm
Jewels: 50
Balance: screw balance with 18 weighted screws in rotating frame
Frequency: 21,600 vph
Balance spring: flat hairspring
Remarks: plate with Glashütte ribbing, blued screws, eccentric, skeletonized, oscillating weight

Caliber 96-01
Automatic; twin spring barrel, bidirectional winding in two speeds via stepped reduction gear; 42-hour power reserve
Functions: hours, minutes (off-center), subsidiary seconds; 2-digit counter (pusher-controlled, forward and backward); flyback chronograph; panorama date
Diameter: 32.2 mm
Height: 8.9 mm
Jewels: 72
Balance: screw balance with 18 weighted screws
Frequency: 28,800 vph
Balance spring: flat hairspring swan-neck fine adjustment,
Remarks: separate wheel bridges for winding and chronograph, finely finished movement

Caliber 100
Automatic; single spring barrel, 55-hour power reserve
Functions: hours, minutes, sweep seconds; 0-reset mechanism for seconds hand activated by case pusher; panorama date
Diameter: 31.15 mm; **Height:** 7.1 mm
Jewels: 59
Balance: screw balance with 18 weighted screws
Frequency: 28,800 vph
Balance spring: flat hairspring swan-neck fine adjustment
Related caliber: 100-01 (power reserve indicator), 100-02 (perpetual calendar), 100-03 (panorama date), 100-04 (moon phases), 100-05 (53 calendar weeks), 100-06 (full calendar)

GRAHAM

In the mid-1990s, unusual creations gave an old English name in watchmaking a brand-new life. In the eighteenth century, George Graham perfected the cylinder escapement and the dead-beat escapement as well as inventing the chronograph. For these contributions and more, Graham certainly earned the right to be considered one of the big wheels in watchmaking history.

Despite his merits in the development of precision timekeeping, it was the mechanism he invented to measure short times—the chronograph—that became the trademark of his wristwatch company. To this day, the fundamental principle of the chronograph hasn't changed at all: A second set of hands can be engaged to or disengaged from the constant flow of energy of the movement. Given the British Masters' aim to honor this English inventor, it is certainly no surprise that the Graham collection includes quite a number of fascinating chronograph variations.

In 2000, the company released the Chronofighter, with its striking thumb-controlled lever mechanism—a modern twist on a function designed for World War II British fighter pilots, who couldn't activate the crown button of their flight chronographs with their thick gloves on. To enhance the retro look and feel, the brand decided to release several models bearing famous World War II pinups. The company has also started a special series to "give back," as it were. Made of a special carbon, this U.S. Navy Seal Chronofighter features a special camo look designed to help hide soldiers from satellite cameras. A part of the sales of these watches will go the nonprofit Navy Seal Foundation.

In recent years, Graham has also added comparatively conventionally designed watches to its collection. For lovers of special pieces, there are the models of the Geo.Graham series. It was the name used by the brilliant watchmaker-inventor. The Tourbillon "Orrery," developed in collaboration with Christophe Claret, is a perfect exemplar of the line, featuring a beautifully finished flying tourbillon with a miniaturized 3D view of the solar system.

Graham
Boulevard des Eplatures 38
CH-2300 La Chaux-de-Fonds
Switzerland

Tel.:
+41-32-910-9888

Fax:
+41-32-910-9889

E-mail:
info@graham1695.com

Website:
www.graham1695.com

Founded:
1995

Number of employees:
approx. 30

Annual production:
5,000–7,000 watches

U.S. distributor:
Graham Watches USA
261 Madison Avenue, 9th Floor
New York, NY 10016
516-526-9092

Most important collections:
Geo.Graham, Chronofighter, Silverstone, Swordfish

Chronofighter Superlight Carbon

Reference number: 2CCBK.B30A
Movement: automatic, Graham Caliber G1747; ø 30 mm, height 8 mm; 25 jewels; 28,800 vph; 48-hour power reserve
Functions: hours, minutes, subsidiary seconds; chronograph; date
Case: carbon fiber, ø 47 mm, height 15 mm; sapphire crystal; transparent case back; crown and pusher with finger lever on left side; water-resistant to 10 atm
Band: rubber, buckle
Price: $9,900

Silverstone RS Racing

Reference number: 2STEA.B16A
Movement: automatic, Graham Caliber G1749 (base ETA 7750); ø 30 mm, height 7.9 mm; 25 jewels; 28,800 vph; 48-hour power reserve
Functions: hours, minutes, subsidiary seconds; chronograph; date, weekday
Case: stainless steel, ø 46 mm, height 16.1 mm; sapphire crystal; transparent case back; water-resistant to 10 atm
Band: rubber, folding clasp
Price: $5,200

Geo.Graham Orrery Tourbillon

Reference number: 2GGAP.U01A
Movement: manually wound, Graham Caliber G1800 (base Christophe Claret); ø 39 mm, height 10.5 mm; 35 jewels; 21,600 vph; 1-minute tourbillon, mechanical model of solar system with 3D planet display, 2 spring barrels, with côtes de Genève; 72-hour power reserve
Functions: hours, minutes (off-center); 100-year calendar with date, month, Zodiac signs, and year (case back); 3D Earth, moon, and Mars
Case: pink gold, ø 48 mm, height 17.6 mm; sapphire crystal; transparent case back; water-resistant to 5 atm
Band: reptile skin, folding clasp
Price: on request; limited to 8 pieces

Chronofighter Grand Vintage
Reference number: 2CVDS.B25A.K134S
Movement: automatic, Graham Caliber G1747; ø 30 mm, height 8 mm; 25 jewels; 28,800 vph; 48-hour power reserve
Functions: hours, minutes, subsidiary seconds; chronograph; date
Case: stainless steel, ø 47 mm; sapphire crystal; transparent case back; crown and pusher and carbon finger lever on left side; water-resistant to 10 atm
Band: calfskin, buckle
Price: $5,950
Variations: various straps and dials

Chronofighter Vintage Nose Art Ltd.
Reference number: 2CVAS.B27A.L127S
Movement: automatic, Graham Caliber G1747; ø 30 mm, height 8 mm; 25 jewels; 28,800 vph; 48-hour power reserve
Functions: hours, minutes, subsidiary seconds; chronograph; date, weekday
Case: stainless steel, ø 44 mm; sapphire crystal; transparent case back; crown and pusher and carbon finger lever on left side; water-resistant to 10 atm
Band: calfskin, buckle
Remarks: various motifs
Price: $5,450; limited to 100 pieces each
Variations: 4 other dial variations

Chronofighter Vintage
Reference number: 2CCBK.V01A
Movement: automatic, Graham Caliber G1747; ø 30 mm, height 8 mm; 25 jewels; 28,800 vph; 48-hour power reserve
Functions: hours, minutes, subsidiary seconds; chronograph; date
Case: carbon fiber composite, ø 47 mm, height 15 mm; sapphire crystal; transparent case back; crown and pusher and carbon finger lever on left side; water-resistant to 10 atm
Band: rubber, buckle
Remarks: carbon dial
Price: $9,900
Variations: various straps and dials

Chronofighter Vintage Pulsometer Ltd.
Reference number: 2CVCS.B20A
Movement: automatic, Graham Caliber G1718; ø 30 mm, height 8 mm; 25 jewels; 28,800 vph; 48-hour power reserve
Functions: hours, minutes, subsidiary seconds; chronograph; date, weekday
Case: stainless steel, ø 44 mm; sapphire crystal; transparent case back; crown and pusher with finger lever on left side; water-resistant to 10 atm
Band: calfskin, buckle
Remarks: pulsometer scale on dial
Price: $4,450; limited to 250 pieces

Chronofighter Target
Reference number: 2CCAC.B33A
Movement: automatic, Graham Caliber G1747; ø 30 mm, height 8 mm; 25 jewels; 28,800 vph; 48-hour power reserve
Functions: hours, minutes, subsidiary seconds; chronograph; date, weekday
Case: stainless steel, ø 47 mm; ceramic bezel; sapphire crystal; transparent case back; crown and pusher and carbon finger lever on left side; water-resistant to 10 atm
Band: calfskin, buckle
Price: $6,900

Chronofighter Target
Reference number: 2CCAU.B32A
Movement: automatic, Graham Caliber G1747; ø 30 mm, height 8 mm; 25 jewels; 28,800 vph; 48-hour power reserve
Functions: hours, minutes, subsidiary seconds; chronograph; date, weekday
Case: stainless steel with black PVD coating, ø 47 mm; ceramic bezel; sapphire crystal; transparent case back; crown and pusher and carbon finger lever on left side; water-resistant to 10 atm
Band: calfskin, buckle
Price: $6,900

GRAND SEIKO

In the hustle and bustle of Baselworld 2017, the news may have been somewhat muffled, but it was a big step for the traditional brand Seiko: Shinji Hattori, president of the Seiko Watch Company, announced that the Grand Seiko line had become a separate *manufacture* brand. The Grand Seiko watches always had their own dedicated personality, he said, and their design required a separate approach to building the movements.

Indeed, the models released in this line always existed in a segment of their own and had become something of a focus for collectors. For the brand's fiftieth anniversary in 2010, the Grand Seiko collection was given a host of new models and started being sold on the European market.

What makes the Grand Seiko collection special is the "Spring Drive" technology, invented by a Seiko engineer, that took twenty-eight years to perfect and, according to the company, six hundred prototypes. Essentially, it consists of a complex combination of mostly mechanical parts with a small but crucial electronic regulating element to tame the energy from the mainspring. These watches also boast some classical mechanical hijinks, such as the "High-Beat" balance with 36,000 vibrations per hour. It's no surprise that classic watch fans have welcomed the Grand Seikos into their midst. The range of models has been widened with a number of sportive diver's watches that are giving established Swiss brands some stiff competition when it comes to price and amenities. In terms of segments, the Seiko Watch Corporation also covers the highest end of the spectrum with its brand Credor, which, however, is not sold outside of Japan.

Seiko Holdings
Ginza, Chuo, Tokyo
Japan

Website:
www.grand-seiko.com

Founded:
1881

Number of employees:
90,000 (for the entire holding)

U.S. distributor:
Grand Seiko Corporation of America
1111 Macarthur Boulevard
Mahwah, NJ 07430
201-529-5730
info@grand-seiko.us.com
www.grand-seiko.com

Most important collections/price range:
Grand Seiko / approx. $5,000 to $59,000

Spring Drive Chronograph
Reference number: SBGC201
Movement: manually wound, Seiko Caliber 9R86; ø 30 mm, height 7.6 mm; 50 jewels; electromagnetic Tri-Synchro Regulator escapement system with sliding wheel; antimagnetic protection 4,800 A/m; 72-hour power reserve
Functions: hours, minutes, subsidiary seconds; additional 24-hour display (2nd time zone), power reserve indicator; chronograph; date
Case: stainless steel, ø 43.5 mm, height 16.1 mm; sapphire crystal; transparent case back; screw-in crown; water-resistant to 10 atm
Band: stainless steel, folding clasp
Price: $8,200
Variations: titanium ($9,600)

Spring Drive GMT
Reference number: SBGE201
Movement: automatic, Seiko Caliber 9R66; ø 30 mm, height 5.1 mm; 30 jewels; electromagnetic Tri-Synchro Regulator escapement system with sliding wheel; antimagnetic up to 4,800 A/m; 72-hour power reserve
Functions: hours, minutes, sweep seconds; additional 24-hour display (2nd time zone), power reserve indicator; date
Case: stainless steel, ø 43.5 mm, height 14.7 mm; bidirectional bezel with 0-24 scale; sapphire crystal; transparent case back; screw-in crown; water-resistant to 20 atm
Band: stainless steel, folding clasp
Price: $5,800

Spring Drive
Reference number: SBGA293
Movement: manually wound, Seiko Caliber 9R65; ø 30 mm, height 5.1 mm; 30 jewels; electromagnetic Tri-Synchro Regulator escapement system with sliding wheel; antimagnetic up to 4,800 A/m; 72-hour power reserve
Functions: hours, minutes, sweep seconds; power reserve indicator; date
Case: stainless steel, ø 40.2 mm, height 12.5 mm; sapphire crystal; transparent case back; screw-in crown; water-resistant to 10 atm
Band: reptile skin, folding clasp, with safety lock
Price: $5,200

GRAND SEIKO

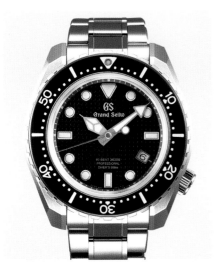

Hi-Beat Professional Diver's
Reference number: SBGH255
Movement: automatic, Seiko Caliber 9S85; ø 28.4 mm, height 5.9 mm; 37 jewels; 36,000 vph; antimagnetic up to 16,000 A/m; 55-hour power reserve
Functions: hours, minutes, sweep seconds; date
Case: titanium, ø 46.9 mm, height 17 mm; unidirectional bezel with 0-60 scale; sapphire crystal; screw-in crown; water-resistant to 60 atm
Band: titanium, folding clasp, with safety lock and extension link
Price: $9,600

Automatic Hi-Beat 36,000 Limited Edition
Reference number: SBGH267
Movement: automatic, Seiko Caliber 9S85; ø 28.4 mm, height 5.9 mm; 37 jewels; 36,000 vph; antimagnetic up to 4,800 A/m; 55-hour power reserve
Functions: hours, minutes, sweep seconds; date
Case: stainless steel, ø 39.5 mm, height 13 mm; sapphire crystal; transparent case back; water-resistant to 10 atm
Band: stainless steel, folding clasp
Price: $6,300; limited to 1,500 pieces

9S Mechanical
Reference number: SBGW231
Movement: manually wound, Seiko Caliber 9S64; ø 28.4 mm, height 4.9 mm; 24 jewels; 28,800 vph; antimagnetic up to 4,800 A/m; 72-hour power reserve
Functions: hours, minutes, sweep seconds
Case: stainless steel, ø 37.3 mm, height 11.6 mm; sapphire crystal; transparent case back; water-resistant to 3 atm
Band: reptile skin, folding clasp
Price: $4,300

High-Precision Quartz 9F 25th Anniversary
Reference number: SBGV238
Movement: quartz, Seiko Caliber 9F82; twin pulse control motor, antimagnetic up to 4,800 A/m
Functions: hours, minutes, sweep seconds; date
Case: stainless steel, ø 40 mm, height 10 mm; bezel yellow gold; sapphire crystal; water-resistant to 10 atm
Band: stainless steel, folding clasp

High-Precision Quartz
Reference number: SBGV225
Movement: quartz, Seiko Caliber 9F82; twin pulse control motor, antimagnetic up to 4,800 A/m
Functions: hours, minutes, sweep seconds; date
Case: stainless steel, ø 40 mm, height 10 mm; sapphire crystal; water-resistant to 10 atm
Band: stainless steel, folding clasp
Price: $2,500

High-Precision Quartz
Reference number: SBGX259
Movement: quartz, Seiko Caliber 9F82; twin pulse control motor, 4,800 A/m
Functions: hours, minutes, sweep seconds; date
Case: stainless steel, ø 37 mm, height 10 mm; sapphire crystal; water-resistant to 10 atm
Band: stainless steel, folding clasp
Price: $2,200

GREUBEL FORSEY

Each year, at the SIHH, the journalists visit brands. But they congregate at the Greubel Forsey booth to take part in something close to a religious experience, an initiation into the esoteric art of *ultra-haute horlogerie*.

In 2004, when Alsatian Robert Greubel and Englishman Stephen Forsey presented a new movement at Baselworld, eyes snapped open: Their watch featured not one but *two* tourbillon carriages working at a 30° incline. In their design, Forsey and Greubel not only took up the basic Abraham-Louis Breguet idea of canceling out the deviations of the balance by the continuous rotation of the tourbillon cage, but they went further, creating a quadruple tourbillon.

In 2010, Greubel Forsey moved into new facilities at a renovated farmhouse between Le Locle and La Chaux-de-Fonds and a brand-new modern building. After capturing an Aiguille D'Or for the magical Double Tourbillon 30° and the Grand Prix d'Horlogerie in Geneva, these two specialists snatched up the top prize at the International Chronometry Competition in Le Locle for the Double Tourbillon 30°.

Greubel and Forsey continue to stun the highest-end fans with some spectacular pieces, like the Quadruple Tourbillon Secret, which shows the complex play of the tourbillons through the case back, and the Greubel Forsey GMT with the names of world cities and a huge floating globe. Their first Art Piece came out in 2013, a most natural collaboration with British miniaturist Willard Wigan, who can sculpt the head of a pin.

Meanwhile, Greubel and Forsey continue to push for tiny increments in chronometric precision. This involves building works of engineering art with, for instance, two inclined oscillators "linked by a spherical differential that averages rating differences and ensures an optimal performance at all times whether in stabilised (horizontal or vertical) positions or dynamically (on the wrist)," to quote their website.

Greubel Forsey SA
Eplatures-Grise 16
CH-2301 La Chaux-de-Fonds
Switzerland

Tel.:
+41-32-925-4545

Fax:
+41-32-925-4502

E-mail:
info@greubelforsey.com
press@greubelforsey.com

Website:
www.greubelforsey.com

Founded:
2004

Number of employees:
approx. 100

Annual production:
approx. 100 watches

U.S. distributor:
Time Art Distribution
550 Fifth Avenue
New York, NY 10036
212-221-8041
info@timeartdistribution.com

Remarks:
Prices given only in Swiss francs (before taxes). Use daily exchange rate for conversion.

QP à Équation
Reference number: P556
Movement: manually wound, Caliber GF07; ø 36.4 mm, height 9.6 mm; 75 jewels; 21,600 vph; 24-second tourbillon, with balance with variable inertia axially inclined at 25°; 2 series-coupled barrels, Phillips end curve, German silver plates and bridges; 72-hour power reserve
Functions: hours, minutes, subsidiary seconds; additional 24-hour display (2nd time zone), power reserve display, function selector; perpetual calendar with large date, weekday, month, leap year
Case: white gold, ø 43.5 mm, height 16 mm; sapphire crystal; transparent case back; water-resistant to 3 atm
Band: reptile skin, folding clasp
Price: CHF 670,000

Double Balancier
Reference number: P193
Movement: manually wound, Caliber GF04; ø 36.4 mm, height 8.15 mm; 50 jewels; 21,600 vph; 2 oscillators inclined at 30°, escapement by constantly rotating differential (4-minute rotation); 72-hour power reserve
Functions: hours, minutes, subsidiary seconds (on differential); power reserve indicator
Case: pink gold, ø 43 mm, height 13.4 mm; sapphire crystal; transparent case back
Band: reptile skin, folding clasp
Price: CHF 350,000

GMT Mouvement 5N
Reference number: P436
Movement: manually wound, Caliber GF05; ø 36.4 mm, height 9.8 mm; 50 jewels; 21,600 vph; 24-second tourbillon; German silver plate with rose gold PVD coating; 72-hour power reserve
Functions: hours, minutes (off-center), subsidiary seconds; additional 12-hour display (2nd time zone), world time indicator on 3D globe with day/night indicator, power reserve indicator
Case: platinum, ø 43.5 mm, height 16.14 mm; sapphire crystal; transparent case back; water-resistant to 3 atm
Band: reptile skin, folding clasp
Price: CHF 550,000

STAY PRECISE, STAY TIMELESS

WOLF1834.COM

WATCH WINDERS | WATCH BOXES
JEWELLERY BOXES | TRAVEL ACCESSORIES

AXIS

H. MOSER & CIE.

H. Moser & Cie. has been making a name for itself in the industry as a serious watchmaker, though not averse to flashes of humor, like the Swiss Mad (sic) Watch made of Vacherin Mont d'Or cheese it presented at the SIHH 2017 (the cheese for the case is mixed with a hardening resin). And there's the Swiss Alp watch, made to look like an Apple Watch, but with all the essential Moser codes.

The company was originally founded by one Heinrich Moser (1805–1874) from Schaffhausen, where he served as "city watchmaker." He moved to Le Locle and, in 1825, founded his company at age twenty-one. Soon after, he moved to Saint Petersburg, Russia, where ambitious watchmakers were enjoying a good market. In 1828, H. Moser & Cie. was brought to life—a brand resuscitated in modern times by a group of investors and watch experts together with Moser's great-grandson, Roger Nicholas Balsiger.

With the help of CTO Jürgen Lange, H. Moser & Cie. has focused on the fundamentals. The company has made movements that contain a separate, removable escapement module supporting the pallet lever, escape wheel, and balance. The latter is fitted with the Straumann spring, made by Precision Engineering, another one of the Moser Group companies.

This small company has considerable technical know-how, which is probably what attracted MELB Holding, owners of Hautlence, and now majority owners of H. Moser shares. Under a new CEO, the brand set out to streamline and refocus its energy on the core look and feel: understatement, soft tones, and subtle technicity. They have been divided over three core collections named Endeavour, Venturer, and Pioneer, all with deceptively simple esthetics. The dials are kept "clean," in solid colors and with a minimum of distractions. The month on the Endeavour, for instance, is a small sweep hand ending in an arrowhead that points to the hours, which double as the months.

H. Moser & Cie.
Rundbuckstrasse 10
CH-8212 Neuhausen am Rheinfall
Switzerland

Tel.:
+41-52-674-0050

Fax:
+41-52-674-0055

E-mail:
info@h-moser.com

Website:
www.h-moser.com

Founded:
1828

Number of employees:
60

Annual production:
approx. 1,500 watches

U.S. distributor:
H. Moser & Cie. distributor:
Horology Works LLC
11 Flagg Road
West Hartford CT 06117
860-986-9676
mmargolis@horologyworks.com
Westime
254 North Rodeo Drive
90210
Beverly Hills, CA 90210
310-271-0000
info@westime.com

Most important collections/price range:
Endeavour / approx. $17,200 to $110,000;
Pioneer / approx. $11,900 to $49,900;
Venturer / approx. $19,500 to $100,000

Swiss Alp Watch Minute Repeater Tourbillon

Reference number: 5901-0200
Movement: manually wound, Caliber HMC 901; 30 × 35 mm, height 6.25 mm; 29 jewels; 21,600 vph; escapement with flying minutes-tourbillon, skeletonized bridges; 90-hour power reserve
Functions: hours, minutes; minute repeater
Case: white gold, 39.8 × 45.8 mm, height 11 mm; sapphire crystal; transparent case back
Band: reptile skin, buckle
Price: $292,000

Endeavour Tourbillon Concept

Reference number: 1804-0200
Movement: automatic, Moser Caliber HMC 804; ø 32 mm, height 5.5 mm; 21,600 vph; interchangeable escapement with flying 1-minute tourbillon, Straumann double hairspring, skeletonized bridges, oscillating mass in pink gold; 72-hour power reserve
Functions: hours, minutes
Case: white gold, ø 42 mm, height 11.6 mm; sapphire crystal; transparent case back
Band: antelope leather, folding clasp
Price: $69,000; limited to 20 pieces
Variations: smoky dial

Pioneer Centre Seconds

Reference number: 3200-1200
Movement: automatic, Moser Caliber HMC 200; ø 32 mm, height 5.5 mm; 18,000 vph; escapement with Straumann hairspring; 72-hour power reserve
Functions: hours, minutes, sweep seconds
Case: stainless steel, ø 42.8 mm, height 11.3 mm; sapphire crystal; transparent case back; water-resistant to 12 atm
Band: rubber, buckle
Price: $11,900

H. MOSER & CIE.

Endeavour Perpetual Calendar Purity Cosmic Green
Reference number: 1800-0202
Movement: manually wound, Moser Caliber HMC 800; ø 34 mm, height 6.3 mm; 32 jewels; 18,000 vph; escapement with Straumann hairspring, "flash calendar," correctable forward and backward; 168-hour power reserve
Functions: hours, minutes, subsidiary seconds; power reserve indicator; perpetual calendar with large date and small sweep month display, leap year (movement side)
Case: white gold, ø 42 mm, height 11.9 mm; sapphire crystal; transparent case back
Band: antelope leather, folding clasp
Price: $60,000; limited to 50 pieces
Variations: midnight blue smoky dial ($60,000)

Venturer Small Seconds XL
Reference number: 2327-0203
Movement: manually wound, Moser Caliber HMC 327; ø 32 mm, height 4.5 mm; 29 jewels; 18,000 vph; escapement with Straumann hairspring; 72-hour power reserve
Functions: hours, minutes, subsidiary seconds; power reserve indicator (movement side)
Case: white gold, ø 43 mm, height 12.6 mm; sapphire crystal; transparent case back
Band: antelope leather, buckle
Price: $23,800
Variations: pink gold ($25,500) or steel ($23,500)

Venturer Big Date
Reference number: 2100-0401
Movement: manually wound, Moser Caliber HMC 100; ø 34 mm, height 6.3 mm; 31 jewels; 18,000 vph; interchangeable escapement with Straumann hairspring; "flash calendar" correctable forward and backward; 168-hour power reserve
Functions: hours, minutes, subsidiary seconds; power reserve indicator (movement side); large date
Case: pink gold, ø 41.5 mm, height 14.5 mm; sapphire crystal; transparent case back
Band: reptile skin, buckle
Price: $29,000
Variations: various dials; white gold ($29,900)

Caliber HMC 341
Manually wound; exchangeable escape with beveled wheels, hardened gold pallet fork and escapement wheel; screw-mounted gold chatons; double spring barrel, 168-hour power reserve
Functions: hours, minutes, subsidiary seconds; power reserve indicator; perpetual calendar with large date and small sweep month hand, leap year (dial side)
Diameter: 34 mm
Height: 5.8 mm
Jewels: 28
Balance: glucydur with white gold screws
Frequency: 18,000 vph
Balance spring: Straumann with Breguet endcurve
Shock protection: Incabloc
Remarks: double-pull crown mechanism for easy switching of crown position

Caliber HMC 802
Automatic; exchangeable 1-minute tourbillon, double hairspring; single spring barrel, 72-hour power reserve
Functions: hours, minutes; additional 12-hour display (2nd time zone, display on request)
Diameter: 34 mm
Height: 6.5 mm
Jewels: 33
Balance: glucydur with white gold screws
Frequency: 21,600 vph
Balance spring: Straumann double hairspring
Shock protection: Incabloc
Remarks: rotor with gold oscillating mass

Caliber HMC 100
Manually wound; exchangeable escapement, gold pallet fork and escapement wheel; double mainspring barrel, 168-hour power reserve
Functions: hours, minutes, subsidiary seconds; power reserve indicator (on movement side); large date
Diameter: 34 mm
Height: 6.3 mm
Jewels: 31
Balance: glucydur with white gold screws
Frequency: 18,000 vph
Balance spring: Straumann with Breguet endcurve
Shock protection: Incabloc
Remarks: double-pull crown mechanism for easy switching of crown position

HABRING²

Fine mechanical works of art are created with smaller and larger complications in a small workshop in Austria's Völkermarkt, where the name Habring² stands for an unusual joint project. "We only come in a set," Maria Kristina Habring jokes. Her husband, Richard, adds with a grin, "You get double for your money here." The couple's first watch labeled with their own name came out in 2004: a simple, congenial three-handed watch based on a refined and unostentatiously decorated ETA pocket watch movement, the Unitas 6498-1. In connoisseur circles the news spread like wildfire that exceptional quality down to the smallest detail was hidden behind its inconspicuous specifications.

Since then, they have put their efforts into such projects as completely revamping the Time Only, powered by brand-new base movement Caliber A09. All the little details that differentiate this caliber are either especially commissioned or are made in-house. Caliber A09 is available both as a manually wound movement (A09M) and a bidirectionally wound automatic with an exclusive gear system. Its sporty version drives a pilot's watch. Also more or less in-house are the components of the Seconde Foudroyante. Because the drive needs a lot of energy, the foudroyante mechanism has been given its own spring barrel. In the Caliber A07F, the eighth of a second is driven by a gear train directly coupled with the movement without surrendering any reliability, power reserve, or amplitude.

For the twentieth anniversary of the IWC double chronograph, Habring² built a limited, improved edition. The movement, based on the ETA 7750 "Valjoux," was conceived in 1991/1992 with an additional module between the chronograph and automatic winder. And for the brand's tenth anniversary, in 2017, the company developed a new manually wound movement, quite a feat for such a small organization. This was already true for the previous generation of movements (A09), for which they only purchased a few parts from other vendors. In addition, rather than short-cut to modern materials, like silicon, and high-tech processes, the Habrings prefer to stay with traditional materials, like classic steel.

Habring Uhrentechnik OG
Hauptplatz 16
A-9100 Völkermarkt
Austria

Tel.:
+43-4232-51-300

Fax:
+43-4232-51-300-4

E-mail:
info@habring.com

Website:
www.habring2.com; www.habring.com

Founded:
1997

Number of employees:
3

Annual production:
200 watches

U.S. retailers:
Martin Pulli (USA-East)
215-508-4610
www.martinpulli.com
Passion Fine Jewelry (USA-West)
858-794-8000
www.passionfinejewelry.com

Most important collections/price range:
Felix / from $5,400; Jumping Second / from $6,300; Doppel 3 / from $9,000; Chrono COS / from $8,300

Doppel Felix

Reference number: Doppel Felix
Movement: manually wound, Habring Caliber A11R; ø 30 mm, height 8.4 mm; 27 jewels; 28,800 vph; Triovis fine adjustment; 48-hour power reserve
Functions: hours, minutes, subsidiary seconds; flyback chronograph
Case: stainless steel, ø 42 mm, height 13 mm; sapphire crystal; transparent case back; water-resistant to 5 atm
Band: calfskin, buckle
Price: $9,000
Variations: various dials

Doppel Felix Date

Reference number: Doppel Felix date
Movement: manually wound, Habring Caliber A11RD; ø 30 mm, height 8.4 mm; 27 jewels; 28,800 vph; Triovis fine adjustment; 48-hour power reserve
Functions: hours, minutes, subsidiary seconds; flyback chronograph; date
Case: stainless steel, ø 42 mm, height 13 mm; sapphire crystal; transparent case back; water-resistant to 5 atm
Band: calfskin, buckle
Price: $9,650
Variations: various dials

Doppel 3

Reference number: Doppel 3
Movement: manually wound, Habring Caliber A11R; ø 30 mm, height 8.4 mm; 27 jewels; 28,800 vph; Triovis fine adjustment; 48-hour power reserve
Functions: hours, minutes, subsidiary seconds; flyback chronograph
Case: stainless steel, ø 42 mm, height 13 mm; sapphire crystal; transparent case back; water-resistant to 5 atm
Band: calfskin, buckle
Price: $9,000
Variations: various dials

Felix

Reference number: Felix
Movement: manually wound, Habring Caliber A11B; ø 30 mm, height 4.2 mm; 18 jewels; 28,800 vph; Triovis fine adjustment, finely finished movement; 48-hour power reserve
Functions: hours, minutes, subsidiary seconds
Case: stainless steel, ø 38.5 mm, height 7 mm; sapphire crystal; transparent case back; water-resistant to 3 atm
Band: calfskin, buckle
Price: $5,400

Erwin

Reference number: Erwin
Movement: manually wound, Habring Caliber A11MS; ø 30 mm, height 5.3 mm; 21 jewels; 28,800 vph; Triovis fine adjustment, finely finished movement; 48-hour power reserve
Functions: hours, minutes, sweep seconds (jumping)
Case: stainless steel, ø 38.5 mm, height 9 mm; sapphire crystal; transparent case back; water-resistant to 3 atm
Band: calfskin, buckle
Price: $6,600

Jumping Second Pilot

Reference number: Jumping Second Pilot
Movement: manually wound, Habring Caliber A11MS; ø 36.6 mm, height 7 mm; 20 jewels; 28,800 vph; Triovis fine adjustment
Functions: hours, minutes, sweep seconds (jumping)
Case: stainless steel, ø 42 mm, height 13 mm; sapphire crystal; transparent case back
Band: calfskin, buckle
Price: $6,300
Variations: various dials; with automatic movement ($6,800)

Jumping Second Pilot Date

Reference number: Jumping Second Pilot Date
Movement: automatic, Habring Caliber A11SD; ø 36.6 mm, height 7.9 mm; 24 jewels; 28,800 vph; Triovis fine adjustment; 48-hour power reserve
Functions: hours, minutes, sweep seconds, (jumping); date
Case: stainless steel, ø 42 mm, height 13 mm; sapphire crystal; transparent case back
Band: calfskin, buckle
Price: $7,400

Foudroyante

Movement: manually wound, Habring Caliber A11MF; ø 30 mm, height 7 mm; 20 jewels; 28,800 vph; Triovis fine adjustment
Functions: hours, minutes, sweep seconds (jumping); eighth of a second display (flashing second or "foudroyante")
Case: stainless steel, ø 42 mm, height 13 mm; sapphire crystal; transparent case back; water-resistant to 5 atm
Band: calfskin, buckle
Price: $7,600
Variations: with automatic winding ($8,000); various dials

Repetition

Reference number: Repetition
Movement: manually wound, Habring Caliber 11B (base with Dubois Dépraz D90 module); ø 36 mm, height 7.85 mm; 18 jewels; 28,800 vph; Triovis fine regulation, finely finished movement; 48-hour power reserve
Functions: hours, minutes, subsidiary seconds; five-minute repeater
Case: titanium, ø 42 mm, height 13.5 mm; sapphire crystal; transparent case back
Band: calfskin, buckle
Price: $21,500

HAGER WATCHES

Keeping it simple and smart is a Hager specialty. The company, owned and operated by American service veteran Pierre "Pete" Brown, is named after the town where the company was started in 2009. The business model is equally streamlined: create high-quality and affordable automatic watches accessible to those who have never experienced the joy of owning a mechanical watch. The look: rugged and refined, for individuals with a bit of adventure in their bones.

The timepieces are designed by Brown and his small team in Hagerstown. All the cues are there for the watch connoisseur, the brushed and polished cases with beveled edges, two-tiered stadium dial, brass markers and hands outlined in black and coated with Superluminova, domed sapphire crystal. Also enhanced with Superluminova are 120-click and 24-click GMT ceramic bezels. The cases are rated a sporty 20 atm, meaning the timepieces are good for more than just washing the dishes. Inside them beats one of a variety of automatic winding Swiss and Japanese mechanical movements that are both installed and regulated in the United States. The goal is to balance quality and affordability. "Although we assemble on-site, we use components from third-party suppliers," says Brown. "We have several suppliers that we work with in Switzerland and Hong Kong that supply us with the best components that I'm seeking." The spirit of elegant adventure is in each of the collections, be it the tough-looking Commando Professional, the Traveler GMT with ceramic inserts and DLC coating, or the classic Flieger (pilot's watch).

Brown is well aware of the foibles of the watch industry, one major complaint being customer service. His personal boast since launching the company is that Hager has never charged a customer for a repair yet, even when it's clear that the customer is at fault. "We aren't just selling watches," says Brown, "we are selling the experience of owning a luxury timepiece. That's not to say that at some point we may have to reverse this because of overall costs, but it's been a hallmark of our brand and it builds brand loyalty."

Hager Watches
36 South Potomac Street
Suite 204
Hagerstown, MD 21740

Tel.:
240-329-0071

E-mail:
info@hagerwatches.com

Website:
www.hagerwatches.com

Founded:
2009

Number of employees:
2

Annual production:
500–1,000 watches

Most important collections/price range:
Commando Professional, GMT Traveler, Skymaster, Flieger, Diplomat / $350 to $3,525

U2 Chronograph
Reference number: 0519G
Movement: quartz; Seiko VK64 "Meca-Quartz," ø 29 mm; height 7.9 mm
Functions: hours, minutes; chronograph with 60-minute and 24-hour subsidiary dials; date
Case: stainless steel, ø 41 mm, height 14.7 mm; sapphire crystal; screwed-down case back; tachymeter scale on bezel; water-resistant to 20 atm
Band: stainless steel, buckle
Price: $650

Diplomat
Reference number: 0389G
Movement: automatic, Soprod Caliber C110, ø 25.60 mm; height 5.1 mm; 29 jewels; 28,800 vph; 42-hour power reserve
Functions: hours, minutes, subsidiary seconds
Case: stainless steel, ø 40 mm, height 12.8 mm; screw-in crown; unilateral bezel with black stainless steel insert; sapphire crystal; screwed-down case back; water-resistant to 30 atm
Band: calfskin, folding clasp
Price: $750
Variations: various dials and case colors

GMT Traveler Blue Insert
Reference number: 16738
Movement: automatic, HGR 60 (base ETA 2836); ø 25.6 mm; height 5.67 mm; 26 jewels; 28,800 vph; 40-hour power reserve
Functions: hours, minutes, sweep seconds; date; 2nd time zone
Case: stainless steel with DLC coating, ø 42 mm, height 14.5 mm; screw-in crown; bidirectional bezel with black ceramic insert; sapphire crystal; screwed-down case back; water-resistant to 20 atm
Band: stainless steel, buckle
Price: $700
Variations: black ceramic insert ($700); DLC with black ceramic insert ($800); DLC with blue ceramic insert ($800)

Hamilton International Ltd.
Mattenstrasse 149
CH-2503 Biel/Bienne
Switzerland

Tel.:
+41-32-343-4004

Fax:
+41-32-343-4006

E-mail:
info@hamiltonwatch.com

Website:
www.hamiltonwatch.com
https://shop.hamiltonwatch.com/

Founded:
1892

U.S. distributor:
Hamilton
Swatch Group (US), Inc.
703 Waterford Way, Suite 450
Miami, FL 33126
800-234-8463
Chelsea.pillsbury@swatchgroup.com

Price range:
between approx. $500 and $2,500

HAMILTON

The Hamilton Watch Co. was founded in 1892 in Lancaster, Pennsylvania, and, within a very brief period, grew into one of the world's largest *manufactures*. Around the turn of the twentieth century, every second railway employee in the United States was carrying a Hamilton watch in his pocket, not only to make sure the trains were running punctually, but also to assist in coordinating them and organizing schedules. And during World War II, the American army officers' kits included a service Hamilton.

Hamilton is the sole survivor of the large U.S. watchmakers—if only as a brand within the Swiss Swatch Group. At one time, Hamilton had itself owned a piece of the Swiss watchmaking industry in the form of the Büren brand in the 1960s and 1970s. As part of a joint venture with Heuer-Leonidas, Breitling, and Dubois Dépraz, Hamilton-Büren also made a significant contribution to the development of the automatic chronograph. Just prior in its history, the tuning fork watch pioneer was all the rage when it took the new movement technology and housed it in a modern case created by renowned industrial designer Richard Arbib. The triangular Ventura hit the watch-world ground running in 1957, in what was truly a frenzy of innovation that benefited the brand especially in the U.S. market. The American spirit of freedom and belief in progress this model embodies, something evoked in Hamilton's current marketing, are taken quite seriously by its designers—even those working in Biel, Switzerland. Today's collections are more inspired by adventure and aviator watches. The brand also continues to focus on revamped remakes of its classics, and prices have come down a bit as a reaction to the unstable markets.

Khaki X-Wind Auto Chrono Limited Edition
Reference number: H77796535
Movement: automatic, Hamilton Caliber H-21 Si (base ETA 7750); ø 30 mm, height 7.9 mm; 25 jewels; 28,800 vph; silicon hairspring; 60-hour power reserve; COSC-certified chronometer
Functions: hours, minutes, subsidiary seconds; chronograph; date, weekday
Case: stainless steel, ø 45 mm, height 14.85 mm; crown-controlled scale ring, with slide rule to calculate drift angle with side winds; sapphire crystal; transparent case back; screw-in crown; water-resistant to 10 atm
Band: calfskin, buckle
Price: $2,595

Khaki Pilot Day Date
Reference number: H64605531
Movement: automatic, Hamilton Caliber H-40 (base ETA 2836-2); ø 25.6 mm, height 5.05 mm; 25 jewels; 21,600 vph; 80-hour power reserve
Functions: hours, minutes, sweep seconds; date, weekday
Case: stainless steel with brown PVD coating, ø 42 mm, height 12.15 mm; sapphire crystal; water-resistant to 10 atm
Band: calfskin, buckle
Price: $995

Khaki Pilot Day Date
Reference number: H64645131
Movement: automatic, Hamilton Caliber H-40 (base ETA 2836-2); ø 25.6 mm, height 5.05 mm; 25 jewels; 21,600 vph; 80-hour power reserve
Functions: hours, minutes, sweep seconds; date, weekday
Case: stainless steel, ø 42 mm, height 12.15 mm; sapphire crystal; water-resistant to 10 atm
Band: stainless steel, folding clasp
Price: $945

HAMILTON

Khaki Pilot Chrono Quartz
Reference number: H76712751
Movement: quartz
Functions: hours, minutes, subsidiary seconds; chronograph; date
Case: stainless steel, ø 44 mm, height 11.05 mm; sapphire crystal; water-resistant to 10 atm
Band: calfskin, buckle
Price: $895

Khaki Navy Scuba
Reference number: H82315331
Movement: automatic, Hamilton Caliber H-10 (base ETA 2824-2); ø 25.6 mm, height 4.6 mm; 25 jewels; 21,600 vph; 80-hour power reserve
Functions: hours, minutes, sweep seconds; date
Case: stainless steel, ø 40 mm, height 12.95 mm; unidirectional bezel, with 0-60 scale; sapphire crystal; transparent case back; screw-in crown; water-resistant to 10 atm
Band: rubber, buckle
Price: $745

Khaki Field 50 mm
Reference number: H69809730
Movement: manually wound, Hamilton Caliber H-50; ø 25.6 mm, height 3.37 mm; 17 jewels; 21,600 vph; 80-hour power reserve
Functions: hours, minutes, sweep seconds; date
Case: stainless steel with black PVD coating, ø 50 mm, height 12.55 mm; sapphire crystal; water-resistant to 10 atm
Band: calfskin, buckle
Price: $1,245

Khaki Field Day Date
Reference number: H70535081
Movement: automatic, Hamilton Caliber H-30 (base ETA 2834-2); ø 25.6 mm, height 5.05 mm; 25 jewels; 21,600 vph; 80-hour power reserve
Functions: hours, minutes, sweep seconds; date, weekday
Case: stainless steel, ø 42 mm, height 11.6 mm; sapphire crystal; water-resistant to 10 atm
Band: textile, buckle
Price: $895

Khaki Field Mechanical
Reference number: H69429901
Movement: manually wound, ETA Caliber 2801-2; ø 25.6 mm, height 3.35 mm; 17 jewels; 28,800 vph; 42-hour power reserve
Functions: hours, minutes, sweep seconds
Case: stainless steel, ø 38 mm, height 9.5 mm; sapphire crystal; water-resistant to 5 atm
Band: textile, buckle
Price: $475

Jazzmaster Thinline
Reference number: H38525881
Movement: automatic, ETA Caliber 2892-A2; ø 25.6 mm, height 3.6 mm; 21 jewels; 28,800 vph; 42-hour power reserve
Functions: hours, minutes; date
Case: stainless steel, ø 40 mm, height 8.45 mm; sapphire crystal; transparent case back; water-resistant to 5 atm
Band: calfskin, buckle
Price: $945

HANHART

Hanhart 1882 GmbH
Hauptstrasse 33
D-78148 Gütenbach
Germany

Tel.:
+49-7723-93-44-0

Fax:
+49-7723-93-44-40

E-mail:
info@hanhart.com

Website:
www.hanhart.com

Founded:
1882 in Diessenhofen, Switzerland;
in Germany since 1902

Number of employees:
22

Annual production:
approx. 1,000 chronographs and 30,000 stopwatches

U.S. distributor:
BluePointe, LLC
207 W. Millbrook Road
Raleigh, NC 27609
(888) 333-4895

Most important collections/price range:
Mechanical stopwatches / from approx. $600;
Pioneer / from approx. $1,090; Primus / from approx. $2,600

The reputation of this rather special company really goes back to the twenties and thirties. At the time, the brand manufactured affordable and robust stopwatches, pocket watches, and chronograph wristwatches. These core timepieces were what the fans of instrument watches wanted, and so they were thrilled as the company slowly abandoned its quartz dabbling of the eighties and reset its sights on the brand's rich and honorable tradition. A new collection was in the wings, raising expectations of great things to come. Support by the shareholding Gaydoul Group provided the financial backbone to get things moving.

Hanhart managed to rebuild a name for itself with a foot in Switzerland and the other in Germany, but it began to drift after the 2009 recession. Following bankruptcy, the company reorganized under the name Hanhart 1822 GmbH and moved everything to its German hometown. It has also returned to its stylistic roots: The characteristic red start/stop pusher graces the new collections, even on the bi-compax chronos of the Racemasters, which come with a smooth bezel. Pilots' chronographs have never lost any of their charm, either, and Hanhart was already making them in the 1930s, notably the Caliber 41 and the Tachy Tele, with asymmetrical pushers and the typical red pusher. These timepieces have to survive extreme conditions, like shocks and severe temperature fluctuations. Obviously, the company has supported a number of personalities in power sports—people like Artur Kielak, the world's number one aerobatic pilot, who flies with an XtremeAir XA41. Hanhart's long tradition and expertise with flyers' chronographs struck a chord with the Austrian Army. It ordered a special edition of the Primus series decorated with the coat of arms of the Austrian Air Force on the dial and certified by the military.

Pioneer One

Reference number: 762.210
Movement: automatic, Sellita Caliber SW200-1; ø 25.6 mm, height 4.6 mm; 26 jewels; 28,800 vph; 38-hour power reserve
Functions: hours, minutes, sweep seconds; date
Case: stainless steel, ø 42 mm, height 12 mm; bidirectional bezel with reference markers; sapphire crystal; water-resistant to 10 atm
Band: calfskin, buckle
Price: $890

Primus Monochrome Pilot

Reference number: 740.240
Movement: automatic, Caliber HAN3809 (base ETA 7750); ø 30 mm, height 10.4 mm; 28 jewels; 28,800 vph; 42-hour power reserve
Functions: hours, minutes, subsidiary seconds; chronograph; date
Case: stainless steel, ø 44 mm, height 15 mm; sapphire crystal; transparent case back; screw-in crown; water-resistant to 10 atm
Band: textile, folding clasp
Remarks: movable lugs
Price: $2,970
Variations: various bands and dials

Primus Race Winner "Black Falcon"

Reference number: 741.511
Movement: automatic, Caliber HAN3809 (base ETA 7750); ø 30 mm, height 10.4 mm; 28 jewels; 28,800 vph; 42-hour power reserve
Functions: hours, minutes, subsidiary seconds; chronograph; date
Case: stainless steel with black DLC coating, ø 44 mm, height 15 mm; sapphire crystal; transparent case back; screw-in crown; water-resistant to 10 atm
Band: rubber, folding clasp
Remarks: movable lugs
Price: $3,580; limited to 111 pieces

HARRY WINSTON

Swatch Group's purchase of the luxury brand Harry Winston in early 2013 for $1 billion came as something of a surprise. But considering the upward flow of money worldwide, banking on a proven high-end luxury brand would seem obvious. On his many travels, founder Harry Winston (1896–1978) bought, recut, and set some of the twentieth century's greatest gems. He was succeeded by his son Ronald, a gifted craftsman himself with several patents in precious metals processing.

It was Ronald who added watches to the company's portfolio, inaugurating two lines: one showcasing the finest precious gems to dovetail with the company's overall focus and one containing clever, complicated timepieces. The result was the stunning Opus line launched by Harry Winston Rare Timepieces. Each of the thirteen models has been developed in conjunction with one exceptional independent watchmaker per year in very small series and containing an exclusive *manufacture* movement. The roster of artist-engineers who have participated reads like a *Who's Who* of independent watchmaking, including François-Paul Journe, Vianney Halter, Felix Baumgartner, and Greubel Forsey all the way to Denis Giguet and Emmanuel Bouchet.

With Nayla Hayek, daughter of Swatch founder Nicolas Hayek, as CEO, the brand has been benefiting from synergies with other Swatch brands, notably Blancpain. While the Opus line has become dormant, Harry Winston continues to explore extreme complications in its Histoire de Tourbillon and Z (for the alloy Zalium) collections. At the same time, jewel watches, some with mechanical movements, continue to conquer wrists, especially of women. The Travel Time brings luxury to the traveler's bedside table.

Harry Winston
701 Fifth Avenue
New York, NY 10022
Tel.: 212-399-1000

Website:
www.harrywinston.com

Founded:
1932

Most important collections:
Avenue, Emerald, Midnight, Premier, Ocean, Opus, Project Z, Histoire de Tourbillon

Histoire de Tourbillon 9
Reference number: HCOMTT47WW001
Movement: manually wound, Harry Winston Caliber HW4504; ø 40.4, height 10.6 mm (without tourbillon); 58 jewels; 21,600 vph; triaxial tourbillon (45, 75, and 300 seconds rotation), variable inertia balance with gold adjustment screws; Phillips end curve, 2 serially coupled spring barrels, 50-hour power reserve
Functions: hours (retrograde), minutes (retrograde dragging); power reserve indicator
Case: white gold; ø 46.5, height 20.84 mm; white gold and rubber crown; different sapphire crystals on hours and tourbillon; transparent case back; water-resistant to 3 atm
Band: reptile skin, white gold buckle
Price: on request; limited to 20 pieces

Harry Winston Midnight Stalactites Automatic 36 mm
Reference number: MIDAHM36RR001
Movement: automatic, Harry Winston Caliber HW2008; ø 26.2 mm, height 3.37 mm; 28 jewels; 28,800 vph; balance spring, skeletonized rotor in white gold, fine finishing with circular côtes de Genève
Functions: hours, minutes (off-center); date (retrograde); moon phase
Case: white gold; ø 36 mm, height 9 mm; sapphire crystal; transparent case back; water-resistant to 3 atm
Remarks: 84 diamonds on case, mother-of-pearl dial with 161 brilliant-cut diamonds in "stalactite" décor
Band: reptile skin, buckle
Price: $34,400
Variations: pink gold

Project Z12
Reference number: OCEAHR42ZZ001
Movement: automatic, Harry Winston Caliber HW3306; ø 34 mm, height 5.37 mm; 35 jewels; 28,800 vph; silicon hairspring, skeletonized rotor in white gold, fine finishing with circular côtes de Genève
Functions: hours, minutes retrograde; date
Case: Zalium, ø 44.2 mm, height 11.27 mm; sapphire crystal; transparent case back; water-resistant to 10 atm
Band: calfskin with denim effect, buckle
Price: $24,800; limited to 300 pieces

HARRY WINSTON

Ocean Biretrograde Automatic 36mm

Reference number: OCEABI36WW035
Movement: automatic, Harry Winston Caliber HW3302, ø 30 mm, height 4.57 mm; 35 jewels, 28,800 vph; silicon hairspring, skeletonized rotor in white gold, fine finishing with circular côtes de Genève
Functions: hours, minutes off-center, retrograde seconds; retrograde date
Case: rose gold, ø 40 mm, height 9.5 mm; bezel set with 57 brilliant-cut diamonds, sapphire crystal; diamond on crown; water-resistant to 10 atm
Band: reptile skin, buckle, set with 42 diamonds
Remarks: burgundy mother-of-pearl dial set with 97 tourmalines, 71 diamonds
Price: $51,100; **Variations:** blue dial

Ocean 20th Anniversary Biretrograde Automatic 36mm

Reference number: OCEABI36WW052
Movement: automatic, Harry Winston Caliber HW3302, ø 30 mm, height 4.57 mm; 35 jewels, 28,800 vph; silicon hairspring, skeletonized rotor in white gold, fine finishing with circular côtes de Genève
Functions: hours, minutes off-center, retrograde seconds; retrograde date
Case: white gold, ø 36 mm, height 9.45 mm; bezel set with 57 brilliant-cut diamonds, sapphire crystal; diamond on crown; water-resistant to 10 atm
Band: reptile skin, buckle, set with 42 diamonds
Remarks: burgundy mother-of-pearl dial set with 32 diamonds
Price: on request; **Variations:** blue or burgundy dial

Premier Precious Butterfly Automatic 36 mm

Reference number: PRNAHM36WW004
Movement: automatic, Harry Winston Caliber HW2008; ø 26.2 mm, height 3.37 mm; 28 jewels; 28,800 vph; balance spring, skeletonized rotor in white gold, fine finishing with circular côtes de Genève, rhodium plating
Functions: hours, minutes
Case: white gold, ø 36 mm, height 8.4 mm; bezel set with 57 brilliant-cut diamonds, sapphire crystal; diamond on crown; water-resistant to 3 atm
Band: satin, buckle set with 17 diamonds
Remarks: dial with "Chrysiridia Madagascariensis" butterfly marquetry
Price: $42,500

Harry Winston Avenue Dual Time Automatic

Reference number: AVEATZ37RR001
Movement: automatic, Harry Winston Caliber HW3502; ø 32 mm, height 5.2; 32 jewels; 28,800 vph; flat silicon spring, white gold rotor, with côtes de Genève; circular grain, beveled bridges
Functions: hours, minutes; 2nd time zone (retrograde); day/night indicator; date
Case: Sedna gold, 53.8 × 35.8 mm, height 10.7 mm; sapphire crystal; transparent case back; water-resistant to 3 atm
Remarks: smoky white sapphire crystal dial, emerald appliques
Band: reptile skin, Sedna gold buckle
Price: $38,300
Variations: Zalium, band with Zalium buckle ($22,200)

Travel Time by Harry Winston

Reference number: HJTQAL66WW001
Movement: quartz, Harry Winston Caliber HW5301; ø 23.9, height 4.1 mm
Functions: hours, minutes
Case: white gold, 66 × 39 mm, height 13 mm; set with 168 brilliant-cut diamonds, sapphire crystal; diamond on crown; water-resistant to 3 atm
Price: on request

Harry Winston Avenue C Art Deco

Reference number: AVCQHM19WW130
Movement: quartz
Functions: hours, minutes
Case: white gold, 19 × 39.5 mm, height 7.8 mm; bezel set with 43 brilliant-cut diamonds, sapphire crystal; diamond on crown; water-resistant to 3 atm
Band: reptile skin, folding clasp
Price: $34,800
Variations: rose gold; various bands and dials

HERMÈS

La Montre Hermès
Erlenstrasse 31A
CH-2555 Brügg
Switzerland

Tel.:
+41-32-366-7100

Fax:
+41-32-366-7101

E-mail:
info@montre-hermes.ch

Website:
www.hermes.ch

Founded:
1978

Number of employees:
150

Annual production:
not specified

U.S. distributor:
Hermès of Paris, Inc.
55 East 59th Street
New York, NY 10022
800-441-4488
www.hermes.com

Most important collections/price range:
Arceau, Cape Cod, Clipper, Dressage, Faubourg, Heure H, Klikti, Kelly, Medor, Slim / $2,400 to $500,000

Thierry Hermès's timing was just right. When he founded his saddlery in Paris in 1837, France's middle class was booming and spending money on beautiful things and activities like horseback riding. Hermès became a household name and a symbol of good taste—not too flashy, not trendy, useful. The advent of the automobile gave rise to luggage, bags, headgear, and soon Hermès, still in family hands today, diversified its range of products—foulards, fashion, porcelain, glass, perfume, and gold jewelry are active parts of its portfolio.

Watches were a natural, especially with the advent of the wristwatch in the years prior to World War I. Hermès even had a timepiece that could be worn on a belt. But some time passed before the company engaged in "real" watchmaking. In 1978, La Montre Hermès opened its watch manufactory in Biel.

Rather than just produce fluffy lifestyle timepieces, Hermès has gone to the trouble to get an in-depth grip on the business. "Our philosophy is all about the quality of time," says Laurent Dordet, who took over as CEO from Luc Perramond in March 2015. "It's about imagination; we want people to dream." "Poetic complications" is what allowed the company to navigate between classy but plain watches and muscular tool timepieces bristling with complications. On the one hand there was the esthetics: the lively leaning numerals of the Arceau series or the bridoon recalling the company's equine roots at 12 o'clock for holding the strap. As for in-house complications, they are produced in collaboration with external designers, notably Jean-Marc Wiederrecht and his company, Agenhor. In the Slim line, one finds a thin perpetual calendar with modern numerals that raise it above the standard retro watch. The clever "Temps Suspendu" lets the wearer stop time for a moment. The "Heure Masquée" hides the hour hand behind hand movements with varying speeds, an option to "suspend" time for a moment, or one to hide time.

Slim d'Hermès GMT
Reference number: CA5.860.230/MM88
Movement: automatic, Hermès Caliber H1950 with GMT module; ø 33.6 mm, height 4 mm; 38 jewels; 21,600 vph; microrotor; 42-hour power reserve
Functions: hours, minutes, date; additional 12-hour display (2nd time zone)
Case: palladium, ø 39.5 mm, height 9.48 mm; sapphire crystal; transparent case back; water-resistant to 3 atm
Band: reptile skin, buckle
Price: $14,700; limited to 120 pieces

Arceau "Le Temps Suspendu"
Reference number: AR8.97A.222/MM41
Movement: automatic, ETA Caliber 2892 (modified); ø 26 mm, height 5.6 mm; 28,800 vph; hands can be parked at 12:30 and then started again at the current time with a pusher; double spring barrel; 42-hour power reserve
Functions: hours, minutes; date (retrograde)
Case: rose gold, ø 43 mm; sapphire crystal; transparent case back; water-resistant to 3 atm
Band: reptile skin, folding clasp
Price: $45,900
Variations: stainless steel ($21,750)

Slim d'Hermès L'Heure Impatiente
Reference number: CA4.870.220/MM7K
Movement: automatic, Hermès Caliber H1912 (base with "L'Heure Impatiente" module); ø 31.96 mm, height 5.9 mm; 36 jewels; 28,800 vph; mainplate and bridges with snail and côtes de Genève decoration; 50-hour power reserve
Functions: hours, minutes; "L'Heure Impatiente" alarm with 60-minute countdown
Case: rose gold, ø 40.5 mm, height 10.67 mm; sapphire crystal; transparent case back; water-resistant to 3 atm
Band: reptile skin, buckle
Price: $39,900

Carré H
Reference number: TI2.710.230/VB34
Movement: automatic, Hermès Caliber H1912; ø 23.3 mm, height 3.7 mm; 28 jewels; 28,800 vph; 50-hour power reserve
Functions: hours, minutes, sweep seconds
Case: stainless steel, 38 × 38 mm, height 10.8 mm; sapphire crystal; transparent case back; water-resistant to 3 atm
Band: calfskin, buckle
Price: $7,125

Arceau Chronographe Titane
Reference number: AR3.941.330/VB34
Movement: automatic, ETA Caliber 2894-2; ø 28.6 mm, height 6.1 mm; 37 jewels; 28,800 vph; 42-hour power reserve
Functions: hours, minutes, subsidiary seconds; chronograph; date
Case: titanium, ø 41 mm, height 14.1 mm; sapphire crystal; water-resistant to 3 atm
Band: calfskin, buckle
Price: $4,950

Cape Cod Chaîne d'Ancre
Movement: quartz
Functions: hours, minutes
Case: stainless steel, 29 × 29 mm, height 8.8 mm; sapphire crystal; water-resistant to 3 atm
Band: calfskin, buckle
Price: $3,500

Nantucket TPM "Jeté de Diamants"
Reference number: NA2.131.221/MM18
Movement: quartz
Functions: hours, minutes
Case: stainless steel, set with 55 diamonds, 17 × 23 mm, height 8.5 mm; sapphire crystal; water-resistant to 3 atm
Band: reptile skin, buckle
Price: $5,675
Variations: various strap colors

Carré Cuir
Reference number: CU2.210.330/LL89
Movement: quartz
Functions: hours, minutes
Case: stainless steel, 24 × 24 mm, height 8.5 mm; sapphire crystal; water-resistant to 3 atm
Band: lizard leather, buckle
Remarks: lizard leather dial; the Carré Cuir is available only in Europe
Price: on request
Variations: set with diamonds

Klikti
Reference number: LL1.292.130/ZZ951
Movement: quartz
Functions: hours, minutes
Case: white gold, 17 × 16 mm, height 6.6 mm; sapphire crystal; water-resistant to 3 atm
Band: reptile skin with white gold elements, buckle
Remarks: case and strap parts set with a total of 178 diamonds
Price: $38,200
Variations: rose gold

HUBLOT

Ever since Hublot moved into a new, modern, spacious factory building in Nyon, near Geneva—in the midst of a recession, no less—the brand has evolved with stunning speed. The growth has been such that Hublot has built a second factory, which is even bigger than the first. The groundbreaking ceremony took place on March 3, 2014, and the man holding the spade was Hublot chairman Jean-Claude Biver, who is also head of LVMH's Watch Division.

Hublot grew and continues to grow thanks to a combination of innovative watchmaking and extremely vigorous communication. It was together with current CEO Ricardo Guadalupe that Biver developed the idea of fusing different and at times incompatible materials in a watch: carbon composite and gold, ceramic and steel, denim and diamonds. In 2011, the brand introduced the first scratchproof precious metal, an alloy of gold and ceramic named "Magic Gold." In 2014, Hublot came out with a watch whose dial is made of osmium, one of the world's rarest metals. Using a new patented process, Hublot has also implemented a unique concept of cutting wafer-thin bits of glass that are set in the open spaces of a skeletonized movement plate.

The "art of fusion" tagline drove the brand into all sorts of technical and scientific partnerships and created a phenomenal buzz that is ongoing, apparently, regardless of the economic environment. Even fashion house Sartoria Rubinacci was picked up by the Hublot radar for a Classic Fusion Chronograph with a hound's-tooth pattern. And during the past decade, Hublot also managed to verticalize. Most of the brand's models are now run on the in-house Unico movements.

Hublot's communication concept also needs mentioning. It consists in being everywhere including on soccer pitches, a sport that was considered too popular for an industry that prides itself on exclusiveness. This particular love affair began in 2006, when the brand sponsored the Swiss national team. It has since signed on numerous other clubs, including the FC Bayern München, Manchester United, and even the lesser-known San Lorenzo de Almagro in Argentina.

Hublot SA
Chemin de la Vuarpillière 33
CH-1260 Nyon
Switzerland

Tel.:
+41-22-990-9000

E-mail:
info@hublot.ch

Website:
www.hublot.com

Founded:
1980

Number of employees:
approx. 600

Annual production:
approx. 50,000 watches

U.S. distributor:
Hublot of America, Inc.
100 N. Biscayne Blvd., Suite 1900
Miami, FL 33132
786-405-8677

Most important collections/price range:
Big Bang / $11,000 to $1,053,000; Classic Fusion / $5,200 to $474,000; Manufacture Piece (MP) / $82,000 to $579,000

Techframe Ferrari Tourbillon Chronograph Sapphire White Gold

Reference number: 408.JW.0123.RX
Movement: manually wound, Caliber HUB 6311; ø 34 mm, height 6.7 mm; 27 jewels; 21,600 vph; 1-minute tourbillon; monopusher for control of chronograph functions; 115-hour power reserve
Functions: hours, minutes; chronograph
Case: white gold, affixed to sapphire crystal inner case, ø 45 mm, height 14.8 mm; sapphire crystal; transparent case back; water-resistant to 3 atm
Band: rubber, folding clasp
Remarks: designed in collaboration with the Ferrari Design Center
Price: $179,000; limited to 70 pieces
Variations: carbon ($137,000)

Big Bang Unico Ferrari Magic Gold

Reference number: 402.MX.0138.WR
Movement: automatic, Caliber HUB 1241 "Unico"; ø 30 mm, height 8.05 mm; 38 jewels; 28,800 vph; black-coated mainplate and bridges; 72-hour power reserve
Functions: hours, minutes, subsidiary seconds; flyback chronograph; date
Case: "Magic Gold" (alloy with some ceramic), ø 45 mm, height 16.7 mm; bezel screwed to case with 6 titanium screws; sapphire crystal; transparent case back; water-resistant to 10 atm
Band: calfskin, folding clasp
Remarks: Ferrari design
Price: $36,700; limited to 250 pieces
Variations: carbon ($29,400; limited to 500 pieces)

Big Bang Unico Baguette Blue Sapphire

Reference number: 411.JL.4809.RT.1901
Movement: automatic, Caliber HUB 1242 "Unico"; ø 30 mm, height 8.05 mm; 38 jewels; 28,800 vph; anthracite-coated mainplate and bridges; 72-hour power reserve
Functions: hours, minutes, subsidiary seconds; flyback chronograph; date
Case: sapphire, ø 45 mm, height 15.85 mm; bezel set with baguette diamonds and screwed to case with 6 titanium screws; sapphire crystal; transparent case back; water-resistant to 5 atm
Band: rubber, folding clasp
Price: $110,000
Variations: set with red stones ($110,000)

Big Bang Unico GMT Titanium

Reference number: 471.NX.7112.RX
Movement: automatic, Caliber HUB 1251 "Unico"; ø 30 mm, height 8.05 mm; 41 jewels; 28,800 vph; anthracite-coated mainplate and bridges; 72-hour power reserve
Functions: hours, minutes, sweep seconds; additional 12-hour display (2nd time zone), day/night indicator
Case: titanium, ø 45 mm, height 15.85 mm; bezel screwed to case with 6 titanium screws; sapphire crystal; transparent case back; water-resistant to 10 atm
Band: rubber, folding clasp
Price: $19,900
Variations: carbon ($23,100)

Big Bang Unico Sang Bleu Titanium Blue

Reference number: 415.NX.7179.VR.MXM18
Movement: automatic, Caliber HUB 1213 "Unico"; ø 30 mm, height 8.05 mm; 28 jewels; 28,800 vph; 72-hour power reserve
Functions: hours, minutes, sweep seconds (disk display with corner pointers)
Case: titanium, ø 45 mm, height 15.55 mm; bezel screwed to case with 6 titanium screws; sapphire crystal; transparent case back; water-resistant to 10 atm
Band: calfskin, folding clasp
Price: $18,800; limited to 200 pieces
Variations: various cases

Big Bang Meca-10 Ceramic Blue

Reference number: 414.EX.5123.RX
Movement: manually wound, Caliber HUB 1201; ø 34.8 mm, height 6.8 mm; 24 jewels; 21,600 vph; bridges with blue coating; 240-hour power reserve
Functions: hours, minutes, subsidiary seconds; power reserve indicator
Case: ceramic, ø 45 mm, height 15.95 mm; bezel screwed to case with 6 titanium screws; sapphire crystal; transparent case back; water-resistant to 10 atm
Band: rubber, folding clasp
Price: $22,000
Variations: rose gold ($40,900)

Big Bang Meca-10 Titanium

Reference number: 414.NI.1123.RX
Movement: manually wound, Caliber HUB 1201; ø 34.8 mm, height 6.8 mm; 24 jewels; 21,600 vph; bridges with black coating; 240-hour power reserve
Functions: hours, minutes, subsidiary seconds; power reserve indicator
Case: titanium, ø 45 mm, height 15.95 mm; bezel screwed to case with 6 titanium screws; sapphire crystal; transparent case back; water-resistant to 10 atm
Band: rubber, folding clasp
Price: $19,900
Variations: various cases

Big Bang Unico Golf Carbon

Reference number: 416.YS.1120.VR
Movement: automatic, Caliber HUB 1580; ø 30 mm, height 8.05 mm; 43 jewels; 28,800 vph; 72-hour power reserve
Functions: hours, minutes; golf counter for strokes per hole, total strokes and holes
Case: carbon and texalium, ø 45 mm, height 18.1 mm; bezel screwed to case with 6 titanium screws; sapphire crystal; transparent case back; water-resistant to 10 atm
Band: rubber calfskin overlay, folding clasp
Price: $31,500

Big Bang Unico King Gold Ceramic

Reference number: 441.OM.1180.RX
Movement: automatic, Caliber HUB 1280; ø 30 mm, height 8.05 mm; 43 jewels; 28,800 vph; mainplate and bridges with gray coating; 72-hour power reserve
Functions: hours, minutes, subsidiary seconds; flyback chronograph; date
Case: rose gold, ø 42 mm, height 14.5 mm; ceramic bezel screwed to case with 6 titanium screws; sapphire crystal; transparent case back; water-resistant to 10 atm
Band: rubber, folding clasp
Price: $33,600
Variations: various cases

HUBLOT

Big Bang Steel Ceramic
Reference number: 341.SB.131.RX
Movement: automatic, Caliber HUB 4300 (base ETA 2894-2); ø 30 mm, height 6.9 mm; 37 jewels; 28,800 vph; 42-hour power reserve
Functions: hours, minutes, subsidiary seconds; chronograph; date
Case: stainless steel, ø 41 mm, height 12.75 mm; ceramic bezel screwed to case with 6 titanium screws; sapphire crystal; transparent case back; water-resistant to 10 atm
Band: rubber, folding clasp
Price: $12,500
Variations: various cases

Classic Fusion Chronograph Italia Independent "Prince of Wales"
Reference number: 521.OX.2709.NR.ITI18
Movement: automatic, Caliber HUB 1143; ø 30 mm, height 6.9 mm; 59 jewels; 28,800 vph; 42-hour power reserve
Functions: hours, minutes, subsidiary seconds; chronograph; date
Case: rose gold, ø 45 mm, height 13.05 mm; bezel screwed to case with 6 titanium screws; sapphire crystal; transparent case back; water-resistant to 5 atm
Band: textile, folding clasp
Remarks: textile dial
Price: $35,100; limited to 50 pieces

Classic Fusion Chronograph Berluti Scritto Bordeaux
Reference number: 521.OX.050V.VR.BER18
Movement: automatic, Caliber HUB 1143; ø 30 mm, height 6.9 mm; 59 jewels; 28,800 vph; 42-hour power reserve
Functions: hours, minutes, subsidiary seconds; chronograph
Case: rose gold, ø 45 mm, height 13.4 mm; bezel screwed to case with 6 titanium screws; sapphire crystal; transparent case back; water-resistant to 5 atm
Band: calfskin, folding clasp
Remarks: leather dial
Price: $36,700; limited to 100 pieces
Variations: various cases, straps, and dials

Classic Fusion Chronograph Ceramic Blue
Reference number: 521.CM.7170.LR
Movement: automatic, Caliber HUB 1143; ø 30 mm, height 6.9 mm; 59 jewels; 28,800 vph; 42-hour power reserve
Functions: hours, minutes, subsidiary seconds; chronograph; date
Case: ceramic, ø 45 mm, height 13.05 mm; bezel screwed to case with 6 titanium screws; sapphire crystal; transparent case back; water-resistant to 5 atm
Band: reptile skin, folding clasp
Price: $11,900
Variations: various cases

Big Bang Unico Red Magic
Reference number: 411.CF.8513.RX
Movement: automatic, Caliber HUB 1242 "Unico"; ø 30 mm, height 8.05 mm; 38 jewels; 28,800 vph; black-coated mainplate and bridges; 72-hour power reserve
Functions: hours, minutes, subsidiary seconds; flyback chronograph; date
Case: ceramic, ø 45 mm, height 15.45 mm; bezel screwed to case with 6 titanium screws; sapphire crystal; transparent case back; water-resistant to 10 atm
Band: rubber, folding clasp
Price: $26,200; limited to 500 pieces

Classic Fusion Chronograph King Gold Green
Reference number: 521.OX.8980.LR
Movement: automatic, Caliber HUB 1143; ø 30 mm, height 6.9 mm; 59 jewels; 28,800 vph; 42-hour power reserve
Functions: hours, minutes, subsidiary seconds; chronograph; date
Case: rose gold, ø 45 mm, height 13.05 mm; bezel screwed to case with 6 titanium screws; sapphire crystal; transparent case back; water-resistant to 5 atm
Band: reptile skin, folding clasp
Price: $30,800
Variations: titanium ($10,800)

HUBLOT

Spirit of Big Bang Sapphire Rainbow
Reference number: 641.JX.0120.RT.4099
Movement: automatic, Caliber HUB 4700 (base Zenith El Primero); ø 30 mm, height 6.6 mm; 31 jewels; 36,000 vph; 50-hour power reserve
Functions: hours, minutes, subsidiary seconds; chronograph; date
Case: sapphire, 42 × 51 mm, height 15.2 mm; bezel screwed to case with 6 titanium screws and set with 48 amethysts; sapphire crystal; transparent case back; water-resistant to 5 atm
Band: rubber, folding clasp
Price: $115,000; limited to 50 pieces
Variations: sapphire ($79,000; limited to 200 pieces)

Spirit of Big Bang Blue Ceramic Blue
Reference number: 601.CI.7170.LR
Movement: automatic, Caliber HUB 4700 (base Zenith El Primero); ø 30 mm, height 6.6 mm; 31 jewels; 36,000 vph; 50-hour power reserve
Functions: hours, minutes, subsidiary seconds; chronograph; date
Case: ceramic, 45 × 51 mm, height 14.5 mm; bezel screwed to case with 6 titanium screws; sapphire crystal; transparent case back; water-resistant to 10 atm
Band: reptile skin, folding clasp
Price: $26,600
Variations: various cases and dials

Big Bang MP-11 14-Day Power Reserve Sapphire
Reference number: 911.JX.0102.RW
Movement: manually wound, Caliber HUB 9011; ø 34 mm, height 10.95 mm; 39 jewels; 28,800 vph; 7 spring barrels stacked perpendicularly to the movement axes, force transmission via oblique spur gears; 336-hour power reserve
Functions: hours, minutes (off-center); power reserve indicator
Case: sapphire, ø 45 mm, height 14.4 mm; sapphire crystal; transparent case back; water-resistant to 3 atm
Band: rubber, folding clasp
Price: $105,000; limited to 200 pieces
Variations: carbon ($82,000)

MP-09 Tourbillon Bi-Axis 5-Day Power Reserve 3D Carbon
Reference number: 909.QD.1120.RX
Movement: automatic, Caliber HUB 9009.H1.RA; ø 43 mm, height 8.4 mm; 43 jewels; 21,600 vph; double-axis tourbillon with different rotation times (60 and 30 seconds); date with rapid correction by lever on case left; anthracite-coated bridges
Functions: hours, minutes (off-center); power reserve indicator; date
Case: carbon composite, ø 49 mm, height 17.95 mm; bezel screwed to case with 5 titanium screws; sapphire crystal; transparent case back; water-resistant to 3 atm
Band: rubber, folding clasp
Price: $190,000; limited to 50 pieces

Caliber HUB 1240
Automatic; column wheel control of chronograph functions; silicon pallet lever and escapement, removable escapement; double-pawl automatic winding (Pellaton system), rotor with ceramic ball bearing; single spring barrel, 70-hour power reserve
Functions: hours, minutes, subsidiary seconds; flyback chronograph; date
Diameter: 30.4 mm
Height: 8.05 mm
Jewels: 38
Balance: glucydur
Frequency: 28,800 vph

Caliber HUB 1201
Manually wound skeletonized movement; silicon pallet lever and escape wheel; double mainspring barrel, 240-hour power reserve
Functions: hours, minutes, subsidiary seconds; power reserve indicator; date
Diameter: 35.19 mm
Height: 6.8 mm
Jewels: 24
Balance: CuBe
Frequency: 21,600 vph
Balance: flat hairspring with fine adjustment
Shock protection: Incabloc
Remarks: 223 parts

HYT

The earliest timekeepers were water clocks, known as clepsydras. The ancient Greeks had already devised a system by which water was guided from one vessel into another through an orifice of a predetermined diameter. Time was read on a calibrated scale on the second vessel. The three founders of HYT loved this idea of displaying the passage of time with moving fluids, and so they set out to solve the many problems generated if one were to introduce a liquid into a watch.

They developed a closed system made up of a capillary tube that would serve as a time track. It had a special pump-tank at either end that could either receive or pump out liquid. The two tanks had their dedicated space at 6 o'clock. After much research, they realized that the tube needed to be filled with two liquids, not one. The different physical properties of the liquids means they don't mix; thus, they create a sharp line at the point where they meet, which could serve as a pointer.

As for the pump-tanks, they were based on sensors used by NASA, a piston-driven bellows made of ultrathin but robust material that bends easily, but offers a stable surface. This allows very exact amounts of liquid to be pumped from one tank into the other. The system is driven by way of an in-house mechanical movement and cams. But it requires a clever thermal compensation system for the fluid in the tube. HYT became an instant sensation when it came out with its first watch in 2012.

Most important, perhaps, the concept left lots of opportunities for designs and new technological gimmicks. HYT has introduced a number of different models that often include clever innovations, such as a miniature hand-wound generator, or attractive design elements, like luminescent liquids.

HYT SA
Rue de Prébarreau 17
CH-2000 Neuchâtel
Switzerland

Tel.:
+41 32-323-2770

E-mail:
contact@hytwatches.com

Website:
www.hytwatches.com

Founded:
2012

Number of employees:
45 (including the sister company Preciflex)

Annual production:
approx. 350 wristwatches

U.S. retailers:
Contact main office in Neuchâtel for information on U.S. retailers

Most important collections/price range:
Various models with a liquid time display, $45,000 to $320,000

H_0

Reference number: 048-DL-90-GF-RU
Movement: manually wound, HYT Caliber; ø 37.8 mm, height 13.2 mm; 35 jewels; 28,800 vph; module with 2 bellows and capillary tube containing 2 immiscible fluids, the meeting between the 2 liquids is the hour pointer; temperature compensated; 65-hour power reserve
Functions: minutes (off-center), hours (capillary display, retrograde), subsidiary seconds; power reserve indicator
Case: stainless steel, black DLC-coated, ø 48.8 mm, height 18.7 mm; sapphire crystal; crown; water-resistant to 3 atm
Band: rubber, folding clasp
Price: $39,000
Variations: H_0 silver case with blue fluid

H^2O

Reference number: 251-AD-46-GF-RU
Movement: manually wound, HYT Caliber; ø 39.9 mm, height 14 mm (including the fluidic module); 28 jewels; 21,600 vph; 2 bellows and a capillary tube; 192-hour power reserve
Functions: minutes (hands), hours (capillary displays, retrograde); power reserve indicator, crown position indicator, thermal indicator
Case: stainless steel, black DLC-coated, ø 51 mm, height 19.95 mm; sapphire crystal; crown; water-resistant to 3 atm
Band: rubber, folding clasp
Price: $95,000; limited to 25 pieces
Variations: blue and red fluid

H_0X Eau Rouge

Reference number: 048-AD-95-RF-RU
Movement: manually wound, HYT Caliber; ø 37.8 mm, height 13.2 mm (including fluidic module); 35 jewels; 28,800 vph; module with 2 bellows and capillary tube containing 2 immiscible fluids, the meeting between the 2 liquids is the hour pointer; thermal compensator; 65-hour power reserve
Functions: minutes (off-center), hours (capillary display, retrograde), subsidiary seconds; power reserve indicator
Case: stainless steel, anthracite DLC-coated, ø 48.8 mm, height 18.7 mm; sapphire crystal; crown; water-resistant to 3 atm
Band: rubber, folding clasp
Price: $39,000
Variations: stainless steel and satin-finished dial

CRUXIBLE™
Life, Liberty and the Pursuit

America that great crucible of humanity bonded by the ideal of equality and the promise of life, liberty, and the pursuit of happiness for all. Through painful trials of war and cultural conflict the American experiment has alloyed a diversity of ideas and cultures into a whole greater than the sum of its parts. Unbeknownst to many Americans, the Second World War, more than any other conflict or challenge that the US has faced, showcased the inspiring potential inherent in our diversity. This has been the driving inspiration behind the 6 years of development we've poured into resurrecting the iconic American tool watch of World War II detailed in the A-11 specification. Largely unappreciated, we aspired to elevate this design so that it could fulfill its potential to symbolize that event, the shared sacrifice, and the universal bond inherent to the ideals of this nation.

Built for the Pursuit™

For additional information and to order direct visit: www.mkiiwatches.com

ITAY NOY

Israeli watchmaker Itay Noy started his career as a jeweler, so it comes as no surprise that his earlier watches tended to emphasize form, while the functional aspects are left to solid Swiss movements. As a shaper, though, he reveals himself to be a pensive, philosophical storyteller making each timepiece a unique, encapsulated tale of sorts. The City Squares model, for example, gives the time on the backdrop of a map of the owner's favorite or native city, thus creating an intimate connection with, perhaps, a past moment. At Baselworld 2013, Noy showcased a square watch run on a Technotime automatic movement with a face-like dial that changes with the movement of the hands, a reminder of how our life has become dominated by the rectangular frame of mobile gadgets. The Cityscape, square as well, represents a modern urban landscape.

Each year brings new ideas as Noy ratchets up his craft and begins reaching into the engineer's magic box—including working on a bespoke movement with a Swiss firm. In 2015, he designed a special module to create a fascinating, two-part square dial for his Part Time, dividing day and night and providing the time plus the position of the sun and the moon through little apertures. This fascination with the times of day was reiterated in 2016 with the Chrono Gears model, which essentially runs on a large invisible circular gear that drives a.m. and p.m. indicators and more. Time Tone gives the owner the choice of a colored hour that only he or she will know, while the minute hand does its work in the center of the dial. The latest, Full Month, shows the date or the moon appearing through a circle of digits carved into the dial. It's no wonder that Noy's collections find their way into museums and special exhibitions. They are works of kinetic art.

Itay Noy
P.O. Box 16661
Tel Aviv 6116601
Israel

Tel.:
+972-352-47-380

Fax:
+972-352-47-381

E-mail:
studio@itay-noy.com

Website:
www.itay-noy.com

Founded:
2000

Number of employees:
4

Annual production:
150–200 pieces

U.S. distributor:
Please contact Studio Itay Noy for information.
www.itay-noy.com

Most important collections/price range:
Time Tone, Full Month, Chrono Gears, Part Time, Open Mind, X-Ray / $2,400 to $9,800

Full Month—Moon
Reference number: FM-MOON
Movements: automatic, Caliber IN.VMF5400; ø 30 mm, height 3 mm; 29 jewels; 21,600 vph; extra-thin microrotor; 48-hour power reserve
Functions: hours, minutes, subsidiary seconds; date
Case: stainless steel, ø 40 mm × 44 mm, height 7.44 mm; sapphire crystal; transparent case back; water resistant to 5 atm
Band: leather, double folding clasp
Remarks: date projected through digit-shaped slits in dial
Price: $9,800; limited and numbered edition of 18 pieces

Full Month
Reference number: FM-NUM.WT
Movements: automatic, Caliber IN.VMF5400; ø 30 mm, height 3 mm; 29 jewels; 21,600 vph; extra-thin microrotor; 48-hour power reserve
Functions: hours, minutes, subsidiary seconds; date
Case: stainless steel, ø 40 mm × 44 mm, height 7.44 mm; sapphire crystal; transparent case back; water resistant to 5 atm
Band: leather, double folding clasp
Case: stainless steel, ø 40 mm × 44 mm, height 6.24 mm, sapphire crystal, transparent case back, water resistant to 5 atm
Remarks: date projected through digit-shaped slits in dial
Price: $9,800; limited and numbered edition of 18 pieces

Time Tone
Reference number: TT.BL
Movement: manually wound; Caliber IN.IP13; ø 36.6 mm, height 5.5 mm; 20 jewels; 21,600 vph; 42-hour power reserve
Functions: dynamic dial with tone color disk; minutes, seconds
Case: stainless steel, ø 44 mm, height 12 mm; sapphire crystal; transparent case back; water-resistant to 5 atm
Band: leather, double folding clasp
Price: $5,600; limited and numbered edition of 24 pieces
Variations: blue or black

Celestial Time
Reference number: CT.W
Movement: manually wound; Caliber IN.IP13, ø 36.6 mm, height 5.5 mm; 20 jewels; 21,600 vph; 42-hour power reserve
Functions: dynamic dial with zodiac hour disk, minutes, seconds
Case: stainless steel, ø 44 mm, height 12 mm; sapphire crystal; transparent case back; water-resistant to 5 atm
Band: handmade leather band, double folding clasp
Price: $5,200; limited and numbered edition of 24 pieces
Variations: Western or Chinese zodiac signs

Chrono Gears
Reference number: CG.G
Movement: manually wound; Caliber IN.IP13, ø 36.6 mm, height 5.5 mm; 20 jewels; 21,600 vph; 42-hour power reserve
Functions: chronogear hand indicator for am/pm, chronogear hand indicator for 8 time situations, central hours, minutes, seconds
Case: stainless steel, ø 44 mm, height 12 mm; sapphire crystal; transparent case back; water-resistant to 5 atm
Band: leather, double folding clasp
Price: $5,800; limited and numbered edition of 24 pieces
Variations: blue or black

Open Mind
Reference number: OM-S.G
Movement: manually wound, ETA Caliber 6497-1; ø 36.6 mm, height 4.5 mm; 17 jewels; 21,600 vph; 38-hour power reserve
Functions: hours, minutes, subsidiary seconds
Case: stainless steel, ø 44 mm, height 12 mm; sapphire crystal; transparent case back; water-resistant to 5 atm
Band: leather, double folding clasp
Price: $4,400; limited and numbered edition of 99 pieces
Variations: blue or black

X-Ray
Reference number: XRAY6498
Movement: manually wound, ETA Caliber 6498-1, ø 36.6 mm, height 4.5 mm; 17 jewels; 21,600 vph; 38-hour power reserve
Functions: hours, minutes, subsidiary seconds
Case: stainless steel, ø 41.6 × 44.6 mm, height 10 mm; sapphire crystal, screw-down case back; water-resistant to 50 m
Band: leather, double folding clasp
Price: $3,640; limited and numbered edition of 99 pieces
Variations: gold-plated dial ($3,900), black or brown leather strap

Part Time
Reference number: PT-DN.BL
Movement: manually wound; Caliber IN.DD&6498-1, ø 36.6 mm, height 5.2 mm; 17 jewels; 21,600 vph; 38-hour power reserve
Functions: hours from 6 a.m–6 p.m., hours from 6 p.m.–6 a.m., analog hours; moon disk, sun disk; minutes, subsidiary seconds
Case: stainless steel, ø 41.6 mm × 44.6 mm, height 10.6 mm; sapphire crystal; transparent case back; water-resistant to 5 atm
Band: leather, double folding clasp
Remarks: limited and numbered edition of 24 pieces
Price: $5,800
Variations: blue or black

ID-Hebrew
Reference number: ID-HEB.G
Movement: automatic, Miyota Caliber 90S5, ø 25.6 mm, height 3.9 mm; 24 jewels; 28,800 vph, gold-plated dial; 42-hour power reserve
Functions: hours, minutes, sweep seconds
Case: stainless steel, ø 42.4 mm, height 10 mm; sapphire crystal; screw-down case back; water-resistant to 5 atm
Band: leather, double folding clasp
Price: $2,800; limited and numbered edition of 99 pieces
Variations: black or brown leather band

IWC

It was an American who laid the cornerstone for an industrial watch factory in Schaffhausen—now environmentally state-of-the-art facilities. In 1868, Florentine Ariosto Jones, watchmaker and engineer from Boston, crossed the Atlantic to the then low-wage venue of Switzerland to open the International Watch Company Schaffhausen.

Jones was not only a savvy businessperson but also a talented designer who had a significant influence on the development of watch movements. Soon, he gave IWC its own seal of approval, the *Ingenieursmarke* (Engineer's Brand), a standard it still maintains today. IWC is synonymous with excellently crafted watches that meet high technical benchmarks. Not even a large variety of owners over the past 100 years has been able to change the company's course. In 2000, it joined Richemont Group.

George Kern, who headed the company until 2017, mixed tradition with a keen sense for flashiness. He pushed in-house movements, such as those found in the Da Vinci and Ingenieur models, and of course the Portuguese, which is still going strong more than seventy-five years after it was first released.

The firm's 150th anniversary was celebrated in 2018 under a new CEO (Christoph Grainger-Herr). A total of twenty-seven limited edition special models were shown at the beginning of the year, each with "150 Years" stamped on the back. The range covers robust pilots' watches, refined Da Vincis, elegant Portofinos, and complicated Portuguese. And then there is the unabashedly retro Pallweber series, which borrows from a pocket watch from the 1880s showing digital time.

One thing is certain: The mechanics are always top-notch. Movements include the Jones caliber, named for the IWC founder, and the pocket watch caliber 89, introduced in 1946 as the creation of then technical director Albert Pellaton. Four years later, Pellaton created the first IWC automatic movement and, with it, a company monument. The careful balance between tapping into past glories to keep fans and newcomers to the brand happy continues to boost the brand. So much so that the company built itself new, ultra-modern, ultra-ecological facilities just outside its home city, Schaffhausen, in Switzerland.

International Watch Co.
Baumgartenstrasse 15
CH-8201 Schaffhausen
Switzerland

Tel.:
+41-52-635-6565

Fax:
+41-52-635-6501

E-mail:
info@iwc.com

Website:
www.iwc.com

Founded:
1868

Number of employees:
approx. 750

U.S. distributor:
IWC North America
645 Fifth Avenue, 5th Floor
New York, NY 10022
800-432-9330

Most important collections/price range:
Da Vinci, Pilot's, Portuguese, Ingenieur, Aquatimer, Pallweber / approx. $4,000 to $260,000

IWC Tribute to Pallweber Edition "150 Years"

Reference number: IW505001
Movement: manually wound, IWC Caliber 94200; ø 37.8 mm, height 7.28 mm; 54 jewels; 28,800 vph; Breguet hairspring; 60-hour power reserve
Functions: hours, minutes (digital), subsidiary seconds
Case: platinum, ø 45 mm, height 12 mm; sapphire crystal; transparent case back; water-resistant to 3 atm
Band: reptile skin, folding clasp
Price: $57,800; limited to 25 pieces

IWC Tribute to Pallweber Edition "150 Years"

Reference number: IW505002
Movement: manually wound, IWC Caliber 94200; ø 37.8 mm, height 7.28 mm; 54 jewels; 28,800 vph; Breguet hairspring; 60-hour power reserve
Functions: hours, minutes (digital), subsidiary seconds
Case: pink gold, ø 45 mm, height 12 mm; sapphire crystal; transparent case back; water-resistant to 3 atm
Band: reptile skin, folding clasp
Price: $36,600; limited to 250 pieces

IWC Tribute to Pallweber Edition "150 Years"

Reference number: IW505003
Movement: manually wound, IWC Caliber 94200; ø 37.8 mm, height 7.28 mm; 54 jewels; 28,800 vph; Breguet hairspring; 60-hour power reserve
Functions: hours, minutes (digital), subsidiary seconds
Case: stainless steel, ø 45 mm, height 12 mm; sapphire crystal; transparent case back; water-resistant to 3 atm
Band: reptile skin, folding clasp
Price: $23,100; limited to 500 pieces

Portuguese Constant-Force Tourbillon Edition "150 Years"
Reference number: IW590202
Movement: manually wound, IWC Caliber 94805; ø 37.8 mm, height 7.7 mm; 41 jewels; 18,000 vph; 1-minute tourbillon, constant force escapement; 96-hour power reserve
Functions: hours, minutes, subsidiary seconds (on tourbillon cage); power reserve indicator; moon phase
Case: platinum, ø 46 mm, height 13.5 mm; sapphire crystal; transparent case back; water-resistant to 3 atm
Band: reptile skin, folding clasp
Price: $249,000; limited to 15 pieces

Portuguese Constant-Force Tourbillon Edition "150 Years"
Reference number: IW590203
Movement: manually wound, IWC Caliber 94805; ø 37.8 mm, height 7.7 mm; 41 jewels; 18,000 vph; 1-minute tourbillon, constant force escapement; 96-hour power reserve
Functions: hours, minutes, subsidiary seconds (on tourbillon cage); power reserve indicator; moon phase
Case: platinum, ø 46 mm, height 13.5 mm; sapphire crystal; transparent case back; water-resistant to 3 atm
Band: reptile skin, folding clasp
Price: $249,000; limited to 15 pieces

Portuguese Perpetual Calendar Tourbillon Edition "150 Years"
Reference number: IW504501
Movement: automatic, IWC Caliber 51950; ø 37.8 mm, height 9.3 mm; 44 jewels; 19,800 vph; flying 1-minute tourbillon, Breguet hairspring, gold oscillating mass; 168-hour power reserve
Functions: hours, minutes; power reserve indicator; perpetual calendar with date, weekday, month, moon phase, year display (4 digits)
Case: pink gold, ø 45 mm, height 15.3 mm; sapphire crystal; transparent case back; water-resistant to 3 atm
Band: reptile skin, folding clasp
Price: $110,000; limited to 50 pieces

Perpetual Calendar Edition "150 Years"
Reference number: IW503405
Movement: automatic, IWC Caliber 52615; ø 37.8 mm, height 9 mm; 54 jewels; 28,800 vph; Breguet hairspring, gold oscillating mass; 168-hour power reserve
Functions: hours, minutes, subsidiary seconds; power reserve indicator; perpetual calendar with date, weekday, month, moon phase, year display (4 digits)
Case: pink gold, ø 44.2 mm, height 14.9 mm; sapphire crystal; transparent case back; water-resistant to 3 atm
Band: reptile skin, folding clasp
Price: $39,900; limited to 250 pieces

Portuguese Hand-Wound Eight Days Edition "150 Years"
Reference number: IW510212
Movement: manually wound, IWC Caliber 59215; ø 37.8 mm, height 5.8 mm; 30 jewels; 28,800 vph; Breguet hairspring; 192-hour power reserve
Functions: hours, minutes, subsidiary seconds
Case: stainless steel, ø 43 mm, height 12.2 mm; sapphire crystal; transparent case back; water-resistant to 3 atm
Band: reptile skin, buckle
Price: $9,900; limited to 1,000 pieces

Portuguese Chronograph Edition "150 Years"
Reference number: IW371601
Movement: automatic, IWC Caliber 69355; ø 30 mm; 27 jewels; 28,800 vph; 46-hour power reserve
Functions: hours, minutes, subsidiary seconds; chronograph
Case: stainless steel, ø 41 mm, height 13.1 mm; sapphire crystal; transparent case back; water-resistant to 3 atm
Band: reptile skin, folding clasp
Price: $7,150; limited to 2,000 pieces
Variations: white dial

IWC

Portofino Hand-Wound Moon Phase Edition "150 Years"

Reference number: IW516407
Movement: manually wound, IWC Caliber 59800; ø 37.8 mm, height 7.3 mm; 30 jewels; 28,800 vph; Breguet hairspring; 192-hour power reserve
Functions: hours, minutes, subsidiary seconds; power reserve indicator; date, moon phase
Case: pink gold, ø 45 mm, height 13.2 mm; sapphire crystal; transparent case back; water-resistant to 3 atm
Band: reptile skin, buckle
Price: $23,600; limited to 150 pieces
Variations: stainless steel

Portofino Chronograph Edition "150 Years"

Reference number: IW391024
Movement: automatic, IWC Caliber 79320 (base ETA 7750); ø 30 mm, height 7.9 mm; 25 jewels; 28,800 vph; 44-hour power reserve
Functions: hours, minutes, subsidiary seconds; chronograph; date, weekday
Case: stainless steel, ø 42 mm, height 13.6 mm; sapphire crystal; water-resistant to 3 atm
Band: reptile skin, buckle
Price: $5,800; limited to 2,000 pieces
Variations: blue dial

IWC Portofino Automatic Edition "150 Years"

Reference number: IW356518
Movement: automatic, IWC Caliber 35111 (base Sellita SW300-1); ø 35.6 mm, height 3.75 mm; 25 jewels; 28,800 vph; 42-hour power reserve
Functions: hours, minutes, sweep seconds; date
Case: stainless steel, ø 40 mm, height 9.3 mm; sapphire crystal; water-resistant to 3 atm
Band: reptile skin, buckle
Price: $4,700; limited to 2,000 pieces
Variations: silver dial

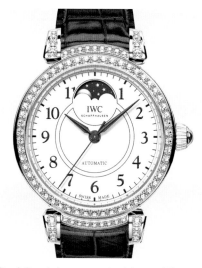

Da Vinci Automatic Edition "150 Years"

Reference number: IW358103
Movement: automatic, IWC Caliber 82200; ø 30 mm, height 6.6 mm; 33 jewels; 28,800 vph; 60-hour power reserve
Functions: hours, minutes, subsidiary seconds
Case: pink gold, ø 40.4 mm, height 12.1 mm; sapphire crystal; transparent case back; water-resistant to 3 atm
Band: reptile skin, buckle
Price: $18,200; limited to 250 pieces
Variations: stainless steel with white or blue dial

Da Vinci Automatic Moon Phase Edition "150 Years"

Reference number: IW459309
Movement: automatic, IWC Caliber 35800; ø 25.6 mm, height 5.35 mm; 25 jewels; 28,800 vph; 42-hour power reserve
Functions: hours, minutes, sweep seconds; moon phase
Case: white gold, ø 36 mm, height 11.7 mm; sapphire crystal; water-resistant to 3 atm
Band: reptile skin, buckle
Remarks: case and lugs set with 206 diamonds
Price: $30,800; limited to 50 pieces
Variations: pink gold

Pilot's Watch Chronograph Edition "150 Years"

Reference number: IW377725
Movement: automatic, IWC Caliber 79320 (base ETA 7750); ø 30 mm, height 7.9 mm; 25 jewels; 28,800 vph; antimagnetic soft iron core; 44-hour power reserve
Functions: hours, minutes, subsidiary seconds; chronograph; date, weekday
Case: stainless steel, ø 43 mm, height 15.2 mm; sapphire crystal; screw-in crown; water-resistant to 6 atm
Band: reptile skin, buckle
Price: $5,150; limited to 1,000 pieces

IWC

Big Pilot's Watch Annual Calendar Edition "150 Years"
Reference number: IW502708
Movement: automatic, IWC Caliber 52850; ø 37.8 mm, height 9.95 mm; 36 jewels; 28,800 vph; Breguet hairspring, gold oscillating mass; 168-hour power reserve
Functions: hours, minutes, subsidiary seconds; power reserve indicator; annual calendar with date, weekday, month
Case: stainless steel, ø 46.2 mm, height 15.5 mm; sapphire crystal; transparent case back; water-resistant to 6 atm
Band: reptile skin, folding clasp
Price: $19,700; limited to 150 pieces

Big Pilot's Watch Big Date Edition "150 Years"
Reference number: IW510503
Movement: manually wound, IWC Caliber 59235; ø 37.8 mm, height 7.3 mm; 30 jewels; 28,800 vph; Breguet hairspring, antimagnetic soft iron core; 192-hour power reserve
Functions: hours, minutes, subsidiary seconds; large date
Case: stainless steel, ø 46.2 mm, height 15.2 mm; sapphire crystal; transparent case back; screw-in crown; water-resistant to 6 atm
Band: reptile skin, folding clasp
Price: $13,800; limited to 100 pieces
Variations: white dial

Pilot's Watch Chronograph
Reference number: IW377709
Movement: automatic, IWC Caliber 79320 (base ETA 7750); ø 30 mm, height 7.9 mm; 25 jewels; 28,800 vph; antimagnetic soft iron core; 44-hour power reserve
Functions: hours, minutes, subsidiary seconds; chronograph; date, weekday
Case: stainless steel, ø 43 mm, height 15.2 mm; sapphire crystal; screw-in crown; water-resistant to 6 atm
Band: textile, buckle
Price: $4,950

Pilot's Watch Mark XVIII Edition "Laureus Sport for Good Foundation"
Reference number: IW324703
Movement: automatic, IWC Caliber 35111 (base Sellita SW300-1); ø 25.6 mm, height 3.75 mm; 25 jewels; 28,800 vph; antimagnetic soft iron core; 42-hour power reserve
Functions: hours, minutes, sweep seconds; date
Case: ceramic, ø 41 mm, height 11 mm; sapphire crystal; screw-in crown; water-resistant to 6 atm
Band: calfskin, buckle
Price: $5,650; limited to 1,500 pieces

Ingenieur Chronograph Sport Edition "76th Goodwood Members' Meeting"
Reference number: IW381201
Movement: automatic, IWC Caliber 69380; ø 30 mm, height 7.9 mm; 33 jewels; 28,800 vph; antimagnetic soft iron core; 46-hour power reserve
Functions: hours, minutes, subsidiary seconds; chronograph; date, weekday
Case: titanium, ø 44.3 mm, height 16.3 mm; sapphire crystal; transparent case back; screw-in crown; water-resistant to 12 atm
Band: calfskin, folding clasp
Price: $8,600; limited to 176 pieces

IWC Tribute to Pallweber Edition "150 Years" Pocket Watch
Reference number: IW505101
Movement: manually wound, IWC Caliber 94200; ø 37.8 mm, height 7.28 mm; 54 jewels; 28,800 vph; Breguet hairspring; 60-hour power reserve
Functions: hours, minutes (digital), subsidiary seconds
Case: pink gold, ø 52 mm, height 14.2 mm; sapphire crystal
Band: pink gold fob
Price: $66,500; limited to 50 pieces

IWC

Caliber 94200

Manually wound; single spring barrel, 60-hour power reserve
Base caliber: 94000
Functions: hours, minutes (digital), subsidiary seconds
Diameter: 37.8 mm
Height: 7.3 mm
Jewels: 54
Frequency: 28,800 vph
Balance spring: Breguet hairspring

Caliber 69375

Automatic; single spring barrel, 46-hour power reserve
Base caliber: 69370
Functions: hours, minutes, subsidiary seconds; chronograph; date
Diameter: 30 mm
Height: 6.9 mm
Jewels: 36
Balance: glucydur
Frequency: 28,800 vph

Caliber 59210

Manually wound; single spring barrel, 192-hour power reserve
Functions: hours, minutes, subsidiary seconds; power reserve indicator; date
Diameter: 37.8 mm
Height: 5.8 mm
Jewels: 30
Balance: glucydur with variable inertia
Frequency: 28,800 vph
Balance spring: Breguet
Shock protection: Incabloc

Caliber 52615

Automatic; double-pawl automatic winding (Pellaton system), with ceramic ball bearings; double spring barrel, 168-hour power reserve
Functions: hours, minutes, subsidiary seconds; power reserve indicator; perpetual calendar with month, weekday, date, double moon phase display (for northern and southern earth hemisphere), year display (4 digits)
Diameter: 37.8 mm
Height: 9 mm
Jewels: 54
Balance: with variable inertia
Frequency: 28,800 vph
Balance spring: Breguet
Shock protection: Incabloc

Caliber 89361

Automatic; double-pawl automatic winding (Pellaton system), column wheel control of chronograph functions; single spring barrel, 68-hour power reserve
Base caliber: 89000
Functions: hours, minutes, subsidiary seconds; flyback chronograph; date
Diameter: 30 mm
Height: 7.46 mm
Jewels: 38
Balance: glucydur with variable inertia
Frequency: 28,800 vph
Balance spring: flat hairspring
Shock protection: Incabloc
Remarks: concentric chronograph totalizer for minutes and hours

Caliber 98295 "Jones"

Manually wound; single spring barrel, 46-hour power reserve
Base caliber: 98000
Functions: hours, minutes, subsidiary seconds
Diameter: 38.2 mm
Height: 5.3 mm
Jewels: 18
Balance: screw balance with fine adjustment cams on balance arms
Frequency: 18,000 vph
Balance spring: Breguet
Shock protection: Incabloc
Remarks: exceptionally long regulator index; three-quarter plate of German silver, hand-engraved balance cock

JAEGER-LECOULTRE

The Jaeger-LeCoultre *manufacture* has a long and tumultuous history. In 1833, Antoine LeCoultre opened his own workshop for the production of gearwheels. Having made his fortune, he then did what many other artisans did: In 1866, he had a large house built and brought together all the craftspeople needed to produce timepieces, from the watchmakers to the turners and polishers. He outfitted the workshop with the most modern machinery of the day, all powered by a steam engine. "La Grande Maison" was the first watch *manufacture* in the Vallée de Joux.

At the start of the twentieth century, the grandson of the company founder, Jacques-David LeCoultre, built slender, complicated watches for the Paris manufacturer Edmond Jaeger. The Frenchman was so impressed with these that, after a few years of fruitful cooperation, he engineered a merger of the two companies.

In the 1970s, the *manufacture* was taken over by the German VDO Group (later Mannesmann). Under the leadership of Günter Blümlein, it weathered the quartz crisis, and during the mechanical watch renaissance in the 1980s, it regained its status as an innovative, high-performance *manufacture*.

In 2000, Mannesmann's watch division (JLC, IWC, A. Lange & Söhne) sold Jaeger-LeCoultre to the Richemont Group. Given the group's strength, Jaeger-LeCoultre continued to grow. Fifty new calibers, including minute repeaters, tourbillons, and other *grandes complications*, a lubricant-free movement, and more than 400 patents tell their own story. Today, it is the largest employer in the Vallée de Joux—just as it was back in the 1860s. With all the complicated watches, though, the most enduring collection produced by the brand is the Reverso, which can swivel around to show a second watch face on the back. Jaeger-LeCoultre's rich history has produced another revived cult watch: the Memovox, originally made in 1968 and with a built-in alarm. A special sounding board in the new 956 caliber gives its rings more power.

Manufacture Jaeger-LeCoultre
Rue de la Golisse, 8
CH-1347 Le Sentier
Switzerland

Tel.:
+41-21-852-0202

Fax:
+41-21-852-0505

E-mail:
info@jaeger-lecoultre.com

Website:
www.jaeger-lecoultre.com

Founded:
1833

Number of employees:
over 1,000

Annual production:
approx. 50,000 watches

U.S. distributor:
Jaeger-LeCoultre
645 Fifth Avenue
New York, NY 10022
800-JLC-TIME
www.jaeger-lecoultre.com

Most important collections/price range:
Atmos starting at $6,600; Duomètre starting at $39,100; Geophysic starting at $9,100; Master starting at $5,700, Rendez-Vous starting at $8,700; Reverso starting at $4,150

Polaris Chronograph
Reference number: 902 81 80
Movement: automatic, JLC Caliber 751H; ø 25.6 mm, height 5.7 mm; 37 jewels; 28,800 vph; skeletonized rotor; 65-hour power reserve
Functions: hours, minutes; chronograph
Case: stainless steel, ø 42 mm, height 11.9 mm; sapphire crystal; transparent case back; water-resistant to 10 atm
Band: stainless steel, double folding clasp
Price: $10,700
Variations: calfskin strap ($9,850)

Polaris Chronograph
Reference number: 902 84 71
Movement: automatic, JLC Caliber 751H; ø 25.6 mm, height 5.7 mm; 37 jewels; 28,800 vph; skeletonized rotor; 65-hour power reserve
Functions: hours, minutes; chronograph
Case: stainless steel, ø 42 mm, height 11.9 mm; sapphire crystal; transparent case back; water-resistant to 10 atm
Band: calfskin, double folding clasp
Price: $9,750
Variations: stainless steel bracelet ($10,800)

Polaris Chronograph
Reference number: 902 24 50
Movement: automatic, JLC Caliber 751H; ø 25.6 mm, height 5.7 mm; 37 jewels; 28,800 vph; skeletonized rotor; 65-hour power reserve
Functions: hours, minutes; chronograph
Case: pink gold, ø 42 mm, height 11.9 mm; sapphire crystal; transparent case back; water-resistant to 10 atm
Band: reptile skin, double folding clasp
Price: $23,900

JAEGER-LECOULTRE

Polaris Automatic
Reference number: 900 84 80
Movement: automatic, JLC Caliber 898E/1; ø 26 mm, height 3.3 mm; 30 jewels; 28,800 vph; 40-hour power reserve
Functions: hours, minutes, sweep seconds
Case: stainless steel, ø 41 mm, height 11.2 mm; crown-activated scale ring, 0-60 scale; sapphire crystal; transparent case back; water-resistant to 10 atm
Band: calfskin, double folding clasp
Price: $6,600
Variations: stainless steel bracelet ($7,600)

Polaris Automatic
Reference number: 900 81 70
Movement: automatic, JLC Caliber 898E/1; ø 26 mm, height 3.3 mm; 30 jewels; 28,800 vph; 40-hour power reserve
Functions: hours, minutes, sweep seconds
Case: stainless steel, ø 41 mm, height 11.2 mm; crown-activated scale ring, 0-60 scale; sapphire crystal; transparent case back; water-resistant to 10 atm
Band: stainless steel, double folding clasp
Price: $7,450
Variations: calfskin strap ($6,700)

Polaris Date
Reference number: 906 86 70
Movement: automatic, JLC Caliber 899A/1; ø 26 mm, height 4.6 mm; 32 jewels; 28,800 vph; 38-hour power reserve
Functions: hours, minutes, sweep seconds; date
Case: stainless steel, ø 42 mm, height 13.1 mm; crown-activated scale ring, 0-60 scale; sapphire crystal; transparent case back; water-resistant to 20 atm
Band: rubber, double folding clasp
Price: $7,600

Polaris Memovox
Reference number: 903 86 70
Movement: automatic, JLC Caliber 956; ø 28 mm, height 7.45 mm; 23 jewels; 28,800 vph; 45-hour power reserve
Functions: hours, minutes, sweep seconds; alarm; date
Case: stainless steel, ø 42 mm, height 15.9 mm; crown-activated scale ring, 0-60 scale; sapphire crystal; water-resistant to 20 atm
Band: rubber, double folding clasp
Price: $12,600

Polaris Chronograph WT
Reference number: 905 T4 80
Movement: automatic, JLC Caliber 752A; ø 28 mm, height 5.7 mm; 37 jewels; 28,800 vph; 2 spring barrels, 65-hour power reserve
Functions: hours, minutes; world time indicator (2nd time zone); chronograph
Case: titanium, ø 44 mm, height 12.5 mm; sapphire crystal; transparent case back; water-resistant to 10 atm
Band: calfskin, double folding clasp
Price: $14,100

Polaris Chronograph WT
Reference number: 905 T4 71
Movement: automatic, JLC Caliber 752A; ø 28 mm, height 5.7 mm; 37 jewels; 28,800 vph; 2 spring barrels, 65-hour power reserve
Functions: hours, minutes; world time indicator (2nd time zone); chronograph
Case: titanium, ø 44 mm, height 12.5 mm; sapphire crystal; transparent case back; water-resistant to 10 atm
Band: reptile skin, double folding clasp
Price: $14,100
Variations: calfskin strap ($14,300)

JAEGER-LECOULTRE

Reverso Classic Large Duoface Small Seconds

Reference number: 384 84 22
Movement: manually wound, JLC Caliber 854A/2; 17.2 × 22 mm, height 3.8 mm; 19 jewels; 21,600 vph; 42-hour power reserve
Functions: hours, minutes, subsidiary seconds; additional 24-hour display (2nd time zone, on rear)
Case: stainless steel, 28.3 × 47 mm, height 10.3 mm; sapphire crystal; water-resistant to 3 atm
Band: calfskin, double folding clasp
Remarks: case turns and swivels 180°
Price: $8,400

Reverso Classic Medium Small Second

Reference number: 243 85 20
Movement: manually wound, JLC Caliber 822/2; height 2.94 mm; 19 jewels; 21,600 vph; 42-hour power reserve
Functions: hours, minutes, subsidiary seconds
Case: stainless steel, 25.5 × 42.9 mm; sapphire crystal; water-resistant to 3 atm
Band: reptile skin, buckle
Remarks: case turns and swivels 180°
Price: $5,900

Reverso Tribute Small Seconds

Reference number: 397 84 80
Movement: manually wound, JLC Caliber 822/2; 17.2 × 22 mm, height 2.94 mm; 19 jewels; 21,600 vph; 42-hour power reserve
Functions: hours, minutes, subsidiary seconds
Case: stainless steel, 27.4 × 45.6 mm, height 8.5 mm; sapphire crystal; water-resistant to 3 atm
Band: calfskin, folding clasp
Price: $7,950

Reverso Tribute Duoface

Reference number: 390 24 20
Movement: manually wound, JLC Caliber 854A/2; 17.2 × 22 mm, height 3.8 mm; 19 jewels; 21,600 vph; 42-hour power reserve
Functions: hours, minutes, subsidiary seconds; additional 24-hour display (2nd time zone) on rear
Case: pink gold, 25.5 × 42.9 mm, height 9.2 mm; sapphire crystal; water-resistant to 3 atm
Band: reptile skin, folding clasp
Remarks: case turns and swivels 180°
Price: $11,400

Reverso Tribute Moon

Reference number: 395 84 20
Movement: manually wound, JLC Caliber 853A; 17.2 × 22 mm, height 5.15 mm; 19 jewels; 21,600 vph; 42-hour power reserve
Functions: hours, minutes; additional 12-hour display (2nd time zone) and day/night indicator (on rear); date, moon phase
Case: stainless steel, 29.9 × 49.4 mm, height 10.9 mm; sapphire crystal; water-resistant to 3 atm
Band: reptile skin, double folding clasp
Remarks: case turns and swivels 180°
Price: $12,700

Reverso Tribute Gyrotourbillon

Reference number: 394 64 20
Movement: manually wound, JLC Caliber 179 × 26.2 × 41 mm, height 5.97 mm; 52 jewels; 21,600 vph; double-axis spherical tourbillon with different rotation times (60 and 12.6 seconds), Gyrolab balance with hemispheric hairspring; certified chronometer according to the German industrial norm (DIN)
Functions: hours, minutes, subsidiary seconds (on tourbillon cage); additional 12-hour display (2nd time zone) and day/night indicator on rear
Case: platinum, 31 × 51.2 mm, height 12.4 mm; sapphire crystal; water-resistant to 3 atm
Band: reptile skin, double folding clasp
Remarks: case turns and swivels 180°
Price: $314,000; limited to 75 pieces

JAEGER-LECOULTRE

Geophysic Universal Time
Reference number: 810 81 20
Movement: automatic, JLC Caliber 772; ø 28.6 mm, height 7.13 mm; 36 jewels; 28,800 vph; jumping seconds drive with remontoir; ringless Gyrolab balance; 40-hour power reserve
Functions: hours, minutes, sweep seconds (jumping); world time indicator (2nd time zone)
Case: stainless steel, ø 41.6 mm, height 11.84 mm; sapphire crystal; transparent case back; water-resistant to 5 atm
Band: stainless steel, double folding clasp
Price: $15,700
Variations: reptile skin strap ($14,300)

Geophysic Tourbillon Universal Time
Reference number: 812 64 20
Movement: automatic, JLC Caliber 948; ø 30 mm, height 11.24 mm; 42 jewels; 28,800 vph; flying 1-minute tourbillon with ringless Gyrolab balance wheel, rotation of world time disk once in 24 hours; 48-hour power reserve
Functions: hours, minutes, subsidiary seconds (on tourbillon cage); world time indicator (2nd time zone)
Case: platinum, ø 43.5 mm, height 14.87 mm; sapphire crystal; transparent case back; water-resistant to 5 atm
Band: reptile skin, double folding clasp
Price: $145,000

Master Control Date
Reference number: 154 85 30
Movement: automatic, JLC Caliber 899/1; ø 26 mm, height 3.3 mm; 32 jewels; 28,800 vph; 38-hour power reserve
Functions: hours, minutes, sweep seconds; date
Case: stainless steel, ø 39 mm, height 8.5 mm; sapphire crystal; transparent case back; water-resistant to 5 atm
Band: reptile skin, buckle
Price: $5,700

Master Ultra-Thin Moon
Reference number: 136 35 40
Movement: automatic, JLC Caliber 925/1; ø 26 mm, height 4.9 mm; 30 jewels; 28,800 vph; 38-hour power reserve
Functions: hours, minutes, sweep seconds; date, moon phase
Case: white gold, ø 39 mm, height 9.9 mm; sapphire crystal; transparent case back; water-resistant to 5 atm
Band: reptile skin, buckle
Price: $18,800

Master Ultra-Thin Perpetual
Reference number: 130 84 70
Movement: automatic, JLC Caliber 868/1; ø 27.8 mm, height 4.72 mm; 46 jewels; 38-hour power reserve
Functions: hours, minutes, sweep seconds; perpetual calendar with date, weekday, month, moon phase, year display (4 digits)
Case: stainless steel, ø 39 mm, height 9.2 mm; sapphire crystal; transparent case back; water-resistant to 5 atm
Band: reptile skin, double folding clasp
Price: $19,000

Master Ultra-Thin Small Seconds
Reference number: 135 84 80
Movement: automatic, JLC Caliber 896/1; ø 26 mm, height 3.98 mm; 32 jewels; 28,800 vph; 43-hour power reserve
Functions: hours, minutes, subsidiary seconds
Case: stainless steel, ø 40 mm, height 8.62 mm; sapphire crystal; transparent case back; water-resistant to 5 atm
Band: reptile skin, double folding clasp
Price: $14,000

JAEGER-LECOULTRE

Caliber 179

Manually wound; double-axis spherical tourbillon with different rotation times (60 and 12.6 seconds); double spring barrel, 40-hour power reserve
Functions: hours, minutes, subsidiary seconds (on tourbillon cage); additional 12-hour display (2nd time zone) and day/night indicator (on rear)
Measurements: 26.2 × 41 mm
Height: 6.85 mm
Jewels: 52
Balance: Gyrolab with hemispheric hairspring
Remarks: 385 parts

Caliber 751

Automatic; column wheel control of chronograph functions; double spring barrel, 65-hour power reserve
Functions: hours, minutes, subsidiary seconds; chronograph
Diameter: 26.2 mm
Height: 5.72 mm
Jewels: 39
Balance: screw balance with 4 weights
Frequency: 28,800 vph
Balance spring: flat hairspring
Shock protection: Kif

Caliber 752

Automatic; column wheel control of chronograph functions; double spring barrel, 65-hour power reserve
Functions: hours, minutes, subsidiary seconds; chronograph; date; world time indicator
Diameter: 26.2 mm
Height: 5.7 mm
Jewels: 41
Balance: screw balance with 4 weights
Frequency: 28,800 vph
Balance spring: flat hairspring
Shock protection: Kif
Remarks: plate with perlage, bridges with côtes de Genève

Caliber 772

Automatic; jumping second hand drive with remontoir; single spring barrel, 40-hour power reserve
Functions: hours, minutes, sweep seconds; world time indicator (2nd time zone)
Diameter: 28.6 mm
Height: 7.13 mm
Jewels: 36
Balance: ringless Gyrolab balance wheel with 4 weights
Frequency: 28,800 vph
Balance spring: flat hairspring
Shock protection: Kif
Remarks: 274 parts

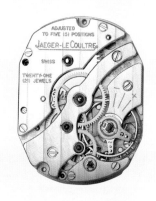

Caliber 822

Manually wound; single spring barrel, 42-hour power reserve
Functions: hours, minutes, subsidiary seconds
Measurements: 17.2 × 22 mm
Height: 2.94 mm
Jewels: 19
Balance: screw balance
Frequency: 21,600 vph
Balance spring: flat hairspring

Caliber 854

Manually wound; single spring barrel, 45-hour power reserve
Functions: hours and minutes with hands on both sides
Measurements: 13 × 15.2 mm
Height: 3.8 mm
Jewels: 21
Balance: glucydur
Frequency: 21,600 vph
Balance spring: flat hairspring
Shock protection: Kif
Remarks: 180 parts

JAEGER-LECOULTRE

Caliber 868
Automatic; single spring barrel, 38-hour power reserve
Functions: hours, minutes, sweep seconds; perpetual calendar with date, weekday, month, moon phase, year display (4 digits)
Diameter: 27.8 mm
Height: 4.72 mm
Jewels: 46
Balance: glucydur
Frequency: 28,800 vph
Remarks: 336 parts

Caliber 896
Automatic; single spring barrel, 43-hour power reserve
Functions: hours, minutes, subsidiary seconds
Diameter: 26 mm
Height: 3.98 mm
Jewels: 32
Frequency: 28,800 vph
Balance spring: flat hairspring
Shock protection: Kif

Caliber 898A
Automatic; single spring barrel, 43-hour power reserve
Functions: hours, minutes, sweep seconds; day/night indicator
Diameter: 26 mm
Height: 3.3 mm
Jewels: 30
Balance: glucydur, 4 regulating screws
Frequency: 28,800 vph
Balance spring: flat hairspring
Shock protection: Kif

Caliber 925
Automatic; single spring barrel, 38-hour power reserve
Functions: hours, minutes, sweep seconds; date, moon phase
Diameter: 26 mm
Height: 4.9 mm
Jewels: 30
Frequency: 28,800 vph
Remarks: 253 parts

Caliber 956
Automatic; automatic winding for time and alarm mechanisms; single spring barrel, 45-hour power reserve
Functions: hours, minutes, sweep seconds; date; alarm
Diameter: 28 mm
Height: 7.45 mm
Jewels: 23
Balance: glucydur
Frequency: 28,800 vph
Balance spring: flat hairspring
Remarks: perlage on mainplate, bridges with côtes de Genève, element for sounding board

Caliber 177
Manually wound; patented spherical tourbillon; double spring barrel, 192-hour power reserve
Functions: hours, minutes, subsidiary seconds (on tourbillon cage); perpetual calendar with date (2 retrograde hands) and month, shows real sun time (equation of time), retrograde leap year on movement side
Diameter: 36.3 mm
Height: 10.85 mm
Jewels: 77
Frequency: 21,600 vph
Remarks: 659 parts, plate with perlage, bridges with côtes de Genève

JAQUET DROZ

Though this watch brand first gained real notice when it was bought by the Swatch Group in 2001, Jaquet Droz looks back on a long tradition. Pierre Jaquet-Droz (1721–1790) was actually supposed to be a pastor, but instead followed the call to become a mechanic and a watchmaker. In the mid-eighteenth century, he began to push the limits of micromechanics, and his enthusiasm for it quickly led him to work on watch mechanisms and more complicated movements, which he attempted to operate through purely mechanical means.

Jaquet-Droz became famous in Europe for his automatons. More than once, he had to answer to religious institutions, whose guardians of public morals suspected there might be some devil's work and witchcraft behind his mechanical children, scribes, and organists. He even designed prostheses. A small enterprise in La Chaux-de-Fonds still produces items of this applied art, proof that the name Jaquet-Droz is alive and well in the Jura mountains, and its watches combined with automatons continue to thrill collectors and enthusiasts alike.

The Swatch Group has developed an esthetically and technically sophisticated collection based on an outstanding Frédéric Piguet movement. In recent years, the classically beautiful watch dials have taken on a slightly modern look without losing any of their identity. The spirit of the maverick founder of the brand still hovers about. Among the leading items in the current portfolio are automatons featuring twittering birds, an ancient technique revived and perfected by the brand. This focus on quality and inventiveness has made Jaquet Droz a top representative in the Swatch Group's portfolio. Its CEO is Marc A. Hayek, who is also CEO of Breguet and Blancpain.

Montres Jaquet Droz SA
CH-2300 La Chaux-de-Fonds
Switzerland

Tel.:
+41-32-924-2888

Fax:
+41-32-924-2882

E-mail:
info@jaquet-droz.com

Website:
www.jaquet-droz.com

Founded:
1738

U.S. Distribution:
The Swatch Group (U.S.), Inc.
1200 Harbor Boulevard
Weehawken, NJ 07086
201-271-1400
www.swatchgroup.com

Most important collections:
Les Ateliers d'Art, Grande Seconde, Grande Seconde SW, Astrale Collection, Petite Heure Minute, Lady 8

Grande Seconde Tribute
Reference number: J003031200
Movement: automatic, Jaquet Droz Caliber 2663. Si; ø 26.2 mm; 30 jewels; 28,800 vph; double spring barrel, silicon pallet forks and hairspring; oscillating mass in yellow gold; 68-hour power reserve
Functions: hours, minutes (off-center), subsidiary seconds
Case: yellow gold, ø 43 mm, height 11.48 mm; sapphire crystal; water-resistant to 3 atm
Band: reptile skin, buckle
Remarks: enamel dial
Price: $21,400; limited to 88 pieces

Grande Seconde Skelet-One Red Gold
Reference number: J003523240
Movement: automatic, Jaquet Droz Caliber 2663 SQ; ø 26.2 mm; 30 jewels; 28,800 vph; double spring barrel, silicon pallet forks and hairspring; skeletonized movement, oscillating mass in red gold; 68-hour power reserve
Functions: hours, minutes (off-center), subsidiary seconds
Case: red gold, ø 41 mm, height 12.3 mm; sapphire crystal; transparent case back; water-resistant to 3 atm
Band: reptile skin, buckle
Remarks: sapphire dial
Price: $33,600

Petite Heure Minute Smalta Clara Tiger
Reference number: J005504500
Movement: automatic, Jaquet Droz Caliber 6150; 29 jewels; 21,600 vph; 38-hour power reserve
Functions: hours and minutes (off-center)
Case: white gold, ø 35 mm, height 10.85 mm; bezel and lugs set with 100 diamonds; sapphire crystal; water-resistant to 3 atm
Band: satin, buckle
Remarks: white gold dial with "plique-à-jour" enamel
Price: $54,600; limited to 28 pieces

Jörg Schauer
c/o Stowa GmbH & Co. KG
Gewerbepark 16
D-75331 Engelsbrand
Germany

Tel.:
+49-7082-9306-0

Fax:
+49-7082-9306-2

E-mail:
info@schauer-germany.com

Website:
www.schauer-germany.com

Founded:
1990

Number of employees:
20

Annual production:
approx. 500 watches

Distribution:
direct sales; please contact the address in Germany

JÖRG SCHAUER

Jörg Schauer's watches are first and foremost cool. The cases have been carefully worked, the look is planned to draw the eye. After all, he is a perfectionist and leaves nothing to chance. He works on every single case himself, polishing and performing his own brand of magic for as long as it takes to display his personal touch. This time-consuming process is one that Schauer believes is absolutely necessary. "I do this because I place a great deal of value on the fact that my cases are absolutely perfect," he explains. "I can do it better than anyone, and I would never let anyone else do it for me."

Schauer, a goldsmith by training, has been making watches since 1990. He began by doing one-off pieces in precious metals for collectors and then opened his business and simultaneously moved to stainless steel. His style is to produce functional, angular cases with visibly screwed-down bezels and straightforward dials in plain black or white. Forget finding any watch close to current trends in his collection; Schauer only builds timepieces that he genuinely likes.

Purchasing a Schauer is not that easy. He has chosen a strategy of genuine quality over quantity and produces only about 500 watches annually. This includes special watches like the One-Hand Durowe, with a movement from one of Germany's movement manufacturers, Durowe, which Schauer acquired in 2002. His production structure is a vital part of his success and includes prototyping, movement modification, finishing, case production, dial painting, and printing—all done in Schauer's own workshop in Engelsbrand. Any support he needs from the outside he prefers to search out among regional specialists.

Edition 9
Reference number: Ed9
Movement: automatic, ETA Caliber 7751; ø 30 mm, height 7.9 mm; 28 jewels; 28,800 vph; finished with ornamental stripes and blued screws, exclusive engraved "Schauer" rotor; 42-hour power reserve
Functions: hours, minutes, subsidiary seconds; additional 24-hour display (2nd time zone); chronograph; full calendar with date, weekday, month, moon phase
Case: stainless steel, ø 42 mm, height 15 mm; bezel fixed with 12 screws; sapphire crystal; transparent case back; water-resistant to 5 atm
Band: stainless steel Milanese mesh, folding clasp
Price: $4,830
Variations: reptile skin strap ($4,530)

Edition 10
Reference number: Ed10
Movement: automatic, ETA Caliber 7753; ø 30 mm, height 7.9 mm; 27 jewels; 28,800 vph; finished with ornamental stripes and blued screws, exclusive engraved "Schauer" rotor; 48-hour power reserve
Functions: hours, minutes, subsidiary seconds; chronograph
Case: stainless steel, ø 42 mm, height 15 mm; bezel fixed with 12 screws; sapphire crystal; transparent case back; water-resistant to 5 atm
Band: calfskin, double folding clasp
Price: $3,820
Variations: stainless steel bracelet ($4,110); reptile skin strap ($3,775); manually wound movement ($4,200)

Edition 12
Reference number: Ed12
Movement: automatic, ETA Caliber 7753; ø 30 mm, height 7.9 mm; 27 jewels; 28,800 vph; finished with ornamental stripes and blued screws, exclusive engraved "Schauer" rotor; 48-hour power reserve
Functions: hours, minutes, subsidiary seconds; chronograph
Case: stainless steel, ø 41 mm, height 15 mm; bezel fixed with 12 screws; sapphire crystal; transparent case back; water-resistant to 5 atm
Band: calfskin, double folding clasp
Price: $4,325
Variations: stainless steel bracelet ($4,725); reptile skin strap ($4,325); manually wound movement ($4,675)

JUNGHANS

Germany has another horological success story besides Glashütte. Schramberg, a recondite town crouching in a valley between wooded slopes on the old trade route from Strasbourg in Alsace south to Lake Constance is where Erhard Junghans founded a watchmaking factory in 1861. His son Arthur then developed it into a large-scale production site built on American models. And so Schramberg became the hub of the watchmaking world. Nearly three thousand men and women worked at that factory and made nine thousand wall clocks and alarm clocks daily.

In the boom years after World War II, the company, with its logo featuring a star, manufactured wristwatches. It went on to ring in modern times with its own solar and radio-controlled watches. Junghans was twice the official timekeeper at the Olympic Games, and for a long time it remained the largest chronometer maker in the world. When quality became less of an issue in the consumer's consciousness, things started quieting down in the Black Forest factory.

Mechanical watches made their comeback in the 1990s, which is when many people remembered the precision timekeepers from the Black Forest. Among them was Dr. Hans-Jochem Steim, a successful entrepreneur from Schramberg, who decided to boost the rebirth of the Schramberg brand. Steim and his son purchased the company and decided to take on the financing of the necessary structural measures themselves. Just in time for the company's 150th anniversary (2011), Junghans devised a new production and distribution schedule. Today, the brand is proud of its extensive collection of high-quality wristwatches, ranging from genuine icons of design to major classics, all the way to sporty chronographs. The Junghans success is also driving the growth of a number of suppliers. Schramberg is once again a big name in the region, and its fame has even spread throughout Germany and beyond. The latest achievement is the opening of a museum devoted to watch- and clockmaking in the restored Terrrassebau, a century-old terraced construction that allowed Junghans employees to work with outstanding natural lighting.

Uhrenfabrik Junghans
GmbH & Co. KG
Geisshaldenstrasse 49
D-78713 Schramberg

Tel.:
+49-742-218-0

Fax:
+49-742-218-665

E-mail:
info@junghans.de

Website:
www.junghans.de

Founded:
1861

Number of employees:
127

Annual production:
approx. 60,000 watches

U.S. distributor:
DKSH Luxury & Lifestyle North America Inc.
9-D Princess Road
Lawrenceville, NJ 08648
609-750-8800

Most important collections/price range:
Junghans Meister; Max Bill by Junghans; Junghans MEGA; from approx. $750 to $2,500, special pieces up to $14,000

Meister Chronoscope Terrassenbau

Reference number: 027/4729.00
Movement: automatic, Caliber J880.1 (base ETA 7750); ø 30 mm, height 7.9 mm; 25 jewels; 28,800 vph; rhodium-plated movement, blued screws, rotor with côtes de Genève; 48-hour power reserve
Functions: hours, minutes, subsidiary seconds; chronograph; date, weekday
Case: stainless steel, ø 40.7 mm, height 13.9 mm; Plexiglas; water-resistant to 3 atm
Band: reptile skin, buckle
Remarks: hardened Plexiglas, antiscratch coating; anniversary model for 100th anniversary of Junghans's terraced industrial buildings
Price: $2,195; limited to 1,000 pieces

Meister Chronoscope Terrassenbau

Reference number: 027/9700.00
Movement: automatic, Caliber J880.1 (base ETA 7750); ø 30 mm, height 7.9 mm; 25 jewels; 28,800 vph; rhodium-plated movement, blued screws, rotor with côtes de Genève; 48-hour power reserve
Functions: hours, minutes, subsidiary seconds; chronograph; date, weekday
Case: rose gold, ø 40.7 mm, height 13.9 mm; sapphire crystal; water-resistant to 3 atm
Band: reptile skin, folding clasp
Remarks: engraved case back; anniversary model for 100th anniversary of Junghans's terraced construction
Price: $9,995; limited to 100 pieces

Meister Driver Chronoscope

Reference number: 027/3684.00
Movement: automatic, Caliber J880.3 (base ETA 2892-A2 with Dubois Dépraz module); ø 30 mm, height 6.5 mm; 45 jewels; 28,800 vph; rhodium-plated movement, blued screws, rotor with stripe finish; 42-hour power reserve
Functions: hours, minutes, subsidiary seconds; chronograph
Case: stainless steel, ø 40.8 mm, height 12.6 mm; Plexiglas; transparent case back; water-resistant to 3 atm
Band: calfskin, buckle
Remarks: hardened Plexiglas with antiscratch coating
Price: $2,095; **Variations:** with gray dial and stainless steel bracelet ($2,090)

JUNGHANS

Meister Telemeter
Reference number: 027/3380.00
Movement: automatic, Caliber J880.3 (base ETA 2892-2 with Dubois Dépraz module); ø 30 mm, height 6.5 mm; 45 jewels; 28,800 vph; rhodium-plated movement, blued screws, rotor with stripe finish; 42-hour power reserve
Functions: hours, minutes, subsidiary seconds; chronograph
Case: stainless steel, ø 40.8 mm, height 12.6 mm; Plexiglas; transparent case back; water-resistant to 3 atm
Band: calfskin, buckle
Remarks: hardened Plexiglas with antiscratch coating
Price: $2,295

Meister Calendar
Reference number: 027/7203.00
Movement: automatic, Caliber J800.3 (base ETA 2824-2 or Sellita SW200-1 with Dubois Dépraz module); ø 25.6 mm, height 6.2 mm; 25 or 26 jewels; 28,800 vph; 42-hour power reserve
Functions: hours, minutes, sweep seconds; full calendar with date, weekday, month, moon phase
Case: stainless steel with rose gold PVD coating, ø 40.4 mm, height 12 mm; Plexiglas; transparent case back; water-resistant to 3 atm
Band: reptile skin, buckle
Remarks: hardened Plexiglas, with antiscratch coating
Price: $2,295

Meister Chronoscope
Reference number: 027/4526.00
Movement: automatic, Caliber J880.1 (base ETA 7750); ø 30 mm, height 7.9 mm; 25 jewels; 28,800 vph; rhodium-plated movement, blued screws, rotor with côtes de Genève; 48-hour power reserve
Functions: hours, minutes, subsidiary seconds; chronograph; date, weekday
Case: stainless steel, ø 40.7 mm, height 13.9 mm; Plexiglas; transparent case back; water-resistant to 3 atm
Band: horse leather, buckle
Remarks: hardened Plexiglas, with antiscratch coating
Price: $1,895
Variations: various straps and dials

Form A
Reference number: 027/4730.00
Movement: automatic, Caliber J800.2 (base ETA 2824-2 or Sellita SW200-1); ø 25.6 mm, height 4.6 mm; 25 or 26 jewels; 28,800 vph; 38-hour power reserve
Functions: hours, minutes, sweep seconds; date
Case: stainless steel, ø 39.3 mm, height 9.5 mm; sapphire crystal; transparent case back; water-resistant to 5 atm
Band: calfskin, buckle
Price: $995
Variations: various straps and dials; quartz chronograph movement ($449)

Form A
Reference number: 027/4832.00
Movement: automatic, Caliber J800.2 (base ETA 2824-2 or Sellita SW200-1); ø 25.6 mm, height 4.6 mm; 25 or 26 jewels; 28,800 vph; 38-hour power reserve
Functions: hours, minutes, sweep seconds; date
Case: stainless steel, ø 39.3 mm, height 9.5 mm; sapphire crystal; transparent case back; water-resistant to 5 atm
Band: calfskin, buckle
Price: $995

Max Bill Manually Wound
Reference number: 027/3701.00
Movement: manually wound, Caliber J805.1 (base ETA 2801-2); ø 25.6 mm, height 3.35 mm; 17 jewels; 28,800 vph; 42-hour power reserve
Functions: hours, minutes, sweep seconds
Case: stainless steel, ø 34 mm, height 9 mm; Plexiglas
Band: calfskin, buckle
Remarks: hardened Plexiglas, scratch-resistant coating
Price: $695
Variations: various straps and dials

KOBOLD

Kobold Watch Company, LLC
1801 Parkway View Drive
Pittsburgh, PA 15205

Tel.:
724-533-3000

E-mail:
info@koboldwatch.com

Website:
www.koboldwatch.com

Founded:
1998

Number of employees:
20

Annual production:
maximum 2,500 watches

Distribution:
factory-direct, select retailers

Most important collections/price range:
Soarway / $1,950 to $35,000

Like many others in the field, Michael Kobold had already developed an interest in the watch industry in childhood. As a young man, he found a mentor in Chronoswiss founder Gerd-Rüdiger Lang, who encouraged him to start his own brand. This he did in 1998—at the age of nineteen while he was still a student at Carnegie Mellon University.

Today, Kobold Watch Company is headquartered in a big red farm in Amish country, Pennsylvania. There, it manufactures cases, movement components, dials, hands, and even straps.

The company's motto—Embrace Adventure—is reflected in the adventure-themed watches it turns out, worn by explorers such as Sir Ranulph Fiennes, whom *Guinness World Records* describes as "the world's greatest living explorer." The brand's centerpiece is the Soarway collection and the fabled Soarway case, which was originally created in 1999 by Sir Ranulph, master watchmaker and Chronoswiss founder Lang, as well as company founder Kobold, himself an avid mountain climber.

Kobold's love of the Himalayas has driven his commitment to the people of Nepal. He produces leather accessories and straps there and uses the operation to offer women vocational training. In 2015, he launched the Soarway Foundation to help Nepal in the event of earthquakes. Coincidentally, a few weeks later the first of two struck, devastating the country and the subsidiary. To help in such emergencies and others, he has started an initiative to get fire trucks to the country, hence the making of fire truck–themed watches. Kobold Nepal also works with Maiti Nepal to reintegrate trafficked women into society.

Kobold has contributed to the renaissance of American watchmaking and originally set its sights even higher, namely, on an in-house U.S.-made movement. Things have changed, though. "We make tough, rugged watches and so the case plays a more important role than the movement," says Kobold. "So for now, we're concentrating on making the toughest cases possible. One day, we'll tackle making in-house movements." The Soarway collection includes several novelties, such as the Soarway Transglobe, a watch with a second time zone that displays minutes as well as hours.

Soarway Transglobe
Reference number: KN 266853
Movement: automatic, Caliber K.793 (base ETA 2892-A2); ø 36 mm, height 4.95 mm; 26 jewels; 28,800 vph; 46-hour power reserve
Functions: hours, minutes, sweep seconds; date; 2nd time zone with hours, minutes
Case: stainless steel, ø 44 mm, height 14.3 mm; sapphire crystal; screw-down case back; water-resistant to 30 atm
Band: canvas, signed buckle
Price: $4,750

Himalaya
Reference number: KN 880121
Movement: automatic, ETA Caliber 2824-A2; ø 30.4 mm, height 10.35 mm; 25 jewels; 28,800 vph; 42-hour power reserve
Functions: hours, minutes, sweep seconds
Case: stainless steel, ø 44 mm, height 11.3 mm; antireflective sapphire crystal; screw-down case back; water-resistant to 10 atm
Band: calfskin, buckle
Price: $3,650
Variations: Arctic Blue, black, or white dial

Intrepid Automatic
Reference number: KD 130852
Movement: automatic, ETA Caliber 2824-A2; ø 30.4 mm, height 10.35 mm; 25 jewels; 28,800 vph; 42-hour power reserve
Functions: hours, minutes, sweep seconds
Case: stainless steel, ø 44 mm, height 11.3 mm; antireflective sapphire crystal; screw-down case back; water-resistant to 10 atm
Band: calfskin, buckle
Price: $3,650
Variations: Arctic blue, white

KOBOLD

Phantom Tactical Chronograph
Reference number: KD 924451
Movement: automatic, ETA Valjoux Caliber 7750; ø 30 mm, height 8.1 mm; 25 jewels; 28,800 vph; 46-hour power reserve; côtes de Genève, perlage, engraved and skeletonized gold-plated rotor
Functions: hours, minutes, subsidiary seconds; date, day; chronograph
Case: PVD-coated stainless steel, made in USA, ø 41 mm, height 15.3 mm; unidirectional bezel with 0-60 scale; screw-in crown/buttons; sapphire crystal; screw-down back; water-resistant to 300 m
Band: PVD-coated stainless steel, folding clasp
Price: $4,250

Intrepid Firetruck Expedition Special Edition
Reference number: KN 664583
Movement: automatic, ETA Caliber 2824-A2; ø 30.4 mm, height 10.35 mm; 25 jewels; 28,800 vph; 42-hour power reserve
Functions: hours, minutes, sweep seconds
Case: stainless steel, ø 44 mm, height 11.3 mm; antireflective sapphire crystal; screw-down case back; water-resistant to 10 atm
Band: alligator, buckle
Price: $3,950; limited to 100 pieces

Seal Firetruck Expedition Special Edition
Reference number: KD 1152147
Movement: automatic, ETA Caliber 2824-A2; ø 30.4 mm, height 10.35 mm; 25 jewels; 28,800 vph; 42-hour power reserve
Functions: hours, minutes, sweep seconds
Case: stainless steel, ø 44.25 mm, height 17.5 mm; antireflective sapphire crystal; screw-down case back; water-resistant to 100 atm
Band: reptile skin, buckle
Price: $4,250; limited to 100 pieces

Soarway Gurkhas
Reference number: KD 1142164
Movement: automatic, ETA Caliber 2892-A2; ø 28 mm, height 3.6 mm; 25 jewels; 28,800 vph; 46-hour power reserve
Functions: hours, minutes, sweep seconds; date
Case: stainless steel, made in USA; ø 43.5 mm, height 15 mm; unidirectional bezel; soft iron core; antireflective sapphire crystal; screw-down case back; screw-in crown; water-resistant to 50 atm
Band: reptile skin and canvas strap, buckle
Price: $4,250
Variations: limited edition of 100 watches

Soarway Diver
Reference number: KD 212441
Movement: automatic, ETA Caliber 2892-A2; ø 28 mm, height 3.6 mm; 25 jewels; 28,800 vph; 46-hour power reserve
Functions: hours, minutes, sweep seconds; date
Case: stainless steel, made in USA; ø 43.5 mm, height 15 mm; unidirectional bezel; soft iron core; antireflective sapphire crystal; screw-down case back; screw-in crown; water-resistant to 50 atm
Band: canvas, buckle
Price: $3,850
Variations: standard, non-California dial; Arctic blue dial

SMG-2
Reference number: KD 956853
Movement: automatic, Caliber ETA 2893-A2; ø 26.2 mm, height 6.1 mm; 21 jewels; 28,800 vph; 40-hour power reserve
Functions: hours, minutes, sweep seconds; 2nd time zone; date
Case: stainless steel, made in USA; ø 43 mm, height 12.75 mm; unidirectional bezel; soft iron core; antireflective sapphire crystal; screw-down case back; screw-in crown; water-resistant to 30 atm
Band: canvas, buckle
Price: $4,250

KUDOKE

Stefan Kudoke, a watchmaker from Frankfurt/Oder, has made a name for himself as an extremely skilled and imaginative creator of timepieces. He apprenticed with two experienced watchmakers and graduated as the number one trainee in the state of Brandenburg. This earned him a stipend from a federal program promoting gifted individuals. He then moved on to one of the large *manufactures* in Glashütte, where he refined his skills in its workshop for complications and prototyping. At the age of twenty-two, with a master's diploma in his pocket, he decided to get an MBA and then devote himself to building his own company.

His guiding principle is individuality, and that is not possible to find in a serial product. So Kudoke began building unique pieces. By realizing the special wishes of customers, he manages to reflect each person's uniqueness in each watch. And he has produced some out-of-the-ordinary pieces, like the ExCentro1 and 2, or more recently a watch with an octopus that seems to be climbing out of the case.

His specialties include engraving and goldsmithing. Within his creations bridges may in fact be graceful bodies, or the fine skeletonizing of a plate fragment, a world of figures and garlands. His recent creations include a skull watch done with characteristic care, and the minimalistic Kurt.

Kudoke Uhren
Tannenweg 5
D-15236 Frankfurt (Oder)
Germany

Tel.:
+49-335-280-0409

E-mail:
info@kudoke.eu

Website:
www.kudoke.eu

Founded:
2007

Number of employees:
1

Annual production:
30–50 watches

Distribution
Contact the brand directly for information.

Most important collections/price range:
between approx. $4,500 and $11,500

Kudoke 1
Movement: manually wound, Kudoke Caliber 1; ø 30 mm, height 4.3 mm; 18 jewels; 28,800 vph; hand-engraved and -finished
Functions: hours, minutes, subsidiary seconds
Case: stainless steel, ø 39 mm, height 10.5 mm; sapphire crystal; transparent case back
Band: reptile skin, buckle
Price: $9,270

Free KudOktopus
Movement: manually wound, ETA Caliber 6498 (modified); ø 36.6 mm, height 4.5 mm; 17 jewels; 18,000 vph; hand-engraved and -finished movement; 46-hour power reserve
Functions: hours, minutes
Case: stainless steel, ø 42 mm, height 10.5 mm; sapphire crystal; transparent case back
Band: reptile skin, buckle
Remarks: case entirely engraved by hand—the stylized octopus seems to be breaking out of the mechanism and reaching over the edge of the case
Price: $14,640

Kurt
Movement: manually wound, ETA Caliber 6498 (modified); ø 36.6 mm, height 4.5 mm; 17 jewels; 18,000 vph; hand-skeletonized movement; 46-hour power reserve
Functions: hours, minutes, subsidiary seconds
Case: stainless steel, ø 42 mm, height 10.5 mm; sapphire crystal; transparent case back
Band: calfskin, buckle
Price: $5,650

Laurent Ferrier
Route de Saint Julien 150
CH-1228 Plan-les-Ouates
Switzerland

Tel.:
+41-22-716-3388

E-mail:
info@laurentferrier.ch

Website:
www.laurentferrier.ch

Founded:
2010

Number of employees:
10

Annual production:
120

U.S. distributor:
Totally Worth It, LLC
76 Division Avenue
Summit, NJ 07901-2309
201-894-4710
info@totallyworthit.com

Most important collections/price range:
Variations of the Galet / from $40,000 to $345,000

LAURENT FERRIER

A rock rolling along a riverbed or being buffeted by coastal surf will, over time, achieve a kind of perfect shape, streamlined, flowing, smooth. It will usually become a comfortable touchstone for the human hand, a beautiful pebble, or *galet* in French. And that is the name given to the watches made by Laurent Ferrier in Geneva, Switzerland. The name refers to the special look and feel of the cases, which are just one hallmark of this very unusual, yet classical, watch brand.

The metaphor of the rock in water optimizing its form in a slow but consistent process could apply to Laurent Ferrier himself. He is a real person, the offspring of a watchmaking family from the Canton of Neuchâtel, and a trained watchmaker. As a young man he had a passion for cars, too, and even raced seven times at the 24 Hours of Le Mans. In 2009, after thirty-five years of employment at Patek Philippe working on new movements, Ferrier decided he had been shaped enough by his industry. He gathered up his deep experience and founded his own enterprise. He was joined by his son, Christian Ferrier, a watchmaker in his own right, and fellow former race driver, François Sérvanin.

One of the first watches was a tourbillon, and it was a winner. The geeks loved the technical wizardry of the natural escapement with the double hairspring ensuring greater accuracy (a technical idea going back to Breguet). The traditionalists found the tourbillon concealed on the movement side (as it used to be) very intriguing and effective, since it kept the dial free of clutter. Purists and estheticians understandably went for precisely that minimal dial, the spear hands, and the drop markers. Ferrier had not missed a single detail.

The brand has since evolved, but the DNA is well in place. The flagship Galet keeps evolving, this year with a minute repeater and an annual calendar. In spite of the complications, the dial of the latter piece remains very clear and organized. The watchmaker projecting his need for order onto the dial.

Galet Micro-Rotor with Ice-Blue Dial

Reference number: LCF004.AC.CG7
Movement: automatic, Laurent Ferrier Caliber FBN229.01; ø 31.6 mm, height 4.35 mm; 21,600 vph; 34 jewels; pawl-fitted microrotor; silicon escapement with dual direct impulse on balance; finely decorated bridges and mainplate; 72-hour power reserve
Functions: hours, minutes, subsidiary seconds
Case: stainless steel, ø 40 mm, height 10.70 mm; sapphire crystal; transparent case back; water-resistant to 3 atm; ball-shaped crown
Band: reptile skin, buckle or folding clasp
Price: $40,000

Galet Montre Ecole Annual Calendar

Reference number: LCF025.AC.A2W
Movement: manually wound, Laurent Ferrier Caliber LF126.01; ø 31.6 mm, height 5.80 mm; 21,600 vph; 23 jewels; Swiss lever escapement, finely decorated bridges and mainplate; semi-instantaneous calendar with correction forward or backward; 80-hour power reserve
Functions: hours, minutes, subsidiary seconds; day, date; power reserve indicator
Case: stainless steel, ø 40 mm, height 10.10 mm; ball-shaped crown; sapphire crystal front, transparent case back; water-resistant to 3 atm
Band: reptile skin, buckle or folding clasp
Price: $58,000

Galet Montre Ecole Minute Repeater

Reference number: LCF030.AC.R1G
Movement: manually wound, Laurent Ferrier Caliber LF707.01; ø 32.40 mm, height 5.36 mm; 21,600 vph; 32 jewels; Swiss lever escapement, finely decorated bridges and mainplate; 80-hour power reserve
Functions: hours, minutes, seconds
Case: stainless steel, ø 40 mm, height 10.40 mm; ball-shaped crown; sapphire crystal front, transparent case back; water-resistant to 3 atm
Band: reptile skin, buckle or folding clasp
Remarks: red gold with vertical satin-brushed finish; 11 indexes in white gold
Price: $345,000

LONGINES

The Longines winged hourglass logo is the world's oldest trademark, according to the World Intellectual Property Organization (WIPO). Since its founding in 1832, the brand has manufactured somewhere in the region of 35 million watches, making it one of the genuine heavyweights of the Swiss watch world. In 1983, Nicolas G. Hayek merged the two major Swiss watch manufacturing groups ASUAG and SIHH into what would later become the Swatch Group. Longines, the leading ASUAG brand, barely missed capturing the same position in the new concern; that honor went to Omega, the SIHH front-runner. However, from a historical and technical point of view, this brand has what it takes to be at the helm of any group. Was it not Longines that equipped polar explorer Roald Amundsen and air pioneer Charles Lindbergh with their watches? It has also been the timekeeper at many Olympic Games and, since 2007, the official timekeeper for the French Open at Roland Garros. In fact, this brand is a major sponsor at many sports events, from riding to archery.

It is not surprising then to find that this venerable Jura company also has an impressive portfolio of in-house calibers in stock, from simple manual winders to complicated chronographs. This broad technological base has benefited the company. As a genuine "one-stop shop," the brand can supply the Swatch Group with anything from cheap, thin quartz watches to heavy gold chronographs and calendars with quadruple retrograde displays. Longines does have one particular specialty, besides elegant ladies' watches and modern sports watches, in that it often has the luxury of rebuilding the classics from its own long history.

Longines Watch Co.
Rue des Noyettes 8
CH-2610 St.-Imier
Switzerland

Tel.:
+41-32-942-5425

Fax:
+41-32-942-5429

E-mail:
info@longines.com

Website:
www.longines.com

Founded:
1832

Number of employees:
worldwide approx. 900

U.S. distributor:
Longines
The Swatch Group (U.S.), Inc.
Longines Division
703 Waterford Way, Ste. 450
Miami, FL 33126
800-897-9477
www.longines.com

Most important collections/price range:
The Longines Master Collection, Longines DolceVita, Conquest V.H.P., HydroConquest, Heritage Collection / from approx. $1,000 to $6,500

HydroConquest

Reference number: L37814566
Movement: automatic, Longines Caliber L888.2 (base ETA A31.L01); ø 25.6 mm, height 3.85 mm; 21 jewels; 25,200 vph; 64-hour power reserve
Functions: hours, minutes, sweep seconds; date
Case: stainless steel, ø 41 mm, height 11.9 mm; unidirectional bezel with ceramic insert, 0-60 scale; sapphire crystal; screw-in crown; water-resistant to 30 atm
Band: stainless steel, double folding clasp, with safety lock and extension link
Price: $1,600
Variations: various straps and dials

HydroConquest

Reference number: L37814966
Movement: automatic, Longines Caliber L888.2 (base ETA A31.L01); ø 25.6 mm, height 3.85 mm; 21 jewels; 25,200 vph; 64-hour power reserve
Functions: hours, minutes, sweep seconds; date
Case: stainless steel, ø 41 mm, height 11.9 mm; unidirectional bezel with ceramic insert, 0-60 scale; sapphire crystal; screw-in crown; water-resistant to 30 atm
Band: stainless steel, double folding clasp, with safety lock and extension link
Price: $1,600
Variations: various straps and dials

HydroConquest

Reference number: L37814769
Movement: automatic, Longines Caliber L888.2 (base ETA A31.L01); ø 25.6 mm, height 3.85 mm; 21 jewels; 25,200 vph; 64-hour power reserve
Functions: hours, minutes, sweep seconds; date
Case: stainless steel, ø 41 mm, height 11.9 mm; unidirectional bezel with ceramic insert, 0-60 scale; sapphire crystal; screw-in crown; water-resistant to 30 atm
Band: rubber, double folding clasp, with safety lock
Price: $1,600
Variations: various straps and dials

LONGINES

HydroConquest Chronograph
Reference number: L3.744.4.96.6
Movement: automatic, Longines Caliber L688 (base ETA A08.L01); ø 30 mm, height 7.9 mm; 27 jewels; 28,800 vph; 54-hour power reserve
Functions: hours, minutes, subsidiary seconds; chronograph; date
Case: stainless steel, ø 41 mm, height 15.6 mm; unidirectional bezel with aluminum insert, 0-60 scale; sapphire crystal; screw-in crown; water-resistant to 30 atm
Band: stainless steel, double folding clasp, with safety lock and extension link
Price: $2,050

The Lindbergh Hour Angle Watch
Reference number: L26784110
Movement: automatic, Longines Caliber L699 (base ETA A07.L01); ø 36.6 mm, height 7.9 mm; 24 jewels; 28,800 vph; 46-hour power reserve
Functions: hours, minutes, sweep seconds, rotatable inner dial to synchronize seconds hand with radio time signals
Case: stainless steel, ø 47.5 mm, height 16.3 mm; bidirectional bezel with scale for time synchronization; sapphire crystal; water-resistant to 3 atm
Band: reptile skin, buckle
Price: $5,000

Record
Reference number: L28208922
Movement: automatic, Longines Caliber L888.4 (base ETA A31.L11); ø 25.6 mm, height 3.85 mm; 21 jewels; 25,200 vph; 64-hour power reserve; COSC-certified chronometer
Functions: hours, minutes, sweep seconds; date
Case: rose gold, ø 38.5 mm, height 10.7 mm; sapphire crystal; transparent case back; water-resistant to 3 atm
Band: reptile skin, folding clasp
Price: $5,850

Record
Reference number: L28215727
Movement: automatic, Longines Caliber L888.4 (base ETA A31.L11); ø 25.6 mm, height 3.85 mm; 21 jewels; 25,200 vph; 64-hour power reserve; COSC-certified chronometer
Functions: hours, minutes, sweep seconds; date
Case: stainless steel, ø 40 mm, height 10.8 mm; bezel and crown with rose gold PVD coating; sapphire crystal; transparent case back; water-resistant to 3 atm
Band: stainless steel with rose gold elements, folding clasp
Price: $3,450

Record
Reference number: L23215877
Movement: automatic, Longines Caliber L592.4 (base ETA A20.L11); ø 19.4 mm, height 4.1 mm; 22 jewels; 28,800 vph; 40-hour power reserve; COSC-certified chronometer
Functions: hours, minutes, sweep seconds; date
Case: stainless steel, ø 30 mm, height 10.7 mm; bezel and crown in rose gold; sapphire crystal; transparent case back; water-resistant to 3 atm
Band: stainless steel with rose gold elements, folding clasp
Remarks: mother-of-pearl dial set with 13 diamonds
Price: $3,200

The Longines Master Collection Moonphase
Reference number: L26734783
Movement: automatic, Longines Caliber L687 (base ETA A08.L91); ø 30 mm, height 7.9 mm; 25 jewels; 28,800 vph; 54-hour power reserve
Functions: hours, minutes, subsidiary seconds; additional 24-hour display; chronograph; full calendar with date, weekday, month, moon phase
Case: stainless steel, ø 40 mm, height 14.3 mm; sapphire crystal; transparent case back; water-resistant to 3 atm
Band: reptile skin, folding clasp
Price: $3,325
Variations: stainless steel bracelet ($3,325)

LONGINES

The Longines Master Collection Annual Calendar
Reference number: L29104786
Movement: automatic, Longines Caliber LL897.2 (base ETA A31.L81); ø 25.6 mm, height 5.2 mm; 21 jewels; 25,200 vph; 64-hour power reserve
Functions: hours, minutes, sweep seconds; annual calendar with date, month
Case: stainless steel, ø 40 mm, height 10.8 mm; sapphire crystal; transparent case back; water-resistant to 3 atm
Band: stainless steel, folding clasp
Price: $2,425

The Longines Master Collection Annual Calendar
Reference number: L29104783
Movement: automatic, Longines Caliber L897.2 (base ETA A31.L81); ø 25.6 mm, height 5.2 mm; 21 jewels; 25,200 vph; 64-hour power reserve
Functions: hours, minutes, sweep seconds; annual calendar with date, month
Case: stainless steel, ø 40 mm, height 10.8 mm; sapphire crystal; transparent case back; water-resistant to 3 atm
Band: reptile skin, folding clasp
Price: $2,425

The Longines Elegant Collection
Reference number: L49104116
Movement: automatic, Longines Caliber L888 (base ETA A31.L01); ø 25.6 mm, height 3.85 mm; 21 jewels; 25,200 vph; 64-hour power reserve
Functions: hours, minutes, sweep seconds; date
Case: stainless steel, ø 39 mm, height 8.6 mm; sapphire crystal; transparent case back; water-resistant to 3 atm
Band: stainless steel, folding clasp
Price: $1,900

The Longines Elegant Collection
Reference number: L49104922
Movement: automatic, Longines Caliber L888 (base ETA A31.L01); ø 25.6 mm, height 3.85 mm; 21 jewels; 25,200 vph; 64-hour power reserve
Functions: hours, minutes, sweep seconds; date
Case: stainless steel, ø 39 mm, height 8.6 mm; sapphire crystal; transparent case back; water-resistant to 3 atm
Band: reptile skin, buckle
Price: $1,900

The Longines Legend Diver Watch
Reference number: L37742509
Movement: automatic, Longines Caliber L888 (base ETA A31.L01); ø 25.6 mm, height 4.6 mm; 25 jewels; 25,200 vph; 64-hour power reserve
Functions: hours, minutes, sweep seconds; date
Case: stainless steel with black PVD coating, ø 42 mm, height 12.7 mm; crown-activated scale ring, 0-60 scale; sapphire crystal; screw-in crown; water-resistant to 30 atm
Band: rubber, double folding clasp with extension link
Price: $2,700

The Longines Legend Diver Watch
Reference number: L37742509
Movement: automatic, Longines Caliber L888 (base ETA A31.L01); ø 25.6 mm, height 4.6 mm; 25 jewels; 25,200 vph; 64-hour power reserve
Functions: hours, minutes, sweep seconds; date
Case: stainless steel, ø 42 mm, height 12.7 mm; crown-activated scale ring, 0-60 scale; sapphire crystal; screw-in crown; water-resistant to 30 atm
Band: stainless steel Milanese mesh, folding clasp
Price: $2,400

Les Ateliers Louis Moinet SA
Rue du Temple 1
CH-2072 Saint-Blaise
Switzerland

Tel.:
+41-32-753-6814

E-mail:
info@louismoinet.com

Website:
www.louismoinet.com

Founded:
2005

U.S. distributor:
Fitzhenry Consulting
1029 Peachtree Parkway, #346
Peachtree City, GA 30269
561-212-6812
Don@fitzhenry.com

Most important collections:
Memoris, Sideralis, 20-Second Spacewalker, Mecanograph, Space Mystery

LOUIS MOINET

In the race to be the first to invent something new, Louis Moinet (1768–1853) emerged as a notable winner: In 2013, a *Compteur de tierces* from 1816 was shown to the public, a chronograph that counts one-sixtieth of a second with a frequency of 216,000 vph. It was proudly signed by Moinet. This professor at the Academy of Fine Arts in Paris and president of the Société Chronométrique was in fact one of the most inventive, multitalented men of his time. He worked with such eminent watchmakers as Breguet, Berthoud, Winnerl, Janvier, and Perrelet. Among his accomplishments is an extensive two-volume treatise on horology.

Following in such footsteps is hardly an easy task, but Jean-Marie Schaller and Micaela Bertolucci decided that their idiosyncratic creations were indeed imbued with the spirit of the great Frenchman. They work with a team of independent designers, watchmakers, movement specialists, and suppliers to produce the most unusual wristwatches filled with clever functions and surprising details. The Jules Verne chronographs have hinged levers, for example, and the second hand on the Tempograph changes direction every ten seconds.

Some watches tell industrial tales, like the Derrick Tourbillon or the Gaz Derrick. Increasingly, this independent-minded brand is exploring the space-time continuum. The dial of the chronograph watch Memoris, for the centenary of the invention of the chronograph by Louis Moinet, is dotted with stars. The epic inverted double tourbillon Sideralis, presented at Baselworld in 2016, features fragments of the famous Rosetta stone and dust from Mars and the moon. Another double tourbillon joins the roster now, with a dial of Acasta gneiss, an ancient Earth-born stone. And space this year is honored by a watch made in collaboration with a gentleman named Alexey Leonov, the first person to leave a capsule to take a little spacewalk . . .

SpaceWalker
Reference number: LM-62.50G.25
Movement: manually wound, Louis Moinet Caliber LM 48; ø 37.65 mm, height 10.33 mm; 20 jewels; 21,600 vph; 72-hour power reserve; "satellite" 13.59-mm tourbillon balanced by a planet on the cage; rhodium-plated, satin-brushed mainplate, bridges with côtes de Genève
Functions: hours, minutes
Case: rose gold, ø 47.4 mm, height 16.9 mm; sapphire crystal; transparent case back; water-resistant to 5 atm
Band: reptile skin, double folding clasp
Price: $199,500; limited to 12 pieces

Metropolis
Reference number: LM-45.50.55
Movement: automatic, Louis Moinet Caliber LM45; ø 30.4 mm, height 6.7 mm; 22 jewels; 28,800 vph; 48-hour power reserve; côtes de Genève decoration
Functions: hours, minutes, subsidiary seconds
Case: rose gold, ø 43.2 mm, height 14.8 mm; sapphire crystal; screwed-in transparent back; water-resistant to 5 atm
Band: reptile skin, folding clasp
Remarks: skeletonized dial and hour markers
Price: $29,900; limited to 60 pieces
Variations: stainless steel ($12,900; limited to 60 pieces); "magic" blue dial pattern

Acasta
Reference number: LM-52.50.AC
Movement: manually wound, Louis Moinet Caliber LM 52; ø 37.6 mm, height 9.85 mm; 36 jewels; 21,600 vph; 52-hour power reserve; double tourbillon; côtes de Genève on bridges
Functions: hours, minutes, automaton
Case: gold, ø 47.4 mm, height 14.92 mm; hand-engraved casing with sapphire crystal; transparent case back; water-resistant to 3 atm
Remarks: dial with Acasta gneiss, a rock from the early earth crust
Band: reptile skin, folding clasp
Price: $325,000; limited to 3 pieces

LOUIS VUITTON

The philosophy of this over 150-year-old brand states that any product bearing the name Louis Vuitton must be manufactured in the company's own facilities. That is why Louis Vuitton has allowed itself the luxury of building its own workshop in Switzerland, specifically in La Chaux-de-Fonds, at the technology center of LVMH (Louis Vuitton, Moët & Hennessy).

Designing is carried out in Paris at the company headquarters, and it is obvious that it would not suit an upscale watch to simply cobble together various parts supplied by outside workshops. The cases and dials with all the details and the hands are all exclusive Louis Vuitton designs, as are other components, such as the pushers and the band clasps—in other words, all that is needed to ensure a unique look. In 2011, Louis Vuitton purchased the dial maker Léman Cadran and the movement specialist Fabrique du Temps (both in Geneva), giving the company a great deal of independence vis-à-vis other brands in the group. And helping it clinch a Geneva Seal for its brand-new tourbillon, whose transparency almost makes it "mysterious." A year after comes the Tambour Moon Mystérieuse Flying Tourbillon driven by a minimalist movement that floats between sapphire crystals without any visible contact to the crown or the case.

For a small *manufacture*, Louis Vuitton has been brave. It has produced the LV Fifty Five collection, the Tambour, and the unconventional Escale world-time watch featuring hand-painted scale fields in the style of the monograms that Louis Vuitton uses to mark its bags. The new Tambour Moons bring together high fashion and "quartz" watchmaking. And for the incurably connected, the brand has developed the Tambour Horizon with a variety of colorful dials.

Louis Vuitton Malletier
2, rue du Pont Neuf
75001
France

Tel.:
+33-1-55-80-41-40

Fax:
+33-1-55-80-41-40

Website:
www.vuitton.com

Founded:
1854

U.S. distributor:
Louis Vuitton
1-866-VUITTON
www.louisvuitton.com

Most important collection/price range:
Tambour / Voyage / LV Fifty Five / Escale / starting at $3,250

Tambour All Black Chronograph
Reference number: QAAA18
Movement: automatic, ETA Caliber 2894-2; ø 28.6 mm, height 6.3 mm; 28 jewels; 28,800 vph; 42-hour power reserve
Functions: hours, minutes, subsidiary seconds; chronograph; date
Case: stainless steel with black PVD coating, ø 46 mm, height 14.7 mm; sapphire crystal; water-resistant to 10 atm
Band: reptile skin, buckle
Price: $7,695

Tambour All Black Petite Seconde
Reference number: QAAA19
Movement: automatic, ETA Caliber 2895-2; ø 25.6 mm, height 4.35 mm; 27 jewels; 28,800 vph; 42-hour power reserve
Functions: hours, minutes, subsidiary seconds; date
Case: stainless steel with black PVD coating, ø 41.5 mm, height 11 mm; sapphire crystal; transparent case back; water-resistant to 10 atm
Band: reptile skin, buckle
Price: $5,195

Tambour Essential Grey Chronograph
Reference number: QA025Z
Movement: automatic, ETA Caliber 2894-2; ø 28.6 mm, height 6.3 mm; 28 jewels; 28,800 vph; 42-hour power reserve
Functions: hours, minutes, subsidiary seconds; chronograph; date
Case: stainless steel, ø 44 mm, height 14.8 mm; sapphire crystal; water-resistant to 10 atm
Band: reptile skin, buckle
Price: $5,600

LOUIS VUITTON

Tambour Moon Chronograph

Reference number: QAAA14
Movement: automatic, ETA Caliber 2894-2; ø 28.6 mm, height 6.1 mm; 37 jewels; 28,800 vph; 42-hour power reserve
Functions: hours, minutes, subsidiary seconds; chronograph; date
Case: stainless steel, ø 44 mm, height 12.02 mm; sapphire crystal; transparent case back; water-resistant to 50 atm
Band: rubber, buckle
Price: $7,150

Tambour Moon Mystérieuse Flying Tourbillon

Movement: manually wound, LV Caliber 110; 13 × 34 mm; 17 jewels; 21,600 vph; flying 1-minute tourbillon without visible drive under sapphire bridges; 80-hour power reserve; Geneva Seal, Qualité Fleurier
Functions: hours, minutes
Case: platinum, ø 45 mm, height 10.6 mm; sapphire crystal; transparent case back; water-resistant to 50 atm
Band: reptile skin, double folding clasp
Price: $256,000

Tambour Moon Blue Chronograph

Movement: automatic, LV Caliber 277 (base Zenith "El Primero"); ø 30 mm, height 6.6 mm; 31 jewels; 36,000 vph; 50-hour power reserve
Functions: hours, minutes, subsidiary seconds; chronograph
Case: rose gold, ø 44 mm, height 12.6 mm; sapphire crystal; water-resistant to 50 atm
Band: reptile skin, buckle
Price: $26,700

Tambour Horizon

Reference number: QAAA67
Movement: quartz, android- and iOS-compatible smartwatch
Functions: hours, minutes; world time indicator (2nd time zone), personalized dials, flight info services, special app for city tours, meteorological information, pace counter; perpetual calendar with date, weekday, month
Case: stainless steel with black PVD coating, ø 42 mm, height 12.55 mm; sapphire crystal; water-resistant to 30 atm
Band: synthetic, buckle
Price: $2,900

Escale Worldtime Blue

Reference number: Q5EK50
Movement: automatic, LV Caliber 106; ø 37 mm, height 6.65 mm; 26 jewels; 28,800 vph; 38-hour power reserve
Functions: hours, minutes; world time indicator (2nd time zone) for 24 time zones on a hand-painted rotating disk
Case: titanium, ø 41 mm, height 9.75 mm; bezel, lugs, and crown in white gold; sapphire crystal; water-resistant to 3 atm
Band: reptile skin, folding clasp
Price: $57,000

Escale Spin Time Blue

Reference number: Q5EG20
Movement: automatic, LV Caliber 77; ø 35.2 mm, height 7.2 mm; 27 jewels; 28,800 vph; 42-hour power reserve
Functions: hours (jumping) on dice elements, minutes
Case: titanium, ø 41 mm, height 11.2 mm; bezel, lugs, and crown in white gold; sapphire crystal; water-resistant to 3 atm
Band: reptile skin, double folding clasp
Price: $44,500

LUMINOX

Watches, as the old industry axiom goes, are jewelry for men. And some men do like watches that express masculinity in no uncertain terms, or that feel like real tools, or that recall the cockpits of fast-moving vehicles. So when Barry Cohen came across the tiny tritium gas–filled luminescent tubes made by the Swiss company mb-microtech, he spotted an opportunity. Here was a way to give sports watches the kind of illumination that would make them dependable and practical time-givers at night. The radioactive tritium, which has a half-life of 12.32 years, makes a coating on the inside of the tubes glow for up to 25 years.

Cohen and his business partner Richard Timbo called their brand Luminox, derived from the Latin "light" and "night." They created a collection of rugged-looking sports timepieces that soon found a loyal following. In 1992 came the first big breakthrough, when a Luminox watch prevailed in a tough competition to become a mission watch for Navy SEALs. The brand now established a reputation, and soon other law enforcement agencies and organizations began ordering watches, notably F-117 Nighthawk and Stealth pilots.

As the brand grew and expanded beyond American borders, it continued developing its product. A new lightweight carbon compound case with a matte finish was developed that is insensitive to outside temperatures. The latest versions of this case can withstand dives of up to 300 meters. A special mineral crystal was also developed that is highly scratch resistant.

Luminox also partnered with the Swiss company Mondaine, famous for its Railroad Watch and the Helvetica, to manufacture watches in their premises in Switzerland. A new company, called Lumondi Inc., was founded to oversee the two brands after Mondaine purchased the 50 percent of remaining shares from Barry Cohen in November 2016.

Luminox watches are unabashedly muscular and outdoorsy. Wherever extreme sports or activities are being performed, that is where Luminox finds its fans. The Scott Cassell Deep Dive Automatic, for example, was made for explorer and deep-seas diver Scott Cassell as part of his "essential gear." The watches are run either on Swiss quartz or on mechanical movements.

Lumondi Inc. (Luminox Watches)
27 W. 24th Street, Suite 804
New York, NY 10010

Tel.:
917-522-3600

E-mail:
info@luminoxusa.com

Website:
www.luminox.com
shop.luminox.com

Founded:
1989

Most important collections:
Navy SEAL 3500, Leatherback Sea, Turtle Giant, XCOR Aerospace

Scott Cassell Deep Dive Automatic
Reference number: 1523
Movement: automatic, ETA Caliber 2826-2; ø 25.6 mm, height 6.2 mm; 25 jewels; 28,800 vph; 38-hour power reserve
Functions: hours, minutes, sweep seconds; large date
Case: stainless steel, ø 44 mm, height 17 mm; unidirectional bezel with blue aluminum ring, bezel locker at 3 o'clock; sapphire crystal with antireflective coating; helium release valve; stainless steel screw-in case back; water-resistant to 50 atm
Band: rubber strap, stainless steel buckle
Remarks: constant glow for up to 25 years in any light condition
Price: $2,000
Variations: black/yellow dial with black rubber strap

XCOR Aerospace Automatic Valjoux Chronograph
Reference number: 5261
Movement: automatic, ETA Caliber 7750 Valjoux; ø 30 mm, height 7.9; 25 jewels; 28,800 vph; 42-hour power reserve
Functions: hours, minutes, chronograph; date
Case: titanium and black PVD, ø 45.5 mm, height 18 mm; sapphire crystal; titanium screw-down case back; water-resistant to 20 atm
Band: calfskin, titanium buckle
Remarks: constant glow for up to 25 years in any light condition
Price: $3,000

F-117 Nighthawk
Reference number: 6422
Movement: quartz, Ronda Caliber 515; ø 26.2 mm, height 3 mm
Functions: hours, minutes; date; 2nd time zone
Case: stainless steel IP gun metal, ø 44 mm, height 12.6 mm; bidirectional bezel; sapphire crystal; water-resistant to 20 atm
Band: stainless steel and buckle with black PVD
Remark: constant glow for up to 25 years in any light condition
Price: $1,400
Variations: Kevlar strap with black stitching, black leather lining, and black PVD buckle

MANUFACTURE ROYALE

Manufacture Royale SA
ZI Le Day
CH-1337 Vallorbe
Switzerland

Tel.:
+41-21-843-01-01

E-mail:
info@manufacture-royale.com

Website:
www.manufacture-royale.com

Founded:
by Voltaire in 1770, revived in 2010

Number of employees:
5

Annual production:
150 watches

Distribution:
Retail and online

In the U.S.:
Café Time, LLC
917-907-1127

Most important collections/price range:
Androgyne / from $47,200; 1770 / from $52,800; Opera / $385,000

François-Marie Arouet (1694–1778), known as Voltaire, was a brilliant playwright, a historian, a freewheeling philosopher, and an all-around thinker. He was also one of the richest men in Europe, which allowed him to express his very progressive views quite freely. In Geneva, where he had often found refuge from the French king, he went further: The local established bourgeoisie steadfastly refused to give political and economic rights to a class of craftsmen known as the *natifs*, whose origins were not local and who made up nearly half the population. In 1770, at his estate in neighboring Ferney-Voltaire (France), Voltaire opened the "Manufacture Royale," which produced very respectable watches.

In 2010, four highly experienced and related watch executives, Gérard, David, and Alexis Gouten, and Marc Guten, decided to revive the brand. Their basic idea: high-end complications, affordable prices, *manufacture* movements assembled in-house.

They set up shop in Vallorbe in the foothills of the Jura, and soon had three respected collections on the market. The 1770 is the most classic piece, with a simple dial as backdrop to a flying tourbillon. The Androgyne, an edgy timepiece with flexible lugs, a screwed-down bezel, and a generally steampunkish look, offers insight into the movement thanks to extensive skeletonizing. Finally, there is the Opera, a minute repeater with a cleverly designed case made of sixty parts that unfolds to form a kind of shell that amplifies the sound of the chimes and looks a little like the Opera House in Sydney, Australia.

The brand has gained lots of experience since its foundation and the self-confidence to sally into more experimental realms, but without ever disturbing the fundamental classicism of the timepieces. The 1770 Haute Voltige series is definitely twenty-first century, with its second time zone cowering under the mysterious bridges that rise from the dial to hold the balance wheel over the dial. The Micromégas Revolution comes in three strong colors. As for the ADN, it cuts to the chase, literally, with extreme skeletonization.

ADN Spirit
Reference number: ADN46.04CS04.DC
Movement: manually wound, Caliber MR10; ø 32.6 mm, height 7.40 mm; 1-minute flying tourbillon on ceramic ball bearings, silicon escapement wheel and pallet; 21 jewels, 48-hour power reserve, 28,800 vph; CVD-coated plates and bridges
Functions: hours (jumping), minutes, subsidiary seconds
Case: stainless steel and forged carbon, ø 46 mm, height 12 mm; sapphire crystal, screwed-down transparent sapphire case back, water-resistant to 3 atm
Band: reptile skin, triple folding clasp
Price: $31,400
Variations: stainless steel ($26,800)

1770 Micromégas Revolution
Reference number: 1770MR45.10.BL
Movement: automatic, MR08 Caliber; ø 36 mm, height 8.7 mm; 26 jewels; 1-minute flying tourbillon (28,800 vph) and 6-second tourbillon (21,600 vph), both with silicon escapement wheel and levers; 40-hour power reserve; skeletonized movement
Functions: hours (off-center), minutes
Case: titanium, ø 45 mm, height 11.8 mm; transparent case back, water-resistant to 3 atm
Band: reptile skin, titanium buckle
Price: $149,100
Variations: pink gold ($172,200)

1770 Haute Voltige
Reference number: 1770HVT45.01.BL
Movement: automatic, MR07 Caliber; 31 jewels; ø 36 mm, height 9.45 mm; 21,600 vph; 40-hour power reserve; red anodized aluminum balance wheel
Functions: hours, minutes, completely independent 2nd time zone
Case: stainless steel, ø 46 mm, height 14 mm; black sunray dial, sapphire crystal; water-resistant to 3 atm
Remarks: raised bridge on dial holds balance wheel over dial; raised subsidiary dial for 2nd time zone
Band: reptile skin, buckle
Price: $36,800
Variations: black sunray dial; "Dragon" with green sunray dial

MAURICE LACROIX

Maurice Lacroix watches are found in sixty countries. The heart of the company, however, remains the production facilities in the highlands of the Jura, in Saignelégier and Montfaucon, where the brand built La Manufacture des Franches-Montagnes SA (MFM) outfitted with state-of-the-art CNC technology for the production of very specific individual parts and movement components.

The watchmaker can thank the clever interpretations of "classic" pocket watch characteristics for its steep ascent in the 1990s. Since then, the *manufacture* has redesigned the complete collection, banning every lick of Breguet-like bliss from its watch designs. In the upper segment, *manufacture* models such as the chronograph and the retrograde variations on Unitas calibers set the tone. In the lower segment, modern "little" complications outfitted with module movements based on ETA and Sellita are the kings. The brand is mainly associated with the hypnotically turning square wheel, the "roue carrée." The idea was used for the latest ladies' watch, the Power of Love, which has three turning hearts forming the word "love" at regular intervals.

Maurice Lacroix's drive to freshen up its look has earned the brand a great deal of recognition in the past years, notably eleven Red Dot awards.

In 2011, DKSH (Diethelm Keller & SiberHegner) took over the brand. This Swiss holding company specializing in international market expansions with 770 establishments throughout the world has ensured Maurice Lacroix a strong position in all major markets, with flagship stores and its own boutiques. A special partnership with the Barcelona football club is bound to have an impact on sales as well, justifying the production of 90,000 watches per year.

Maurice Lacroix SA
Rüschlistrasse 6
CH-2502 Biel/Bienne
Switzerland

Tel.:
+41-44-209-1111

E-mail:
info@mauricelacroix.com

Website:
www.mauricelacroix.com

Founded:
1975

Number of employees:
about 250 worldwide

Annual production:
Approx. 90,000 watches

U.S. distributor:
DKSH Luxury & Lifestyle North America Inc.
9-D Princess Road
Lawrenceville, NJ 08648
609-750-8800

Most important collections/price range:
Aikon / $890 to $2,900; Les Classiques / $950 to $4,300; Fiaba (ladies) / $980 to $2,900; Pontos / $1,750 to $7,900; Masterpiece *manufacture* models / $6,800 to $14,900

Masterpiece Square Wheel Retrograde

Reference number: MP6058-SS001-110-1
Movement: automatic, Caliber ML 258 (base Sellita SW200 with ML module); ø 34 mm, height 8.6 mm; 37 jewels; 28,800 vph; 36-hour power reserve
Functions: hours, minutes, subsidiary seconds; date (retrograde)
Case: stainless steel, ø 43 mm, height 15 mm; sapphire crystal; transparent case back; water-resistant to 10 atm
Band: reptile skin, folding clasp
Price: $7,490

Masterpiece Calendar Retrograde

Reference number: MP6568-SS001-132-1
Movement: automatic, Caliber ML 190; ø 36.6 mm, height 9.9 mm; 50 jewels; 18,000 vph; 52-hour power reserve
Functions: hours, minutes, subsidiary seconds; power reserve indicator; date (retrograde)
Case: stainless steel, ø 43 mm, height 15 mm; sapphire crystal; transparent case back; water-resistant to 5 atm
Band: reptile skin, folding clasp
Price: $4,500

Masterpiece Gravity

Reference number: MP6118-SS001-434-1
Movement: automatic, Caliber ML 230; ø 37.2 mm, height 9.05 mm; 35 jewels; 18,000 vph; inverted movement construction with escapement on dial; silicon pallet lever and pallet fork; 48-hour power reserve
Functions: hours, minutes (off-center), subsidiary seconds
Case: stainless steel, ø 43 mm, height 16.2 mm; sapphire crystal; transparent case back; water-resistant to 5 atm
Band: reptile skin, folding clasp
Price: $13,900
Variations: various cases and dials

MAURICE LACROIX

Masterpiece Chronograph Skeleton
Reference number: MP6028-PVC01-002-1
Movement: automatic, Caliber ML 206; ø 30.4 mm, height 8.4 mm; 25 jewels; 28,800 vph; skeletonized movement, bridges black gold-plated; 48-hour power reserve
Functions: hours, minutes, subsidiary seconds; chronograph
Case: stainless steel with blue PVD coating, ø 45 mm, height 16.4 mm; sapphire crystal; screw-in crown; water-resistant to 10 atm
Band: reptile skin, folding clasp
Remarks: skeletonized dial
Price: $7,900

Masterpiece Double Retrograde
Reference number: MP6578-SS001-331-1
Movement: automatic, Caliber ML 191; ø 36.6 mm, height 8.2 mm; 74 jewels; 18,000 vph; 52-hour power reserve
Functions: hours, minutes, subsidiary seconds; additional 24-hour display (2nd time zone, retrograde); date (retrograde)
Case: stainless steel, ø 43 mm, height 15.4 mm; sapphire crystal; transparent case back; water-resistant to 5 atm
Band: reptile skin, folding clasp
Price: $4,500

Aikon Automatic Chronograph LE
Reference number: AI6018-PVB01-331-1
Movement: automatic, Caliber ML 157 (base ETA 7753); ø 30 mm, height 7.9 mm; 27 jewels; 28,800 vph; 48-hour power reserve
Functions: hours, minutes, subsidiary seconds; chronograph; date
Case: stainless steel with black PVD coating, ø 44 mm, height 15 mm; sapphire crystal; transparent case back; screw-in crown; water-resistant to 20 atm
Band: calfskin, double folding clasp
Price: $3,690; limited to 500 pieces

Aikon Automatic Chronograph
Reference number: AI6038-SS001-131-1
Movement: automatic, Caliber ML 112 (base ETA 7750); ø 30 mm, height 7.9 mm; 25 jewels; 28,800 vph; 48-hour power reserve
Functions: hours, minutes, subsidiary seconds; chronograph; date, weekday
Case: stainless steel, ø 44 mm, height 15 mm; sapphire crystal; transparent case back; screw-in crown; water-resistant to 20 atm
Band: calfskin, double folding clasp
Price: $2,890

Aikon Automatic
Reference number: AI6008-SS001-430-1
Movement: automatic, Caliber ML 115 (base Sellita SW200-1); ø 25.6 mm, height 4.6 mm; 26 jewels; 28,800 vph; 38-hour power reserve
Functions: hours, minutes, sweep seconds; date
Case: stainless steel, ø 42 mm, height 11 mm; sapphire crystal; transparent case back; screw-in crown; water-resistant to 20 atm
Band: calfskin, double folding clasp
Price: $1,890

Aikon Automatic Skeleton
Reference number: AI6028-SS001-030-1
Movement: automatic, Caliber ML 234; ø 36.6 mm, height 8.7 mm; 34 jewels; 18,000 vph; movement completely skeletonized; 52-hour power reserve
Functions: hours, minutes, subsidiary seconds
Case: stainless steel, ø 45 mm, height 13 mm; sapphire crystal; transparent case back; screw-in crown; water-resistant to 10 atm
Band: calfskin, double folding clasp
Price: $5,990

MAURICE LACROIX

Aikon Automatic
Reference number: AI6008-SS002-330-1
Movement: automatic, Caliber ML 115 (base Sellita SW200-1); ø 25.6 mm, height 4.6 mm; 26 jewels; 28,800 vph; 38-hour power reserve
Functions: hours, minutes, sweep seconds; date
Case: stainless steel, ø 42 mm, height 11 mm; sapphire crystal; transparent case back; screw-in crown; water-resistant to 20 atm
Band: stainless steel, double folding clasp
Price: $1,990

Les Classiques Date
Reference number: LC6098-SS001-320-1
Movement: automatic, Caliber ML 115 (base Sellita SW200-1); ø 25.6 mm, height 4.6 mm; 26 jewels; 28,800 vph; 38-hour power reserve
Functions: hours, minutes, sweep seconds; date
Case: stainless steel, ø 40 mm, height 10.5 mm; sapphire crystal; transparent case back; water-resistant to 3 atm
Band: reptile skin, folding clasp
Price: $1,480
Variations: stainless steel bracelet

Les Classiques Moonphase
Reference number: LC6168-SS001-122-1
Movement: automatic, Caliber ML 37 (base ETA 2892 with special module); ø 25.6 mm, height 5.4 mm; 25 jewels; 28,800 vph; 38-hour power reserve
Functions: hours, minutes, sweep seconds; date, moon phase
Case: stainless steel, ø 40 mm, height 12 mm; sapphire crystal
Band: reptile skin, folding clasp
Price: $2,650

Caliber ML 106-2
Manually wound; column wheel control of chronograph functions; single spring barrel, 42-hour power reserve
Functions: hours, minutes, subsidiary seconds; chronograph
Diameter: 36.6 mm
Height: 6.9 mm
Jewels: 22
Balance: screw balance with weights
Frequency: 18,000 vph
Balance spring: with swan-neck fine adjustment
Shock protection: Kif Elastor
Remarks: anodized movement, bridges with grand colimaçon snailing, 2 gold chatons

Caliber ML 191
Automatic; single spring barrel, 52-hour power reserve; COSC-certified chronometer
Functions: hours, minutes, subsidiary seconds; additional 24-hour display (2nd time zone), power reserve indicator; date (retrograde)
Diameter: 36.6 mm
Height: 8.2 mm
Jewels: 74
Balance: glucydur
Frequency: 18,000 vph
Balance spring: Nivarox
Shock protection: Incabloc
Remarks: rhodium plating, brushed bridges, winding rotor with grand colimaçon snailing

Caliber ML 230
Automatic; inverted movement construction with escapement on dial; silicon pallet lever and pallet fork; single spring barrel, 50-hour power reserve
Functions: hour, minutes (off-center), subsidiary seconds
Diameter: 37.2 mm
Height: 9.05 mm
Jewels: 35
Balance: glucydur
Frequency: 18,000 vph
Remarks: rhodium-plated three-quarter plate (movement side) with côtes de Genève; dial assembled to mainplate

MB&F
Boulevard Helvétique 22
Case postale 3466
CH-1211 Geneva 3
Switzerland

Tel.:
+41-22-786-3618

Fax:
+41-22-786-3624

E-mail:
info@mbandf.com

Website:
www.mbandf.com

Founded:
2005

Number of employees:
24

Annual production:
approx. 250 watches

U.S. Distributors:
Westime Los Angeles and Miami
310-470-1388 (Los Angeles); 786-347-5353 (Miami)
info@westime.com
Provident Jewelry, Florida
561-747-4449; nick@providentjewelry.com
Stephen Silver, Redwood City (California)
650-325-9500; www.shsilver.com
Cellini, New York
212-888-0505
contact@cellinijewelers.com

Most important collections/price range:
Horological Machines / from $63,000; Legacy Machines / from $64,000

MB&F

Maximilian Büsser & Friends goes beyond the standard idea of a brand. Perhaps calling it a tribe would be better: one aiming to create unique works of horology. MB&F is doing something unconventional in an industry that usually takes its innovation in small doses.

After seeing the Opus projects to fruition at Harry Winston, Büsser decided it was time to set the creators free. At MB&F he acts as initiator and coordinator. His Horological Machines are developed and realized in cooperation with highly specialized watchmakers, inventors, and designers in an "idea collective" creating unheard-of mechanical timepieces of great inventiveness, complication, and exclusivity. The composition of this collective varies as much as each machine. Number 5 ("On the Road Again") is an homage to the 1970s, when streamlining rather than brawn represented true strength. The display in the lateral window is reflected by a prism. The "top" of the watch opens to let in light to charge the Superluminova numerals on the disks. As for the Space Pirate, Number 6, it is a talking piece that makes a genial nod to sci-fi moviemakers, and all the talk was real: The model won a coveted Red Dot "Best of the Best" award in 2015. Contrasting sharply with the modern productions are the Legacy Machines, which reach into horological history and reinterpret past mechanical feats.

The spirit of Büsser is always present in each new watch, but it is now vented freely in the M.A.D. Gallery in Geneva, where "mechanical art objects" on display are beautiful, intriguing, technically impeccable, and sometimes perfectly useless. They have their own muse and serve as worthy companions to the sci-fi-inspired table clocks that MB&F produces with L'épée 1938. One is shaped like a spaceship; the other is a huge spider; yet another like a tank with real treads. MB&F has been presenting at the SIHH as well lately, carrying his genuinely disruptive ideas to the more traditional brands and shaking things up a bit. All these pieces tell the time very accurately and mechanically.

MoonMachine 2
Reference number: 81.BTL.B
Movement: automatic, MB&F Caliber HM8 (base Girard-Perregaux); ø 35.8 mm, height 13.5 mm; 30 jewels; 28,800 vph; vertical display on head side via reflecting sapphire prism; 42-hour power reserve; with additional module for moon phase indication
Functions: hours (digital, jumping), minutes (disk display); moon phase
Case: titanium, black coating, 51.5 × 49 mm, height 19.5 mm; sapphire crystal; transparent case back; water-resistant to 3 atm
Band: reptile skin, folding clasp
Price: $88,000; limited to 12 pieces
Variations: pink gold (limited to 12 pieces), natural titanium (limited to 12 pieces)

Legacy Machine Perpetual
Reference number: 03.WL.B
Movement: manually wound, MB&F Caliber LM Perpetual; ø 36.6 mm, height 12.6 mm; 41 jewels; 18,000 vph; double spring barrel, inverted movement design with one balance floating over dial; finely finished with côtes de Genève; 72-hour power reserve
Functions: hours, minutes (off-center); power reserve indicator; perpetual calendar with date, weekday, month, leap year (backward counting)
Case: white gold, ø 44 mm, height 17.5 mm; sapphire crystal; transparent case back; water-resistant to 3 atm
Band: reptile skin, folding clasp
Price: $158,000

HM7 Aquapod Ti Green
Reference number: 70.TGL.B
Movement: automatic, MB&F Caliber HM7; ø 31.4 mm, height 17.45 mm; 35 jewels; 18,000 vph; flying 1-minute tourbillon, 3D vertical architecture, titanium and platinum rotor; 72-hour power reserve
Functions: hours, minutes (on spherical titanium-aluminum disk)
Case: titanium, ø 53.8 mm, height 21.3 mm; unidirectional bezel in sapphire crystal, with 0-60 scale; sapphire crystal; transparent case back; crowns for winding and time-setting; water-resistant to 5 atm
Band: rubber, folding clasp
Price: $108,000; limited to 50 pieces

MEISTERSINGER

In 2014, MeisterSinger completed a long process of reorientation, setting the German brand in redux mode. At Baselworld 2014, it presented a portfolio of exclusively one-hand watches, the actual core of the brand. These watches express a relaxed and self-determined approach to the perception of time apparent in the special diurnal rituals that everyone knows, young, old, in private, or at work. These rituals actually divide up and define certain moments. And it is the reiteration of these moments which leads to order, or at least avoiding chaos.

Founder Martin Brassler launched his little collection of stylistically neat one-hand dials at the beginning of the new millennium. Looking at these ultimately simplified dials does tempt one to classify the one-hand watch as an archetype. The single hand simply cannot be reduced any further, and the 144 minutes for 12 hours around the dial do have a normative function of sorts. In a frenetic era when free time has become so rare, these watches slow things down a little. The most recent one-hander does provide the hour, jumping very precisely in a window under 12 o'clock—hence its Italian name "Salthora," or jumping hour.

Minimalism, however, does not mean les dynamism. For 2018, Brassler, who designs all the watches himself, has put out a pleasant range of watches. There is the fun Metris, on the one hand, or the rigorously elegant Lunascope moon phase. The Black Line, for its part, pays homage to the color black and its potential depths.

Design, product planning, service, and management all happen in Münster, Germany, but the watches are Swiss made, with ETA and Sellita movements. The Circularis, however, is the brand's first model with an in-house movement, a manually wound caliber with two barrel springs developed in collaboration with the Swiss firm Synergies Horlogères.

MeisterSinger GmbH & Co. KG
Hafenweg 46
D-48155 Münster
Germany

Tel.:
+49-251-133-4860

E-mail:
info@meistersinger.de

Website:
www.meistersinger.de

Founded:
2001

Number of employees:
13

Annual production:
approx. 10,000 watches

U.S. distributor:
Duber Time
1920 Dr. MLK Jr. Street North
Suite #D
St. Petersburg, FL 33704
727-202-3262
damir@meistersingertime.com

Most important collections/price range:
from approx. $1,200 to $7,000

Circularis Power Reserve Black Line
Reference number: CCP302BL
Movement: manually wound, MeisterSinger Caliber MSH02; ø 32.7 mm, height 5.4 mm; 31 jewels; 28,800 vph; 2 spring barrels; finely finished movement; 120-hour power reserve
Functions: hours (each scale line is 5 minutes); power reserve indicator; date
Case: stainless steel with black DLC coating, ø 43 mm, height 12.5 mm; sapphire crystal; transparent case back; water-resistant to 5 atm
Band: reptile skin, double folding clasp
Price: $5,825

Lunascope
Reference number: LS901
Movement: automatic, ETA Caliber 2836-2 with MeisterSinger module; ø 25.6 mm, height 5.05 mm; 25 jewels; 28,800 vph; 38-hour power reserve
Functions: hours (each scale line is 5 minutes); date, moon phase
Case: stainless steel, ø 40 mm, height 12 mm; sapphire crystal; transparent case back; water-resistant to 5 atm
Band: calfskin, buckle
Price: $3,845

N° 03 40 mm
Reference number: DM908
Movement: automatic, ETA Caliber 2824-2 or Sellita SW200-1; ø 25.6 mm, height 4.6 mm; 25 or 26 jewels; 28,800 vph; with côtes de Genève; 38-hour power reserve
Functions: hours (each scale line is 5 minutes); date
Case: stainless steel, ø 40 mm, height 10.3 mm; sapphire crystal; transparent case back; water-resistant to 5 atm
Band: calfskin, buckle
Price: $2,195

MEISTERSINGER

Perigraph
Reference number: AM1007OR
Movement: automatic, ETA Caliber 2824-2 or Sellita SW200-1; ø 25.6 mm, height 4.6 mm; 25 or 26 jewels; 28,800 vph; with côtes de Genève; 38-hour power reserve
Functions: hours (each scale line is 5 minutes); date
Case: stainless steel, ø 43 mm, height 11.5 mm; sapphire crystal; transparent case back; water-resistant to 5 atm
Band: calfskin, buckle
Price: $2,295

Metris
Reference number: ME908
Movement: automatic, ETA Caliber 2824-2; ø 25.6 mm, height 4.6 mm; 25 jewels; 28,800 vph; 38-hour power reserve
Functions: hours (each scale line is 5 minutes); date
Case: stainless steel, ø 38 mm, height 11.1 mm; sapphire crystal; water-resistant to 20 atm
Band: textile, buckle
Price: $1,795
Variations: various dial colors

Urban
Reference number: UR902
Movement: automatic, Miyota Caliber 8245; ø 25.6 mm, height 5.67 mm; 21 jewels; 21,600 vph; 42-hour power reserve
Functions: hours (each scale line is 5 minutes)
Case: stainless steel, ø 40 mm, height 12 mm; sapphire crystal; transparent case back; water-resistant to 5 atm
Band: calfskin, buckle
Price: $995
Variations: various dial colors

Salthora Meta X
Reference number: SAMX902
Movement: automatic, ETA Caliber 2824-2 or Sellita SW200-1 with in-house module for jumping hours display; ø 25.6 mm, height 6.9 mm; 25 or 26 jewels; 28,800 vph; with côtes de Genève; 38-hour power reserve
Functions: hours (digital, jumping), minutes
Case: stainless steel, ø 43 mm, height 14.2 mm; unidirectional ceramic bezel, with 0-60 scale; sapphire crystal; screw-in crown; water-resistant to 20 atm
Band: stainless steel Milanese mesh, folding clasp
Price: $3,820
Variations: green indices; blue dial and bezel; reptile skin ($3,595)

Pangaea
Reference number: PM903
Movement: automatic, ETA Caliber 2892-A2 or Sellita SW300-1; ø 25.6 mm, height 3.6 mm; 21 or 25 jewels; 28,800 vph; 42-hour power reserve
Functions: hours (each scale line is 5 minutes)
Case: stainless steel, ø 40 mm, height 10.1 mm; sapphire crystal; transparent case back; water-resistant to 5 atm
Band: calfskin, buckle
Price: $2,595
Variations: various dial colors; reptile skin strap ($2,950); stainless steel Milanese mesh band ($2,950)

N° 01 40 mm
Reference number: DM317
Movement: manually wound, Sellita Caliber SW210; ø 25.6 mm, height 3.4 mm; 19 jewels; 28,800 vph; 42-hour power reserve
Functions: hours (each scale line is 5 minutes)
Case: stainless steel, ø 40 mm, height 11.5 mm; sapphire crystal; transparent case back; water-resistant to 5 atm
Band: calfskin, buckle
Price: $1,695
Variations: various dial colors

MIDO

Among the legacies of World War I was the popularization of the wristwatch, which had freed up soldiers' and aviators' hands to fight and steer, respectively, and permitted artillery officers to coordinate barrages. And, not surprisingly, this led to a kind of re-industrialization of the watch industry. Among the earliest companies to appear on the scene was Mido, which was founded on November 11, 1918—Armistice Day—by Georges Schaeren in Solothurn, Switzerland. The name means "I measure" in Spanish.

At first, the brand produced colorful and imaginative watches that were well-suited to the Roaring Twenties. In the 1930s, however, Mido began making more serious, robust, sportive timepieces better suited for everyday use.

For the watch fan of today, water resistance and self-winding are normal. Mido, however, was already offering this functionality in the 1930s with the introduction of the Multifort, which really put the company on the map at the time. This Swiss manufacturer was equally innovative with its movements. It developed a number of very practical novelties, like the Radiotime model (1939) and the Multicenterchrono (1941), which today have become genuine collectors' items.

In 1971 the Schaeren family sold the company to the General Watch Co. Ltd., a holding company belonging to ASUAG, which, in turn became the SMH and, ultimately, Swatch Group.

Mido continues to produce mostly mechanical watches with about one-quarter of its production devoted to quartz movements. In 1998, the company decided to revive some of its older watchmaking values, a strategy that ultimately worked out well. The marching orders were simply "manufacturing robust and water-resistant state-of-the-art automatic watches with timeless design." The Multifort, Commander, Battalion, and Baroncelli collections are each in their own way expressions of that mission. Nothing in-your-face, just affordable timepieces with the basic hallmarks of a good Swiss watch, like côtes de Genève on the rotors and, in some cases, even COSC certification.

Mido SA
Chemin des Tourelles 17
CH-2400 Le Locle
Switzerland

Tel.:
+41 32 933 35 11

Fax:
+41 32 933 55 00

Website:
www.mido.ch

Founded:
1918

Number of employees:
50 (estimated)

Annual production:
over 100,000

U.S. distributor:
Mido, division of The Swatch Group (U.S.) Inc.
703 Waterford Way, Suite 450
Miami, FL 33126
www.midowatches.com

Most important collections/price range:
Baroncelli, Commander, Commander II, Multifort; $800 to $2,300

Commander Big Date

Reference number: M021.626.11.061.00
Movement: automatic, Mido Caliber 80.651 (base ETA C07.651); ø 29.4 mm, height 5.77 mm; 25 jewels; 21,600 vph; rotor with côtes de Genève; 80-hour power reserve
Functions: hours, minutes, sweep seconds; large date
Case: stainless steel, ø 42 mm, height 11.97 mm; sapphire crystal; transparent case back; water-resistant to 5 atm
Band: stainless steel, folding clasp
Price: $950

Commander Icon

Reference number: M031.631.11.031.00
Movement: automatic, Mido Caliber 80.821 (base ETA C07.821); ø 25.6 mm, height 5.22 mm; 25 jewels; 21,600 vph; silicon hairspring; 80-hour power reserve; COSC-certified chronometer
Functions: hours, minutes, sweep seconds; date, weekday
Case: stainless steel, ø 42 mm, height 11.42 mm; sapphire crystal; transparent case back; water-resistant to 5 atm
Band: stainless steel Milanese mesh, folding clasp
Price: $1,240

Commander Chronograph

Reference number: M016.414.11.041.00
Movement: automatic, Mido Caliber 60 (base ETA A05.H21); ø 30 mm, height 7.9 mm; 27 jewels; 28,800 vph; rotor with côtes de Genève; 60-hour power reserve
Functions: hours, minutes, subsidiary seconds; chronograph; date, weekday
Case: stainless steel, ø 42.5 mm, height 14.89 mm; sapphire crystal; transparent case back; water-resistant to 5 atm
Band: stainless steel, double folding clasp
Price: $1,960

MIDO

Baroncelli Chronometer Si
Reference number: M027.408.11.011.00
Movement: automatic, Mido Caliber 80 (base ETA C07.821); ø 25.6 mm, height 5.22 mm; 25 jewels; 21,600 vph; silicon hairspring; 80-hour power reserve; COSC-certified chronometer
Functions: hours, minutes, sweep seconds; date
Case: stainless steel, ø 40 mm, height 9.43 mm; sapphire crystal; transparent case back; water-resistant to 3 atm
Band: stainless steel, folding clasp
Price: $1,190

Baroncelli Heritage Gent
Reference number: M027.407.36.260.00
Movement: automatic, Mido Caliber 1192 (base ETA 2892-A2); ø 25.6 mm, height 3.6 mm; 21 jewels; 28,800 vph; rotor with côtes de Genève; 42-hour power reserve
Functions: hours, minutes, sweep seconds; date
Case: stainless steel with rose gold PVD coating, ø 39 mm, height 6.95 mm; sapphire crystal; transparent case back; water-resistant to 3 atm
Band: calfskin, buckle
Price: $1,090

Multifort Escape
Reference number: M032.607.36.050.99
Movement: automatic, Mido Caliber 80.611 (base ETA C07.611); ø 25.6 mm, height 4.47 mm; 25 jewels; 21,600 vph; 80-hour power reserve
Functions: hours, minutes, sweep seconds; date
Case: stainless steel with gray PVD coating, ø 44 mm, height 11.88 mm; sapphire crystal; transparent case back; water-resistant to 10 atm
Band: calfskin, buckle
Remarks: comes with second calfskin strap
Price: $1,090

Multifort Chronograph Adventure
Reference number: M025.627.36.061.10
Movement: automatic, Mido Caliber 60 (base ETA A05.H31); ø 30 mm, height 7.9 mm; 27 jewels; 28,800 vph; rotor with côtes de Genève; 60-hour power reserve
Functions: hours, minutes, subsidiary seconds; chronograph; date
Case: stainless steel with black PVD coating, ø 44 mm, height 14.7 mm; sapphire crystal; transparent case back; screw-in crown; water-resistant to 10 atm
Band: calfskin, folding clasp
Price: $2,130

Commander II
Reference number: M021.431.16.051.00
Movement: automatic, Mido Caliber 80.621 (base ETA C07.621); ø 25.6 mm, height 5.22 mm; 25 jewels; 21,600 vph; rotor with côtes de Genève; 80-hour power reserve; COSC-certified chronometer
Functions: hours, minutes, sweep seconds; date, weekday
Case: stainless steel, ø 40 mm, height 11.3 mm; sapphire crystal; transparent case back; water-resistant to 5 atm
Band: calfskin, folding clasp
Price: $1,170

Ocean Star Captain Titanium
Reference number: M026.430.44.061.00
Movement: automatic, Mido Caliber 80 (base ETA C07.621); ø 25.6 mm, height 5.22 mm; 25 jewels; 21,600 vph; rotor with côtes de Genève; 80-hour power reserve
Functions: hours, minutes, sweep seconds; date, weekday
Case: titanium, ø 42.5 mm, height 11.75 mm; unidirectional bezel, with 0-60 scale; sapphire crystal; screw-in crown; water-resistant to 20 atm
Band: titanium, folding clasp, with extension link
Price: $1,040

MK II

If vintage and unserviceable watches had their say, they would probably be naturally attracted to Mk II for the name alone, which is a military designation for the second generation of equipment. The company, which was founded by watch enthusiast and maker Bill Yao in 2002, not only puts retired designs back into service, but also modernizes and customizes them. Before the screwed-down crown, diving watches were not nearly as reliably sealed, for example. And some beautiful old pieces were made with plated brass cases or featured Bakelite components, which are either easily damaged or have aged poorly. The company substitutes proven modern materials as well as modern manufacturing methods and techniques to ensure a better outcome.

These are material issues that the team at Mk II handles with great care. They will not, metaphorically speaking, airbrush a Model-T. As genuine watch lovers themselves, they make sure that the final design is in the spirit of the watch itself, which still leaves a great deal of leeway for many iterations, given a sufficient number of parts. In the company's output, vintage style and modern functionality are key. The watches are assembled by hand at the company's workshop in Pennsylvania—and subjected to a rigorous regime of testing. The components are individually inspected, the cases tested at least three times for water resistance, and at the end the whole watch is regulated in six positions. Looking to the future, Mk II aspires to carry its clean vintage style into the development of what it hopes will be future classics of its own.

Mk II Corporation
303 W. Lancaster Avenue, #283
Wayne, PA 19087

E-mail:
info@mkiiwatches.com

Website:
www.mkiiwatches.com

Founded:
2002

Number of employees:
3

Annual production:
800 watches

Distribution:
direct sales and select retail

Most important collections/price range:
Ready-to-Wear Collection / $500 to $995;
Bencrafted Collection / $1,000 to $2,000

Paradive
Reference number: CD04.1-1012N
Movement: automatic (hack setting), Caliber SII NE15; ø 27.40 mm, height 5.32 mm; 24 jewels; 21,600 vph; 50-hour power reserve; rotor decorated with côtes de Genève
Functions: hours, minutes, sweep seconds, date
Case: stainless steel, ø 41.2 mm, height 15.50 mm; 120-click unidirectional bezel; high domed sapphire crystal with antireflective coating; screw-down case back; screw-in crown; water-resistant to 20 atm
Band: nylon, pin buckle
Price: $895
Variations: without date, dive bezel

Hawkinge AGL
Reference Number: CG05-3001N
Movement: automatic (hack setting), Caliber SII NE15; ø 27.40 mm, height 5.32 mm; 24 jewels; 21,600 vph; 50-hour power reserve; rotor decorated with côtes de Genève
Functions: hours, minutes, sweep seconds
Case: stainless steel; ø 37.80 mm, height 12.75 mm; domed sapphire crystal with antireflective coating; screw-down case back; screw-in crown; water-resistant to 10 atm
Band: nylon, pin buckle
Price: $595
Variations: leather strap

Cruxible Type A-11
Reference Number: CG06-1001N
Movement: automatic (hack setting), Caliber SII NE15; ø 27.40 mm, height 5.32 mm; 24 jewels; 21,600 vph; 50-hour power reserve; rotor decorated with côtes de Genève
Functions: hours, minutes, sweep seconds
Case: stainless steel; ø 39 mm, height 13.55 mm; domed sapphire crystal with antireflective coating; screw-down case back; screw-in crown; water-resistant to 10 atm
Band: nylon, pin buckle
Price: $649
Variations: date, leather strap

Montblanc Montre SA
10, chemin des Tourelles
CH-2400 Le Locle
Switzerland

Tel.:
+41-32-933-8888

Fax:
+41-32-933-8880

E-mail:
service@montblanc.com

Website:
www.montblanc.com

Founded:
1997 (1906 in Hamburg)

Number of employees:
worldwide approx. 3,000

U.S. distributor:
Montblanc North America
645 Fifth Avenue, 7th Floor
New York, NY 10022
Ph.: 1-800-995-4810
Fax: 1-908-464-3722
www.montblanc.com

Most important collections:
Heritage Chronométrie, Heritage Spirit, Meisterstück, Star, Nicolas Rieussec, 4810, TimeWalker, Collection Villeret, 1858 Collection

MONTBLANC

It was with great skill and cleverness that Nicolas Rieussec (1781–1866) used the invention of a special chronograph—the "Time Writer," a device that released droplets of ink onto a rotating sheet of paper—to make a name for himself. Montblanc, once famous only for its exclusive writing implements, borrowed that name on its way to becoming a distinguished watch brand. Within a few years, it had created an impressive range of chronographs driven by in-house calibers: from simple automatic stopwatches to flagship pieces with two independent spring barrels for time and "time-writing."

The Richemont Group, owner of Montblanc, has placed great trust in its "daughter" company, having put the little *manufacture* Minerva, which it purchased at the beginning of 2007, at the disposal of Montblanc. Minerva, which was founded in Villeret in 1858, was already building keyless pocket watches in the 1880s, and by the early twentieth century was producing monopusher chronographs with a reputation for precision. Today, the Minerva Institute serves as a kind of think tank for the future, a place where young watchmakers can absorb the old traditions and skills, as well as the wealth of experience and mind-set of the masters.

Montblanc is continuing the Minerva tradition today with three leading collections: the 1858, the Star Legacy, and the TimeWalker line.

Star Legacy Suspended Exo Tourbillon

Reference number: 116829
Movement: manually wound, Montblanc Caliber MB M16.68; ø 38.3 mm, height 10.6 mm; 19 jewels; 18,000 vph; flying 1-minute tourbillon with external hairspring, screw balance; 50-hour power reserve
Functions: hours and minutes (off-center), subsidiary seconds (on tourbillon cage)
Case: pink gold, ø 44.8 mm, height 15.03 mm; sapphire crystal; transparent case back; water-resistant to 3 atm
Band: reptile skin, triple folding clasp
Price: on request; limited to 58 pieces

Star Legacy Nicolas Rieussec Chronograph

Reference number: 118537
Movement: automatic, Montblanc Caliber MB R200; ø 31 mm, height 8.46 mm; 40 jewels; 28,800 vph; 2 spring barrels; monopusher column-wheel control of chronograph functions; 72-hour power reserve
Functions: hours, minutes, subsidiary seconds; additional 12-hour display (2nd time zone); chronograph; date
Case: stainless steel, ø 44.8 mm, height 15.02 mm; sapphire crystal; transparent case back; water-resistant to 3 atm
Band: reptile skin, double folding clasp
Price: $8,000

Star Legacy Automatic Chronograph

Reference number: 118514
Movement: automatic, Montblanc Caliber MB 25.02; ø 30 mm, height 7.9 mm; 27 jewels; 28,800 vph; 46-hour power reserve
Functions: hours, minutes, subsidiary seconds; chronograph; date
Case: stainless steel, ø 42 mm, height 14.23 mm; sapphire crystal; transparent case back; water-resistant to 3 atm
Band: reptile skin, triple folding clasp
Price: $4,300

MONTBLANC

Star Legacy Full Calendar
Reference number: 118516
Movement: automatic, Montblanc Caliber MB 29.12 (base ETA 7751); ø 30 mm, height 7.9 mm; 25 jewels; 28,800 vph; 42-hour power reserve
Functions: hours, minutes, sweep seconds; full calendar with date, weekday, month, moon phase
Case: stainless steel, ø 42 mm, height 11.43 mm; sapphire crystal; transparent case back; water-resistant to 3 atm
Band: reptile skin, triple folding clasp
Price: $4,600

Star Legacy Moonphase
Reference number: 116508
Movement: automatic, Montblanc Caliber MB 29.14 (base ETA 7751); ø 30 mm, height 7.9 mm; 25 jewels; 28,800 vph; 42-hour power reserve
Functions: hours, minutes; date, moon phase
Case: stainless steel, ø 42 mm, height 10.88 mm; sapphire crystal; transparent case back; water-resistant to 3 atm
Band: reptile skin, triple folding clasp
Price: $4,200
Variations: various straps

Star Legacy Automatic Date
Reference number: 117575
Movement: automatic, Montblanc Caliber MB 24.01 (base ETA 2892-A2); ø 25.6 mm, height 3.6 mm; 21 jewels; 28,800 vph; 42-hour power reserve
Functions: hours, minutes, sweep seconds; date
Case: stainless steel, ø 42 mm, height 9.58 mm; sapphire crystal; transparent case back; water-resistant to 3 atm
Band: reptile skin, triple folding clasp
Price: $2,990
Variations: various straps

Star Legacy Small Second
Reference number: 118532
Movement: automatic, Montblanc Caliber MB 24.08 (base ETA 2895-2); ø 25.6 mm, height 4.35 mm; 27 jewels; 28,800 vph; 42-hour power reserve
Functions: hours, minutes, subsidiary seconds
Case: pink gold, ø 32 mm, height 9.39 mm; sapphire crystal; transparent case back; water-resistant to 3 atm
Band: reptile skin, buckle
Remarks: second ring set with 28 diamonds
Price: $7,300

Star Legacy Ladies' Small Second
Reference number: 118508
Movement: automatic, Montblanc Caliber MB 24.16 (base Sellita SW260-1); ø 25.6 mm, height 4.35 mm; 31 jewels; 28,800 vph; 38-hour power reserve
Functions: hours, minutes, subsidiary seconds
Case: stainless steel, ø 36 mm, height 10.75 mm; bezel set with 76 diamonds; sapphire crystal; transparent case back; water-resistant to 3 atm
Band: reptile skin, buckle
Remarks: second ring set with 28 diamonds
Price: $5,400

1858 Monopusher Chronograph LE 100
Reference number: 117834
Movement: manually wound, Montblanc Caliber MB M13.21 (base Minerva); ø 29.5 mm, height 6.4 mm; 22 jewels; 18,000 vph; screw balance, monopusher for column wheel control of chronograph functions; 55-hour power reserve
Functions: hours, minutes, subsidiary seconds; chronograph
Case: stainless steel, ø 40 mm, height 12.15 mm; sapphire crystal; transparent case back; water-resistant to 10 atm
Band: reptile skin, buckle
Price: on request; limited to 100 pieces

MONTBLANC

1858 Geosphere LE 1858
Reference number: 119347
Movement: automatic, Montblanc Caliber MB 29.25; 26 jewels; 28,800 vph; 42-hour power reserve
Functions: hours, minutes; additional 12-hour display (2nd time zone), synchronously counter-rotating world time indicators for northern and southern hemispheres; date
Case: bronze, ø 42 mm, height 12.8 mm; bezel with ceramic inlay; sapphire crystal; transparent case back; water-resistant to 10 atm
Band: calfskin, folding clasp
Price: $6,300; limited to 1,858 pieces
Variations: stainless steel ($6,300)

1858 Geosphere
Reference number: 119286
Movement: automatic, Montblanc Caliber MB 29.25; 26 jewels; 28,800 vph; 42-hour power reserve
Functions: hours, minutes; additional 12-hour display (2nd time zone), synchronously counter-rotating world time indicators for northern and southern hemispheres; date
Case: stainless steel, ø 42 mm, height 12.8 mm; bezel with ceramic inlay; sapphire crystal; transparent case back; water-resistant to 10 atm
Band: calfskin, folding clasp
Price: $5,600
Variations: bronze ($5,600, limited to 1,858 pieces)

1858 Automatic Chronograph
Reference number: 117835
Movement: automatic, Montblanc Caliber MB 25.11 (base ETA 7753); ø 30 mm, height 7.9 mm; 27 jewels; 28,800 vph; 48-hour power reserve
Functions: hours, minutes, subsidiary seconds; chronograph
Case: stainless steel, ø 42 mm, height 14.55 mm; sapphire crystal; transparent case back; water-resistant to 10 atm
Band: textile, buckle
Price: $4,300

1858 Automatic
Reference number: 117832
Movement: automatic, Montblanc Caliber MB 24.15 (base Sellita SW200-1); ø 25.6 mm, height 4.6 mm; 26 jewels; 28,800 vph; 38-hour power reserve
Functions: hours, minutes
Case: stainless steel, ø 40 mm, height 11.07 mm; bezel in bronze; sapphire crystal; water-resistant to 10 atm
Band: calfskin, buckle
Price: $2,670
Variations: textile strap; champagne-colored dial

TimeWalker Manufacture Chronograph
Reference number: 118490
Movement: automatic, Montblanc Caliber MB 25.10; ø 30.15 mm, height 7.9 mm; 33 jewels; 28,800 vph; column wheel control of chronograph functions, screw balance; 46-hour power reserve
Functions: hours, minutes, subsidiary seconds; chronograph; date
Case: stainless steel, ø 43 mm, height 15.2 mm; ceramic bezel; sapphire crystal; transparent case back; water-resistant to 10 atm
Band: stainless steel, triple folding clasp
Price: $5,700
Variations: calfskin strap ($5,700)

TimeWalker Manufacture Chronograph
Reference number: 118488
Movement: automatic, Montblanc Caliber MB 25.10; ø 30.15 mm, height 7.9 mm; 33 jewels; 28,800 vph; column wheel control of chronograph functions, screw balance; 46-hour power reserve
Functions: hours, minutes, subsidiary seconds; chronograph; date
Case: stainless steel, ø 43 mm, height 15.2 mm; ceramic bezel; sapphire crystal; transparent case back; water-resistant to 10 atm
Band: calfskin, triple folding clasp
Price: $5,400
Variations: stainless steel bracelet ($5,400)

MONTBLANC

Caliber MB R200

Automatic; monopusher column-wheel control, vertical chronograph clutch, stop-seconds mechanism; double spring barrel, 72-hour power reserve
Functions: hours, minutes, subsidiary seconds; additional 12-hour display (2nd time zone), chronograph; date
Diameter: 31 mm
Height: 8.46 mm
Jewels: 40
Balance: screw balance
Frequency: 28,800 vph
Balance spring: flat hairspring
Remarks: rhodium-plated mainplate with perlage, bridges with côtes de Genève

Caliber MB M16.68

Manually wound; exo-tourbillon with external balance spring; mainplate of German silver; single spring barrel, 50-hour power reserve
Functions: hours, minutes (off-center), subsidiary seconds (on tourbillon cage)
Diameter: 38.3 mm
Height: 10.6 mm
Jewels: 19
Balance: screw balance
Balance spring: flat hairspring with Phillips end curve
Remarks: 218 parts; finely finished, with côtes de Genève

Caliber MB M13.21

Manually wound; column-wheel control of chronograph functions; single spring barrel, 60-hour power reserve
Functions: hours, minutes, subsidiary seconds; chronograph
Diameter: 29.5 mm
Height: 6.4 mm
Jewels: 22
Balance: screw balance with weights
Frequency: 18,000 vph
Balance spring: with Phillips end curve
Shock protection: Incabloc
Remarks: rhodium-plated mainplate and bridges of German silver, partially decorated with perlage and hand-beveled

Caliber MB M16.24

Manually wound; monopusher for column-wheel control of chronograph functions; single spring barrel, 50-hour power reserve
Functions: hours, minutes; chronograph
Diameter: 38.4 mm
Height: 6.3 mm
Jewels: 22
Balance: screw balance
Frequency: 18,000 vph
Balance spring: with Phillips end curve
Remarks: rhodium-plated mainplate with perlage, gold-plated wheelworks

Caliber MB 25.10

Automatic; monopusher for column-wheel control of chronograph functions, horizontal chronograph clutch, stop-seconds mechanism; single spring barrel, 46-hour power reserve
Functions: hours, minutes, subsidiary seconds; chronograph
Diameter: 30.15 mm
Height: 7.9 mm
Jewels: 33
Balance: screw balance
Frequency: 28,800 vph
Balance spring: flat hairspring
Remarks: 232 parts; rhodium-plated mainplate with perlage, bridges with côtes de Genève

Caliber MB M29.24

Automatic; 1-minute tourbillon with external hairspring; double spring barrel, microrotor; 50-hour power reserve
Functions: hours, minutes
Diameter: 30.6 mm
Height: 4.5 mm
Jewels: 29
Balance: screw balance with 18 weighted screws
Frequency: 21,600 vph
Balance spring: flat hairspring
Remarks: gold microrotor, bridges with côtes de Genève

Mühle Glashütte GmbH
Nautische Instrumente und Feinmechanik
Altenberger Strasse 35
D-01768 Glashütte
Germany

Tel.:
+49-35053-3203-0

Fax:
+49-35053-3203-136

E-mail:
info@muehle-glashuette.de

Website:
www.muehle-glashuette.de

Founded:
first founding 1869; second founding 1993

Number of employees:
47

U.S. distributor:
Mühle Glashütte
Old Northeast Jewelers
1131 4th Street North
Saint Petersburg, FL 33701
800-922-4377
www.muehle-glashuette.com

Most important collections/price range:
mechanical wristwatches / approx. $1,399 to $5,400

MÜHLE GLASHÜTTE

Mühle Glashütte has survived all the ups and downs of Germany's history. The firm Rob. Mühle & Sohn was founded by its namesake in 1869. At that time, the company made precision measuring instruments for the local watch industry and the German School of Watchmaking. In the early 1920s, the firm established itself as a supplier for the automobile industry, making speedometers, automobile clocks, tachometers, and other measurement instruments.

Having manufactured instruments for the military during the war, the company was not only bombarded by the Soviet air force, but was also nationalized in 1945, as it was in the eastern part of the country. After the fall of the Iron Curtain, it was reestablished as a limited liability corporation. In 2007, Thilo Mühle took over the helm from his father, Hans-Jürgen Mühle.

The company's wristwatch business was launched in 1996 and now overshadows the nautical instruments that had made the name Mühle Glashütte famous. Its collection comprises mechanical wristwatches at entry and mid-level prices. For these, the company uses Swiss base movements that are equipped with such in-house developments as a patented woodpecker-neck regulation and the Mühle rotor. The modifications are so extensive that they have led to the calibers having their own names. The traditional line named "R. Mühle & Sohn," introduced in 2014, is equipped with the RMK 1 and RMK 2 calibers. And there are other, somewhat less nautically inspired timepieces, like the Lunova series or the 29ers, which are simply elegant in an unspectacular way.

Panova Blau

Reference number: M1-40-72-NB
Movement: automatic, Sellita Caliber SW200-1, version Mühle; ø 25.6 mm, height 4.6 mm; 26 jewels; 28,800 vph; woodpecker-neck regulator, Mühle rotor, carefully reworked with special Mühle finish; 38-hour power reserve
Functions: hours, minutes, sweep seconds
Case: stainless steel, ø 40 mm, height 10.4 mm; sapphire crystal; screw-in crown; water-resistant to 10 atm
Band: textile, buckle
Price: $999

29er Big

Reference number: M1-25-33-MB
Movement: automatic, Sellita Caliber SW200-1, version Mühle; ø 25.6 mm, height 4.6 mm; 26 jewels; 28,800 vph; woodpecker-neck regulator, Mühle rotor, carefully reworked with special Mühle finish; 38-hour power reserve
Functions: hours, minutes, sweep seconds; date
Case: stainless steel, ø 42.4 mm, height 11.3 mm; sapphire crystal; transparent case back; screw-in crown; water-resistant to 10 atm
Band: stainless steel, folding clasp
Price: $2,199
Variations: calfskin strap ($2,099); rubber strap ($2,099)

29er Tag/Datum

Reference number: M1-25-34-NB
Movement: automatic, Sellita Caliber SW220-1, version Mühle; ø 25.6 mm, height 5.05 mm; 26 jewels; 28,800 vph; woodpecker-neck regulator, Mühle rotor, carefully reworked with special Mühle finish; 38-hour power reserve
Functions: hours, minutes, sweep seconds; date, weekday
Case: stainless steel, ø 42.4 mm, height 12.2 mm; sapphire crystal; transparent case back; screw-in crown; water-resistant to 10 atm
Band: textile, buckle
Price: $2,299
Variations: stainless steel bracelet ($2,399)

MÜHLE GLASHÜTTE

S.A.R. Rescue-Timer
Reference number: M1-41-03-MB
Movement: automatic, Sellita Caliber SW 200-1, version Mühle; ø 25.6 mm, height 4.6 mm; 26 jewels; 28,800 vph; with woodpecker-neck regulator, Mühle rotor, carefully reworked with special Mühle finish; 38-hour power reserve
Functions: hours, minutes, sweep seconds; date
Case: stainless steel, ø 42 mm, height 13.5 mm; bezel with rubber ring; sapphire crystal; screw-in crown; water-resistant to 100 atm
Band: stainless steel, folding clasp, with extension link
Price: $2,699
Variations: rubber strap ($2,799)

S.A.R. Flieger-Chronograph
Reference number: M1-41-33-KB
Movement: automatic, Caliber MU 9413; ø 30 mm, height 7.9 mm; 25 jewels; 28,800 vph; woodpecker-neck regulator, three-quarter plate, Mühle rotor, carefully reworked with special Mühle finish; 48-hour power reserve
Functions: hours, minutes, subsidiary seconds; chronograph; date
Case: stainless steel, ø 45 mm, height 16.2 mm; bidirectional 60-minute bezel; sapphire crystal; transparent case back; screw-in crown; water-resistant to 10 atm
Band: rubber, folding clasp, with extension link
Price: $4,299
Variations: stainless steel bracelet ($4,499)

ProMare Go
Reference number: M1-42-32-NB
Movement: automatic, Sellita Caliber SW200-1, version Mühle; ø 25.6 mm, height 4.6 mm; 26 jewels; 28,800 vph; woodpecker-neck regulator, Mühle rotor, carefully reworked with special Mühle finish; 38-hour power reserve
Functions: hours, minutes, sweep seconds; date
Case: stainless steel, ø 42 mm, height 12.2 mm; bidirectional 60-minute bezel; sapphire crystal; transparent case back; screw-in crown; water-resistant to 30 atm
Band: rubber with leather overlay, buckle
Price: $2,100

ProMare Chronograph
Reference number: M1-42-04-NB
Movement: automatic, Mühle Caliber MU 9408; ø 30 mm, height 7.9 mm; 25 jewels; 28,800 vph; woodpecker-neck regulator, Glashütte three-quarter plate, Mühle rotor, carefully reworked with special Mühle finish; 48-hour power reserve
Functions: hours, minutes, subsidiary seconds; chronograph; date
Case: stainless steel, ø 44 mm, height 15.4 mm; sapphire crystal; transparent case back; screw-in crown; water-resistant to 30 atm
Band: rubber with leather overlay, buckle
Price: $4,299
Variations: stainless steel bracelet ($4,399)

Teutonia Sport II
Reference number: M1-29-73-NB
Movement: automatic, Sellita Caliber SW290-1, version Mühle; ø 25.6 mm, height 5.6 mm; 31 jewels; 28,800 vph; woodpecker-neck regulator, finely reworked with special Mühle finish; 38-hour power reserve
Functions: hours, minutes, subsidiary seconds; date
Case: stainless steel, ø 41.6 mm, height 12.8 mm; sapphire crystal; transparent case back; screw-in crown; water-resistant to 10 atm
Band: rubber with leather overlay, buckle
Price: $2,549

Terrasport II
Reference number: M1-37-44-LB
Movement: automatic, Sellita Caliber SW200-1, version Mühle; ø 25.6 mm, height 4.6 mm; 26 jewels; 28,800 vph; woodpecker-neck regulator, Mühle rotor, carefully reworked with special Mühle finish; 38-hour power reserve
Functions: hours, minutes, sweep seconds; date
Case: stainless steel, ø 40 mm, height 10 mm; sapphire crystal; transparent case back; screw-in crown; water-resistant to 10 atm
Band: calfskin, buckle
Price: $1,799
Variations: stainless steel bracelet ($1,999)

MÜHLE GLASHÜTTE

Teutonia II Chronograph
Reference number: M1-30-95-LB
Movement: automatic, Mühle Caliber MU 9413; ø 30 mm, height 7.9 mm; 25 jewels; 28,800 vph; woodpecker-neck regulator, Glashütte three-quarter plate, Mühle rotor, carefully reworked with special Mühle finish; 48-hour power reserve
Functions: hours, minutes, subsidiary seconds; chronograph; date, weekday
Case: stainless steel, ø 42 mm, height 15.5 mm; sapphire crystal; transparent case back; screw-in crown; water-resistant to 10 atm
Band: reptile skin, double folding clasp
Price: $4,599
Variations: stainless steel bracelet ($4,799)

Teutonia II Weltzeit
Reference number: M1-33-82-LB
Movement: automatic, Sellita Caliber SW330-1, version Mühle; ø 25.6 mm, height 4.1 mm; 25 jewels; 28,800 vph; woodpecker-neck regulator, Mühle rotor, carefully reworked with special Mühle finish; 42-hour power reserve
Functions: hours, minutes, sweep seconds; world time indicator (2nd time zone); date
Case: stainless steel, ø 41 mm, height 13 mm; crown-activated scale ring, with reference city names; sapphire crystal; transparent case back; water-resistant to 10 atm
Band: reptile skin, double folding clasp
Price: $2,899
Variations: stainless steel bracelet ($2,999)

Lunova Tag/Datum
Reference number: M1-43-26-LB
Movement: automatic, Sellita Caliber SW220-1, version Mühle; ø 25.6 mm, height 5.05 mm; 26 jewels; 28,800 vph; woodpecker-neck regulator, Mühle rotor, carefully reworked with special Mühle finish; 38-hour power reserve
Functions: hours, minutes, sweep seconds; date, weekday
Case: stainless steel, ø 42.3 mm, height 11 mm; sapphire crystal; transparent case back; screw-in crown; water-resistant to 10 atm
Band: reptile skin, buckle
Price: $2,349

Lunova Lady
Reference number: M1-43-36-LM
Movement: automatic, Sellita Caliber SW200-1, version Mühle; ø 25.6 mm, height 4.6 mm; 26 jewels; 28,800 vph; woodpecker-neck regulator, carefully reworked with special Mühle finish; 38-hour power reserve
Functions: hours, minutes, sweep seconds; date
Case: stainless steel, ø 35 mm, height 9.6 mm; sapphire crystal; transparent case back; screw-in crown; water-resistant to 10 atm
Band: calfskin, buckle
Price: $1,949
Variations: calfskin strap without jewels ($1,949)

Robert Mühle Zeigerdatum
Reference number: M1-11-46-LB
Movement: manually wound, Robert Mühle Caliber RMK 03; ø 36.6 mm, height 8.35 mm; 33 jewels; 21,600 vph; engraved balance cock, with woodpecker-neck regulator, three-fifth plate with Glashütte long-slot click; 3 screw-mounted gold chatons; 56-hour power reserve
Functions: hours, minutes, subsidiary seconds; power reserve indicator; date
Case: stainless steel, ø 44 mm, height 12.56 mm; sapphire crystal; transparent case back; water-resistant to 10 atm
Band: reptile skin, folding clasp
Price: $8,799; limited to 100 pieces

RMK 03
Manually wound; woodpecker-neck regulator; single spring barrel, 56-hour power reserve
Functions: hours, minutes, subsidiary seconds; power reserve indicator; date
Diameter: 36.6 mm
Height: 8.35 mm
Jewels: 33, including 3 screw-mounted gold chatons
Balance: glucydur
Frequency: 21,600 vph
Balance spring: Nivarox
Remarks: three-fifth plate, Glashütte stopwork, hand-engraved balance cock

NIVREL

In 1891, master goldsmith Friedrich Jacob Kraemer founded a jewelry and watch shop in Saarbrücken that proved to be the place to go for fine craftsmanship. Gerd Hofer joined the family business in 1956, carrying it on into the fourth generation. However, his true passion was for watchmaking. In 1993, he and his wife, Gitta, bought the rights to use the Swiss name Nivrel, a brand that had been established in 1936, and integrated production of these watches into their German-based operations.

Today, Nivrel is led by the Hofers' daughter Anja, who is keeping both lineages alive. Mechanical complications with Swiss movements of the finest technical level and finishing as well as gold watches in the high-end design segment of the industry are manufactured with close attention to detail and an advanced level of craftsmanship. In addition to classic automatic watches, the brand has introduced everything from complicated chronographs and skeletonized watches to perpetual calendars and tourbillons. The movements and all the "habillage" of the watches—case, dial, crystal, crown, etc.—are made in Switzerland. Watch design, assembly, and finishing are done in Saarbrücken.

Nivrel watches are a perfect example of how quickly a watch brand incorporating a characteristic style and immaculate quality can make a respected place for itself in the industry. Affordable prices also play a significant role in this brand's success, but they do not keep the brand from innovating. Nivrel has teamed up with the Department of Metallic Materials of Saarland University to develop a special alloy for repeater springs that is softer and does not need as much energy to press.

Nivrel Uhren
Gerd Hofer GmbH
Kossmannstrasse 3
D-66119 Saarbrücken
Germany

Tel.:
+49-681-584-6576

Fax:
+49-681-584-6584

E-mail:
info@nivrel.com

Website:
www.nivrel.com

Founded:
1978

Number of employees:
10, plus external staff members

Distribution:
Please contact headquarters for enquiries.

Most important collections/price range:
mechanical watches, most with complications / approx. $600 to $45,000

Réplique Manuelle
Reference number: N 322.001 CAAES
Movement: manually wound, ETA Caliber 6498-1; ø 36.6 mm, height 4.5 mm; 17 jewels; 18,000 vph; blued screws, côtes de Genève; 46-hour power reserve
Functions: hours, minutes, subsidiary seconds
Case: stainless steel, ø 44 mm, height 13 mm; sapphire crystal; transparent case back; water-resistant to 5 atm
Band: calfskin, buckle
Price: $990

Réplique Manuelle Squelette
Reference number: N 322.001 CKKKS
Movement: manually wound, ETA Caliber 6497; ø 36.6 mm, height 4.5 mm; 17 jewels; 18,000 vph; hand-engraved and skeletonized movement; 46-hour power reserve
Functions: hours, minutes
Case: stainless steel, ø 44 mm, height 13 mm; sapphire crystal; transparent case back; water-resistant to 5 atm
Band: calfskin, buckle
Remarks: skeletonized dial
Price: $2,900

Red Voyager
Reference number: N 148.001 AASDS
Movement: automatic, ETA Caliber 2824-2; ø 25.6 mm, height 4.6 mm; 25 jewels; 28,800 vph; 42-hour power reserve
Functions: hours, minutes, sweep seconds; date
Case: stainless steel, ø 43 mm, height 14.5 mm; unidirectional bezel, with 0-60 scale; sapphire crystal; screw-in crown; water-resistant to 20 atm
Band: calfskin, buckle
Remarks: comes with additional silicon strap
Price: $799; limited to 100 pieces

Jubilé III

Reference number: N 121.001 AAWAS
Movement: automatic, ETA Caliber 2824-2; ø 25.6 mm, height 4.6 mm; 25 jewels; 28,800 vph; 38-hour power reserve
Functions: hours, minutes, sweep seconds; date
Case: stainless steel, ø 40 mm, height 10 mm; sapphire crystal; transparent case back; water-resistant to 5 atm
Band: calfskin, buckle
Price: $699
Variations: black dial; stainless steel Milanese mesh bracelet ($750)

Jubilé III

Reference number: N 121.001 AASAM
Movement: automatic, ETA Caliber 2824-2; ø 25.6 mm, height 4.6 mm; 25 jewels; 28,800 vph; 38-hour power reserve
Functions: hours, minutes, sweep seconds; date
Case: stainless steel, ø 40 mm, height 10 mm; sapphire crystal; transparent case back; water-resistant to 5 atm
Band: stainless steel Milanese mesh, folding clasp
Price: $750
Variations: white dial; calfskin strap ($699)

Deep Ocean Black

Reference number: N 145.001
Movement: automatic, ETA Caliber 2824-2; ø 25.6 mm, height 4.6 mm; 25 jewels; 28,800 vph; 38-hour power reserve
Functions: hours, minutes, sweep seconds; date
Case: stainless steel, ø 43 mm, height 13.5 mm; unidirectional bezel, 0-60 scale; sapphire crystal; screw-in crown; water-resistant to 50 atm
Band: stainless steel, folding clasp, with safety lock, with extension link
Remarks: comes with additional silicon strap
Price: $980

Coeur de la Sarre: Sarreguemines

Reference number: N 130.001 CABDS
Movement: automatic, ETA Caliber 2824-2; ø 25.6 mm, height 4.6 mm; 25 jewels; 28,800 vph; rotor with côtes de Genève; 38-hour power reserve
Functions: hours, minutes, sweep seconds
Case: stainless steel, ø 42 mm, height 12.5 mm; mineral sapphire; transparent case back; water-resistant to 5 atm
Band: calfskin, buckle
Price: $740

Héritage Grand Chronographe

Reference number: N 512.001 AAAAS
Movement: automatic, ETA Caliber 7750; ø 30 mm, height 7.9 mm; 25 jewels; 28,800 vph; fine finishing; 42-hour power reserve
Functions: hours, minutes, subsidiary seconds; chronograph; date, weekday
Case: stainless steel, ø 42 mm, height 13.5 mm; sapphire crystal; transparent case back; screw-in crown; water-resistant to 5 atm
Band: calfskin, buckle
Price: $2,350

Héritage Perpétuel

Reference number: N 401.001
Movement: automatic, ETA 2892-A2 with calendar module; ø 25.6 mm; 21 jewels; 28,800 vph; fine finishing, rotor with côtes de Genève; 42-hour power reserve
Functions: hours, minutes, sweep seconds; perpetual calendar with date, weekday, month, moon phase, leap year
Case: stainless steel, ø 38 mm, height 10 mm; sapphire crystal; transparent case back; water-resistant to 5 atm
Band: reptile skin, buckle
Price: $11,500
Variations: black reptile skin strap

NOMOS

Still waters run deep, and discreet business practices at times travel far. Nomos, founded in 1990, has suddenly become a full-fledged *manufacture* with brand-new facilities and a smart policy of only so much growth as the small team gathered around the founder, Roland Schwertner, and his associate Uwe Ahrendt can easily absorb.

The collection has grown to thirteen model families in a short period of time, with around one hundred variations. The number of calibers available is growing at an impressive rate, including two luxury manually wound movements with fine finishings. Nomos even produced an in-house escapement with a spring "made in Germany," the DUW 4401 (Deutsche Uhrenwerke Nomos Glashütte), which is gradually being used in all the movements, including the new, automatic ultrathin DUW 3001. The DUW 6101 features a safe and easy date correction, and it is a mere 3.6 millimeters high including the date, which fits well in the company's design efforts.

Speaking of design, the 300 people working at Nomos include about 40 design and communication staff at the Berlinblau in-house design studio and in the United States, where Nomos has offices (in New York) and about fifty points of sale. The key strategy: outstanding watches at an affordable price, a simple look full of subtle details, and marketing that is bold and humorous. Nomos, visibly, is a member of the deutscher Werkbund, precursor to the Bauhaus school, meaning pared-down industrial design, with a touch of Berlin's biting humor. This esthetic scrim, as it were, has produced the swimmer's watch Ahoi (as in "ship ahoy!"), which comes with an optional synthetic strap like those that carry locker keys at Germany's public swimming pools, or the related, colorful, Aqua series. Nomos has also been addressing the young and chic with the highly affordable Campus models. In 2018 came the Autobahn, designed in collaboration with Werner Aisslinger. And, yes, it definitely reminds one of a speedometer, with a fun open circle of luminescent segments.

NOMOS Glashütte/SA
Roland Schwertner KG
Ferdinand-Adolph-Lange-Platz 2
01768 Glashütte
Germany

Tel.:
+49-35053-4040

Fax:
+49-35053-40480

E-mail:
nomos@glashuette.com

Website:
nomos-glashuette.com

Founded:
1990

Number of employees:
300

U.S. distributor:
For the U.S. market, please contact:
NOMOS Glashuette USA Inc.
347 W. 36th St., Suite 600
New York, NY 10018
212-929-2575
E-mail: contact@nomos-watches.com

Most important collections/price range:
Ahoi / $4,020 to $4,660; Autobahn / $4,800; Club / $1,500 to $3,550; Lambda / $17,000 to $20,000; Ludwig / $1,700 to $4,000; Metro / $3,480 to $3,780; Orion / $1,920 to $4,200; Tangente / $1,760 to $4,100; Tangomat / $3,280 to $4,920; Tetra / $1,980 to $3,980; Zürich / $4,480 to $6,100

Autobahn Neomatik 41 Date

Reference number: 1301
Movement: automatic, Nomos Caliber DUW 6101; ø 35.2 mm, height 3.6 mm; 27 jewels; 21,600 vph; 42-hour power reserve
Functions: hours, minutes, subsidiary seconds; date
Case: stainless steel, ø 41 mm, height 10.5 mm; sapphire crystal; transparent case back; water-resistant to 10 atm
Band: textile, buckle
Price: $4,800

Autobahn Neomatik 41 Date Midnight Blue

Reference number: 1302
Movement: automatic, Nomos Caliber DUW 6101; ø 35.2 mm, height 3.6 mm; 27 jewels; 21,600 vph; 42-hour power reserve
Functions: hours, minutes, subsidiary seconds; date
Case: stainless steel, ø 41 mm, height 10.5 mm; sapphire crystal; transparent case back; water-resistant to 10 atm
Band: textile, buckle
Price: $4,800

Autobahn Neomatik 41 Date Sports Gray

Reference number: 1303
Movement: automatic, Nomos Caliber DUW 6101; ø 35.2 mm, height 3.6 mm; 27 jewels; 21,600 vph; 42-hour power reserve
Functions: hours, minutes, subsidiary seconds; date
Case: stainless steel, ø 41 mm, height 10.5 mm; sapphire crystal; transparent case back; water-resistant to 10 atm
Band: textile, buckle
Price: $4,800

NOMOS

Tangente Neomatik 41 Update
Reference number: 180
Movement: automatic, Nomos Caliber DUW 6101; ø 35.2 mm, height 3.6 mm; 27 jewels; 21,600 vph; 42-hour power reserve
Functions: hours, minutes, subsidiary seconds; date
Case: stainless steel, ø 40.5 mm, height 7.9 mm; sapphire crystal; transparent case back; water-resistant to 5 atm
Band: horse leather, buckle
Price: $4,100

Orion Neomatik 41 Date
Reference number: 360
Movement: automatic, Nomos Caliber DUW 6101; ø 35.2 mm, height 3.6 mm; 27 jewels; 21,600 vph; 42-hour power reserve
Functions: hours, minutes, subsidiary seconds; date
Case: stainless steel, ø 40.5 mm, height 9.35 mm; sapphire crystal; transparent case back; water-resistant to 5 atm
Band: horse leather, buckle
Price: $4,200

Ludwig Neomatik 41 Date
Reference number: 260
Movement: automatic, Nomos Caliber DUW 6101; ø 35.2 mm, height 3.6 mm; 27 jewels; 21,600 vph; 42-hour power reserve
Functions: hours, minutes, subsidiary seconds; date
Case: stainless steel, ø 40.5 mm, height 7.7 mm; sapphire crystal; transparent case back; water-resistant to 5 atm
Band: horse leather, buckle
Price: $4,000

Metro Date Power Reserve
Reference number: 1101
Movement: manually wound, Nomos Caliber DUW 4401; ø 32.1 mm, height 2.8 mm; 23 jewels; 21,600 vph; 42-hour power reserve
Functions: hours, minutes, subsidiary seconds; power reserve indicator; date
Case: stainless steel, ø 37 mm, height 7.65 mm; sapphire crystal; transparent case back; water-resistant to 3 atm
Band: horse leather, buckle
Price: $3,780

Metro Rose Gold Neomatik 39
Reference number: 1180
Movement: automatic, Nomos Caliber DUW 3001; ø 28.8 mm, height 3.2 mm; 27 jewels; 21,600 vph; 43-hour power reserve
Functions: hours, minutes, subsidiary seconds
Case: rose gold, ø 38.5 mm, height 8.35 mm; sapphire crystal; transparent case back; water-resistant to 3 atm
Band: horse leather, buckle
Price: $9,700

Metro Neomatik 39 Silvercut
Reference number: 1114
Movement: automatic, Nomos Caliber DUW 3001; ø 28.8 mm, height 3.2 mm; 27 jewels; 21,600 vph; 43-hour power reserve
Functions: hours, minutes, subsidiary seconds
Case: stainless steel, ø 38.5 mm, height 8.35 mm; sapphire crystal; transparent case back; water-resistant to 5 atm
Band: horse leather, buckle
Price: $4,280

NOMOS

Club 38 Campus Night
Reference number: 736
Movement: manually wound, Nomos Caliber Alpha; ø 23.3 mm, height 2.6 mm; 17 jewels; 21,600 vph; 43-hour power reserve
Functions: hours, minutes, subsidiary seconds
Case: stainless steel, ø 38.5 mm, height 8.45 mm; sapphire crystal; water-resistant to 10 atm
Band: suede, buckle
Price: $1,650

Ahoi
Reference number: 550
Movement: automatic, Nomos Caliber Epsilon; ø 31 mm, height 4.3 mm; 26 jewels; 21,600 vph; 43-hour power reserve
Functions: hours, minutes, subsidiary seconds
Case: stainless steel, ø 40.3 mm, height 10.54 mm; sapphire crystal; transparent case back; screw-in crown; water-resistant to 20 atm
Band: Textile, buckle
Price: $4,060

Zürich World Time Midnight Blue
Reference number: 807
Movement: automatic, Nomos Caliber DUW 5201; ø 31 mm, height 5.7 mm; 26 jewels; 21,600 vph; 42-hour power reserve
Functions: hours, minutes, subsidiary seconds; world time indicator (2nd time zone)
Case: stainless steel, ø 39.9 mm, height 10.85 mm; sapphire crystal; transparent case back; water-resistant to 3 atm
Band: horse leather, buckle
Price: $6,100

Tetra Azure
Reference number: 496
Movement: manually wound, Nomos Caliber Alpha; ø 23.3 mm, height 2.6 mm; 17 jewels; 21,600 vph; 43-hour power reserve
Functions: hours, minutes, subsidiary seconds
Case: stainless steel, 29.5 × 29.5 mm, height 6.25 mm; sapphire crystal; water-resistant to 3 atm
Band: suede, buckle
Price: $2,080

Tetra Matcha
Reference number: 495
Movement: manually wound, Nomos Caliber Alpha; ø 23.3 mm, height 2.6 mm; 17 jewels; 21,600 vph; 43-hour power reserve
Functions: hours, minutes, subsidiary seconds
Case: stainless steel, 29.5 × 29.5 mm, height 6.25 mm; sapphire crystal; water-resistant to 3 atm
Band: suede, buckle
Price: $2,080

Tetra Neomatik 39 Silvercut
Reference number: 423
Movement: automatic, Nomos Caliber DUW 3001; ø 28.8 mm, height 3.2 mm; 27 jewels; 21,600 vph; 43-hour power reserve
Functions: hours, minutes, subsidiary seconds
Case: stainless steel, 33 × 33 mm, height 7.3 mm; sapphire crystal; transparent case back; water-resistant to 3 atm
Band: horse leather, buckle
Price: $3,980

Tangente
Reference number: 139
Movement: manually wound, Nomos Caliber Alpha; ø 23.3 mm, height 2.6 mm; 17 jewels; 21,600 vph; 43-hour power reserve
Functions: hours, minutes, subsidiary seconds
Case: stainless steel, ø 35 mm, height 6.6 mm; sapphire crystal; transparent case back; water-resistant to 3 atm
Band: horse leather, buckle
Price: $2,180

Orion 33 Rose
Reference number: 325
Movement: manually wound, Nomos Caliber Alpha; ø 23.3 mm, height 2.6 mm; 17 jewels; 21,600 vph; 43-hour power reserve
Functions: hours, minutes, subsidiary seconds
Case: stainless steel, ø 32.8 mm, height 8.54 mm; sapphire crystal; transparent case back; water-resistant to 3 atm
Band: suede, buckle
Price: $2,240

Ludwig Neomatik Champagne
Reference number: 283
Movement: automatic, Nomos Caliber DUW 3001; ø 28.8 mm, height 3.2 mm; 27 jewels; 21,600 vph; 43-hour power reserve
Functions: hours, minutes, subsidiary seconds
Case: stainless steel, ø 36 mm, height 6.95 mm; sapphire crystal; transparent case back; water-resistant to 3 atm
Band: calfskin, buckle
Price: $3,480

Caliber DUW 2002
Manually wound; swan-neck fine adjustment; double spring barrel, 84-hour power reserve
Functions: hours, minutes, subsidiary seconds
Measurement: 22.6 × 32.6 mm
Height: 3.6 mm
Jewels: 23, including 5 screw-mounted gold chatons
Balance: screw balance
Frequency: 21,600 vph
Balance spring: Nivarox 1A
Shock protection: Incabloc
Remarks: hand-engraved balance cock, beveled and angled edges, movement surfaces with Glashütte sunburst brushing and perlage

Caliber DUW 6101
Automatic; single spring barrel, 43-hour power reserve
Functions: hours, minutes, subsidiary seconds; date
Diameter: 35.2 mm
Height: 3.6 mm
Jewels: 27
Balance: in-house manufacturing
Frequency: 21,600 vph
Balance: in-house manufacture, heat-blued
Shock protection: Incabloc
Remarks: three-quarter plate, rhodium-plated movement with Glashütte ribbing and perlage, gold-plated engraving

Caliber DUW 3001
Automatic; single spring barrel, 43-hour power reserve
Functions: hours, minutes, subsidiary seconds
Diameter: 28.8 mm
Height: 3.2 mm
Jewels: 27
Balance: in-house manufacture
Frequency: 21,600 vph
Balance spring: in-house manufacture, heat-blued
Shock protection: Incabloc
Remarks: three-quarter plate, rhodium-plated movement with Glashütte ribbing and perlage

OMEGA

Omega SA
Jakob-Stämpfli-Strasse 96
CH-2502 Biel/Bienne
Switzerland

Tel.:
+41-32-343-9211

E-mail:
info@omegawatches.com

Website:
www.omegawatches.com

Founded:
1848

U.S. distributor:
Omega
A division of The Swatch Group (U.S.), Inc.
1200 Harbor Boulevard
Weehawken, NJ 07086
201-271-1400
www.omegawatches.com

Omega is not the Swatch Group's most important brand. But it may be the one with the greatest reach. For one thing, it has appeared in James Bond movies, and was the official timekeeper at the 2010 Winter Olympics in Vancouver. It also played a major role historically in Switzerland's industry history. The brand was founded in 1848. In 1930, it merged with Tissot to form SIHH, which in turn merged with watch conglomerate ASUAG to form the Swatch Group in 1983, of which Omega was the leading brand. In the 1990s, the brand managed to expand incrementally into the Chinese market and thus established a firm foothold in Asia. This also led to a steep growth in production numbers, putting it neck and neck with Rolex.

Today Omega is back in the competition, with technology as the key. It introduced the innovative coaxial escapement to several collections, which has pushed the brand back among the technological front-runners in its market segment. Swatch Group subsidiary Nivarox-FAR has finally mastered the production of the difficult, oil-free parts of the system designed by Englishman George Daniels, although the escapement continues to include lubrication, as the long-term results of "dry" coaxial movements are less than satisfactory. Thus, the most important plus for this escapement design remains high rate stability after careful regulation. Omega has even revived the Ladymatic, adding a silicon spring and the trademark coaxial escapement.

A few years ago, Omega officially unveiled a brand-new 15,000-gauss amagnetic movement, a technological innovation that is gradually being added to all new movements. Indeed, in recent years, Omega has presented no fewer than eight new "Master Chronometer" movements. These not only meet the stringent requirements set out by the COSC but also have to pass the tests developed by Switzerland's Federal Institute of Metrology (METAS). The testing and certification process is performed in the new production building at the entirely renovated Swatch Group premises in Bienne/Biel. After promulgating the benefits of decentralization for years, Omega appears to be returning to the good old *manufacture* system of all crafts under a single roof.

Seamaster Aqua Terra Master Chronometer

Reference number: 220.10.41.21.01.001
Movement: automatic, Omega Caliber 8900; ø 29 mm, height 5.5 mm; 39 jewels; 25,200 vph; 2 spring barrels, coaxial escapement, silicon balance and hairspring, antimagnetic to 15,000 gauss; METAS-certified chronometer; 60-hour power reserve
Functions: hours, minutes, sweep seconds; date
Case: stainless steel, ø 41 mm, height 13.2 mm; sapphire crystal; transparent case back; screw-in crown; water-resistant to 15 atm
Band: stainless steel, folding clasp
Price: $5,500
Variations: calfskin or reptile skin strap

Seamaster Aqua Terra Master Chronometer

Reference number: 220.13.41.21.03.002
Movement: automatic, Omega Caliber 8900; ø 29 mm, height 5.5 mm; 39 jewels; 25,200 vph; 2 spring barrels, coaxial escapement, silicon balance and hairspring, antimagnetic to 15,000 gauss; METAS-certified chronometer; 60-hour power reserve
Functions: hours, minutes, sweep seconds; date
Case: stainless steel, ø 41 mm, height 13.2 mm; sapphire crystal; transparent case back; screw-in crown; water-resistant to 15 atm
Band: reptile skin, folding clasp
Price: $5,400
Variations: rubber strap

Seamaster Aqua Terra Master Chronometer

Reference number: 220.12.38.20.02.001
Movement: automatic, Omega Caliber 8800; ø 26 mm, height 4.6 mm; 35 jewels; 25,200 vph; 2 spring barrels, coaxial escapement, silicon balance and hairspring, antimagnetic to 15,000 gauss; METAS-certified chronometer; 55-hour power reserve
Functions: hours, minutes, sweep seconds; date
Case: stainless steel, ø 38 mm, height 12.26 mm; sapphire crystal; transparent case back; screw-in crown; water-resistant to 15 atm
Band: rubber, folding clasp
Price: $5,400
Variations: various dial colors

OMEGA

Seamaster Diver 300M

Reference number: 210.32.42.20.06.001
Movement: automatic, Omega Caliber 8800; ø 26 mm, height 4.6 mm; 35 jewels; 25,200 vph; 2 spring barrels, coaxial escapement, silicon balance and hairspring, antimagnetic to 15,000 gauss; METAS-certified chronometer; 55-hour power reserve
Functions: hours, minutes, sweep seconds; date
Case: stainless steel, ø 42 mm, height 13.56 mm; unidirectional bezel with ceramic insert, 0-60 scale; sapphire crystal; transparent case back; screw-in crown; helium valve; water-resistant to 30 atm
Band: rubber, folding clasp
Price: $4,750

Seamaster Diver 300M

Reference number: 210.30.42.20.01.001
Movement: automatic, Omega Caliber 8800; ø 26 mm, height 4.6 mm; 35 jewels; 25,200 vph; 2 spring barrels, coaxial escapement, silicon balance and hairspring, antimagnetic to 15,000 gauss; METAS-certified chronometer; 55-hour power reserve
Functions: hours, minutes, sweep seconds; date
Case: stainless steel, ø 42 mm, height 13.56 mm; unidirectional bezel with ceramic insert, 0-60 scale; sapphire crystal; transparent case back; screw-in crown; helium valve; water-resistant to 30 atm
Band: stainless steel, folding clasp
Price: $4,850

Seamaster Diver 300M

Reference number: 210.30.42.20.03.001
Movement: automatic, Omega Caliber 8800; ø 26 mm, height 4.6 mm; 35 jewels; 25,200 vph; 2 spring barrels, coaxial escapement, silicon balance and hairspring, antimagnetic to 15,000 gauss; METAS-certified chronometer; 55-hour power reserve
Functions: hours, minutes, sweep seconds; date
Case: stainless steel, ø 42 mm, height 13.56 mm; unidirectional bezel with ceramic insert, 0-60 scale; sapphire crystal; transparent case back; screw-in crown; helium valve; water-resistant to 30 atm
Band: stainless steel, folding clasp
Price: $4,850

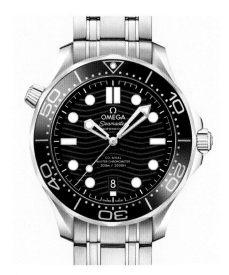

Seamaster Diver 300M

Reference number: 210.20.42.20.01.002
Movement: automatic, Omega Caliber 8800; ø 26 mm, height 4.6 mm; 35 jewels; 25,200 vph; 2 spring barrels, coaxial escapement, silicon balance and hairspring, antimagnetic to 15,000 gauss; METAS-certified chronometer; 55-hour power reserve
Functions: hours, minutes, sweep seconds; date
Case: stainless steel, ø 42 mm, height 13.56 mm; unidirectional bezel in yellow gold with ceramic inlay, 0-60 scale; sapphire crystal; transparent case back; screw-in crown, yellow gold; helium valve; water-resistant to 30 atm
Band: stainless steel with yellow gold elements, folding clasp
Price: $9,700

Seamaster Aqua Terra Annual Calendar

Reference number: 231.10.43.22.02.003
Movement: automatic, Omega Caliber 8602; ø 29 mm, height 6.5 mm; 39 jewels; 25,200 vph; coaxial escapement, silicon balance and hairspring; 60-hour power reserve; COSC-certified chronometer
Functions: hours, minutes, sweep seconds; annual calendar with date, month
Case: stainless steel, ø 43 mm, height 14.3 mm; sapphire crystal; transparent case back; screw-in crown; water-resistant to 15 atm
Band: stainless steel, folding clasp
Price: $8,600
Variations: various dials; reptile skin strap; pink gold bezel; in pink gold

Seamaster Aqua Terra GMT

Reference number: 231.13.43.22.01.001
Movement: automatic, Omega Caliber 8605; ø 29 mm, height 6 mm; 38 jewels; 25,200 vph; 2 spring barrels, coaxial escapement, silicon balance wheel and hairspring, antimagnetic to 15,000 gauss; METAS-certified chronometer; 60-hour power reserve
Functions: hours, minutes, sweep seconds; additional 24-hour display (2nd time zone); date
Case: stainless steel, ø 43 mm, height 14.15 mm; sapphire crystal; transparent case back; screw-in crown; water-resistant to 15 atm
Band: reptile skin, folding clasp
Price: $7,800
Variations: various dials; stainless steel bracelet; pink gold bezel; in pink gold

OMEGA

Seamaster Planet Ocean Master Chronometer
Reference number: 215.30.44.21.01.002
Movement: automatic, Omega Caliber 8900; ø 29 mm, height 5.5 mm; 39 jewels; 25,200 vph; 2 spring barrels, coaxial escapement, silicon balance and hairspring, antimagnetic to 15,000 gauss; METAS-certified chronometer; 60-hour power reserve
Functions: hours, minutes, sweep seconds; date
Case: stainless steel, ø 43.5 mm, height 16.04 mm; unidirectional bezel with ceramic insert, 0-60 scale; sapphire crystal; transparent case back; screw-in crown; helium valve; water-resistant to 60 atm
Band: stainless steel, folding clasp
Price: $6,550
Variations: rubber or reptile skin strap

Seamaster Planet Ocean Deep Black Master Chronometer
Reference number: 215.92.46.22.01.001
Movement: automatic, Omega Caliber 8906; ø 29 mm, height 6 mm; 38 jewels; 25,200 vph; 2 spring barrels, coaxial escapement, silicon balance and hairspring, antimagnetic to 15,000 gauss; METAS-certified chronometer; 60-hour power reserve
Functions: hours, minutes, sweep seconds; additional 24-hour display (2nd time zone); date
Case: ceramic, ø 43.5 mm, height 17.04 mm; unidirectional bezel, 0-60 scale; sapphire crystal; screw-in crown; helium valve; water-resistant to 60 atm
Band: rubber with textile overlay, folding clasp
Price: $11,700

Seamaster Aqua Terra Railmaster
Reference number: 220.10.40.20.01.001
Movement: automatic, Omega Caliber 8806; ø 26 mm, height 4.6 mm; 35 jewels; 25,200 vph; coaxial escapement, silicon balance and hairspring, antimagnetic to 15,000 gauss; METAS-certified chronometer; 55-hour power reserve
Functions: hours, minutes, sweep seconds
Case: stainless steel, ø 40 mm, height 12.65 mm; sapphire crystal; water-resistant to 15 atm
Band: stainless steel, folding clasp
Price: $5,000
Variations: various dials; calfskin strap; textile strap

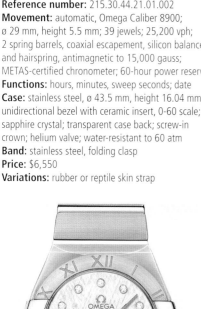

Constellation
Reference number: 127.10.27.20.52.001
Movement: automatic, Omega Caliber 8700; ø 20 mm, height 5.3 mm; 28 jewels; 25,200 vph; coaxial escapement, silicon balance and hairspring, antimagnetic to 15,000 gauss; METAS-certified chronometer; 50-hour power reserve
Functions: hours, minutes, sweep seconds; date
Case: stainless steel, ø 27 mm, height 12.25 mm; sapphire crystal; water-resistant to 10 atm
Band: stainless steel, folding clasp
Remarks: dial set with 11 diamonds
Price: $7,350

Globemaster Master Chronometer
Reference number: 130.33.39.21.03.001
Movement: automatic, Omega Caliber 8900; ø 29 mm, height 5.5 mm; 39 jewels; 25,200 vph; coaxial escapement, silicon balance and hairspring, antimagnetic to 15,000 gauss; METAS-certified chronometer; 60-hour power reserve
Functions: hours, minutes, sweep seconds; date
Case: stainless steel, ø 39 mm, height 12.53 mm; sapphire crystal; water-resistant to 10 atm
Band: reptile skin, folding clasp
Price: $6,900
Variations: various dials; stainless steel bracelet; pink gold bezel; in pink gold

De Ville Prestige
Reference number: 424.13.40.20.02.005
Movement: automatic, Omega Caliber 2500; ø 25.6 mm, height 4.1 mm; 27 jewels; 25,200 vph; coaxial escapement; 48-hour power reserve; COSC-certified chronometer
Functions: hours, minutes, sweep seconds; date
Case: stainless steel, ø 39.5 mm, height 10.1 mm; sapphire crystal; water-resistant to 3 atm
Band: reptile skin, folding clasp
Price: $3,600
Variations: stainless steel bracelet; pink gold or yellow gold bezel

OMEGA

Speedmaster Dark Side of the Moon "Apollo 8"
Reference number: 311.92.44.30.01.001
Movement: manually wound, Omega Caliber 1869; ø 27 mm, height 6.87 mm; 19 jewels; 21,600 vph; 48-hour power reserve
Functions: hours, minutes, subsidiary seconds; chronograph
Case: ceramic, ø 44.25 mm, height 13.8 mm; sapphire crystal; transparent case back; water-resistant to 5 atm
Band: calfskin, buckle
Remarks: mainplate and bridges with carefully replicated moon surface
Price: $12,000

Speedmaster Ladies Co-Axial Chronometer
Reference number: 324.30.38.50.06.001
Movement: automatic, Omega Caliber 3330; ø 30 mm, height 7.9 mm; 31 jewels; 28,800 vph; coaxial escapement, silicon balance and hairspring; 52-hour power reserve; COSC-certified chronometer
Functions: hours, minutes, subsidiary seconds; chronograph; date
Case: stainless steel, ø 38 mm, height 14.7 mm; bezel with ceramic inlay; sapphire crystal; water-resistant to 10 atm
Band: stainless steel, folding clasp
Price: $4,900

Speedmaster Moonphase
Reference number: 304.33.44.52.03.001
Movement: automatic, Omega Caliber 9904; ø 32.5 mm, height 8.35 mm; 54 jewels; 28,800 vph; 2 spring barrels, coaxial escapement, silicon balance and hairspring, antimagnetic to 15,000 gauss; METAS-certified chronometer; 60-hour power reserve
Functions: hours, minutes, subsidiary seconds; chronograph; date, moon phase
Case: stainless steel, ø 44.25 mm, height 16.85 mm; bezel with ceramic inlay; sapphire crystal; water-resistant to 10 atm
Band: reptile skin, folding clasp
Price: $10,600

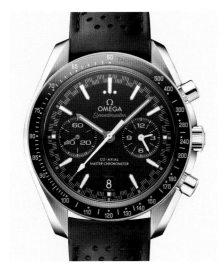

Speedmaster Racing Master Chronometer
Reference number: 329.32.44.51.01.001
Movement: automatic, Omega Caliber 9900; ø 32.5 mm, height 7.6 mm; 54 jewels; 28,800 vph; 2 spring barrels, coaxial escapement, silicon balance and hairspring, antimagnetic to 15,000 gauss; METAS-certified chronometer; 60-hour power reserve
Functions: hours, minutes, subsidiary seconds; chronograph; date
Case: stainless steel, ø 44.25 mm, height 14.9 mm; ceramic bezel; sapphire crystal; water-resistant to 5 atm
Band: calfskin, folding clasp
Price: $8,450
Variations: various dials; reptile skin strap; stainless steel bracelet

Speedmaster Racing Master Chronometer
Reference number: 329.32.44.51.06.001
Movement: automatic, Omega Caliber 9900; ø 32.5 mm, height 7.6 mm; 54 jewels; 28,800 vph; 2 spring barrels, coaxial escapement, silicon balance and hairspring, antimagnetic to 15,000 gauss; METAS-certified chronometer; 60-hour power reserve
Functions: hours, minutes, subsidiary seconds; chronograph; date
Case: stainless steel, ø 44.25 mm, height 14.9 mm; ceramic bezel; sapphire crystal; water-resistant to 5 atm
Band: calfskin, folding clasp
Price: $8,450
Variations: various dials; reptile skin strap; stainless steel bracelet

Speedmaster '57
Reference number: 331.10.42.51.01.002
Movement: automatic, Omega Caliber 9300; ø 32.5 mm, height 7.6 mm; 54 jewels; 28,800 vph; coaxial escapement, silicon balance and hairspring; 60-hour power reserve; COSC-certified chronometer
Functions: hours, minutes, subsidiary seconds; chronograph; date
Case: stainless steel, ø 41.5 mm, height 16.17 mm; sapphire crystal; water-resistant to 10 atm
Band: stainless steel, folding clasp
Price: $9,000
Variations: various dials; calfskin strap; in pink gold; in yellow gold

OMEGA

Caliber 8800

Automatic; coaxial escapement; antimagnetic up to 15,000 gauss; METAS-certified chronometer; single spring barrel, 55-hour power reserve
Functions: hours, minutes, sweep seconds; date
Diameter: 26 mm
Height: 4.6 mm
Jewels: 35
Balance: silicon, without regulator
Frequency: 25,200 vph
Balance spring: silicon
Shock protection: Nivachoc
Remarks: blackened screws

Caliber 8801

Automatic; coaxial escapement; antimagnetic up to 15,000 gauss; METAS-certified chronometer; single spring barrel, 55-hour power reserve
Functions: hours, minutes, subsidiary seconds; date
Jewels: 35
Balance: silicon, without regulator
Frequency: 25,200 vph
Balance spring: silicon
Shock protection: Nivachoc
Remarks: gold rotor, gold balance wheel bridge, blackened screws

Caliber 8807

Automatic; coaxial escapement; antimagnetic up to 15,000 gauss; METAS-certified chronometer; single spring barrel, 55-hour power reserve
Functions: hours, minutes, sweep seconds
Diameter: 26 mm
Height: 4.6 mm
Jewels: 35
Balance: silicon, without regulator
Frequency: 25,200 vph
Balance spring: silicon
Shock protection: Nivachoc
Remarks: gold rotor, gold balance wheel bridge, blackened screws

Caliber 8900

Automatic; coaxial escapement; antimagnetic up to 15,000 gauss; METAS-certified chronometer; double spring barrel, 60-hour power reserve
Functions: hours, minutes, sweep seconds; date
Diameter: 29 mm
Height: 5.5 mm
Jewels: 39
Balance: silicon, without regulator
Frequency: 25,200 vph
Balance spring: silicon
Shock protection: Nivachoc
Remarks: mainplate, bridges and rotor with "arabesque" côtes de Genève, rhodium-plated, spring barrels, blackened balance wheel and screws

Caliber 9900

Automatic; coaxial escapement; column wheel control of chronograph functions; antimagnetic up to 15,000 gauss; METAS-certified chronometer; double spring barrel, 60-hour power reserve
Functions: hours, minutes, subsidiary seconds; chronograph; date
Diameter: 32.5 mm
Height: 7.6 mm
Jewels: 54
Balance: silicon, without regulator
Frequency: 28,800 vph
Balance spring: silicon
Shock protection: Nivachoc
Remarks: mainplate, bridges, and rotor with "arabesque" côtes de Genève

Caliber 1861

Manually wound; single spring barrel, 48-hour power reserve
Base caliber: Lémania 1873
Functions: hours, minutes, subsidiary seconds; chronograph
Diameter: 27 mm
Height: 6.87 mm
Jewels: 18
Frequency: 21,600 vph
Balance spring: flat hairspring
Remarks: rhodium-plated, gold-plated engravings

Oris SA
Ribigasse 1
CH-4434 Hölstein
Switzerland

Tel.:
+41-61-956-1111

E-mail:
info@oris

Website:
www.oris.ch

Founded:
1904

Number of employees:
90

U.S. distributor:
Oris Watches USA
50 Washington Street, Suite 412
Norwalk, CT 06854
203-857-4769; 203-857-4782 (fax)

Most important collections/price range:
Diver, Big Crown, Artelier, Aquis, Williams /
approx. $1,100 to $5,500

ORIS

Oris has been producing mechanical watches in the little town of Hölstein in northwestern Switzerland, near Basel, since 1904, so 2014 was a celebratory year. The brand's strategy has always been to keep prices low and quality high, so Oris has managed to expand in a segment relinquished by other big-name competitors as they sought their fortune in the higher-end markets. The result has been growing international success for Oris, whose portfolio is divided up into four "product worlds," each with its own distinct identity: aviation, motor sports, diving, and culture. In utilizing specific materials—a tungsten bezel for the divers, for example—and functions based on these types, Oris makes certain that each will fit perfectly into the world for which it was designed. Yet the heart of every watch houses a small, high-quality "high-mech" movement identifiable by the brand's standard red rotor.

The brand surprised everyone for its 110th birthday by signing off on the in-house Caliber 110, a plain, but technically efficient, manually wound movement. It was made together with the engineers from the Technical College of Le Locle, and features a massive barrel spring with a 6-foot (1.8-m) spring. It was followed by the Caliber 111 with an optimized spring that could provide ten days of power of even torque. The power reserve indicator on the right of the dial does not move evenly, however, due to the transmission ratio. Toward the end, the markers are somewhat longer to give a more accurate idea of the remaining power in the spring. Ever since, the company has come out with a caliber a year. The 112 has GMT function and day/night indication. The fourth in-house caliber, 113, was equipped with a clever sweep hand indication of calendar weeks that also shows the month. Add to that the apertures for date and day of the week, and you have a complete calendar for businesspeople and others who need to stay dialed into the date. And in 2018 came the 114, which is used not in the Artelier series, but rather in the Big Crown ProPilot series.

Artelier Caliber 112

Reference number: 112 7726 6351
Movement: manually wound, Oris Caliber 112; ø 34 mm, height 6.4 mm; 40 jewels; 21,600 vph; 240-hour power reserve
Functions: hours, minutes, subsidiary seconds; additional 12-hour display (2nd time zone), day/night indicator, power reserve indicator; date
Case: stainless steel, ø 43 mm, height 12.7 mm; bezel in rose gold; sapphire crystal; transparent case back; water-resistant to 5 atm
Band: reptile skin, folding clasp
Price: $8,200
Variations: stainless steel ($6,700)

Artelier Caliber 113

Reference number: 113 7738 4031
Movement: manually wound, Oris Caliber 113; ø 34 mm, height 6.65 mm; 40 jewels; 21,600 vph; 240-hour power reserve
Functions: hours, minutes, subsidiary seconds; power reserve indicator; full calendar with date, weekday, calendar week, month
Case: stainless steel, ø 43 mm, height 13.05 mm; sapphire crystal; transparent case back; water-resistant to 5 atm
Band: reptile skin, folding clasp
Price: $6,300
Variations: anthracite dial; with stainless steel bracelet

Artelier Caliber 113

Reference number: 113 7738 4063
Movement: manually wound, Oris Caliber 113; ø 34 mm, height 6.65 mm; 40 jewels; 21,600 vph; 240-hour power reserve
Functions: hours, minutes, subsidiary seconds; power reserve indicator; full calendar with date, weekday, calendar week, month
Case: stainless steel, ø 43 mm, height 13.05 mm; sapphire crystal; transparent case back; water-resistant to 5 atm
Band: reptile skin, folding clasp
Price: $6,300
Variations: opaline dial; stainless steel bracelet

ORIS

Big Crown Pointer Date
Reference number: 754 7741 4065
Movement: automatic, Oris Caliber 754 (base Sellita SW200-1); ø 25.6 mm, height 4.6 mm; 26 jewels; 28,800 vph; 38-hour power reserve
Functions: hours, minutes, sweep seconds; date
Case: stainless steel, ø 40 mm, height 11.8 mm; sapphire crystal; transparent case back; water-resistant to 5 atm
Band: stainless steel, folding clasp
Price: $1,800
Variations: calfskin strap ($1,500)

Big Crown ProPilot Caliber 114
Reference number: 01 114 7746 4063
Movement: manually wound, Oris Caliber 114; ø 34 mm, height 6 mm; 40 jewels; 21,600 vph; 240-hour power reserve
Functions: hours, minutes, subsidiary seconds; additional 24-hour display (2nd time zone), power reserve indicator; date
Case: stainless steel, ø 44 mm, height 14 mm; sapphire crystal; transparent case back; screw-in crown; water-resistant to 10 atm
Band: reptile skin, folding clasp
Price: $6,100
Variations: textile strap ($5,800); stainless steel bracelet ($5,900)

Big Crown ProPilot GMT
Reference number: 748 7710 4063
Movement: automatic, Oris Caliber 748 (base Sellita SW500-1); ø 32.2 mm, height 5.5 mm; 28 jewels; 28,800 vph; 38-hour power reserve
Functions: hours, minutes, subsidiary seconds; additional 24-hour display (2nd time zone); date
Case: stainless steel, ø 45 mm, height 12.9 mm; sapphire crystal; transparent case back; screw-in crown; water-resistant to 10 atm
Band: calfskin, folding clasp
Price: $2,300
Variations: textile strap ($2,300); stainless steel bracelet ($2,500)

Big Crown ProPilot Worldtimer
Reference number: 01 690 7735 4063
Movement: automatic, Oris Caliber 690 (base ETA 2836-2); ø 25.6 mm, height 5.05 mm; 30 jewels; 28,800 vph; 38-hour power reserve
Functions: hours, minutes, subsidiary seconds; additional 12-hour display (2nd time zone); date
Case: stainless steel, ø 44.7 mm, height 13.1 mm; sapphire crystal; transparent case back; water-resistant to 10 atm
Band: calfskin, folding clasp
Price: $3,600

Divers Sixty-Five
Reference number: 01 733 7707 4354
Movement: automatic, Oris Caliber 733 (base Sellita SW200-1); ø 25.6 mm, height 4.6 mm; 26 jewels; 28,800 vph; 38-hour power reserve
Functions: hours, minutes, sweep seconds; date
Case: stainless steel, ø 40 mm, height 12.8 mm; bezel in bronze, unidirectional bezel, with 0-60 scale; sapphire crystal; screw-in crown; water-resistant to 10 atm
Band: calfskin, buckle
Price: $2,000
Variations: various colors; textile strap ($2,000); rubber strap ($2,000); stainless steel bracelet ($2,300)

Regulateur "Master Diver"
Reference number: 749 7734 7154
Movement: automatic, Oris Caliber 749 (base Sellita SW220-1); ø 25.6 mm, height 5.05 mm; 28 jewels; 28,800 vph; 38-hour power reserve
Functions: hours (off-center), minutes, subsidiary seconds; date
Case: titanium, ø 43.5 mm, height 12.65 mm; unidirectional bezel, with 0-60 scale; sapphire crystal; screw-in crown; helium valve; water-resistant to 30 atm
Band: titanium, folding clasp
Remarks: comes with additional rubber strap
Price: $3,350

Williams Engine Date
Reference number: 733 7740 4154
Movement: automatic, Oris Caliber 733 (base Sellita SW200-1); ø 25.6 mm, height 4.6 mm; 26 jewels; 28,800 vph; 38-hour power reserve
Functions: hours, minutes, sweep seconds; date
Case: stainless steel, ø 42 mm, height 11.55 mm; sapphire crystal; water-resistant to 10 atm
Band: rubber, folding clasp
Remarks: skeletonized dial
Price: $1,650
Variations: stainless steel bracelet ($1,850)

Chronoris Date
Reference number: 01 733 7737 4054
Movement: automatic, Oris Caliber 733 (base Sellita SW200-1); ø 25.6 mm, height 4.6 mm; 26 jewels; 28,800 vph; 38-hour power reserve
Functions: hours, minutes, sweep seconds; date
Case: stainless steel, ø 39 mm, height 12.4 mm; crown-activated scale ring, with 0-60 scale; sapphire crystal; water-resistant to 10 atm
Band: calfskin, buckle
Price: $1,750
Variations: stainless steel bracelet ($1,950); textile or rubber strap ($1,750)

Williams 40th Anniversary Limited Edition
Reference number: 673 7739 4084
Movement: automatic, Oris Caliber 673 (base ETA 7750); ø 30 mm, height 7.9 mm; 25 jewels; 28,800 vph; 48-hour power reserve
Functions: hours, minutes; chronograph; date
Case: stainless steel, ø 40 mm, height 15.4 mm; sapphire crystal; water-resistant to 10 atm
Band: stainless steel, folding clasp
Price: $3,950
Variations: textile strap ($3,700); rubber strap ($3,700); calfskin strap ($3,700)

Aquis Titanium Date
Reference number: 01 733 7730 7153
Movement: automatic, Oris Caliber 733 (base Sellita SW200-1); ø 25.6 mm, height 4.6 mm; 26 jewels; 28,800 vph; 38-hour power reserve
Functions: hours, minutes, sweep seconds; date
Case: titanium, ø 43.5 mm, height 12.7 mm; unidirectional bezel, with 0-60 scale; sapphire crystal; transparent case back; screw-in crown; water-resistant to 30 atm
Band: rubber, folding clasp
Price: $2,000

Aquis Clipperton Limited Edition
Reference number: 733 7730 4185
Movement: automatic, Oris Caliber 733 (base Sellita SW200-1); ø 25.6 mm, height 4.6 mm; 26 jewels; 28,800 vph; 38-hour power reserve
Functions: hours, minutes, sweep seconds; date
Case: stainless steel, ø 43.5 mm, height 12.7 mm; unidirectional bezel, with 0-60 scale; sapphire crystal; water-resistant to 30 atm
Band: rubber, buckle
Price: $2,000; limited to 2,000 pieces
Variations: stainless steel bracelet ($2,200)

Aquis Source of Life Limited Edition
Reference number: 733 7730 4125
Movement: automatic, Oris Caliber 733 (base Sellita SW200-1); ø 25.6 mm, height 4.6 mm; 26 jewels; 28,800 vph; 38-hour power reserve
Functions: hours, minutes, sweep seconds; date
Case: stainless steel, ø 43.5 mm, height 12.9 mm; unidirectional bezel, with 0-60 scale; sapphire crystal; water-resistant to 30 atm
Band: stainless steel, folding clasp
Price: $2,400; limited to 2,343 pieces
Variations: rubber strap ($2,200)

PANERAI

Officine Panerai (in English: Panerai Workshops) joined the Richemont Group in 1997. Since then, it has made an unprecedented rise from an insider niche brand to a lifestyle phenomenon. The company, founded in 1860 by Giovanni Panerai, supplied the Italian navy with precision instruments. In the 1930s, the Florentine engineers developed a series of waterproof wristwatches that could be used by commandos under especially extreme and risky conditions. After 1997, under the leadership of Angelo Bonati, the company came out with a collection of oversize wristwatches, both stylistically and technically based on these historical models.

In 2002, Panerai opened a *manufacture* in Neuchâtel, and by 2005 it was already producing its own movements (caliber family P.2000). In 2009, the new "little" Panerai *manufacture* movements (caliber family P.9000) were released. From the start, the idea behind them was to provide a competitive alternative to the base movements available until a couple of years ago. In 2014, a new *manufacture* was inaugurated in Neuchâtel to handle development, manufacturing, assembly, and quality control under one roof.

Parallel to consolidating, the brand has been steadily expanding its portfolio of new calibers. Fairly early on, it came out with an automatic chronograph with a flyback function, the P.9100. This was followed by a string of new calibers, almost one per year, to gradually replace "foreign" movements. Notorious is the P.4000, with an off-center winding rotor. At 3.95 millimeters, it is very thin for Panerai, but then again, it was developed for a new set of models.

In 2016, the company updated its cases and added the P.4001 and P.4002 calibers to the P.4000 family. The new calibers featured a date function and a power reserve display. The latest caliber, the P.6000 is 100 percent Panerai made in Switzerland. Interestingly, it comes with a deadbeat second and will be fit into the Luminors.

Officine Panerai
Viale Monza, 259
I-20126 Milan
Italy

Tel.:
+39-02-363-138

Fax:
+39-02-363-13-297

Website:
www.panerai.com

Founded:
1860 in Florence, Italy

Number of employees:
approx. 250

U.S. distributor:
Panerai
645 Fifth Avenue
New York, NY 10022
877-PANERAI
concierge.usa@panerai.com; www.panerai.com

Most important collections/price range:
Luminor / $5,000 to $25,000; Luminor 1950 / $8,000 to $30,000; Radiomir / $7,000 to $25,000; Radiomir 1940 / $8,000 to $133,000; special editions / $10,000 to $125,000; clocks and instruments / $20,000 to $250,000

Lo Scienziato—Luminor 1950 Tourbillon GMT Titanio

Reference number: PAM00767
Movement: manually wound, Panerai Caliber 2005/T; ø 36.6 mm, height 10.05 mm; 31 jewels; 28,800 vph; 3 spring barrels, 1-minute tourbillon; skeletonized movement; 44-hour power reserve
Functions: hours, minutes, subsidiary seconds; additional 12-hour display (2nd time zone), day/night indicator, power reserve indicator (on back)
Case: titanium, ø 47 mm, height 17.66 mm; sapphire crystal; transparent case back; crown protector with hinged lever; water-resistant to 10 atm
Band: calfskin, buckle
Price: $143,000; limited to 100 pieces

Radiomir 3 Days Acciaio

Reference number: PAM00721
Movement: manually wound, Panerai Caliber P.3000; ø 37.2 mm, height 5.3 mm; 21 jewels; 21,600 vph; 2 spring barrels 72-hour power reserve
Functions: hours, minutes
Case: stainless steel, ø 47 mm, height 15.97 mm; sapphire crystal; transparent case back; water-resistant to 10 atm
Band: calfskin, buckle
Remarks: sandwich dial with lower layer of luminous mass
Price: $9,200

Luminor Base Logo 3 Days Acciaio

Reference number: PAM00774
Movement: manually wound, Panerai Caliber P.6000; ø 34.96 mm, height 4.5 mm; 19 jewels; 21,600 vph; 72-hour power reserve
Functions: hours, minutes
Case: stainless steel, ø 44 mm, height 13.05 mm; sapphire crystal; crown protector with hinged lever; water-resistant to 10 atm
Band: textile, buckle
Price: $4,750

Luminor Base Logo 3 Days Acciaio

Reference number: PAM00775
Movement: manually wound, Panerai Caliber P.6000; ø 34.96 mm, height 4.5 mm; 19 jewels; 21,600 vph; 72-hour power reserve
Functions: hours, minutes
Case: stainless steel, ø 44 mm, height 13 mm; sapphire crystal; crown protector with hinged lever; water-resistant to 10 atm
Band: calfskin, buckle
Price: $5,000

Luminor Marina Logo 3 Days Acciaio

Reference number: PAM00777
Movement: manually wound, Panerai Caliber P.6000; ø 34.96 mm, height 4.5 mm; 19 jewels; 21,600 vph; 72-hour power reserve
Functions: hours, minutes, subsidiary seconds
Case: stainless steel, ø 44 mm, height 13.05 mm; sapphire crystal; crown protector with hinged lever; water-resistant to 10 atm
Band: textile, buckle
Price: $15,300

Luminor Due 3 Days Automatic Oro Rosso

Reference number: PAM00756
Movement: automatic, Panerai Caliber OP XXXIV; ø 28.19 mm, height 4.2 mm; 22 jewels; 28,800 vph; 72-hour power reserve
Functions: hours, minutes, subsidiary seconds; date
Case: pink gold, ø 38 mm, height 11.2 mm; sapphire crystal; crown protector with hinged lever; water-resistant to 3 atm
Band: reptile skin, buckle
Price: $15,300
Variations: stainless steel ($6,000)

Luminor Due 3 Days Automatic Acciaio

Reference number: PAM00904
Movement: automatic, Panerai Caliber OP XXXIV; ø 28.19 mm, height 4.2 mm; 22 jewels; 28,800 vph; 72-hour power reserve
Functions: hours, minutes, subsidiary seconds; date
Case: stainless steel, ø 42 mm, height 10.7 mm; sapphire crystal; crown protector with hinged lever; water-resistant to 3 atm
Band: calfskin, buckle
Remarks: sandwich dial with lower layer of luminous mass
Price: $6,400

Luminor Due 3 Days Automatic Acciaio

Reference number: PAM00906
Movement: automatic, Panerai Caliber OP XXXIV; ø 28.19 mm, height 4.2 mm; 22 jewels; 28,800 vph; 72-hour power reserve
Functions: hours, minutes, subsidiary seconds; date
Case: stainless steel, ø 42 mm, height 10.7 mm; sapphire crystal; crown protector with hinged lever; water-resistant to 3 atm
Band: calfskin, buckle
Price: $6,400
Variations: pink gold ($21,900)

Luminor Due 3 Days Automatic Acciaio

Reference number: PAM00943
Movement: automatic, Panerai Caliber P.4001; ø 31 mm, height 5.04 mm; 31 jewels; 28,800 vph; 2 spring barrels, microrotor; 72-hour power reserve
Functions: hours, minutes, subsidiary seconds; power reserve indicator (on back); date
Case: stainless steel, ø 45 mm, height 11.4 mm; sapphire crystal; transparent case back; crown protector with hinged lever; water-resistant to 3 atm
Band: reptile skin, buckle
Remarks: sandwich dial with lower layer of luminous mass
Price: $10,100

PANERAI

Luminor Due 3 Days GMT Power Reserve Automatic Acciaio

Reference number: PAM00944
Movement: automatic, Panerai Caliber P.4002; ø 31 mm, height 4.8 mm; 31 jewels; 28,800 vph; 2 spring barrels, microrotor; 72-hour power reserve
Functions: hours, minutes, subsidiary seconds; additional 12-hour display (2nd time zone), power reserve indicator; date
Case: stainless steel, ø 45 mm; sapphire crystal; transparent case back; crown protector with hinged lever; water-resistant to 3 atm
Band: reptile skin, buckle
Remarks: sandwich dial with lower layer of luminous mass
Price: $11,100

Luminor Submersible 1950 3 Days Automatic Acciaio

Reference number: PAM00682
Movement: automatic, Panerai Caliber P.9010; ø 31.1 mm, height 6 mm; 31 jewels; 28,800 vph; 72-hour power reserve
Functions: hours, minutes, subsidiary seconds; date
Case: stainless steel, ø 42 mm; unidirectional bezel, with 0-60 scale; sapphire crystal; crown protector with hinged lever; water-resistant to 30 atm
Band: rubber, buckle
Price: $8,700
Variations: pink gold with ceramic bezel ($26,700)

Luminor Submersible 1950 BMG-TECH 3 Days Automatic

Reference number: PAM00692
Movement: automatic, Panerai Caliber P.9010; ø 31.1 mm, height 6 mm; 31 jewels; 28,800 vph; 72-hour power reserve
Functions: hours, minutes, subsidiary seconds; date
Case: composite material, BMG technology (alloy of zirconium, copper, aluminum, titanium, nickel), ø 47 mm; unidirectional bezel, with 0-60 scale; sapphire crystal; crown protector with hinged lever; water-resistant to 30 atm
Band: rubber, buckle
Price: $13,300

Luminor Marina 1950 3 Days Automatic Acciaio

Reference number: PAM01312
Movement: automatic, Panerai Caliber P.9010; ø 31 mm, height 6 mm; 31 jewels; 28,800 vph; 72-hour power reserve
Functions: hours, minutes, subsidiary seconds; date
Case: stainless steel, ø 44 mm; sapphire crystal; transparent case back; crown protector with hinged lever; water-resistant to 30 atm
Band: reptile skin, buckle
Remarks: sandwich dial with lower layer of luminous mass
Price: $7,500

Luminor Marina 1950 3 Days Automatic Acciaio

Reference number: PAM01499
Movement: automatic, Panerai Caliber P.9010; ø 31 mm, height 6 mm; 31 jewels; 28,800 vph; 72-hour power reserve
Functions: hours, minutes, subsidiary seconds; date
Case: stainless steel, ø 44 mm; sapphire crystal; transparent case back; crown protector with hinged lever; water-resistant to 30 atm
Band: calfskin, buckle
Price: $7,500

Radiomir Black Seal 8 Days Acciaio

Reference number: PAM00609
Movement: manually wound, Panerai Caliber P.5000; ø 34.97 mm, height 4.5 mm; 21 jewels; 21,600 vph; 2 spring barrels, screw balance; 192-hour power reserve
Functions: hours, minutes, subsidiary seconds
Case: stainless steel, ø 45 mm; sapphire crystal; transparent case back; water-resistant to 10 atm
Band: calfskin, buckle
Price: $6,000

Caliber P.2005/S

Manually wound; 30-second tourbillon, rotation along long axis, skeletonized mainplate and bridges; triple serial spring barrel, 144-hour power reserve
Functions: hours, minutes, subsidiary seconds; additional 24-hour display (2nd time zone), power reserve indicator (on rear)
Diameter: 36.6 mm
Height: 10.05 mm
Jewels: 31
Balance: glucydur
Frequency: 28,800 vph
Shock protection: Kif
Remarks: 277 parts

Caliber P.3000

Manually wound; double spring barrel, double serial spring barrel, 72-hour power reserve
Functions: hours, minutes
Diameter: 37.2 mm
Height: 5.3 mm
Jewels: 21
Balance: glucydur
Frequency: 21,600 vph
Remarks: 160 parts

Caliber P.4002

Automatic; microrotor; double serial spring barrel, 72-hour power reserve
Functions: hours, minutes, subsidiary seconds; additional 12-hour display (2nd time zone); date
Diameter: 30 mm
Height: 4.8 mm
Jewels: 31
Balance: glucydur
Frequency: 28,800 vph
Balance spring: flat hairspring
Shock protection: Kif
Remarks: 288 parts

Caliber P.6000

Manually wound; single spring barrel, 72-hour power reserve
Functions: hours, minutes
Diameter: 35 mm
Height: 4.5 mm
Jewels: 19
Balance: glucydur
Frequency: 21,600 vph
Shock protection: Incabloc
Remarks: 110 parts

Caliber P.5000

Manually wound; double serial spring barrel, 192-hour power reserve
Functions: hours, minutes, subsidiary seconds
Diameter: 37.2 mm
Height: 4.5 mm
Jewels: 21
Balance: glucydur with weighted screws
Frequency: 21,600 vph
Balance spring: flat hairspring
Remarks: 127 parts

Caliber P.4001

Automatic; microrotor; stop-seconds mechanism and zero position on pulling crown; double spring barrel, 72-hour power reserve
Functions: hours, minutes, subsidiary seconds; additional 24-hour display (2nd time zone), power reserve indicator (on rear); date
Diameter: 30 mm
Height: 5.04 mm
Jewels: 31
Balance: glucydur
Frequency: 28,800 vph
Balance spring: flat hairspring
Shock protection: Kif
Remarks: 278 parts

PARMIGIANI

What began as the undertaking of a single man—a gifted watchmaker and reputable restorer of complicated vintage timepieces—in the small town of Fleurier in Switzerland's Val de Travers has now grown into an empire of sorts comprising several factories and more than 400 employees.

Michel Parmigiani is in fact just doing what he has done since 1976, when he began restoring vintage works. An exceptional talent, his output soon attracted the attention of the Sandoz Family Foundation, an organization established by a member of one of Switzerland's most famous families in 1964. The foundation bought 51 percent of Parmigiani Mesure et Art du Temps SA in 1996, turning what was practically a one-man show into a full-fledged and fully financed watch *manufacture*.

After the merger, Swiss suppliers were acquired by the partners, furthering the quest for horological autonomy. Atokalpa SA in Alle (Canton of Jura) manufactures parts such as pinions, wheels, and micro components. Bruno Affolter SA in La Chaux-de-Fonds produces precious metal cases, dials, and other specialty parts. Les Artisans Boitiers (LAB) and Quadrance et Habillage (Q&H) in La Chaux-de-Fonds manufacture cases out of precious metals and dials as well. Elwin SA in Moutier specializes in turned parts. In 2003, the movement development and production department officially separated from the rest as Vaucher Manufacture, now an autonomous entity.

Parmigiani has enjoyed great independence and, hence, strong growth, notably in the United States. The brand also set its sights on Latin America, particularly Brazil, where it signed a partnership with the Confederação Brasileira de Futebol. In addition to making watches and unique pieces, like the famed Islamic clock based on a lunar calendar, Parmigiani devotes a part of its premises to restoring ancient timepieces.

Parmigiani Fleurier SA
Rue du Temple 11
CH-2114 Fleurier
Switzerland

Tel.:
+41-32-862-6630

Fax:
+41-32-862-6631

E-mail:
info@parmigiani.ch

Website:
www.parmigiani.ch

Founded:
1996

Number of employees:
520

Annual production:
approx. 4,000 watches

U.S. distributor:
Parmigiani Fleurier Distribution Americas LLC
2655 S. Le Jeune Road
Penthouse 1G
Coral Gables, FL 33134
305-260-7770; 305-269-7770
americas@parmigiani.com

Most important collections/price range:
Kalpa, Tonda, Pershing, Toric, Bugatti / approx. $7,800 to $700,000 for *haute horlogerie* watches; no limit for unique models

Tonda 1950

Reference number: PFC288-0000600-XA3142
Movement: automatic, Parmigiani Caliber PF702; ø 30 mm, height 2.6 mm; 21,600 vph; 48-hour power reserve
Functions: hours, minutes, subsidiary seconds
Case: stainless steel, ø 40 mm, height 8.2 mm; sapphire crystal; water-resistant to 3 atm
Band: reptile skin, folding clasp
Price: $9,900

Tonda Métrographe

Reference number: PFC274-0000100-XC1342
Movement: automatic, Parmigiani Caliber PF315; ø 28 mm, height 6 mm; 46 jewels; 28,800 vph; double spring barrel; finely finished with côtes de Genève; 42-hour power reserve
Functions: hours, minutes, subsidiary seconds; chronograph; date
Case: stainless steel, ø 40 mm, height 11.7 mm; sapphire crystal; transparent case back; water-resistant to 3 atm
Band: calfskin, folding clasp
Price: $11,500
Variations: black dial; stainless steel bracelet ($10,900)

Tonda Calendrier Annuel

Reference number: PFC272-1002401-HA1242
Movement: automatic, Parmigiani Caliber PF339; ø 27.1 mm, height 5.5 mm; 32 jewels; 28,800 vph; double spring barrel, 50-hour power reserve
Functions: hours, minutes, sweep seconds; annual calendar with date (retrograde), weekday, month, moon phase
Case: rose gold, ø 40 mm, height 11.2 mm; sapphire crystal; water-resistant to 3 atm
Band: reptile skin, folding clasp
Price: $29,600

PARMIGIANI

Toric Hémisphères Rétrograde
Reference number: PFC493-1002400-HA1442
Movement: automatic, Parmigiani Caliber PF317; ø 35.6 mm, height 5.45 mm; 28 jewels; 28,800 vph; double spring barrel, 50-hour power reserve
Functions: hours, minutes, subsidiary seconds; 2 additional 12-hour dislays (2nd and 3rd time zones), day/night indicator; date (retrograde)
Case: rose gold, ø 43.8 mm, height 14.2 mm; sapphire crystal; water-resistant to 3 atm
Band: reptile skin, folding clasp
Price: $29,500

Toric Hémisphères Rétrograde
Reference number: PFC493-0001400-XA1442
Movement: automatic, Parmigiani Caliber PF317; ø 35.6 mm, height 5.45 mm; 28 jewels; 28,800 vph; double spring barrel, 50-hour power reserve
Functions: hours, minutes, subsidiary seconds; 2 additional 12-hour displays (2nd and 3rd time zone), day/night indicator; date (retrograde)
Case: stainless steel, ø 43.8 mm, height 14.2 mm; sapphire crystal; water-resistant to 3 atm
Band: reptile skin, folding clasp
Price: $18,500

Toric Chronomètre
Reference number: PFC423-1602400-HA1441
Movement: automatic, Parmigiani Caliber PF331; ø 25.6 mm, height 3.5 mm; 32 jewels; 28,800 vph; double spring barrel, 55-hour power reserve; COSC-certified chronometer
Functions: hours, minutes, sweep seconds; date
Case: rose gold, ø 40.8 mm, height 9.5 mm; sapphire crystal; water-resistant to 3 atm
Band: reptile skin, buckle
Price: $18,500
Variations: white gold ($18,500)

Bugatti Type 390
Reference number: PFH390-1201401-HA1442
Movement: manually wound, Parmigiani Caliber PF390; 25 × 37.5 mm, height 2.6 mm; 32 jewels; 28,800 vph; cylindrical layered movement construction with 1-minute tourbillon on left side, skeletonized hands mechanism; 80-hour power reserve
Functions: hours, minutes; power reserve indicator (roller-shaped)
Case: white gold, 42.2 × 57.7 mm, height 18.4 mm; sapphire crystal; water-resistant to 3 atm
Band: reptile skin, folding clasp
Remarks: 2-part hinged case, can be folded to 12° to better fit wrist
Price: $295,000

Kalpa Qualité Fleurier
Reference number: PFC194-1601400-HA1441
Movement: manually wound, Parmigiani Caliber PF442; 28.7 × 35.9 mm, height 3.7 mm; 29 jewels; 28,800 vph; double spring barrel, 60-hour power reserve; Qualité Fleurier
Functions: hours, minutes, sweep seconds; date
Case: rose gold, 32.1 × 42.3 mm, height 10.1 mm; sapphire crystal; water-resistant to 3 atm
Band: reptile skin, buckle
Remarks: certified according to high Qualité Fleurier standards, covers COSC certification and wearing simulation
Price: $23,500

Kalpagraphe Chronometer
Reference number: PFC193-1002500-XA1442
Movement: manually wound, Parmigiani Caliber PF362; 31.9 × 39.7 mm, height 7 mm; 42 jewels; 36,000 vph; 65-hour power reserve; COSC-certified chronometer
Functions: hours, minutes, subsidiary seconds; chronograph; date
Case: rose gold, 40.4 × 48.2 mm, height 14 mm; sapphire crystal; water-resistant to 3 atm
Band: reptile skin, folding clasp
Price: $35,000

PARMIGIANI

Tonda 1950 Lune
Reference number: PFC284-0000600-XA3242
Movement: automatic, Parmigiani Caliber PF708; ø 30 mm, height 4 mm; 21,600 vph; 48-hour power reserve
Functions: hours, minutes, subsidiary seconds; date, double moon phase
Case: stainless steel, ø 39.1 mm, height 9.6 mm; sapphire crystal; water-resistant to 3 atm
Band: reptile skin, buckle
Price: $12,900

Kalparisma Nova
Reference number: PFC125-1000700-B10002
Movement: automatic, Parmigiani Caliber PF332; ø 25.6 mm, height 3.5 mm; 32 jewels; 28,800 vph; double spring barrel, 55-hour power reserve
Functions: hours, minutes, subsidiary seconds
Case: rose gold, 31.2 × 37.5 mm, height 8.4 mm; sapphire crystal; water-resistant to 3 atm
Band: rose gold, folding clasp
Remarks: ivory dial
Price: $32,100

Kalparisma Agenda
Reference number: PFC123-1000700-HE2421
Movement: automatic, Parmigiani Caliber PF331; ø 25.6 mm, height 3.5 mm; 32 jewels; 28,800 vph; double spring barrel, 55-hour power reserve
Functions: hours, minutes, sweep seconds; date
Case: rose gold, 31.2 × 37.5 mm, height 8.4 mm; sapphire crystal; water-resistant to 3 atm
Band: calfskin, buckle
Remarks: ivory dial
Price: $18,500

Kalpa Donna
Reference number: PFC160-0020701-B00002
Movement: quartz
Functions: hours, minutes
Case: stainless steel, 24.8 × 34.8 mm, height 6.8 mm; sapphire crystal; water-resistant to 3 atm
Band: stainless steel, folding clasp
Remarks: case set with 43 diamonds
Price: $9,400
Variations: various dials

Tonda Métropolitaine
Reference number: PFC273-0000600-B00002
Movement: automatic, Parmigiani Caliber PF310; ø 23.9 mm, height 3.9 mm; 28 jewels; 28,800 vph; double spring barrel; finely finished with côtes de Genève; 50-hour power reserve
Functions: hours, minutes, subsidiary seconds; date
Case: stainless steel, ø 33.1 mm, height 8.65 mm; sapphire crystal; transparent case back; water-resistant to 3 atm
Band: stainless steel, folding clasp
Price: $8,900
Variations: calfskin strap ($8,100); textile strap ($8,000); various dials

Tonda Métropolitaine Selène Galaxy
Reference number: PFC283-0060601-XA3131
Movement: automatic, Parmigiani Caliber PF318; ø 26 mm, height 4.7 mm; 28 jewels; 28,800 vph; double spring barrel, 50-hour power reserve
Functions: hours, minutes, subsidiary seconds; date, moon phase
Case: stainless steel, ø 33.7 mm, height 9.6 mm; bezel with 72 diamonds; sapphire crystal; transparent case back; water-resistant to 3 atm
Band: textile, buckle
Remarks: aventurine dial
Price: $16,200

Caliber PF361

Manually wound; control by 2 column wheels; skeletonized movement; rose gold mainplate and bridges; single spring barrel, 65-hour power reserve
Functions: hours, minutes, subsidiary seconds; flyback chronograph; large date
Diameter: 30.6 mm
Height: 8.5 mm
Jewels: 25
Frequency: 36,000 vph
Remarks: 317 parts; limited edition

Caliber PF702

Automatic; platinum microrotor; single spring barrel, 48-hour power reserve
Functions: hours, minutes, subsidiary seconds
Diameter: 30 mm
Height: 2.6 mm
Jewels: 29
Frequency: 21,600 vph

Caliber PF442

Automatic; double spring barrel, 60-hour power reserve; Qualité Fleurier
Functions: hours, minutes, sweep seconds; date
Measurement: 28.7 × 35.9 mm
Height: 3.7 mm
Jewels: 29
Frequency: 28,800 vph

Caliber PF110

Manually wound; double spring barrel, 192-hour power reserve
Functions: hours, minutes, subsidiary seconds; power reserve indicator; date
Measurement: 29.3 × 23.6 mm
Height: 4.9 mm
Jewels: 28
Frequency: 21,600 vph

Caliber PF390

Manually wound; cylindrical movement design, with 1-minute tourbillon on left-hand side, skeletonized mechanism driving hands; single spring barrel, 80-hour power reserve
Functions: hours, minutes; power reserve indicator (cylindrical)
Measurement: 25 × 37.5 mm
Height: 2.6 mm
Jewels: 32
Frequency: 28,800 vph
Remarks: 302 parts; mainplate and bridges with black PVD coating; limited edition

Caliber PF365

Automatic; rose gold mainplate, bridges, and rotor; column wheel control of chronograph functions; single spring barrel, 65-hour power reserve; COSC-certified chronometer
Functions: hours, minutes, subsidiary seconds; chronograph; date
Measurement: 31.9 × 39.7 mm
Height: 7 mm
Jewels: 42
Frequency: 36,000 vph
Remarks: 348 parts

PATEK PHILIPPE

In the Swiss watchmaking landscape, Patek Philippe has a special status as the last independent family-owned business. The company originated in 1839 with two Polish emigrés to Switzerland, Count Norbert Antoine de Patek and Frantiszek Czapek. In 1845, following the natural end of their contract, Patek sought another partner in the master watchmaker Jean Adrien Philippe, who had developed a keyless winding and time-setting mechanism. Ever since, Patek Philippe has been known for creating high-quality mechanical watches, some with extremely sophisticated complications. Even among its competition, the *manufacture* enjoys the greatest respect.

In 1932, Charles-Henri Stern took over the *manufacture*. His son Henri and grandson Philippe continued the tradition of solid leadership, steering the company through the notorious quartz crisis without ever compromising quality. The next in line, also Henri, heads the enterprise these days.

In 1997, Patek Philippe moved into new quarters, based on the most modern standards. The facility boasts the world's largest assembly of watchmakers under one roof, and yet production figures are comparatively modest. A small section of the building is reserved for restoring old watches either using parts from a large and valuable collection of components or rebuilding them from scratch.

The company opened a highly industrialized second branch between La Chaux-de-Fonds and Le Locle, where case components are manufactured, cases are polished, and gem setting is done. Patek Philippe's main headquarters remain in Geneva, but the *manufacture* no longer has a need for that city's famed seal: All of the company's mechanical watches now feature the "Patek Philippe Seal," the criteria for which far exceed the requirements of the *Poinçon de Genève* and include specifications for the entire watch, not just the movement. Among the most recent creations to make that grade is the World Time Chronograph, a masterful extension of the company's large range of chronographs. To make space, there is no second hand and only a thirty-minute counter. A moving city ring and twenty-four-hour ring have a place on the dial as well, and the whole piece is just over 12 millimeters high.

Patek Philippe SA
Chemin du pont-du-centenaire 141
CH-1228 Plan-les-Ouates
Switzerland

Tel.:
+41-22-884-20-20

Fax:
+41-22-884-20-40

Website:
www.patek.com

Founded:
1839

Number of employees:
approx. 2,000 (estimated)

Annual production:
approx. 60,000 watches worldwide per year

U.S. distributor:
Patek Philippe USA
45 Rockefeller Center, Suite 401
New York, NY 10111
212-218-1240; 212-218-1283 (fax)

Most important collections/price range:
Calatrava, Nautilus, Gondolo, Ellipse, Aquanaut / ladies' timepieces

World Time with Minute Repeater
Reference number: 5531R-001
Movement: automatic, Patek Philippe Caliber R 27 HU; ø 32 mm, height 8.5 mm; 45 jewels; 21,600 vph; chime with traditional gong; microrotor winding; 43-hour power reserve
Functions: hours, minutes; world time display, minute repeater (chimes local time, can be set using pusher)
Case: rose gold, ø 40.2 mm, height 11.49 mm; sapphire crystal; transparent case back
Band: reptile skin, folding clasp
Remarks: dial center with cloisonné enamel motif
Price: on request

Perpetual Calendar with Tourbillon and Minute Repeater
Reference number: 5207G-001
Movement: manually wound, Patek Philippe Caliber R TO 27 PS QI; ø 32 mm, height 9.33 mm; 37 jewels; 21,600 vph; 1-minute tourbillon; chimes with traditional gongs; 38-hour power reserve
Functions: hours, minutes, subsidiary seconds; minute repeater; perpetual calendar with date, weekday, month, moon phase
Case: white gold, ø 41 mm, height 13.81 mm; sapphire crystal; transparent case back
Band: reptile skin, folding clasp
Price: on request

Chronograph Perpetual Calendar with Tourbillon and Minute Repeater
Reference number: 5208R-001
Movement: automatic, Patek Philippe Caliber R CH 27 PS QI; ø 32 mm, height 10.35 mm; 63 jewels; 21,600 vph; 1-minute tourbillon; monopusher control of chronograph functions; microrotor winding; 38-hour power reserve
Functions: hours, minutes, subsidiary seconds; minute repeater; chronograph; perpetual calendar with date, weekday, month, moon phase, leap year
Case: rose gold, ø 42 mm, height 15.11 mm; sapphire crystal; transparent case back
Band: reptile skin, folding clasp
Price: on request

PATEK PHILIPPE

Chronograph Perpetual Calendar
Reference number: 5270P-001
Movement: manually wound, Patek Philippe Caliber CH 29 535 PS Q; ø 32 mm, height 7 mm; 33 jewels; 28,800 vph; 55-hour power reserve
Functions: hours, minutes, subsidiary seconds; day/night indicator; chronograph; perpetual calendar with date, weekday, month, moon phase, leap year
Case: platinum, ø 41 mm, height 12.4 mm; sapphire crystal; transparent case back; water-resistant to 3 atm
Band: reptile skin, folding clasp
Price: $187,114

Chronograph Perpetual Calendar
Reference number: 5270/1R-001
Movement: manually wound, Patek Philippe Caliber CH 29 535 PS Q; ø 32 mm, height 7 mm; 33 jewels; 28,800 vph; 55-hour power reserve
Functions: hours, minutes, subsidiary seconds; day/night indicator; chronograph; perpetual calendar with date, weekday, month, moon phase, leap year
Case: rose gold, ø 41 mm, height 12.4 mm; sapphire crystal; transparent case back; water-resistant to 3 atm
Band: rose gold, folding clasp
Price: $192,784

Perpetual Calendar
Reference number: 5320G-001
Movement: automatic, Patek Philippe Caliber 324 S Q; ø 32 mm, height 4.97 mm; 29 jewels; 28,800 vph; gold rotor; 35-hour power reserve
Functions: hours, minutes, sweep seconds; day/night indicator; perpetual calendar with date, weekday, month, moon phase, leap year
Case: white gold, ø 40 mm, height 11.1 mm; sapphire crystal; transparent case back; water-resistant to 3 atm
Band: reptile skin, folding clasp
Remarks: comes with additional white-gold case back
Price: $87,320

Annual Calendar
Reference number: 5205G-013
Movement: automatic, Patek Philippe Caliber 324 S QA LU 24H; ø 32.6 mm, height 5.78 mm; 34 jewels; 28,800 vph; 35-hour power reserve
Functions: hours, minutes, sweep seconds; additional 24-hour display; annual calendar with date, weekday, month, moon phase
Case: white gold, ø 40 mm, height 11.36 mm; sapphire crystal; transparent case back; water-resistant to 3 atm
Band: reptile skin, buckle
Price: $47,970

Calatrava Pilot Travel Time
Reference number: 5524R-001
Movement: automatic, Patek Philippe Caliber 324 S C FUS; ø 31 mm, height 4.9 mm; 29 jewels; 28,800 vph; silicon Spiromax hairspring; 35-hour power reserve
Functions: hours, minutes, sweep seconds; additional 12-hour display (2nd time zone), day/night indicator; date
Case: rose gold, ø 42 mm, height 10.78 mm; sapphire crystal; transparent case back; water-resistant to 6 atm
Band: calfskin, buckle
Price: $47,600
Variations: white gold ($47,600)

Chronograph Ladies'
Reference number: 7150/250R-001
Movement: manually wound, Patek Philippe Caliber CH 29-535 PS; ø 29.6 mm, height 5.35 mm; 33 jewels; 28,800 vph; column wheel control of chronograph functions; 65-hour power reserve
Functions: hours, minutes, subsidiary seconds; chronograph
Case: rose gold, ø 38 mm, height 10.59 mm; bezel with 72 diamonds; sapphire crystal; transparent case back; water-resistant to 3 atm
Band: reptile skin, buckle with 27 diamonds
Price: $83,918

PATEK PHILIPPE

Calatrava Pilot Travel Time Ladies'
Reference number: 7234R-001
Movement: automatic, Patek Philippe Caliber 324 S C FUS; ø 31 mm, height 4.9 mm; 29 jewels; 28,800 vph; silicon Spiromax hairspring; 35-hour power reserve
Functions: hours, minutes, sweep seconds; additional 12-hour display (2nd time zone), day/night indicator; date
Case: rose gold, ø 37.5 mm, height 10.78 mm; sapphire crystal; transparent case back; water-resistant to 3 atm
Band: calfskin, buckle
Price: $43,093

World Time Chronogaph
Reference number: 5930G-001
Movement: automatic, Patek Philippe Caliber CH 28-520 HU; ø 33 mm, height 7.97 mm; 38 jewels; 28,800 vph; Spiromax silicon hairspring; 50-hour power reserve
Functions: hours, minutes; world time display (2nd time zone); chronograph
Case: white gold, ø 39.5 mm, height 12.86 mm; 24 pusher-controlled time zones on a scale flange with reference city names; sapphire crystal; transparent case back; water-resistant to 3 atm
Band: reptile skin, folding clasp
Price: $73,712

World Time
Reference number: 5230G-001
Movement: automatic, Patek Philippe Caliber 240 HU; ø 27.5 mm, height 3.88 mm; 33 jewels; 21,600 vph; silicon Spiromax hairspring; 48-hour power reserve
Functions: hours, minutes; world time display (2nd time zone)
Case: white gold, ø 38.5 mm, height 10.23 mm; pusher-activated inner bezel with city references; sapphire crystal; transparent case back; water-resistant to 3 atm
Band: reptile skin, folding clasp
Price: $47,629

Annual Calendar Chronograph
Reference number: 5960/01G-001
Movement: automatic, Patek Philippe Caliber CH 28 520 IRM QA 24H; ø 33 mm, height 7.68 mm; 40 jewels; 28,800 vph; 45-hour power reserve
Functions: hours, minutes; power reserve indicator; flyback chronograph; annual calendar with date, weekday, month
Case: white gold, ø 40.5 mm, height 13.5 mm; sapphire crystal; transparent case back; water-resistant to 3 atm
Band: calfskin, buckle
Price: $65,774

Annual Calendar
Reference number: 5396/R-015
Movement: automatic, Patek Philippe Caliber 324 S QA LU 24H/303; ø 33.3 mm, height 5.78 mm; 34 jewels; 28,800 vph; silicon Spiromax hairspring; 35-hour power reserve
Functions: hours, minutes, sweep seconds; additional 24-hour display; annual calendar with date, weekday, month, moon phase
Case: rose gold, ø 38.5 mm, height 11.2 mm; sapphire crystal; transparent case back; water-resistant to 3 atm
Band: reptile skin, folding clasp
Remarks: dial with 12 baguette diamonds
Price: $47,970
Variations: white gold ($47,970)

Annual Calendar
Reference number: 4947G-010
Movement: automatic, Patek Philippe Caliber 324 S QA LU; ø 30 mm, height 5.32 mm; 34 jewels; 28,800 vph; 35-hour power reserve
Functions: hours, minutes, sweep seconds; annual calendar with date, weekday, month, moon phase
Case: white gold, ø 38 mm, height 11 mm; bezel set with 141 diamonds; sapphire crystal; transparent case back; crown set with 14 diamonds; water-resistant to 3 atm
Band: reptile skin, buckle
Price: $49,897

Calatrava
Reference number: 6006G-001
Movement: automatic, Patek Philippe Caliber 240 PS C; ø 30 mm, height 3.43 mm; 27 jewels; 21,600 vph; gold microrotor
Functions: hours, minutes, subsidiary seconds; date
Case: white gold, ø 39 mm, height 8.86 mm; sapphire crystal; transparent case back; water-resistant to 3 atm
Band: reptile skin, folding clasp
Price: $30,619

Gondolo Ellipse d'Or
Reference number: 5738R-001
Movement: automatic, Patek Philippe Caliber 240; ø 27.5 mm, height 2.53 mm; 27 jewels; 21,600 vph; gold microrotor; 48-hour power reserve
Functions: hours, minutes
Case: rose gold, 34.5 × 39.5 mm, height 5.9 mm; sapphire crystal
Band: reptile skin, buckle
Price: $30,846

Nautilus Perpetual Calendar
Reference number: 5740/1G-001
Movement: automatic, Patek Philippe Caliber 240 Q; ø 27.5 mm, height 3.88 mm; 27 jewels; 21,600 vph; 38-hour power reserve
Functions: hours, minutes; additional 24-hour display; perpetual calendar with date, weekday, month, moon phase, leap year
Case: white gold, ø 40 mm, height 8.42 mm; sapphire crystal; transparent case back; screw-in crown; water-resistant to 6 atm
Band: white gold, folding clasp
Price: $119,073

Nautilus Ladies
Reference number: 7118/1A-010
Movement: automatic, Patek Philippe Caliber 324 S C; ø 27 mm, height 3.57 mm; 29 jewels; 28,800 vph; 35-hour power reserve
Functions: hours, minutes, sweep seconds; date
Case: stainless steel, ø 35.2 mm, height 8.62 mm; sapphire crystal; transparent case back; screw-in crown; water-resistant to 6 atm
Band: stainless steel, folding clasp
Price: $24,836

Aquanaut Chronograph
Reference number: 5968A-001
Movement: automatic, Patek Philippe Caliber CH 28-520 C; ø 30 mm, height 6.63 mm; 32 jewels; 28,800 vph; 45-hour power reserve
Functions: hours, minutes; chronograph; date
Case: stainless steel, ø 42.2 mm, height 11.9 mm; sapphire crystal; transparent case back; screw-in crown; water-resistant to 12 atm
Band: rubber, folding clasp
Price: $43,774

Aquanaut Ladies'
Reference number: 5067A-025
Movement: quartz
Functions: hours, minutes, sweep seconds; date
Case: stainless steel, ø 35.6 mm, height 7.7 mm; bezel with 46 diamonds; sapphire crystal; screw-in crown; water-resistant to 12 atm
Band: rubber, folding clasp
Price: $16,217

PATEK PHILIPPE

Caliber R 27 HU
Automatic; gold microrotor, chime with traditional gong activated by lateral slider; single spring barrel, 43-hour power reserve
Functions: hours, minutes; world time display, minute repeater (chimes local time, set by pusher)
Diameter: 32 mm
Height: 8.5 mm
Jewels: 45
Balance: Gyromax
Frequency: 21,600 vph
Balance spring: Spiromax
Remarks: 462 parts

Caliber R TO 27 PS QI
Manually wound; 1-minute tourbillon; chime with traditional gong; single spring barrel, 38-hour power reserve
Functions: hours, minutes, subsidiary seconds; minute repeater; perpetual calendar with date, weekday, month, moon phase
Diameter: 32 mm
Height: 9.33 mm
Jewels: 37
Balance: Gyromax
Frequency: 21,600 vph
Balance spring: Breguet
Remarks: 549 parts

Caliber R CH 27 PS QI
Automatic; 1-minute tourbillon; chime with traditional gong; monopusher control of chronograph functions; microrotor; single spring barrel, 38-hour power reserve
Functions: hours, minutes, subsidiary seconds; minute repeater; chronograph; perpetual calendar with date, weekday, month, moon phase, leap year
Diameter: 32 mm
Height: 10.35 mm
Jewels: 63
Balance: Gyromax
Frequency: 21,600 vph
Balance spring: Spiromax
Remarks: 719 parts

Caliber CH 29-535 PS
Manually wound; column wheel control of chronograph functions, precisely jumping 30-minute totalizer; single spring barrel, 65-hour power
Functions: hours, minutes, subsidiary seconds; flyback chronograph
Diameter: 29.6 mm
Height: 7.1 mm
Jewels: 34
Balance: Gyromax, four-armed, with four regulating weights
Frequency: 28,800 vph
Balance spring: Breguet
Shock protection: Incabloc
Remarks: 312 parts

Caliber CHR 29-535 PS Q
Manually wound; two column wheels for control of chronograph functions, flyback mechanism with isolator; single spring barrel, 65-hour power reserve
Functions: hours, minutes, subsidiary seconds; day/night indicator; flyback chronograph; perpetual calendar with date, weekday, month, moon phase, leap year
Diameter: 32 mm
Height: 8.7 mm
Jewels: 34
Balance: Gyromax, four-armed, with 4 regulating weights
Frequency: 28,800 vph
Balance spring: Breguet
Remarks: 496 parts: 182 for perpetual calendar, 42 for flyback mechanism

Caliber CH 28-520 HU
Automatic; single spring barrel, 50-hour power reserve
Functions: hours, minutes; world time display (2nd time zone); chronograph
Diameter: 33 mm
Height: 7.91 mm
Jewels: 38
Balance: Gyromax
Frequency: 28,800 vph
Balance spring: silicon Spiromax
Remarks: 343 parts; finely finished movement

PATEK PHILIPPE

Caliber 324 S Q

Automatic; gold rotor; single spring barrel, 35-hour power reserve
Functions: hours, minutes, sweep seconds; day/night indicator; perpetual calendar with date, weekday, month, moon phase, leap year
Diameter: 32 mm
Height: 4.97 mm
Jewels: 29
Frequency: 28,800 vph

Caliber 324 SC

Automatic; gold rotor; single spring barrel, 35-hour power reserve
Functions: hours, minutes, sweep seconds; date
Diameter: 27 mm
Height: 3.3 mm
Jewels: 29
Balance: Gyromax
Frequency: 28,800 vph
Balance spring: Breguet

Caliber 324 S C FUS

Automatic; gold rotor; single spring barrel, 35-hour power reserve
Functions: hours, minutes, sweep seconds; additional 12-hour display (2nd time zone), day/night indicator; date
Diameter: 31 mm
Height: 4.82 mm
Jewels: 29
Balance: Gyromax
Frequency: 28,800 vph
Balance spring: silicon Spiromax
Remarks: 294 parts

Caliber 324 S QA LU

Automatic; gold rotor; single spring barrel, 35-hour power reserve
Functions: hours, minutes, sweep seconds; annual calendar with date, weekday, month, moon phase
Diameter: 30 mm
Height: 5.32 mm
Jewels: 34
Balance: Gyromax
Frequency: 28,800 vph
Balance spring: Breguet

Caliber 324 S QA LU 24H-303

Automatic; central rotor in 21-kt gold; single spring barrel, 45-hour power reserve
Functions: hours, minutes, sweep seconds; additional 24-hour display (2nd time zone); annual calendar with date, weekday, month, moon phase
Diameter: 32.6 mm
Height: 5.78 mm
Jewels: 34
Balance: Gyromax
Frequency: 28,800 vph
Balance spring: silicon Spiromax
Remarks: silicon escape wheel; 347 parts

Caliber CH 28-520 IRM QA 24H

Automatic; column wheel control of chronograph functions, central rotor in 21-kt gold; single spring barrel, 55-hour power reserve
Functions: hours, minutes, sweep seconds; day/night indicator, power reserve indicator; chronograph with combined hour and minute counter; annual calendar with date, weekday, month, moon phase
Diameter: 33 mm
Height: 7.68 mm
Jewels: 40
Balance: Gyromax
Frequency: 28,800 vph
Balance spring: Breguet
Remarks: 456 parts

 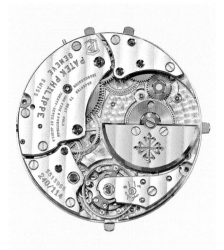 

Caliber 240 HU

Automatic; off-center, ball bearing–mounted, unidirectional gold microrotor in 22-kt gold; single spring barrel, 48-hour power reserve
Functions: hours, minutes; world time display (2nd time zone)
Diameter: 27.5 mm
Height: 3.88 mm
Jewels: 33
Balance: Gyromax
Frequency: 21,600 vph
Remarks: 239 parts

Caliber 240 Q

Automatic; gold microrotor; single spring barrel, 48-hour power reserve
Functions: hours, minutes; additional 24-hour display (2nd time zone); perpetual calendar with date, weekday, month, moon phase, leap year
Diameter: 30 mm
Height: 3.75 mm
Jewels: 27
Balance: Gyromax, with 8 masselotte regulating weights
Frequency: 21,600 vph
Balance spring: flat hairspring
Shock protection: Kif

Caliber 240 SQU

Automatic; gold microrotor; single spring barrel, 48-hour power reserve
Functions: hours, minutes
Diameter: 27.5 mm
Height: 2.53 mm
Jewels: 27
Balance: Gyromax, with 8 masselotte regulating weights
Frequency: 21,600 vph
Balance spring: flat hairspring
Shock protection: Kif
Remarks: movement entirely skeletonized, engraved and decorated by hand

Caliber 240 PS C

Automatic; gold microrotor; single spring barrel, 48-hour power reserve
Functions: hours, minutes, subsidiary seconds; date
Diameter: 30 mm
Height: 3.43 mm
Jewels: 27
Balance: Gyromax
Frequency: 21,600 vph

Caliber R 27 PS

Automatic; gold microrotor, chime with traditional gong activated by slide in flank; single spring barrel
Functions: hours, minutes, subsidiary seconds; minute repeater
Diameter: 28 mm
Height: 5.05 mm
Jewels: 39
Frequency: 21,600 vph

Caliber R TO 27 PS QR

Manually wound; 1-minute tourbillon; single spring barrel, 48-hour power reserve, COSC-certified chronometer
Functions: hours, minutes, subsidiary seconds; minute repeater; perpetual calendar with date (retrograde), weekday, month and leap year (in window), moon phases
Diameter: 32 mm
Height: 8.61 mm
Jewels: 28
Balance: Gyromax
Frequency: 21,600 vph
Balance spring: Breguet
Remarks: 336 parts

QUILL & PAD
KEEPING WATCH ON TIME

Make **time** for **a unique watch experience**

Breaking Stories
Unique Photography
Interesting Angles and Subjects

Quill & Pad is an online platform combining decades of excellence and experience in watch journalism, bringing **you** original stories and photography.

Ian Skellern

Elizabeth Doerr

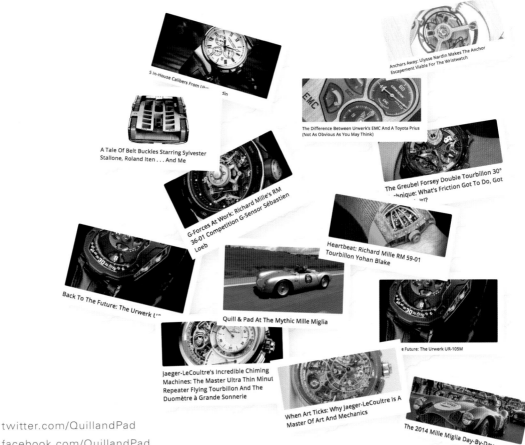

twitter.com/QuillandPad
facebook.com/QuillandPad
quillandpad.tumblr.com
instagram.com/quillandpad

www.QuillandPad.com

PAUL GERBER

Watchmaker Paul Gerber has already developed mechanisms and complications, including calendar movements, alarms, and tourbillons, for numerous renowned watchmakers over the decades. Time and again, this genial watchmaker has astonished the horological world with outrageously complicated mechanisms, which he somehow manages to create by fitting hundreds of additional tiny parts into filigree movements. Gerber is the one who designed the complicated calendar mechanism for the otherwise minimalist MIH watch conceived by Ludwig Oechslin, curator of the International Museum of Horology (MIH) in La Chaux-de-Fonds and himself a watchmaker. To avoid cluttering a dial for a special customer, he recently devised a battery-run moon phase that fits in the watch strap. Twice his work has appeared in *Guiness World Records*.

When his daily work for others lets up, Gerber gets around to building watches bearing his own name with such marvelous features as a retrograde second hand in an elegant thin case and a synchronously, unidirectional rotor system with miniature oscillating weights for his self-winding Retro Twin model. Gerber's works are all limited editions.

After designing a tonneau-shaped manually wound wristwatch with a three-dimensional moon phase display, Gerber created a simple, three-hand watch with an automatic movement conceived and produced completely in-house. It features a 100-hour power reserve and is wound by three synchronically turning gold rotors. Gerber also offers the triple rotor and large date features in a watch with an ETA movement and lightweight titanium case as a classic pilot watch design or in a version with a more modern dial (the Synchron model). The Model 41 has an optional complication that switches the second hand from sweep to dead-beat motion by way of a pusher at 2 o'clock.

Gerber is allegedly retired. But a watchmaker never really does. Besides continuing to produce outstanding pieces, he occasionally gives three-day workshops for people wanting to get a real feel for the work.

Paul Gerber
Uhren-Konstruktionen
Bockhornstrasse 69
CH-8047 Zürich
Switzerland

Tel.:
+41-44-401-4569

E-mail:
info@gerber-uhren.ch

Website:
www.gerber-uhren.ch

Founded:
1976

Number of employees:
n/a

Annual production:
up to 50 watches

U.S. distributor:
Intro Swiss—Michel Schmutz
7615 Estate Circle
Niwot, CO 80503
303-652-1520
introswiss@q.com

Most important collections/price range:
mechanical watches / from approx. $4,900 to $60,000; tourbillon desk clocks / from approx. $48,000 to $70,000

Retro Twin
Reference number: 156
Movement: automatic, Gerber Caliber 15 (base ETA 7001); ø 28 mm, height 5.2 mm; 27 jewels; 21,600 vph; automatic winding with 2 synchronously rotating platinum rotors
Functions: hours, minutes, subsidiary seconds (retrograde)
Case: rose gold, ø 36 mm, height 10.8 mm; sapphire crystal; transparent case back; water-resistant to 3 atm
Band: reptile skin, buckle
Price: $17,600
Variations: yellow gold ($16,600) or white gold ($17,050); with platinum rotors set with brilliants (plus $2,000)

Modell 42
Reference number: 420 DT
Movement: automatic, Gerber Caliber 42 (base ETA 2824); ø 36 mm, height 6.1 mm; 25 jewels; 28,800 vph; automatic winding with 3 synchronously rotating gold rotors
Functions: hours, minutes, sweep seconds; dual time indication; date
Case: titanium, ø 42 mm, height 12 mm; sapphire crystal; transparent case back; screw-in crown; water-resistant to 10 atm
Band: calfskin, buckle
Price: $5,800
Variations: pilot's/synchron dial ($4,950); as Caliber 42 pilot's/synchron day and night ($5,850)

Modell 33
Reference number: 336
Movement: manually wound, Gerber Caliber 33; 28 × 34 × 5 mm; 20 jewels; 21,600 vph; special Gerber escapement
Functions: hours, minutes, seconds, 3D moon phase
Case: white gold, 34 × 40 × 10.2 mm; sapphire crystal; transparent case back; water-resistant to 3 atm
Band: reptile skin, gold buckle
Remarks: moon corrected for 128 years, ø 6 mm, corrected for 128 years; one hemisphere set with 54 diamonds, the other of lapis lazuli
Price: $37,670
Variations: rose gold ($37,670); platinum ($46,300)

Speake-Marin
Chemin en-Baffa 2
CH-1183 Bursins
Switzerland

Tel.:
+41-21-825-5069

E-mail:
info@speake-marin.com

Website:
www.speake-marin.com

Founded:
2002

Number of employees:
5

Annual production:
400 watches

U.S. distributor:
About Time Luxury Group
210 Bellevue Avenue
Newport, RI 02840
401-846-0598
Speake-marin@abouttimeluxury.com
www.abouttimeluxury.com

Most important collections:
HMS, J-Class, London, Resilience, Serpent Calendar, Velsheda

PETER SPEAKE-MARIN

Peter Speake-Marin brings realism, genius, and a sense of romance to his work. As a horological innovator, he could have been a poet or adventurer. He has an outstanding reputation for originality, virtuosity, and being a very friendly and helpful colleague in a highly competitive field. He has also had his skilled fingers in a number of iconic timepieces, like the HM1 of MB&F, the Chapter One for Maîtres du Temps, and the Harry Winston Excenter Tourbillon.

Born in Essex in 1968, Speake-Marin attended Hackney College, London, and WOSTEP in Switzerland, before earning his spurs restoring antique watches at a Somlo in Piccadilly. In 1996, he moved to Le Locle, Switzerland, to work with Renaud et Papi, when he also set about making his own pieces. A dual-train tourbillon (the Foundation Watch) opened the door to the prestigious AHCI.

The recession taught him something crucial: "I was a watchmaker, not an entrepreneur," he shares. "I had to become entrepreneurial to become a watchmaker again." And so he reorganized himself as a brand with three watch families. For his flashes of creative madness, he has the grab-bag Cabinet des Mystères. The Spirit models have a military, adventurous feel. And the J-Class family recalls the discreet elegance of J-Class yachts, such as the Velsheda, launched originally in 1933 and named for Velma, Sheila, and Daphne, the three daughters of the first owner, Woolworth executive William Lawrence Stephenson.

Speake-Marin is a profoundly creative watchmaker, one willing to give a hand even to his competitors. So the surprise at the 2017 announcement that he was actually leaving the brand he had given birth to and shaped for sixteen years was somewhat mitigated. In a brief letter, CEO Christelle Rosnoblet promised to continue the brand's characteristic "British elegance and impertinence." It's there in the bold London chronograph, which sports a red subdial for the minutes.

Resilience
Movement: automatic, Caliber Vaucher 3002; ø 26 mm, height 4.3 mm; 28 jewels; 28,800 vph; 50-hour power reserve
Functions: hours, minutes, sweep seconds
Case: pink gold, ø 38 mm, height 12 mm; sapphire crystal; transparent case back; water-resistant to 3 atm
Band: reptile skin, pink gold buckle
Remarks: enamel dial
Price: $24,800
Variations: with 42-mm case; various dials

Spirit Seafire
Movement: automatic, ETA Caliber 7750 (modified); ø 30.4 mm, height 7.9 mm; 25 jewels; 28,800 vph; 48-hour power reserve
Functions: hours, minutes, subsidiary seconds (rosetta form); chronograph; date
Case: titanium, ø 42 mm, height 15 mm; sapphire crystal; water-resistant to 3 atm
Band: calfskin, buckle
Price: $10,350; limited edition of 28 pieces

London Chronograph
Movement: hand-wound, Caliber Valjoux 92; ø 29.5 mm, height 6 mm; 28 jewels; 18,000 vph; 40-hour power reserve
Functions: hours, minutes, subsidiary seconds; chronograph (bicompax)
Case: titanium, ø 42 mm, height 15 mm; sapphire crystal; transparent case back; water-resistant to 3 atm
Band: rubber, buckle
Price: $18,400; limited to 15 pieces

PIAGET

One of the oldest watch manufacturers in Switzerland, Piaget began making watch movements in the secluded Jura village of La Côte-aux-Fées in 1874. For decades, those movements were delivered to other watch brands. The *manufacture* itself, strangely enough, remained in the background. It wasn't until the 1940s that the Piaget family began to offer complete watches under their own name.

Even today, Piaget, which long ago moved the business side of things to Geneva, still makes its watch movements at its main facility high in the Jura mountains.

In the late fifties, Piaget began investing in the design and manufacturing of ultrathin movements. This lends these watches the kind of understated elegance that became the company's hallmark. In 1957, Valentin Piaget presented the first ultrathin men's watch, the Altiplano, with the manual caliber 9P, which was 2 millimeters high. Shortly after, it came out with the 12P, an automatic caliber that clocked in at 2.3 millimeters.

The Altiplano has faithfully accompanied the brand for sixty years now. The movement has evolved over time. The recent 900P measures just 3.65 mm and is inverted to enable repairs, making the case back the mainplate, with the dial set on the upper side. Striving for the thinnest watch produced the Altiplano Ultimate Concept in 2018, a 2-millimeter, manually wound watch, released—but not for sale—after four years of R&D.

Worthy of note, too, is Piaget's concept watch that combines mechanical and quartz technology. The spring barrel, wound by hand or microrotor, drives a miniature generator that turns at a constant 5.33 rpm and replaces the escapement and balance wheel. It supplies the regulating quartz with power, which in turn regulates the movement. This inverted caliber is just 5.5 millimeters high. All the basic parts are on the dial side and partly visible thanks to skeletonized plates.

Piaget SA
CH-1228 Plan-les-Ouates
Switzerland

Tel.:
+41-32-867-21-21

E-mail:
info@piaget.com

Website:
www.piaget.com

Founded:
1874

Number of employees:
900

Annual production:
watches not specified; plus about 20,000 movements for Richemont Group

U.S. distributor:
Piaget North America
645 5th Avenue, 6th Floor
New York, NY 10022
212-909-4362; 212-909-4332 (fax)
www.piaget.com

Most important collections/price range:
Altiplano / approx. $13,500 to $22,000

Altiplano Ultimate Automatic

Reference number: G0A43120
Movement: automatic, Piaget Caliber 910P; ø 41 mm, height 4.3 mm (case with movement); 30 jewels; 21,600 vph; inverted movement with case and hubless peripheral rotor; 50-hour power reserve
Functions: hours, minutes (off-center)
Case: rose gold, ø 41 mm, height 4.3 mm; sapphire crystal
Band: reptile skin, buckle
Price: $27,300

Altiplano Ultimate Automatic

Reference number: G0A43121
Movement: automatic, Piaget Caliber 910P; ø 41 mm, height 4.3 mm (case with movement); 30 jewels; 21,600 vph; inverted movement with case, hubless peripheral rotor; 50-hour power reserve
Functions: hours, minutes (off-center)
Case: white gold, ø 41 mm, height 4.3 mm; sapphire crystal
Band: reptile skin, buckle
Price: $28,400

Altiplano Ultimate Concept

Reference number: G0A43900
Movement: manually wound, Piaget Caliber 900P-UC; ø 41 mm, height 2 mm (case with movement); 13 jewels; 28,800 vph; inverted movement construction integrated in the case, flying gearwheels mounted (one-sided); 44-hour power reserve
Functions: hours, minutes (off-center)
Case: cobalt-based alloy, ø 41 mm, height 2 mm; sapphire crystal
Band: reptile skin, buckle
Remarks: concept watch, thinnest mechanical wristwatch ever made; probably not for serial production
Price: on request

Altiplano
Reference number: G0A39105
Movement: manually wound, Piaget Caliber 450P; ø 20.5 mm, height 2.1 mm; 18 jewels; 21,600 vph; 43-hour power reserve
Functions: hours, minutes, subsidiary seconds
Case: rose gold, ø 34 mm, height 6.3 mm; sapphire crystal; water-resistant to 3 atm
Band: reptile skin, buckle
Price: $15,100

Altiplano
Reference number: G0A31114
Movement: manually wound, Piaget Caliber 430P; ø 20.5 mm, height 2.1 mm; 18 jewels; 21,600 vph; 43-hour power reserve
Functions: hours, minutes
Case: rose gold, ø 38 mm, height 6.4 mm; sapphire crystal; water-resistant to 3 atm
Band: reptile skin, buckle
Price: $15,200

Altiplano Date
Reference number: G0A38131
Movement: automatic, Piaget Caliber 1205P; ø 29.9 mm, height 3 mm; 27 jewels; 21,600 vph; microrotor in rose gold; côtes de Genève; 44-hour power reserve
Functions: hours, minutes, subsidiary seconds; date
Case: rose gold, ø 40 mm, height 6.36 mm; sapphire crystal; transparent case back; water-resistant to 3 atm
Band: reptile skin, buckle
Price: $23,800
Variations: white gold ($24,700)

Altiplano 900P
Reference number: G0A39112
Movement: manually wound, Piaget Caliber 900P; ø 30.4 mm, height 3.65 mm; 20 jewels; 21,600 vph; inverted movement; 48-hour power reserve
Functions: hours, minutes
Case: white gold, ø 38 mm, height 3.65 mm; bezel set with 78 diamonds; sapphire crystal
Band: reptile skin, buckle
Price: $32,300

Polo S Automatic
Reference number: G0A43001
Movement: automatic, Piaget Caliber 1110P; ø 25.58 mm, height 4 mm; 25 jewels; 28,800 vph; perlage on mainplate, blued screws, finely finished with côtes de Genève; 50-hour power reserve
Functions: hours, minutes, sweep seconds; date
Case: stainless steel, ø 42 mm, height 9.4 mm; sapphire crystal; transparent case back; water-resistant to 10 atm
Band: reptile skin, folding clasp
Price: $8,100

Polo S Chronograph
Reference number: G0A43002
Movement: automatic, Piaget Caliber 1160P; ø 25.58 mm, height 5.72 mm; 35 jewels; 28,800 vph; perlage on mainplate, blued screws, finely finished with côtes de Genève; 50-hour power reserve
Functions: hours, minutes; chronograph; date
Case: stainless steel, ø 42 mm, height 11.2 mm; sapphire crystal; transparent case back; water-resistant to 10 atm
Band: reptile skin, folding clasp
Price: $11,200
Variations: black dial; stainless steel bracelet

PIAGET

Polo S Automatic
Reference number: G0A43010
Movement: automatic, Piaget Caliber 1110P; ø 25.58 mm, height 4 mm; 25 jewels; 28,800 vph; perlage on mainplate, blued screws, finely finished with côtes de Genève; 50-hour power reserve
Functions: hours, minutes, sweep seconds; date
Case: rose gold, ø 42 mm, height 9.4 mm; sapphire crystal; transparent case back; water-resistant to 10 atm
Band: reptile skin, folding clasp
Price: $21,000

Polo S Chronograph
Reference number: G0A43011
Movement: automatic, Piaget Caliber 1160P; ø 25.58 mm, height 5.72 mm; 35 jewels; 28,800 vph; perlage on mainplate, blued screws, finely finished with côtes de Genève; 50-hour power reserve
Functions: hours, minutes; chronograph; date
Case: rose gold, ø 42 mm, height 11.2 mm; sapphire crystal; transparent case back; water-resistant to 10 atm
Band: reptile skin, folding clasp
Price: $27,700

Possession
Reference number: G0A43082
Movement: quartz
Functions: hours, minutes
Case: rose gold, ø 29 mm, height 7.47 mm; bezel set with 42 diamonds; sapphire crystal; water-resistant to 3 atm
Band: reptile skin, buckle
Remarks: dial set with 11 diamonds
Price: $12,600

Possession
Reference number: G0A43090
Movement: quartz
Functions: hours, minutes
Case: stainless steel, ø 34 mm, height 7.47 mm; bezel set with a diamond; sapphire crystal; water-resistant to 3 atm
Band: reptile skin, buckle
Remarks: dial set with 11 diamonds
Price: $4,000

Limelight Gala Milanese Mesh
Reference number: G0A41213
Movement: quartz
Functions: hours, minutes
Case: rose gold, ø 32 mm, height 7.4 mm; bezel with 62 diamonds; sapphire crystal; water-resistant to 3 atm
Band: rose gold Milanese mesh, sliding fastener
Price: $34,500
Variations: white gold ($36,000)

Limelight Stella
Reference number: G0A40123
Movement: automatic, Piaget Caliber 584P; ø 26 mm; 21,600 vph; 42-hour power reserve
Functions: hours, minutes, sweep seconds; moon phase
Case: rose gold, ø 36 mm, height 9.9 mm; bezel set with 126 diamonds; sapphire crystal; transparent case back; water-resistant to 3 atm
Band: reptile skin, buckle
Remarks: dial set with 14 diamonds
Price: $37,200
Variations: white gold ($38,700)

PIAGET

Caliber 910P

Automatic; inverted movement construction as single unit with watch case; hubless peripheral rotor on movement edge; single spring barrel, 50-hour power reserve
Functions: hours, minutes (off-center)
Diameter: 41 mm
Height: 4.3 mm
Jewels: 30
Balance: glucydur
Frequency: 21,600 vph
Balance spring: flat hairspring
Remarks: finely finished movement; 238 parts

Caliber 1110P

Automatic; single spring barrel, 50-hour power reserve
Functions: hours, minutes, sweep seconds; date
Diameter: 25.58 mm
Height: 4 mm
Jewels: 25
Frequency: 28,800 vph
Remarks: perlage on mainplate, blued screws, finely finished with côtes de Genève; 180 parts

Caliber 1160P

Automatic; single spring barrel, 50-hour power reserve
Functions: hours, minutes; chronograph; date
Diameter: 25.58 mm
Height: 5.72 mm
Jewels: 35
Frequency: 28,800 vph
Remarks: perlage on mainplate, blued screws, finely finished with côtes de Genève; 262 parts

Caliber 450P

Manually wound; single spring barrel, 43-hour power reserve
Functions: hours, minutes, subsidiary seconds
Diameter: 20.5 mm
Height: 2.1 mm
Jewels: 18
Balance: glucydur
Frequency: 21,600 vph
Balance spring: flat hairspring
Remarks: finely finished movement; 131 parts

Caliber 1205P

Automatic; microrotor; single spring barrel, 44-hour power reserve; Geneva Seal
Functions: hours, minutes, subsidiary seconds; date
Diameter: 29.9 mm
Height: 3 mm
Jewels: 27
Balance: glucydur
Frequency: 21,600 vph
Balance spring: flat hairspring
Shock protection: Incabloc
Remarks: world's thinnest automatic movement with date from current production; 221 parts

Caliber 1208P

Automatic; microrotor; single spring barrel, 42-hour power reserve; Geneva Seal
Functions: hours, minutes, subsidiary seconds
Diameter: 29.9 mm
Height: 2.35 mm
Jewels: 27
Balance: glucydur
Frequency: 21,600 vph
Balance spring: flat hairspring
Shock protection: Incabloc

PORSCHE DESIGN

Porsche Design has always made sure it was partnering with the best to manufacture its products. In 1978, it was the Schaffhausen-based brand IWC that produced watches under the name Porsche Design through a license agreement with the F.A. Porsche design firm. But when the Porsche family purchased Eterna in 1995, a new era began—for both brands. When the IWC license expired in 1998, Eterna took over manufacturing responsibilities for the designer brand. In March 2014, Eterna and Porsche Design separated. Since September of that year, all Porsche Design watches have been developed by the company subsidiary Porsche Design Timepieces in Solothurn, Switzerland, in collaboration with the well-established design studio in Zell-am-See, Austria.

Porsche Design was founded by Professor Ferdinand Alexander Porsche in 1972—the fountainhead of numerous objects in daily use beyond just watches. The Professor—a title bestowed by the Austrian government—who died in April 2012, created a string of classic objects at his "Studio," but sports car fans will always remember him for the Porsche 911.

In 2003, the Professor decided to found his own company, which is separate from the carmaker. The brand is proud not only of its unusual designs but also of its use of light metals: In the 1970s already, it was using black PVD-coated aluminum and titanium for its watches and cases. The Chronotimer collection harked back to this very avant-garde esthetic statement. The streamlined and rigorous design is also visible in the new 1919 collection, named for the foundation date of the Bauhaus movement.

Porsche Design engineers are obviously not averse to picking up ideas from the automobile industry. The innovative rocker arm that activates the chronograph of the new Monobloc Actuator was inspired from the valve control of high-powered race cars using tappets. It improves ease of use and increases the mechanism's durability.

Porsche Design joined in the sports car's 70th anniversary celebrations with the launch of a special edition 1919 Datetimer. A total of 1,948 units to the 70Y Sports Car Limited Edition are being manufactured, the number a reminder of the company's founding year. It comes in a special box with a commemorative plaque.

Porsche Design Group
Groenerstrasse 5
D-71636 Ludwigsburg
Germany

Tel.:
+49-711-911-0

E-mail:
timepieces@porsche-design.us

Website:
www.porsche-design.com

Founded:
1972

U.S. distributor:
Porsche Design of America, Inc
Plaza Tower
600 Anton Blvd., Suite 1280
Costa Mesa, CA 92626
770-290-7500
770-290-0227 (fax)
timepieces@porsche-design.us
www.porsche-design.com

Most important collections:
Chronotimer Series 1, 1919 Datetimer Eternity, 1919 Globetimer, 1919 Chronotimer, Monobloc Actuator / $3,150 to $7,450

Monobloc Actuator 24H-Chronotimer All Black

Reference number: 4046901818685
Movement: automatic, ETA Caliber 7754; ø 30 mm, height 7.9 mm; 25 jewels; 28,800 vph; 48-hour power reserve
Functions: hours, minutes; rate control; additional 24-hour display (2nd time zone); chronograph; date
Case: titanium with black titanium carbide coating, ø 45.5 mm, height 15.6 mm; sapphire crystal; transparent case back; screw-in crown; water-resistant to 10 atm
Band: titanium with black titanium carbide coating, folding clasp
Price: $7,250

Monobloc Actuator Chronotimer Flyback Limited Edition

Reference number: 4046901810504
Movement: automatic, Porsche Design Caliber Werk 01.200 (base ETA 7750); 48-hour power reserve; COSC-certified chronometer
Functions: hours, minutes; rate control; flyback chronograph; date
Case: titanium with black titanium carbide coating, ø 45.5 mm, height 15.6 mm; sapphire crystal; transparent case back; screw-in crown; water-resistant to 10 atm
Band: rubber, folding clasp
Remarks: large rocker pusher integrated into right case side
Price: $8,500; limited to 251 pieces

1919 Chronotimer Flyback Brown & Leather

Reference number: 4046901809379
Movement: automatic, Porsche Design Caliber Werk 01.200 (base ETA 7750); ø 30 mm, height 7.9 mm; 25 jewels; 28,800 vph; 48-hour power reserve; COSC-certified chronometer
Functions: hours, minutes; rate control; flyback chronograph; date
Case: titanium, ø 42 mm, height 14.9 mm; sapphire crystal; transparent case back; screw-in crown; water-resistant to 10 atm
Band: calfskin, folding clasp
Price: $6,350

PORSCHE DESIGN

Chronotimer Flyback Special Edition

Reference number: 4046901811006
Movement: automatic, Porsche Design Caliber Werk 01.200 (base ETA 7750); ø 30 mm, height 7.9 mm; 25 jewels; 28,800 vph; 48-hour power reserve; COSC-certified chronometer
Functions: hours, minutes; rate control; flyback chronograph; date
Case: titanium with black titanium carbide coating, ø 42 mm, height 14.62 mm; sapphire crystal; transparent case back; screw-in crown; water-resistant to 10 atm
Band: calfskin, folding clasp
Price: $6,700

Monobloc Actuator 24H-Chronotimer Black & Rubber

Reference number: 4046901568047
Movement: automatic, ETA Caliber 7754; ø 30 mm, height 7.9 mm; 25 jewels; 28,800 vph; 48-hour power reserve
Functions: hours, minutes; rate control; additional 24-hour display (2nd time zone); chronograph; date
Case: titanium with black titanium carbide coating, ø 45.5 mm, height 15.6 mm; sapphire crystal; transparent case back; screw-in crown; water-resistant to 10 atm
Band: rubber, folding clasp
Remarks: large rocker pusher integrated into right case side
Price: $6,700

Monobloc Actuator GMT All Titanium

Reference number: 4046901564124
Movement: automatic, ETA Caliber 7754; ø 30 mm, height 7.9 mm; 25 jewels; 28,800 vph; 48-hour power reserve
Functions: hours, minutes; radio control; additional 24-hour display (2nd time zone); chronograph; date
Case: titanium, ø 45.5 mm, height 15.6 mm; sapphire crystal; transparent case back; screw-in crown; water-resistant to 10 atm
Band: titanium, folding clasp
Remarks: large rocker pusher integrated into right case side
Price: $6,900

1919 Datetimer Eternity Brown Alligator Leather

Reference number: 4046901986117
Movement: automatic, Sellita Caliber SW200-1; ø 25.6 mm, height 4.6 mm; 26 jewels; 28,800 vph; 38-hour power reserve
Functions: hours, minutes, sweep seconds; date
Case: titanium, ø 42 mm, height 11.92 mm; sapphire crystal; screw-in crown; water-resistant to 10 atm
Band: reptile skin, folding clasp
Price: $4,200

Chronotimer Series 1 Deep Blue

Reference number: 404901408770
Movement: automatic, ETA Caliber 7750; ø 30 mm, height 7.9 mm; 25 jewels; 28,800 vph; 48-hour power reserve
Functions: hours, minutes, subsidiary seconds; chronograph; date
Case: titanium, ø 42 mm, height 14.62 mm; sapphire crystal; transparent case back; screw-in crown; water-resistant to 5 atm
Band: textile, folding clasp
Price: $4,200

Chronotimer Series 1 Matte Black

Reference number: 4046901408695
Movement: automatic, ETA Caliber 7750; ø 30 mm, height 7.9 mm; 25 jewels; 28,800 vph; 48-hour power reserve
Functions: hours, minutes, subsidiary seconds; chronograph; date
Case: titanium with black titanium carbide coating, ø 42 mm, height 14.62 mm; sapphire crystal; transparent case back; screw-in crown; water-resistant to 5 atm
Band: titanium with black titanium carbide coating, folding clasp
Price: $5,100

PRAMZIUS

Whatever their political affiliations or leanings, no one can deny that Eastern Europe, the Baltic states, and Russia, in particular, exert a considerable fascination on people. It may be the extreme quality of everything that comes from that part of the world that appeals to our need for drama—the long and troubled history; the brutal leaders; the staggeringly talented people, from musicians to chess players; the brooding novels about adultery, complex love, suicide, war, dark morality; the antics of modern-day Russians caught on smartphones and dashcams. That may explain the success of Détente Group and its boisterous watches celebrating Big Mechanics—not for the limp-wristed, by any stretch.

In 2017, Craig Hester, distributor of Vostok-Europe, Sturmanskie, and other brands, parlayed over 20 years' experience in the watch business into the launch of a series of watches that would continue paying tribute to this wild, dangerous, creative, and at times sincerely eccentric part of the world. The name of the brand, Pramzius, is a reference to the Baltic Ruler of Time, an ancient and, appropriately, pagan god. Funding for the project came from a brief but successful Kickstarter campaign.

The first series of Pramzius was devoted to the renowned Trans-Siberian Railway and was inspired by a correspondingly themed pocket watch from back in the day. The modern version is, accordingly, big—48 millimeters in diameter—made of high-grade steel instead of brass. The wide watch face makes for excellent readability even at night, when the three hands sweep the full-lume dial of Superluminova. A relief of the train appears on the case back. The machine inside is a robust Seiko NH23; it keeps time and will not break the bank. The second in the series celebrates the thirtieth anniversary of the fall of the Berlin Wall, with a dial featuring graffiti from the last stretch of the Wall, the open-air East Side Gallery in Berlin, and a bit of the Wall in the crown as a souvenir of the ghastly frontier.

Pramzius Watches
31 Halls Hill Road
Colchester, CT 06415

Website:
www.pramzius.com

E-mail:
info@pramzius.com

Founded:
2017

Number of employees:
4

Annual production:
2,500

Distribution:
direct sales

Most important collections/price range:
Trans-Siberian Railroad $479; Berlin Wall Watch, $649 to $949

Trans-Siberian Railroad Black

Movement: automatic, Seiko Caliber NH38A; ø 27.4 mm, height 5.32 mm; 24 jewels; 21,600 vph; open-heart dial on the balance; 41-hour power reserve
Functions: hours, minutes, sweep seconds
Case: stainless steel, ø 48 mm, height 14.60 mm, K1 mineral glass; 3D locomotive rendering on case back; water-resistant to 3 atm
Band: calfskin bund strap, buckle
Remarks: comes with extra leather and nylon strap
Price: $479
Variation: sapphire crystal ($529)

Berlin Wall Watch

Movement: automatic, Seiko Caliber NH35; ø 27.4 mm, height 5.32 mm; 24 jewels; 21,600 vph; bidirectional rotor; 41-hour power reserve
Functions: hours, minutes, sweep seconds; date
Case: stainless steel, ø 42 mm, height 14 mm, K1 mineral glass; screw-down crown; 3D rendering of Brandenburg Gate on case back; water-resistant to 10 atm
Band: calfskin strap, buckle
Remarks: genuine marble dial featuring original graffiti from Berlin Wall; bits of Berlin Wall in crown; comes with extra leather-nylon strap
Price: $649 leather and $699 bracelet
Variation: ETA 2824 movement ($899–$949); 48-millimeter case ($649–$699)

Berlin Wall Watch

Movement: automatic, Seiko Caliber NH35; ø 27.4 mm, height 5.32 mm; 24 jewels; 21,600 vph; bidirectional rotor; 41-hour power reserve
Functions: hours, minutes, sweep seconds; date
Case: stainless steel, ø 48 mm, height 14 mm, K1 mineral glass; screw-down crown; 3D rendering of Brandenburg Gate on case back; water-resistant to 10 atm
Band: calfskin strap, buckle
Remarks: genuine marble dial featuring original graffiti from Berlin Wall; bits of Berlin Wall in crown; comes with extra leather-nylon strap
Price: $649 leather
Variation: stainless steel bracelet ($699); ETA 2824 movement ($899–$949); 48-millimeter case ($649–$699)

Rado Uhren AG
Bielstrasse 45
CH-2543 Lengnau
Switzerland

Tel.:
+41-32-655-6111

Fax:
+41-32-655-6112

E-mail:
info@rado.com

Website:
www.rado.com
store.us.rado.com

Founded:
1957

Number of employees:
approx. 470

U.S. distributor:
Rado
The Swatch Group (U.S.), Inc.
703 Waterford Way, Suite 450
Miami, FL 33126
786-725-5393

Most important collections/price range:
Hyperchrome / from approx. $1,100; Diamaster / from approx. $1,500; Integral / from approx. $2,000; True / from approx. $1,400; Centrix / from approx. $800; Coupole Classic / from approx. $1,000; Tradition / from approx. $2,000

RADO

Rado is a relatively young brand, especially for a Swiss one. The company, which grew out of the Schlup clockwork factory, launched its first watches in 1957, but it achieved international fame only five years later, in 1962, when it surprised the world with a revolutionary invention. Rado's oval DiaStar was the first truly scratch-resistant watch ever, sporting a case made of the impervious alloy hard metal. In 1985, its parent company, the Swatch Group, decided to put Rado's know-how and extensive experience in developing materials to good use. From then on, the brand intensified its research activities at its home in Lengnau, Switzerland, and continued to produce only watches with extremely hard cases. A record of sorts was even set in 2004, when they managed to create a 10,000-Vickers material, which is as hard as natural diamonds.

Rado also made jewel watches, but over time, it was the high-tech watches and pioneering spirit of the brand's ceramic researchers and engineers that won out. The company already holds more than thirty patents arising from research and production of new case materials. In 2011, for example, they produced the ultra-light Ceramos, which went into the D-Star collection. It has returned in a series of slim automatics branded DiaMaster. This innovative material is made up of 90 percent ceramic and 10 percent of a special metal alloy. It is scratch resistant and comes in a rosy gold or steely hue. Rado also uses a plasma-ceramic process to produce a material in warm metallic tones that keeps its sheen. Finally, there's the HyperChrome, which offers more resilience than regular ceramic, while weighing far less.

HyperChrome Automatic Chronograph

Reference number: R32249152
Movement: automatic, ETA Caliber 2894-S2; ø 28 mm, height 6.1 mm; 37 jewels; 28,800 vph; fully skeletonized movement; 42-hour power reserve
Functions: hours, minutes, subsidiary seconds; chronograph
Case: ceramic, ø 45 mm, height 13 mm; sapphire crystal; transparent case back; water-resistant to 10 atm
Band: ceramic, triple folding clasp
Price: $7,050

DiaMaster Thinline

Reference number: R14068026
Movement: automatic, ETA Caliber A31.L01; ø 25.6 mm, height 3.85 mm; 21 jewels; 25,200 vph; 64-hour power reserve
Functions: hours, minutes; date
Case: Ceramos (ceramic-stainless steel composite), ø 41 mm, height 8.3 mm; sapphire crystal; water-resistant to 5 atm
Band: calfskin, folding clasp
Price: $2,250

Coupole Classic XL Petite Seconde COSC

Reference number: R22880013
Movement: automatic, ETA Caliber C07.881; ø 25.6 mm, height 5.77 mm; 25 jewels; 21,600 vph; 80-hour power reserve; COSC-certified chronometer
Functions: hours, minutes, subsidiary seconds; date
Case: stainless steel, ø 41 mm, height 11.7 mm; sapphire crystal; water-resistant to 5 atm
Band: stainless steel, triple folding clasp
Price: $1,950

RADO

HyperChrome Automatic Chronograph
Reference number: R32168155
Movement: automatic, ETA Caliber 2894-2; ø 28.6 mm, height 6.1 mm; 37 jewels; 28,800 vph; 42-hour power reserve
Functions: hours, minutes, subsidiary seconds; chronograph; date
Case: ceramic with bronze barrel, crown, and pushers, ø 45 mm, height 13 mm; sapphire crystal; transparent case back; water-resistant to 10 atm
Band: calfskin, folding clasp
Price: $4,950

Ceramica
Reference number: R21631202
Movement: quartz
Functions: hours, minutes, sweep seconds
Case: ceramic, 30 × 41.7 mm, height 7.6 mm; sapphire crystal; water-resistant to 5 atm
Band: ceramic, triple folding clasp
Price: on request (not available in U.S.)

True Automatic
Reference number: R27510152
Movement: automatic, ETA Caliber C07.631; ø 25.6 mm, height 4.74 mm; 25 jewels; 21,600 vph; 80-hour power reserve
Functions: hours, minutes, sweep seconds
Case: ceramic, titanium case back, ø 40 mm, height 10.4 mm; sapphire crystal; transparent case back; water-resistant to 5 atm
Band: ceramic, triple folding clasp
Price: $2,200

DiaMaster Automatic
Reference number: R14061106
Movement: automatic, ETA Caliber 2899-S2; ø 25.6 mm, height 5.6 mm; 25 jewels; 28,800 vph; 42-hour power reserve
Functions: hours, minutes (off-center), subsidiary seconds; date
Case: ceramic, ø 43 mm, height 11.8 mm; sapphire crystal; water-resistant to 5 atm
Band: calfskin, folding clasp
Price: $3,050

Coupole Classic XL Petite Seconde COSC
Reference number: R22880205
Movement: automatic, ETA Caliber C07.881; ø 25.6 mm, height 5.77 mm; 25 jewels; 21,600 vph; 80-hour power reserve; COSC-certified chronometer
Functions: hours, minutes, subsidiary seconds; date
Case: stainless steel, ø 41 mm, height 11.7 mm; sapphire crystal; transparent case back; water-resistant to 5 atm
Band: calfskin, folding clasp
Price: $1,900
Variations: stainless steel bracelet and white or gray dial ($1,950)

Tradition 1965 Automatic
Reference number: R33017205
Movement: automatic, ETA Caliber C07.621; ø 25.6 mm, height 5.2 mm; 25 jewels; 21,600 vph; 80-hour power reserve
Functions: hours, minutes, sweep seconds; date, weekday
Case: titanium, 44 × 32 mm, height 11.8 mm; sapphire crystal; water-resistant to 5 atm
Band: calfskin, folding clasp
Price: $2,350

Ressence Watches
Meirbrug 1
2000 Antwerp
Belgium

Tel.:
+32-3-446-0060

E-mail:
hello@ressence.be

Website:
www.ressencewatches.com

Founded:
2011

U.S. distributor:
Totally Worth It, LLC
76 Division Avenue
Summit, NJ 07901-2309
201-894-4710
724-263-2286
info@totallyworthit.com

Most important collections/price range:
Type 1 / from $20,600; Type 3/ at $42,200; Type 5 /at $35,800

RESSENCE

It's rare for anyone to cause a buzz at the great Baselworld watch and jewelry fair based on the presentation of some prototypes. But Belgian Benoit Mintiens had the luck of the newcomer. The vicious 2008–2009 recession had left space available in the Palace pavilion in Basel, one of the crucibles of innovative watchmaking, and an audience curious about novelties in the post-bling world.

He returned in 2011 with the Type 1001. It consisted of a large rotating dial carrying a hand that pointed to a minute track on the bezel. Hours, small second, and a day-night indication rotated on dedicated subsidiary dials. The ballet on the dial mesmerized those who saw it, so he sold all his fifty models off the bat.

The mechanics behind the Ressence watches—the name is a compounding of Renaissance of the Essential—are basically simple: a stripped-down and rebuilt ETA 2824 leaves the minute wheel as the main driver of the other wheels. Mintiens, however, was about to go further.

In 2012 came the Type 0 series, which included a few design changes. Then, in a successful bid to improve readability, he immersed the dial section in oil, giving the displays a very contemporary two-dimensional look, much like an electronic watch. The movement had to be kept separate from the oil, and was connected to the dial using magnets and a set of superconductors and a Faraday cage to protect the movement from magnetism. A series of baffles compensates for the expanding and contracting of the oil due to temperature shifts. The Type 3 also lost the crown, the only obstacle to making a perfectly smooth watch, in favor of a clever setting and winding mechanism controlled by the case back. The watch is an automatic, of course. Not surprisingly, it won the Revelation Prize at the Grand Prix d'Horlogerie in Geneva in 2013.

The variations on the theme have been emerging from the Ressence studio at a regular pace. Besides a diver's watch, the brand has also created the Type 1 Squared, for instance. With its pillow-shaped brushed steel case, it combines the ultramodern display with an almost vintage look.

Type 1 Squared
Movement: automatic, ROCS 1.3 (module with ETA 2824-2 base); ø 32 mm, 40 jewels; 28,800 vph; Ressence Orbital Convex System: rotating minute dial, with rotating satellites for additional displays; winding and hand-setting using case back; 27 gearwheels, 212 components; 36-hour power reserve
Functions: minutes; hours, subsidiary seconds (eccentric, peripheral); weekday (eccentric, peripheral)
Case: stainless steel, 41 × 41 mm, height 11.5 mm; sapphire crystal
Band: reptile skin, buckle
Price: $20,600

Type 5 BB
Movement: automatic, ROCS 5 (module with ETA 2824-2 base); ø 32 mm; 41 jewels; 28,800 vph; Ressence Orbital Convex System: rotating minute dial, with rotating satellites for additional displays; winding and hand-setting using case back; 36-hour power reserve
Functions: minutes; hours, 90-second "runner," oil temperature gauge (eccentric, peripheral)
Case: titanium with black PVD coating, ø 46 mm, height 15.5 mm; sapphire crystal; water-resistant to 10 atm
Band: calfskin, buckle
Remarks: 2 separate, sealed case chambers; dial-side chamber filled with oil; magnetic drive for display disks
Price: $42,200

Type 3
Movement: automatic, ROCS 3 (module with ETA 2824-2 base); ø 32 mm; 47 jewels; 28,800 vph; Ressence Orbital Convex System: rotating minutes dial, with rotating satellites for additional displays; winding and hand-setting using case back; 36-hour power reserve
Functions: minutes; hours, 120-second "runner" (eccentric, peripheral), oil temperature (eccentric, peripheral), date, weekday (eccentric, peripheral)
Case: titanium, ø 44 mm, height 15 mm; sapphire crystal
Band: calfskin, buckle
Remarks: 2 separate, sealed case chambers; dial-side chamber filled with oil; magnetic drive for display disks
Price: $35,800

RGM

If there is any part of the United States that can somehow be considered its "watch valley," it may be the state of Pennsylvania. And one of the big players there is no doubt Roland Murphy, founder of RGM. Murphy, born in Maryland, went through the watchmaker's drill, studying at the Bowman Technical School, then in Switzerland, and finally working with Swatch before launching his own business in 1992.

His first series, Signature, paid homage to local horological genius through vintage pocket watch movements developed by Hamilton. His second big project was the Caliber 801, the first "high-grade mechanical movement made in series in America since Hamilton stopped production of the 992 B in 1969," Murphy grins. The next goal was to manufacture an all-American-made watch, the Pennsylvania Tourbillon.

And so, model by model, Murphy continues to expand his "Made in U.S.A." portfolio. "You cannot compare us to the big brands," says Murphy. "We are small and specialized, the needs are different. We work directly with the customer." In 2012, the brand's twentieth anniversary, RGM went retro with the 801 Aircraft. In 2015 came a baseball-themed watch. As for functionality, RGM makes a fine diver water-resistant to 70 atm, and the series 400 chronograph with a pulsometer and extra-large subdials for visibility.

One of the brand's main creations is the Caliber "20," which revives an old invention once in favor for railroad watches. The motor barrel is a complex but robust system in which the watch is wound by the barrel and the barrel arbor then drives the gear train. Less friction and wear and a slimmer chance of damage to the watch if the mainspring breaks are the two main advantages. Even the finishing on components is done following research into earlier American models.

RGM Watch Company
801 W. Main Street
Mount Joy, PA 17552

Tel.:
717-653-9799

Fax:
717-653-9770

E-mail:
sales@rgmwatches.com

Founded:
1992

Number of employees:
12

Annual production:
200–300 watches

Distribution:
RGM deals directly with customers.

Most important collections/price range:
Pennsylvania Series (completely made in the U.S.) / $2,500 to $125,000 range

Chess in Enamel
Reference number: PS 801 CH
Movement: manually wound, RGM Caliber 801; ø 37 mm; 19 jewels; lever escapement; screw balance; U.S. components: bridges, mainplate, settings, 7-tooth winding click; circular côtes de Genève; 42-hour power reserve
Functions: hours, minutes, sweep seconds
Case: stainless steel, ø 43.3 mm, height 12.3 mm; sapphire crystal
Band: reptile skin, buckle
Remarks: grand feu, white glass enamel dial with chess piece hour markers
Price: $13,900; limited edition
Variations: 18k rose gold ($29,900), platinum ($39,900)

Caliber 20
Reference number: Caliber 20
Movement: manually wound, RGM motor barrel movement; 34.4 × 30.4 mm; 19 jewels; 18,000 vph; perlage and côtes de Genève; 42-hour power reserve
Functions: hours, minutes, subsidiary seconds on disk; moon phase
Case: stainless steel, 42.5 × 38.5 mm, height 9.7 mm; hands of blued steel; sapphire crystal; transparent case back
Band: reptile skin, folding clasp
Remarks: full silver guilloché or skeleton dial available
Price: $29,500
Variations: rose gold ($42,500)

222 Railroad Series—Boxcar Dial
Reference number: 222 RR
Movement: restored, manually wound, Hamilton 923; ø 38.1 mm; 23 jewels; lever escapement; screw balance; U.S. components: bridges, mainplate, settings, 7-tooth winding click; circular côtes de Genève; 42-hour power reserve
Functions: hours, minutes, subsidiary seconds
Case: stainless steel, ø 41 mm, height 12 mm; sapphire crystal; transparent case back; water-resistant to 5 atm
Band: leather, buckle
Remarks: grand feu, white glass enamel "boxcar" dial
Price: $7,900
Variations: Hamilton 921 ($5,900)

Pennsylvania Series 801

Reference number: PS 801 EE
Movement: manually wound, RGM Caliber 801; ø 37 mm; 19 jewels; lever escapement; screw balance; U.S. components: bridges, moiré guilloché mainplate, settings, 7-tooth winding click; circular côtes de Genève; 42-hour power reserve
Functions: hours, minutes, subsidiary seconds
Case: stainless steel, ø 43.3 mm, height 12.3 mm; sapphire crystal; transparent case back; water-resistant to 5 atm
Band: reptile skin, buckle
Remarks: grand feu enamel skeleton dial with Roman numerals
Price: $11,900
Variations: rose gold ($24,700)

Model 25

Reference number: Model 25
Movement: automatic, RGM/ETA Caliber 2892-A2; ø 25.6 mm; 21 jewels; 28,800 vph; rhodium finish with perlage and côtes de Genève; 42-hour power reserve
Functions: hours, minutes, sweep seconds
Case: stainless steel, ø 40 mm, height 11.2 mm; guilloché dial, sapphire crystal; transparent case back; water-resistant to 5 atm
Band: calfskin, buckle
Remarks: several hand-cut dial options
Price: $6,450

Enamel Corps of Engineers

Reference number: 801 COE
Movement: manually wound, RGM Caliber 801; ø 37 mm; 19 jewels; lever escapement; screw balance; U.S. components: bridges, mainplate, settings, 7-tooth winding click; circular côtes de Genève; 42-hour power reserve
Functions: hours, minutes, subsidiary seconds
Case: stainless steel, ø 42 mm, height 10.5 mm; sapphire crystal; transparent case back; water-resistant to 5 atm
Band: leather, buckle
Remarks: grand feu white glass enamel dial with vintage luminous numerals
Price: $9,700
Variations: stainless steel bracelet ($10,450)

Pilot Professional

Reference number: 151 B
Movement: automatic, RGM/ETA Caliber 2892-A2; ø 25.6 mm; 21 jewels; 28,800 vph; rhodium finish with perlage and côtes de Genève; 42-hour power reserve
Functions: hours, minutes, sweep seconds; date
Case: brushed stainless steel, ø 38.5 mm, height 9.9 mm; sapphire crystal; transparent case back
Band: leather, buckle
Remarks: date at 3, 6, or no date
Price: $2,950
Variations: titanium ($3,950)

Time Zone Big Date

Reference number: 350 TZBD
Movement: automatic, RGM/ETA Caliber 2892-A2 with time zone and big date modules; ø 25.6 mm; 21 jewels; 28,800 vph; rhodium finish with perlage and côtes de Genève; 42-hour power reserve
Functions: hours, minutes, sweep seconds; 2nd time zone; big date
Case: brushed stainless steel, ø 38.7 mm, height 12 mm; curved sapphire crystal; transparent case back
Band: leather, buckle
Price: $3,700
Variations: none

Professional Diver

Reference number: 300-2 Series 2
Movement: automatic, modified ETA Caliber 2892; ø 25.6 mm, height 3.6 mm; 21 jewels; 28,800 vph; bridges and plates with perlage and côtes de Genève; 42-hour power reserve
Functions: hours, minutes, sweep seconds; date
Case: brushed stainless steel, ø 43.5 mm, height 17 mm; sapphire crystal (5 mm thick); unidirectional bezel with 0-60 scale (240 clicks); screw-down back; screw-in crown; water-resistant to more than 70 atm
Band: rubber strap, buckle
Price: $3,700; limited to 75 pieces
Variations: stainless steel bracelet ($4,450); without date function

RICHARD MILLE

Mille never stops delivering the wow to the watch world with what he calls his "race cars for the wrist." He is not an engineer, however, but rather a marketing expert who earned his first paychecks in the watch division of the French defense, automobile, and aerospace concern Matra in the early 1980s. This was a time of fundamental changes in technology, and the European watch industry was being confronted with gigantic challenges. "I have no historical relationship with watchmaking whatsoever," says Mille, "and so I have no obligations either. The mechanics of my watches are geared towards technical feasibility."

In the 1990s, Mille had to go to the expert workshop of Audemars Piguet Renaud & Papi (APRP) in Le Locle to find a group of watchmakers and engineers who would take on the Mille challenge. Audemars Piguet even succumbed to the temptation of testing those scandalous innovations—materials, technologies, functions—in a Richard Mille watch before daring to use them in its own collections (Tradition d'Excellence).

Since 2007, Audemars Piguet has also become a shareholder in Richard Mille, and so the three firms are now closely bound. The assembly of the watches is done in the Franches-Montagnes region in the Jura, where Richard Mille opened the firm Horométrie.

Richard Mille timepieces have also found their way onto the wrists of elite athletes, like tennis star Rafael Nadal and sprinter Yohan Blake. To keep its fans happy, the brand never ceases to explore the lunatic fringe of the technically possible, like the collaboration with Airbus Corporate Jets, which gave rise to a case made of a lightweight titanium-aluminum alloy used in turbines. Then there is the superlight and tough material called graphene developed at the University of Manchester and used by McLaren. The 2018 watch for polo star Pablo Mac Donough features a unique, reinforced sapphire crystal. Mille's other side is more romantic, like the Tourbillon Fleur or the Pink Lady Sapphire.

Richard Mille
c/o Horométrie SA
11, rue du Jura
CH-2345 Les Breuleux
Switzerland

Tel.:
+41-32-959-4353

Fax:
+41-32-959-4354

E-mail:
info@richardmille.ch

Website:
www.richardmille.com

Founded:
2000

Annual production:
approx. 4,600 watches

U.S. distributor:
Richard Mille Americas
8701 Wilshire Blvd.
Beverly Hills, CA 90211
310-205-5555

Tourbillon Alain Prost

Reference number: RM 70-01
Movement: manually wound, Richard Mille Caliber RM70; 29.7 × 37.1 mm, height 10.7 mm; 32 jewels; 21,600 vph; 1-minute tourbillon; 5 pusher-activated number rollers; distance tracker; 69-hour power reserve.
Functions: hours, minutes; kilometer counter
Case: TPT carbon; sapphire crystal; transparent case back; crown with torque limiter; water-resistant to 5 atm
Band: silicon, folding clasp
Remarks: homage to passionate cyclist Alain Prost; price includes custom-made Colnago racing bike
Price: $815,500; limited to 30 pieces

Tourbillon Split Seconds Chronograph McLaren F1

Reference number: RM 50-03
Movement: manually wound, Richard Mille Caliber RM52-03 × 31.1 × 32.15 mm, height 9.92 mm; 43 jewels; 21,600 vph; 1-minute tourbillon; titanium and TPT carbon plates and bridges; 70-hour power reserve
Functions: hours, minutes, subsidiary seconds; power reserve, torque and crown position indicator; split-second chronograph
Case: composite material, Graph TPT, 44.5 × 49.65 mm, height 16.1 mm; sapphire crystal; transparent case back
Band: textile, folding clasp
Remarks: collaboration with McLaren F1
Price: $1,030,000; limited to 75 pieces

Tourbillon Split Seconds Chronograph McLaren F1

Reference number: RM 50-03
Movement: manually wound, Richard Mille Caliber RM52-03 × 31.1 × 32.15 mm, height 9.92 mm; 43 jewels; 21,600 vph; 1-minute tourbillon; titanium and TPT carbon plates and bridges; 70-hour power reserve
Functions: hours, minutes, subsidiary seconds; power reserve, torque and crown position indicator; split-second chronograph
Case: composite material, Graph TPT, 44.5 × 49.65 mm, height 16.1 mm; sapphire crystal; transparent case back
Band: rubber, folding clasp
Remarks: cooperation with McLaren F1
Price: $1,030,000; limited to 75 pieces

RICHARD MILLE

Flyback Chronograph Le Mans Classic 2018
Reference number: RM 011-03 LMC
Movement: automatic, Richard Mille Caliber RMAC3; 28.45 × 30.25 mm, height 9 mm; 68 jewels; 28,800 vph; winding rotor with variable geometry
Functions: hours, minutes, subsidiary seconds; flyback chronograph; annual calendar with large date, month
Case: ceramic, 44.5 × 50 mm, height 16.15 mm; sapphire crystal; transparent case back
Band: rubber, folding clasp
Remarks: special 2018 edition for vintage car race "Le Mans Classic" sponsored by Richard Mille
Price: $180,000; limited to 150 pieces

Tourbillon Pablo Mac Donough
Reference number: RM 53-01
Movement: manually wound, Richard Mille Caliber RM53-01; 30.26 × 32 mm, height 6.35 mm; 19 jewels; 21,600 vph; 1-minute tourbillon; movement suspended by cables held in place by 10 pulleys with 2 tensioners at 3 and 9 o'clock; 70-hour power reserve
Functions: hours, minutes
Case: TPT carbon fiber, 44.5 × 50 mm, height 16.15 mm; sapphire crystal; water-resistant to 3 atm
Band: silicon, buckle
Remarks: 3-layered sapphire crystal with elastomer interlayers; homage to polo professional Pablo Mac Donough
Price: $888,000; limited to 30 pieces

Automatic
Reference number: RM 016
Movement: automatic, Richard Mille Caliber RM 005-S (base Vaucher 331); 28.6 × 30.2 mm, height 6.35 mm; 31 jewels; 28,800 vph; skeletonized titanium mainplate and bridges, 2 spring barrels, winding with adjustable rotor geometry and winding power
Functions: hours, minutes; large date
Case: pink gold, 38 × 49.8 mm, height 8.25 mm; sapphire crystal; transparent case back; water-resistant to 5 atm
Band: rubber, folding clasp
Price: on request; not available in U.S.

Automatic Alexander Zverev
Reference number: RM 67-02
Movement: automatic, Richard Mille Caliber CRMA7; 28.4 × 31.2 mm, height 3.6 mm; 25 jewels; 28,800 vph; skeletonized titanium mainplate and bridges, TPT carbon winding rotor with white gold oscillating mass; 50-hour power reserve
Functions: hours, minutes
Case: composite material (quartz TPT), 38 × 49.8 mm, height 8.25 mm; sapphire crystal; transparent case back; water-resistant to 5 atm
Band: rubber, folding clasp
Remarks: homage to tennis pro Alexander Zverev
Price: $132,000

Automatic Alexis Pinturault
Reference number: RM 67-02
Movement: automatic, Richard Mille Caliber CRMA7; 28.4 × 31.2 mm, height 3.6 mm; 25 jewels; 28,800 vph; skeletonized titanium mainplate and bridges, TPT carbon winding rotor with white gold oscillating mass; 50-hour power reserve
Functions: hours, minutes
Case: composite material (quartz TPT), 38 × 49.8 mm, height 8.25 mm; sapphire crystal; transparent case back; water-resistant to 5 atm
Band: rubber, folding clasp
Remarks: homage to ski racer Alexis Pinturault
Price: $132,000

Pink Lady Sapphire
Reference number: RM 07-02
Movement: automatic, Richard Mille Caliber CRMA5; 22 × 29.9 mm, height 5.05 mm; 25 jewels; 28,800 vph; pink gold mainplate and bridges, with diamonds; winding rotor set with diamonds; with variable geometry; 50-hour power reserve
Functions: hours, minutes
Case: pink sapphire, 32.9 × 46.75 mm, height 14.35 mm; sapphire crystal; transparent case back
Band: silicon, folding clasp
Remarks: dial with diamond pavé
Price: $1,034,000

ROGER DUBUIS

Roger Dubuis has always been a *manufacture* committed to luxury and *"très haute horlogerie."* The brand makes some outstanding movement components—parts that, because of their quality and geographical origins, bear the coveted Seal of Geneva. Founder Roger Dubuis, who passed away in 2017, was steeped in watchmaking and the business. In 2008, he sold 60 percent of the company shares to Richemont Group, which benefited from the resulting synergies, especially Cartier, which gets its skeletonized movements from the Roger Dubuis *manufacture*. In early 2016, Richemont went all the way and acquired the remaining 40 percent of the Genevan brand.

Roger Dubuis was founded in 1995 as SOGEM SA (Société Genevoise des Montres) by name-giver Roger Dubuis and financier Carlos Dias. These two exceptional men created a complete collection of unusual watches in no time flat—timepieces with unheard-of dimensions and incomparable complications. The meteoric development of this *manufacture* and the incredible frequency of its new introductions—even technical ones—continue to astound the traditional, rather conservative watch industry. Today, Roger Dubuis develops all of its own movements, currently numbering more than thirty different mechanical calibers. In addition, it produces just about all of its individual components in-house, from base plates to escapements and balance springs. With this heavy-duty technological know-how in its quiver, the brand has been able to build some remarkable movements, like the massive RD101, with four balance springs and all manner of differentials and gear works to drive the Excalibur Quatuor, the equivalent in horology to a monster truck. Even in their more delicate versions, like the Brocéliande, featuring colored ivy leaves embracing the movement, Roger Dubuis watches always seem ready to jump off your wrist.

Recently, the brand has rubbed elbows with car racing. Its partnership with Pirelli has produced models using genuine Formula One rubber as straps on the crowns. The rubber has the original tire color coding for different driving options: wet, medium, soft, supersoft, ultrasoft, and hypersoft.

Manufacture Roger Dubuis
2, rue André-De-Garrini - CP 149
CH-1217 Meyrin 2 (Geneva)
Switzerland

Tel.:
+41-22-783-2828

Fax:
+41-22-783-2882

E-mail:
info@rogerdubuis.com

Website:
www.rogerdubuis.com

Founded:
1995

Annual production:
over 5,000 watches (estimated)

U.S. distributor:
Roger Dubuis New York
545 Madison Ave.
New York, NY 10022
212-651-3773
Roger Dubuis Beverly Hills
9490C Brighton Way
Beverly Hills, CA 90210
310-734-1855

Most important collections:
Excalibur, Velvet / $12,000 to $1,100,000

Excalibur Spider Skelett Pirelli
Reference number: RDDBEX0704
Movement: automatic, Roger Dubuis Caliber RD820SQ; ø 36.1 mm, height 6.38 mm; 35 jewels; 28,800 vph; skeletonized movement, microrotor; 60-hour power reserve; Geneva Seal
Functions: hours, minutes
Case: titanium with black DLC coating, partly rubber-coated, ø 45 mm, height 14.02 mm; sapphire crystal; transparent case back; water-resistant to 5 atm
Band: rubber, folding clasp
Price: $73,500; limited to 88 pieces

Excalibur Aventador S
Reference number: RDDBEX0686
Movement: manually wound, Roger Dubuis Caliber RD103SQ; ø 36.1 mm, height 7.8 mm; 48 jewels; 57,600 vph; double sprung balance with angled balance wheels, 28,800 vph each; transmission power and synchronization via planetary gears; skeletonized movement; 40-hour power reserve; Geneva Seal
Functions: hours, minutes, sweep seconds (jumping); power reserve indicator
Case: carbon fiber with titanium inner case, partly rubber-coated, ø 45 mm, height 14.05 mm; sapphire crystal; transparent case back; crown with rubber covering; water-resistant to 5 atm
Band: rubber with calfskin overlay, folding clasp
Price: $210,000; limited to 88 pieces

Excalibur Spider Skeleton Flying Tourbillon
Reference number: RDDBEX0589
Movement: manually wound, Roger Dubuis Caliber RD10SQ; ø 37.8 mm, height 7.67 mm; 50 jewels; 21,600 vph; flying tourbillon with equalizing differential, skeletonized movement, galvanically blacked, beveled, and with perlage; 52-hour power reserve; Geneva Seal
Functions: hours, minutes
Case: titanium with black DLC coating, partly rubber-coated, ø 47 mm, height 14.95 mm; sapphire crystal; transparent case back; water-resistant to 5 atm
Band: rubber with calfskin overlay, folding clasp
Price: $305,000; limited to 28 pieces

ROGER DUBUIS

Excalibur Quatuor Cobalt MicroMelt
Reference number: RDDBEX0571
Movement: manually wound, Roger Dubuis Caliber RD101; ø 37.9 mm, height 10.6 mm; 113 jewels; 28,800 vph; four coupled escapement systems each with 4-Hz frequency; power transmission and synchronization through 3 satellite differentials; 40-hour power reserve; Geneva Seal
Functions: hours, minutes; power reserve indicator
Case: special alloy (chrome-cobalt), ø 48 mm, height 18.38 mm; sapphire crystal; transparent case back; water-resistant to 5 atm
Band: reptile skin, folding clasp
Price: $400,000; limited to 8 pieces

Excalibur Spider Skeleton
Reference number: RDDBEX0647
Movement: automatic, Roger Dubuis Caliber RD820SQ; ø 36.1 mm, height 6.38 mm; 35 jewels; 28,800 vph; skeletonized movement, microrotor; 60-hour power reserve; Geneva Seal
Functions: hours, minutes
Case: rose gold, with titanium inner case, partly rubber-coated, ø 45 mm, height 14.02 mm; sapphire crystal; transparent case back; water-resistant to 5 atm
Band: rubber with leather overlay, folding clasp
Price: $97,000; limited to 88 pieces
Variations: in carbon ($88,500)

Excalibur Skeleton Automatic
Reference number: RDDBEX0473
Movement: automatic, Roger Dubuis Caliber RD820SQ; ø 36.1 mm, height 6.38 mm; 35 jewels; 28,800 vph; skeletonized movement, microrotor; 60-hour power reserve; Geneva Seal
Functions: hours, minutes
Case: carbon fiber, ø 42 mm, height 11.9 mm; sapphire crystal; transparent case back; water-resistant to 3 atm
Band: reptile skin, folding clasp
Price: $63,500
Variations: white gold ($82,500); rose gold ($77,500)

Excalibur 45 Automatic
Reference number: RDDBEX0602
Movement: automatic, Roger Dubuis Caliber RD830; ø 29.21 mm, height 4 mm; 27 jewels; 28,800 vph; 48-hour power reserve
Functions: hours, minutes, subsidiary seconds; date
Case: titanium, ø 45 mm, height 14.7 mm; sapphire crystal; transparent case back; water-resistant to 5 atm
Band: rubber, folding clasp
Price: $16,600
Variations: various straps, cases, and dials

Excalibur 45 Automatic
Reference number: RDDBEX0566
Movement: automatic, Roger Dubuis Caliber RD830; ø 29.21 mm, height 4 mm; 27 jewels; 28,800 vph; 48-hour power reserve
Functions: hours, minutes, subsidiary seconds; date
Case: rose gold, ø 45 mm, height 14.7 mm; sapphire crystal; transparent case back; water-resistant to 3 atm
Band: reptile skin, folding clasp
Price: $30,500
Variations: various straps, cases, and dials

Excalibur 45 Automatic
Reference number: RDDBEX0567
Movement: automatic, Roger Dubuis Caliber RD830; ø 29.21 mm, height 4 mm; 27 jewels; 28,800 vph; 48-hour power reserve
Functions: hours, minutes, subsidiary seconds; date
Case: titanium with black DLC coating, ø 45 mm, height 14.7 mm; sapphire crystal; transparent case back; water-resistant to 5 atm
Band: rubber, folding clasp
Price: $15,800
Variations: various straps, cases, and dials

ROGER DUBUIS

Excalibur 42 Automatic
Reference number: RDDBEX0538
Movement: automatic, Roger Dubuis Caliber RD830; ø 29.21 mm, height 4 mm; 27 jewels; 28,800 vph; microrotor in rose gold; 48-hour power reserve
Functions: hours, minutes, subsidiary seconds; date
Case: pink gold, ø 42 mm, height 10.18 mm; sapphire crystal; transparent case back; water-resistant to 5 atm
Band: reptile skin, folding clasp
Price: $25,700
Variations: various straps, cases, and dials

Excalibur 42 Automatic
Reference number: RDDBEX0619
Movement: automatic, Roger Dubuis Caliber RD830; ø 29.21 mm, height 4 mm; 27 jewels; 28,800 vph; microrotor in rose gold; 48-hour power reserve
Functions: hours, minutes, subsidiary seconds; date
Case: stainless steel, ø 42 mm, height 10.18 mm; sapphire crystal; transparent case back; water-resistant to 5 atm
Band: stainless steel, folding clasp
Price: $15,200
Variations: various straps, cases, and dials

Excalibur 42 Automatic
Reference number: RDDBEX0537
Movement: automatic, Roger Dubuis Caliber RD830; ø 29.21 mm, height 4 mm; 27 jewels; 28,800 vph; microrotor in rose gold; 48-hour power reserve
Functions: hours, minutes, subsidiary seconds; date
Case: pink gold, ø 42 mm, height 10.18 mm; sapphire crystal; transparent case back; water-resistant to 5 atm
Band: reptile skin, folding clasp
Price: $25,700
Variations: various straps, cases, and dials

Excalibur 36 Automatic
Reference number: RDDBEX0587
Movement: automatic, Roger Dubuis Caliber RD830; ø 29.21 mm, height 4 mm; 27 jewels; 28,800 vph; microrotor in rose gold; 48-hour power reserve
Functions: hours, minutes, subsidiary seconds; date
Case: rose gold, ø 36 mm, height 9.8 mm; sapphire crystal; transparent case back; water-resistant to 5 atm
Band: reptile skin, folding clasp
Price: $22,100
Variations: titanium set with diamonds and sapphires ($17,400)

Velvet Caviar
Reference number: RDDBVE0079
Movement: automatic, Roger Dubuis Caliber RD830; ø 29.21 mm, height 4 mm; 27 jewels; 28,800 vph; microrotor in rose gold; 48-hour power reserve
Functions: hours, minutes
Case: white gold, ø 36 mm, height 9.7 mm; bezel and case set with 86 diamonds; sapphire crystal; transparent case back; water-resistant to 5 atm
Band: textile set with crystals, folding clasp with 14 diamonds
Remarks: mother-of-pearl dial
Price: $42,700

Velvet Automatic
Reference number: RDDBVE0070
Movement: automatic, Roger Dubuis Caliber RD830; ø 29.21 mm, height 4 mm; 27 jewels; 28,800 vph; microrotor in rose gold; 48-hour power reserve
Functions: hours, minutes
Case: white gold, ø 36 mm, height 9.8 mm; sapphire crystal; transparent case back; water-resistant to 5 atm
Band: reptile skin, buckle
Remarks: flange with 64 diamonds
Price: $23,600

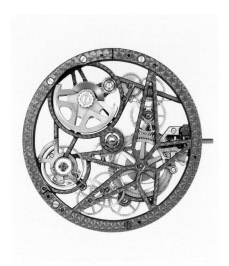

Caliber RD820SQ

Automatic; microrotor; single spring barrel, 60-hour power reserve; Geneva Seal
Functions: hours, minutes
Diameter: 36.1 mm
Height: 6.38 mm
Jewels: 35
Frequency: 28,800 vph
Remarks: skeletonized movement, rhodium-plated, finely finished with côtes de Genève; 167 parts

Caliber RD01SQ

Manually wound; 2 flying 1-minute tourbillons with equalizing differential skeletonized movement; single spring barrel, 48-hour power reserve; Geneva Seal, COSC-certified chronometer
Functions: hours, minutes
Diameter: 37.8 mm
Height: 7.67 mm
Jewels: 28
Balance: glucydur variable inertia balance
Frequency: 21,600 vph
Remarks: 319 components

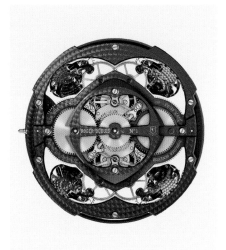

Caliber RD101

Manually wound; 4 radially assembled lever escapements, synchronized with 3 balancing differentials; additional planetary gears for winding and power reserve indicator; skeletonized movement; double spring barrel, 40-hour power reserve; Geneva Seal
Functions: hours, minutes; power reserve indicator
Diameter: 37.9 mm
Height: 10.6 mm
Jewels: 113
Balance: glucydur (4 ×)
Frequency: 28,800 vph
Balance spring: flat hairspring
Remarks: galvanic black movement, beveled and with perlage; 590 parts

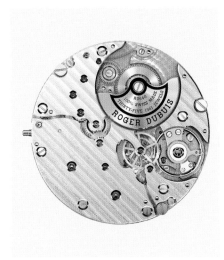

Caliber RD640

Automatic; microrotor; single spring barrel, 52-hour power reserve; Geneva Seal
Functions: hours, minutes, subsidiary seconds; date
Diameter: 31.1 mm
Height: 4.5 mm
Jewels: 35
Balance: glucydur with smooth collar
Frequency: 28,800 vph
Balance spring: flat hairspring
Shock protection: Incabloc
Remarks: finely finished with côtes de Genève; 198 parts

Caliber RD103SQ

Manually wound; double sprung balance with angled balance wheels, 28,800 vph each; power transmission and synchronization via planetary gears; skeletonized movement; single spring barrel, 40-hour power reserve; Geneva Seal
Functions: hours, minutes, sweep seconds (jumping); power reserve indicator
Diameter: 36.1 mm
Height: 7.8 mm
Jewels: 48
Frequency: 57,600 vph

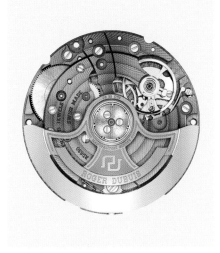

Caliber RD830

Automatic; rotor in rose gold; single spring barrel, 48-hour power reserve
Functions: hours, minutes, subsidiary seconds; date
Diameter: 29.21 mm
Height: 4 mm
Jewels: 27
Frequency: 28,800 vph
Remarks: finely finished with côtes de Genève; 183 parts

ROLEX

Essentially, the Rolex formula for success has always been "what you see is what you get"—and plenty of it. For over a century now, the company has made wristwatch history without a need for *grandes complications*, perpetual calendars, tourbillons, or exotic materials. And its output in sheer quantity is phenomenal, at not quite a million watches per year. But make no mistake about it: The quality of these timepieces is legendary.

For as long as anyone can remember, this brand has held the top spot in the COSC's statistics, and year after year Rolex delivers just about half of all of the official institute's successfully tested mechanical chronometer movements. The brand has also pioneered several fundamental innovations: Rolex founder Hans Wilsdorf invented the hermetically sealed Oyster case in the 1920s, which he later outfitted with a screwed-in crown and an automatic movement wound by rotor. Shock protection, water resistance, the antimagnetic Parachrom hairspring, and automatic winding are some of the virtues that make wearing a Rolex timepiece much more comfortable and reliable. Because Wilsdorf patented his inventions for thirty years, Rolex had a head start on the competition.

Rolex watches and movements were at first produced in two different companies at two different sites. Only in 2004 did Geneva-based Rolex buy and integrate the Rolex movement factory in Biel. Then, in 2008, for its 100th birthday, the company built three gigantic new buildings with loads of steel and dark glass in the industrial suburb of Plan-les-Ouates. The automatic caliber 3255 suggests a brand still in innovation mode: It features new materials (nickel-phosphorus), special micromanufacturing technology (LIGA) to make the pallet fork and balance wheel of the Chronergy escapement, and a barrel spring that can store up more energy than ever. Meanwhile the company has built up representation in nearly one hundred countries in the world, with over thirty subsidiaries with customer service centers. The network also includes around four thousand watchmakers trained according to Rolex standards.

Rolex SA
Rue François-Dussaud 3
CH-1211 Geneva 26
Switzerland

Website:
www.rolex.com

Founded:
1908

Number of employees:
over 2,000 (estimated)

Annual production:
approx. 1,000,000 watches (estimated)

U.S. distributor:
Rolex Watch U.S.A., Inc.
Rolex Building
665 Fifth Avenue
New York, NY 10022-5358
212-758-7700; 212-980-2166 (fax)
www.rolex.com

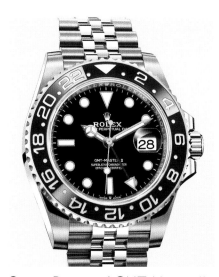

Oyster Perpetual GMT-Master II

Reference number: 126710BLRO
Movement: automatic, Rolex Caliber 3285; ø 28.5 mm, height 6.4 mm; 31 jewels; 28,800 vph; Parachrom hairspring, Chronergy escapement; 70-hour power reserve; COSC-certified chronometer
Functions: hours, minutes, sweep seconds; additional 24-hour display (2nd time zone); date
Case: stainless steel, ø 40 mm, height 13 mm; bidirectional bezel with ceramic inlay, with 24-hour division; sapphire crystal; screw-in crown; water-resistant to 10 atm
Band: Jubilé stainless steel, folding clasp, with safety lock and extension link
Price: $8,400

Oyster Perpetual GMT-Master II

Reference number: 116710BLNR
Movement: automatic, Rolex Caliber 3186; ø 28.5 mm, height 6.4 mm; 31 jewels; 28,800 vph; Parachrom hairspring; 48-hour power reserve; COSC-certified chronometer
Functions: hours, minutes, sweep seconds; additional 24-hour display (2nd time zone); date
Case: stainless steel, ø 40 mm, height 13 mm; bidirectional bezel with ceramic insert, with 24-hour division; sapphire crystal; screw-in crown; water-resistant to 10 atm
Band: Oyster stainless steel, folding clasp, with safety lock and extension link
Price: $8,100

Oyster Perpetual GMT-Master II

Reference number: 126711CHNR
Movement: automatic, Rolex Caliber 3285; ø 28.5 mm, height 6.4 mm; 31 jewels; 28,800 vph; Parachrom hairspring, Chronergy escapement; 70-hour power reserve; COSC-certified chronometer
Functions: hours, minutes, sweep seconds; additional 24-hour display (2nd time zone); date
Case: stainless steel, ø 40 mm, height 13 mm; bezel in rose gold with ceramic inlay, bidirectional bezel with 0-24 scale; sapphire crystal; screw-in crown; water-resistant to 10 atm
Band: Oyster stainless steel with rose gold elements, folding clasp, with safety lock and extension link
Price: $8,950
Variations: Everose gold ($36,750)

Oyster Perpetual Deepsea

Reference number: 126660
Movement: automatic, Rolex Caliber 3235; ø 29.1 mm; 31 jewels; 28,800 vph; Parachrom hairspring, Chronergy escapement; 70-hour power reserve; COSC-certified chronometer
Functions: hours, minutes, sweep seconds; date
Case: stainless steel, ø 44 mm; unidirectional bezel with ceramic insert, 0-60 scale; sapphire crystal; screw-in crown; helium valve; water-resistant to 390 atm
Band: Oyster stainless steel, folding clasp, with safety lock and extension link
Price: $12,250
Variations: D-blue dial ($12,550)

Oyster Perpetual Sea-Dweller

Reference number: 126600
Movement: automatic, Rolex Caliber 3235; ø 29.1 mm; 31 jewels; 28,800 vph; Parachrom hairspring, Chronergy escapement; 70-hour power reserve; COSC-certified chronometer
Functions: hours, minutes, sweep seconds; date
Case: stainless steel, ø 43 mm, height 13.8 mm; unidirectional bezel with ceramic insert, 0-60 scale; sapphire crystal; screw-in crown; helium valve; water-resistant to 122 atm
Band: Oyster stainless steel, folding clasp, with extension link
Price: $11,350

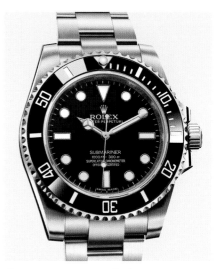

Oyster Perpetual Submariner

Reference number: 114060
Movement: automatic, Rolex Caliber 3130; ø 28.5 mm; 31 jewels; 28,800 vph; Parachrom hairspring; 48-hour power reserve; COSC-certified chronometer
Functions: hours, minutes, sweep seconds
Case: stainless steel, ø 40 mm, height 12.5 mm; unidirectional bezel with ceramic insert, 0-60 scale; sapphire crystal; screw-in crown; water-resistant to 30 atm
Band: Oyster stainless steel, folding clasp, with extension link
Price: $7,500

Oyster Perpetual Air-King

Reference number: 116900
Movement: automatic, Rolex Caliber 3131; ø 28.5 mm; 31 jewels; 28,800 vph; Parachrom hairspring, glucydur balance with microstella regulating bolts; soft-iron cap for antimagnetic protection; 48-hour power reserve; COSC-certified chronometer
Functions: hours, minutes, sweep seconds
Case: stainless steel, ø 40 mm; sapphire crystal; screw-in crown; water-resistant to 10 atm
Band: Oyster stainless steel, folding clasp, with safety lock, with extension link
Price: $6,200

Oyster Perpetual Explorer

Reference number: 214270
Movement: automatic, Rolex Caliber 3132; ø 28.5 mm; 31 jewels; 28,800 vph; Parachrom hairspring, Paraflex shock protection, glucydur balance with microstella regulating bolts; 48-hour power reserve; COSC-certified chronometer
Functions: hours, minutes, sweep seconds
Case: stainless steel, ø 39 mm; sapphire crystal; screw-in crown; water-resistant to 10 atm
Band: Oyster stainless steel, folding clasp, with extension link
Price: $6,550

Oyster Perpetual Cosmograph Daytona

Reference number: 116500LN
Movement: automatic, Rolex Caliber 4130; ø 30.5 mm, height 6.5 mm; 44 jewels; 28,800 vph; Parachrom hairspring, glucydur balance with microstella regulating bolts; 72-hour power reserve; COSC-certified chronometer
Functions: hours, minutes, subsidiary seconds; chronograph
Case: stainless steel, ø 40 mm, height 12.8 mm; Cerachrom bezel; sapphire crystal; screw-in crown and pusher; water-resistant to 10 atm
Band: Oyster stainless steel, folding clasp, with safety lock and extension link
Price: $12,400
Variations: black dial

ROLEX

Oyster Perpetual Yacht-Master II
Reference number: 116680
Movement: automatic, Rolex Caliber 4161 (base Caliber 4130); ø 31.2 mm, height 8.05 mm; 42 jewels; 28,800 vph; Parachrom hairspring; 72-hour power reserve; COSC-certified chronometer
Functions: hours, minutes, subsidiary seconds; programmable regatta countdown with memory
Case: stainless steel, ø 44 mm, height 13.8 mm; bidirectional bezel with ceramic insert to control functions; sapphire crystal; screw-in crown; water-resistant to 10 atm
Band: Oyster stainless steel, folding clasp, with safety lock and extension link
Price: $18,750; **Variations:** yellow gold ($43,550); stainless steel/Everose gold ($25,150)

Oyster Perpetual Yacht-Master 40
Reference number: 116655
Movement: automatic, Rolex Caliber 3135; ø 28.5 mm, height 6 mm; 31 jewels; 28,800 vph; Parachrom hairspring, glucydur balance with microstella regulating bolts; 48-hour power reserve; COSC-certified chronometer
Functions: hours, minutes, sweep seconds; date
Case: rose gold, ø 40 mm, height 11.7 mm; bidirectional bezel with ceramic insert, 0-60 scale; sapphire crystal; screw-in crown; water-resistant to 10 atm
Band: elastomer with metal mesh, folding clasp
Price: $24,950

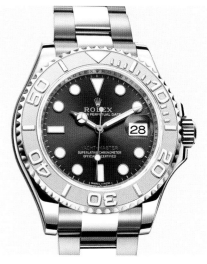

Oyster Perpetual Yacht-Master 40
Reference number: 116621
Movement: automatic, Rolex Caliber 3135; ø 28.5 mm, height 6 mm; 31 jewels; 28,800 vph; Parachrom hairspring, glucydur balance with microstella regulating bolts; 48-hour power reserve; COSC-certified chronometer
Functions: hours, minutes, sweep seconds; date
Case: stainless steel, ø 40 mm, height 11.7 mm; bezel in rose gold, bidirectional bezel with 0-60 scale; sapphire crystal; screw-in crown; water-resistant to 10 atm
Band: Oyster stainless steel and Everose gold, folding clasp
Price: $14,050

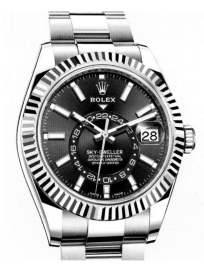

Oyster Perpetual Sky-Dweller
Reference number: 326934
Movement: automatic, Rolex Caliber 9001; ø 33 mm, height 8 mm; 40 jewels; 28,800 vph; 72-hour power reserve; COSC-certified chronometer
Functions: hours, minutes, sweep seconds; additional 24-hour display (2nd time zone); annual calendar with date, month
Case: stainless steel, ø 42 mm, height 14.1 mm; bidirectional white gold bezel to control functions; sapphire crystal; screw-in crown; water-resistant to 10 atm
Band: Oyster stainless steel, folding clasp
Price: $14,400
Variations: stainless steel/yellow gold ($17,150); yellow gold ($46,150); in Everose gold ($48,850)

Oyster Perpetual Day-Date 40
Reference number: 228239
Movement: automatic, Rolex Caliber 3255; ø 29.1 mm, height 5.4 mm; 31 jewels; 28,800 vph; Parachrom hairspring, Paraflex shock protection, Chronergy escapement, glucydur balance with microstella regulating bolts; 70-hour power reserve; COSC-certified chronometer
Functions: hours, minutes, sweep seconds; date, weekday
Case: white gold, ø 40 mm, height 11.6 mm; sapphire crystal; screw-in crown; water-resistant to 10 atm
Band: President white gold, folding clasp
Price: $37,500
Variations: yellow gold ($34,850); Everose gold ($37,550); platinum ($62,500)

Oyster Perpetual Datejust 41
Reference number: 126334
Movement: automatic, Rolex Caliber 3235; ø 29.1 mm; 31 jewels; 28,800 vph; Parachrom hairspring, Paraflex shock protection, Chronergy escapement, glucydur balance with microstella regulating bolts; 70-hour power reserve; COSC-certified chronometer
Functions: hours, minutes, sweep seconds; date
Case: stainless steel, ø 41 mm, height 11.6 mm; white gold bezel; sapphire crystal; screw-in crown; water-resistant to 10 atm
Band: Oyster stainless steel, folding clasp, with extension link
Price: $9,350

ROLEX

Oyster Perpetual Datejust 36
Reference number: 126231
Movement: automatic, Rolex Caliber 3235; ø 29.1 mm; 31 jewels; 28,800 vph; Parachrom hairspring, Paraflex shock protection, Chronergy escapement, glucydur balance with microstella regulating bolts; 70-hour power reserve; COSC-certified chronometer
Functions: hours, minutes, sweep seconds; date
Case: stainless steel, ø 36 mm, height 11.3 mm; rose gold bezel; sapphire crystal; screw-in crown, in rose gold; water-resistant to 10 atm
Band: Oyster stainless steel with rose gold elements, folding clasp, with extension link
Price: $10,950
Variations: available with Jubilé straps

Oyster Perpetual Datejust 31
Reference number: 278289RBR
Movement: automatic, Rolex Caliber 2236; ø 20 mm, height 5.95 mm; 31 jewels; 28,800 vph; Syloxi hairspring, glucydur balance with microstella regulating bolts; 55-hour power reserve; COSC-certified chronometer
Functions: hours, minutes, sweep seconds; date
Case: white gold, ø 31 mm, height 11 mm; bezel with 46 diamonds; sapphire crystal; screw-in crown; water-resistant to 10 atm
Band: President white gold, folding clasp
Price: $40,100

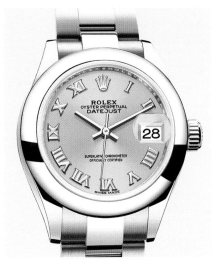

Oyster Lady Datejust 28
Reference number: 279160
Movement: automatic, Rolex Caliber 2236; ø 20 mm, height 5.95 mm; 31 jewels; 28,800 vph; Syloxi hairspring, glucydur balance with microstella regulating bolts; 55-hour power reserve; COSC-certified chronometer
Functions: hours, minutes, sweep seconds; date
Case: stainless steel, ø 28 mm, height 10.5 mm; sapphire crystal; screw-in crown; water-resistant to 10 atm
Band: Oyster stainless steel, folding clasp, with extension link
Price: $6,300
Variations: various straps and dials

Cellini Moonphase
Reference number: 50535
Movement: automatic, Rolex Caliber 3195; ø 28.5 mm; 31 jewels; 28,800 vph; Parachrom hairspring, Paraflex shock protection; 48-hour power reserve; COSC-certified chronometer
Functions: hours, minutes, sweep seconds; date, moon phase
Case: rose gold, ø 39 mm; sapphire crystal; screw-in crown; water-resistant to 5 atm
Band: reptile skin, folding clasp
Remarks: blue-enameled disk with rhodium-plated full moon
Price: $26,750

Cellini Date
Reference number: 50519
Movement: automatic, Rolex Caliber 3165 (base Rolex Caliber 3187); ø 28.5 mm; 31 jewels; 28,800 vph; Parachrom Breguet hairspring; 48-hour power reserve; COSC-certified chronometer
Functions: hours, minutes, sweep seconds; date
Case: white gold, ø 39 mm; sapphire crystal; screw-in crown; water-resistant to 5 atm
Band: reptile skin, buckle
Price: $17,800
Variations: various dials; rose gold

Cellini Dual Time
Reference number: 50525
Movement: automatic, Rolex Caliber 3180 (base Rolex Caliber 3187 with module); 28,800 vph; Parachrom hairspring; 48-hour power reserve; COSC-certified chronometer
Functions: hours, minutes, sweep seconds; additional 12-hour display (2nd time zone), day/night indicator
Case: rose gold, ø 39 mm; sapphire crystal; screw-in crown; water-resistant to 5 atm
Band: reptile skin, buckle
Price: $19,400
Variations: various dials

ROLEX

Caliber 3255
Automatic; optimized Chronergy escapement, nickel phosphorus pallet lever and escape wheel made using LIGA process; single spring barrel, 70-hour power reserve; COSC-certified chronometer
Functions: hours, minutes, sweep seconds; date, weekday
Diameter: 29.1 mm
Height: 5.4 mm
Jewels: 31
Balance: glucydur with microstella regulating bolts
Frequency: 28,800 vph
Balance spring: Parachrom Breguet hairspring
Shock protection: Paraflex
Remarks: used in the Day-Date 40

Caliber 3235
Automatic; optimized Chronergy escapement, nickel phosphorus pallet lever and escape wheel made using LIGA process; single spring barrel, 70-hour power reserve; COSC-certified chronometer
Functions: hours, minutes, sweep seconds; date
Diameter: 28.5 mm
Height: 6 mm
Jewels: 31
Balance: glucydur with microstella regulating bolts
Frequency: 28,800 vph
Balance spring: Parachrom Breguet hairspring
Shock protection: Paraflex
Remarks: used in the Datejust

Caliber 2236
Automatic; single spring barrel, 55-hour power reserve; COSC-certified chronometer
Functions: hours, minutes, sweep seconds; date
Diameter: 20 mm
Height: 5.95 mm
Jewels: 31
Balance: glucydur with microstella regulating bolts
Frequency: 28,800 vph
Balance spring: Parachrom flat hairspring
Shock protection: Kif
Remarks: used in the Lady Datejust

Caliber 4130
Automatic; single spring barrel, 72-hour power reserve; COSC-certified chronometer
Functions: hours, minutes, subsidiary seconds; chronograph
Diameter: 30.5 mm
Height: 6.5 mm
Jewels: 44
Balance: glucydur with microstella regulating bolts
Frequency: 28,800 vph
Balance spring: Parachrom Breguet hairspring
Shock protection: Kif
Remarks: used in the Daytona

Caliber 4161
Automatic; single spring barrel, 72-hour power reserve; COSC-certified chronometer
Base caliber: Caliber 4130
Functions: hours, minutes, subsidiary seconds; programmable regatta countdown with memory
Diameter: 31.2 mm
Height: 8.05 mm
Jewels: 42
Balance: glucydur with microstella regulating bolts
Frequency: 28,800 vph
Balance spring: Parachrom Breguet hairspring
Shock protection: Kif
Remarks: used in the Yacht-Master II

Caliber 9001
Automatic; single spring barrel, 72-hour power reserve; COSC-certified chronometer
Functions: hours, minutes, sweep seconds; additional 24-hour display (2nd time zone); annual calendar with date, month
Diameter: 33 mm
Height: 8 mm
Jewels: 40
Balance: glucydur with microstella regulating bolts
Frequency: 28,800 vph
Balance spring: Parachrom Breguet hairspring
Shock protection: Kif
Remarks: used In the Sky-Dweller

The Ideal Shipmate for Every Wrist: the Yacht-Timer BRONZE

With its blue face, blue textile strap and bronze case with a – you guessed it – blue bezel, the Yacht-Timer BRONZE is a prime example of how colour can reveal the true character of a watch. The latest special edition to be launched by Nautische Instrumente Mühle-Glashütte uses its stunning blue design to immediately tell its wearers and admirers where it feels right at home: out on the water. This model is not only the perfect companion for a casual boating jaunt but also an outstanding shipmate for more challenging adventures at sea. **Limited edition: 500 pieces**

For more information please contact:

West: Arizona Fine Time – Scottsdale, AZ / Feldmar Watch Company – Los Angeles, CA / Ravits – San Francisco, CA / Right Time Denver – Denver CO / Right Time Highlands Ranch – Highlands Ranch, CO / Time Spot – Thousand Oaks, CA and Fox's – Seattle WA
Midwest: Abt – Glenview, IL / Chalmers Jewelers – Middleton, WI and IW Marks – Houston, TX
East: Joseph Edwards – New York, NY / Little Treasury – Gambrills, MD and Martin Pulli – Philadelphia, PA
Florida: Exquisite Timepieces – Naples, FL / Old Northeast Jewelers – St. Petersburg, FL / Old Northeast Jewelers – Tampa FL / and Orlando Watch Co. – Winter Park, FL

SCHAUMBURG WATCH

Frank Dilbakowski is the owner of this small watchmaking business in Rinteln, Westphalia, which has been producing very unusual yet affordable timepieces since 1998. The name Schaumburg comes from the surrounding region. The firm has gained a reputation for high-performance timepieces for rugged sports and professional use. The chronometer line Aquamatic, with water resistance to 1,000 meters, and the Aquatitan models, secure to 2,000 meters, confirm the company's maxim that form, function, and performance are inseparable from one another.

By the same token, traditional watchmaking is also high on the agenda. The Rinteln workbenches produce the plates and bridges and provide all the finishing as well (perlage, engraving, skeletonizing). Some of the bracelets, cases, and dials are even manufactured here, but the base movements come from Switzerland. Besides unadorned one-hand watches, the current portfolio of timepieces includes such outstanding creations as a special moon phase, which, rather than simply showing a moon, has a "shadow" crossing over an immobile photo-like reproduction of the moon. And, of course, the Schaumburg workshop also produces unique pieces.

The latest design, one that is bound to raise eyebrows and generate talk, is the artificial aging of cases, movements, and components. The Steam Punk collection looks as if it had been buried in someone's garden and has now been unearthed, cleaned up, but not restored. The watches work, of course.

Schaumburg Watch
Lindburgh & Benson
Kirchplatz 5 and 6
D-31737 Rinteln
Germany

Tel.:
+49-5751-923-351

E-mail:
info@lindburgh-benson.com

Website:
www.schaumburgwatch.com

Founded:
1998

Number of employees:
7

Annual production:
not specified

Distribution:
retail

U.S. distributor:
Schaumburg Watch
About Time Luxury Group
210 Bellevue Avenue
Newport, RI 02840
401-846-0598
nicewatch@aol.com

Most important collections/price range:
mechanical wristwatches / approx. $1,500 to $13,000

GT SuperCup Chronograph
Reference number: SWGTSCC
Movement: automatic, SW Caliber 50c (base ETA 7750); ø 30 mm, height 7.9 mm; 25 jewels; 28,800 vph; 42-hour power reserve
Functions: hours, minutes, subsidiary seconds; chronograph; date
Case: stainless steel, ø 45 mm, height 15 mm; sapphire crystal; screw-in crown; water-resistant to 10 atm
Band: calfskin, buckle
Price: $2,500

Steam Punk
Movement: manually wound, SW Caliber 07 (base ETA 6498); ø 36.6 mm, height 4.5 mm; 17 jewels; 21,600 vph; 48-hour power reserve
Functions: hours, minutes, subsidiary seconds
Case: stainless steel, ø 42 mm, height 11.8 mm; sapphire crystal; transparent case back; water-resistant to 5 atm
Band: calfskin, buckle
Remarks: dial made up of indices on artificially aged metal in typical steampunk look
Price: $2,200

Moon Blue Nebula
Movement: automatic, SW Caliber 11 (base MAB 88); ø 25.6 mm, height 3.6 mm; 25 jewels; 28,800 vph; very large, astronomically precise moon phase display
Functions: hours, minutes; date, moon phase
Case: stainless steel, ø 43 mm, height 12 mm; sapphire crystal; transparent case back; water-resistant to 5 atm
Band: reptile skin, folding clasp
Remarks: mother-of-pearl dial
Price: $5,200

Schwarz Etienne SA
Route de L'Orée-du-Bois 5
CH-2300 La Chaux-de-Fonds
Switzerland

Tel.:
+41-32-967-9420

E-mail:
info@schwarz-etienne.ch

Website:
www.schwarz-etienne.com

Founded:
1902

Number of employees:
20

Annual production:
300 to 500

U.S. distributor:
Right Time
1485 S. Colorado Blvd.
Denver, CO 80222
877-470-TIME

Most important collections:
La Chaux-de-Fonds, Roma, Roswell

SCHWARZ ETIENNE

When Raffaello Radicchi talks about his business, you might think he was talking about a little shop he set up in Neuchâtel. Maybe it's the charming, lilting French still tinged with his mother tongue, even though he left his native Perugia decades earlier. Maybe it's the quick laugh, the smiling, vivacious eyes, the earthiness he exudes. It all seems easy. For example, ask him why he went into watchmaking, he'll answer: "I was allergic to the metal and could only wear a gold watch." Subtext: He could not afford a gold watch, so he founded a watch company.

Radicchi is a genuine maverick and a lone figure in this somewhat hermetic industry. He arrived in Switzerland at 18, a mason. Unable to continue his work, he retrained as a carpenter and started renovating homes, then buying and renovating, and soon he was earning some serious money. Easy-peasy. In the early aughts, an acquaintance bought up a watch brand in La Chaux-de-Fonds and suggested that Radicchi buy the building that came with it. The brand, once a big name in the industry and a supplier of movements (to Chanel, among others), had originally been founded by Paul Schwarz and Olga Etienne.

By 2008, Radicchi owned the whole package. He understood that the company needed independence to survive. Having a number of outstanding suppliers locally to partner with was a good start. But Schwarz Etienne needed movements. By 2013, he had two, and a third came in 2015. These calibers drive a series of watches, including a tourbillon, that are classical in look, yet very modern-technical, thanks to the inverted movement construction that puts the off-center microrotor on the dial.

For all its traditionalism, the brand still maintains a feeling of youthful creativity. The Roswell's case, for example, is shaped a little like a flying vessel. And if you have seven figures to spend, you can purchase a special box of seven watches honoring the days of the week, their planets, and their astrological sign.

Roswell 08 Date
Reference number: WRW11TJ32SS21AA-A
Movement: automatic, Schwarz Etienne Caliber ISE 100.11; ø 30.40 mm, height 6.30 mm; 36 jewels; 21,600 vph; inverted movement, irreversible engraved microrotor; visible date module; 86-hour power reserve
Functions: hours, minutes, subsidiary seconds; date (at 2 o'clock)
Case: stainless steel, ø 45 mm, height 12.24 mm; sapphire crystal; water-resistant to 5 atm
Band: calfskin, folding clasp
Price: $15,500

La Chaux-de-Fonds Flying Tourbillon
Reference number: WCF09TSE16SS04AA
Movement: automatic, Schwarz Etienne Caliber TSE 121.00; ø 30.40 mm, height 6.35 mm; 40 jewels; 21,600 vph; inverted movement, microrotor on dial side; 1-minute flying tourbillon; plate/bridges sandblasted and chamfered; 70-hour power reserve
Functions: hours, minutes (off-center)
Case: stainless steel, ø 44 mm, height 13.70 mm; sapphire crystal; transparent case back; water-resistant to 5 atm
Band: reptile skin, folding clasp
Price: on request

Roma Manufacture Small Second
Reference number: WRO15MA25SS01AA
Movement: automatic, Schwarz Etienne Caliber ASE 100.00; ø 30.4 mm, height 5.35 mm; 34 jewels; 21,600 vph; 96-hour power reserve
Functions: hours, minutes, subsidiary seconds
Case: stainless steel, ø 42 mm, height 12.13 mm; sapphire crystal; transparent case back; water-resistant to 5 atm
Band: reptile skin, buckle
Remarks: silver-colored dial with rhodium-blue hands
Price: $9,995
Variations: various dials

SEIKO

The Japanese watch giant is a part of the Seiko Holding Company, but the development and production of its watches are fully self-sufficient. Seiko makes every variety of portable timepiece and offers mechanical watches with both manual and automatic winding, quartz watches with battery and solar power or with the brand's own mechanical "Kinetic" power generation, as well as the groundbreaking "Spring Drive" hybrid technology. This intelligent mix of mechanical energy generation and electronic regulation is reserved for Seiko's top models.

Also in the top segment of the brand is the Grand Seiko line, a group of watches that enjoys cult status among international collectors. Only recently did the Tokyo-based company offer a large collection to the global market. Today, there are several watches with the Spring Drive technology, but most new Grand Seikos (see page 156) are conventional, mechanical hand-wound and automatic watches.

Classic Seikos are designed for tradition-conscious buyers. The Astron, however, with its automatic GPS-controlled time setting, suggests the watch of the future. In its second incarnation, the Astron is 30 percent more compact, and the energy required by the GPS system inside is supplied by a high-tech solar cell on the dial. As for the new Prospex collection, released for the fiftieth anniversary of the first Seiko diver's watches, it has an unmistakably modern look. The old protective case of the Marinemaster is now of ceramic instead of plastic.

Seiko Holdings
Ginza, Chuo, Tokyo
Japan

Website:
www.seikowatches.com

Founded:
1881

U.S. distributor:
Seiko Corporation of America
1111 Macarthur Boulevard
Mahwah, NJ 07430
201-529-5730
custserv@seikousa.com
www.seikousa.com

Most important collections/price range:
Ananta / approx. $2,400 to $8,500; Astron / approx. $1,850 to $3,400; Seiko Elite (Sportura, Premier, Velatura, Arctura) / approx. $430 to $1,500; Presage / approx. $490 to $2,600; Prospex / approx. $395 to $6,000

Astron GPS Solar Dual Time

Reference number: SSE167J1
Movement: quartz, Seiko Caliber 8X53; autonomous energy from solar cells on dial
Functions: hours, minutes, sweep seconds; additional 12-hour display (2nd time zone), world time function (40 time zones), flight mode, signal reception display, summer time display; perpetual calendar with date, weekday
Case: titanium (with hard coating), ø 46.7 mm, height 14.5 mm; ceramic bezel and lateral protection; sapphire crystal; water-resistant to 20 atm
Band: silicon, folding clasp
Remarks: comes with additional calfskin strap
Price: $2,400

Astron GPS Solar Chronograph

Reference number: SSE003J1
Movement: quartz, Seiko Caliber 8X82; autonomous energy from solar cells on dial
Functions: hours, minutes, sweep seconds; world time display (2nd time zone, GPS alignment of 40 time zones), flight mode, signal reception indicator, power reserve indicator, daylight savings indicator; chronograph; date
Case: titanium, ø 45 mm, height 13.5 mm; ceramic bezel; sapphire crystal; water-resistant to 10 atm
Band: titanium, folding clasp
Price: $2,300
Variations: various cases and dials

Astron GPS Solar World Time

Reference number: SSE161J1
Movement: quartz, Seiko Caliber 8X22; autonomous energy from solar cells on dial
Functions: hours, minutes, sweep seconds, world time function (40 time zones), flight mode, signal reception display, summer time display; perpetual calendar with date
Case: stainless steel (with hardened coating), ø 45.4 mm, height 12.4 mm; ceramic bezel; sapphire crystal; water-resistant to 10 atm
Band: stainless steel, folding clasp
Price: on request

SEIKO

Prospex Automatic Diver's Save the Ocean Special Edition
Reference number: SRPC93K1
Movement: automatic, Seiko Caliber 4R35; ø 27 mm, height 4.95 mm; 23 jewels; 21,600 vph; 41-hour power reserve
Functions: hours, minutes, sweep seconds; date
Case: stainless steel, ø 43.8 mm, height 13.4 mm; unidirectional bezel, with 0-60 scale; Plexiglas; screw-in crown; water-resistant to 20 atm
Band: stainless steel, folding clasp, with safety lock and strap extension
Price: $525

Prospex Automatic Diver's
Reference number: SRP777K1
Movement: automatic, Seiko Caliber 4R36; ø 27 mm, height 5.32 mm; 24 jewels; 21,600 vph; 41-hour power reserve
Functions: hours, minutes, sweep seconds; date, weekday
Case: stainless steel, ø 44.3 mm, height 13 mm; unidirectional bezel, with 0-60 scale; Plexiglas; screw-in crown; water-resistant to 20 atm
Band: silicon, buckle
Price: $495

Prospex Automatic Diver's PADI Special Edition
Reference number: SPB071J1
Movement: automatic, Seiko Caliber 6R15; ø 27.4 mm, height 4.95 mm; 23 jewels; 21,600 vph; antimagnetic up to 4,800 A/m; 50-hour power reserve
Functions: hours, minutes, sweep seconds; date
Case: stainless steel (with hardened coating), ø 42.6 mm, height 13.4 mm; unidirectional bezel, with 0-60 scale; sapphire crystal; screw-in crown; water-resistant to 20 atm
Band: silicon, buckle
Price: $850

Presage Automatic Chronograph
Reference number: SRQ025J1
Movement: automatic, Seiko Caliber 8R48; ø 28 mm, height 7.5 mm; 34 jewels; 28,800 vph; chronograph control by vertical clutch and column wheel; 45-hour power reserve
Functions: hours, minutes, subsidiary seconds; chronograph; date
Case: stainless steel, ø 42 mm, height 15.2 mm; sapphire crystal; transparent case back; water-resistant to 10 atm
Band: reptile skin, folding clasp
Price: $1,900

Presage Automatic Limited Edition
Reference number: SJE073J1
Movement: automatic, Seiko Caliber 6L35; height 3.7 mm; 26 jewels; 28,800 vph; 45-hour power reserve
Functions: hours, minutes, sweep seconds; date
Case: stainless steel (with hardened coating), ø 40.7 mm, height 9.8 mm; sapphire crystal; transparent case back; water-resistant to 10 atm
Band: stainless steel, folding clasp
Price: $2,200

Presage Automatic
Reference number: SRPB41J1
Movement: automatic, Seiko Caliber 4R35; ø 27 mm, height 4.95 mm; 23 jewels; 21,600 vph; antimagnetic up to 4,800 A/m; 41-hour power reserve
Functions: hours, minutes, sweep seconds; date
Case: stainless steel, ø 40.5 mm, height 11.8 mm; Plexiglas; transparent case back; water-resistant to 5 atm
Band: stainless steel, folding clasp
Price: $450

SINN

Pilot and flight instructor Helmut Sinn began manufacturing watches in Frankfurt am Main because he thought the pilot's watches on the market were too expensive. The resulting combination of top quality, functionality, and a good price-performance ratio turned out to be an excellent sales argument. There is hardly another source that offers watch lovers such a sophisticated and reasonable collection of sporty watches, many conceived to survive in extreme conditions by conforming to German DIN industrial norms.

In 1994, Lothar Schmidt took over the brand, and his product developers began looking for inspiration in other industries and the sciences. They did so out of a practical technical impulse. Research and development are consistently aimed at improving the functionality of the watches. This includes application of special Sinn technology such as moisture-proofing cases by pumping in an inert gas like argon. Other Sinn innovations include the Diapal (a lubricant-free lever escapement), the Hydro (an oil-filled diver's watch), and tegiment processing (for hardened steel and titanium surfaces). Schmidt also negotiated a partnership with the Aachen Technical University to create the Technischer Standard Fliegeruhren (TESTAF, or Technical Standards for Aviator Watches), which is housed at the Eurocopter headquarters.

Sinn recently joined forces with two German watch companies, the Sächsische Uhrentechnologie Glashütte (SUG) and the Uhren-Werke-Dresden (UWD). The latter produced the outstanding UWD 33.1 caliber with Sinn as chaperone. That movement was then used to drive the Meisterbund I, which translates as "master alliance."

All these moves have brought the company enough wherewithal to open new headquarters in the Sossenheim district of Frankfurt. The building offers nearly 25,000 square feet of space, most of which is devoted to assembly and manufacturing.

At the heart of the two-story construction is a grandiose atrium with a skylight offering lots of natural light. The roof was also turned into an open-air terrace. Customers can come to the new building to admire the entire collection of Sinn watches in beautiful surroundings, and make a purchase as well.

Sinn Spezialuhren GmbH
Im Füldchen 5-7
D-60489 Frankfurt / Main
Germany

Tel.:
+49-69-9784-14-200

Fax:
+49-69-9784-14-201

E-mail:
info@sinn.de

Website:
www.sinn.de

Founded:
1961

Number of employees:
approx. 100

Annual production:
approx. 12,500 watches

U.S. distributor:
WatchBuys
888-333-4895
www.watchbuys.com

Most important collections/price range:
Financial District, U-Models, Diapal / from approx. $700 to $27,500

936

Reference number: 936.010
Movement: automatic, Sinn Caliber SZ 05 (base ETA 7750); ø 30 mm, height 7.9 mm; 26 jewels; 28,800 vph; antimagnetic according to German Industrial Norm (DIN), up to 80,000 A/m; 42-hour power reserve
Functions: hours, minutes, subsidiary seconds; chronograph; date
Case: tegimented stainless steel, ø 43 mm, height 15 mm; sapphire crystal; screw-in crown; water-resistant to 10 atm
Band: calfskin, buckle
Price: $2,980
Variations: stainless steel bracelet ($3,240)

836

Reference number: 836.010
Movement: automatic, ETA Caliber 2892-A2; ø 25.6 mm, height 3.6 mm; 21 jewels; 28,800 vph; shock-resistant and antimagnetic according to German Industrial Norm (DIN); antimagnetic up to 80,000 A/m; 42-hour power reserve
Functions: hours, minutes, sweep seconds; date
Case: tegimented stainless steel, ø 43 mm, height 10.6 mm; sapphire crystal; screw-in crown; water-resistant to 10 atm
Band: tegimented stainless steel, folding clasp, with safety lock with extension link
Price: $2,050
Variations: calfskin strap ($1,790)

3006 Hunting Watch

Reference number: 3006.010
Movement: automatic, ETA Caliber 7751; ø 30 mm, height 7.9 mm; 25 jewels; 28,800 vph; shock-resistant and antimagnetic according to German Industrial Norm (DIN); 42-hour power reserve
Functions: hours, minutes, subsidiary seconds; additional 24-hour display (2nd time zone); chronograph; date, weekday, month, moon phase
Case: tegimented stainless steel, ø 44 mm, height 15.5 mm; sapphire crystal; transparent case back; screw-in crown; water-resistant to 20 atm
Band: calfskin, buckle
Remarks: dehumidifying technology (special gas)
Price: $3,980
Variations: stainless steel bracelet ($4,340); silicon strap ($4,100)

EZM 12

Reference number: 112.010
Movement: automatic, ETA Caliber 2836-2; ø 25.6 mm, height 5.05 mm; 25 jewels; 28,800 vph; protected from magnetic fields up to 80,000 A/m; 38-hour power reserve
Functions: hours, minutes, sweep seconds; date, weekday
Case: tegimented stainless steel, black hard coating, ø 44 mm, height 14 mm; bidirectional bezel with 0-60 scale (counting downward); crown-activated inner ring with 0-60 scale (counting upward); sapphire crystal; water-resistant to 20 atm
Band: silicon, folding clasp with extension link
Remarks: developed for emergency medics; comes with pocketknife; dehumidifying technology
Price: $3,560

EZM 10 TESTAF

Reference number: 950.011
Movement: automatic, Sinn Caliber SZ 01 (base ETA 7750); ø 30 mm, height 7.9 mm; 29 jewels; 28,800 vph; sweep minute counter, lubrication-free escapement (Diapal), shockproof and antimagnetic
Functions: hours, minutes, subsidiary seconds; additional 24-hour display; chronograph; date
Case: tegimented titanium, ø 46.5 mm, height 15.6 mm; bidirectional bezel with 0-60 scale; sapphire crystal; screw-in crown; water-resistant to 20 atm
Band: calfskin, buckle
Remarks: certified according to Technical Standard for Flyers' Watches (TESTAF); dehumidifying technology (protective gas)
Price: $5,480

857 UTC VFR

Reference number: 857.0401
Movement: automatic, ETA Caliber 2893-2; ø 25.6 mm, height 4.1 mm; 21 jewels; 28,800 vph; shockproof and antimagnetic (DIN-norm); 42-hour power reserve
Functions: hours, minutes, sweep seconds; additional 24-hour display (2nd time zone); date
Case: tegimented stainless steel, ø 43 mm, height 12 mm; sapphire crystal; screw-in crown
Band: silicon, folding clasp with safety lock and extension link
Remarks: certified according to DIN 8830 for pilots' watches; dehumidifying technology (protective gas)
Price: $2,490
Variations: without DIN certification ($2,180); without UTC ($2,080)

910 SRS

Reference number: 910.020
Movement: automatic, ETA Caliber 7750 (modified); ø 30 mm, height 8.4 mm; 25 jewels; 28,800 vph; column wheel control of chronograph functions; shock-resistant and antimagnetic according to German Industrial Norm (DIN); 46-hour power reserve
Functions: hours, minutes, subsidiary seconds; flyback chronograph; date
Case: stainless steel, ø 41.5 mm, height 15.5 mm; sapphire crystal; transparent case back; water-resistant to 10 atm
Band: calfskin, buckle
Price: $3,980
Variations: stainless steel bracelet ($4,280)

356 Sa Flieger III

Reference number: 356.0721
Movement: automatic, Sellita Caliber SW500 (modified); ø 30.4 mm, height 7.9 mm; 25 jewels; 28,800 vph; shockproof and antimagnetic (DIN-norm); finely finished movement; 42-hour power reserve
Functions: hours, minutes, subsidiary seconds; chronograph; date, weekday
Case: stainless steel, ø 38.5 mm, height 15 mm; sapphire crystal; transparent case back; water-resistant to 10 atm
Band: calfskin, buckle
Price: $2,650
Variations: acrylic crystal ($2,960); black dial ($2,580); copper-colored dial ($2,660)

104 St Sa I W

Reference number: 104.012
Movement: automatic, Sellita Caliber SW220-1; ø 25.6 mm, height 5.05 mm; 26 jewels; 28,800 vph; shockproof and antimagnetic (DIN-norm); 38-hour power reserve
Functions: hours, minutes, sweep seconds; date, weekday
Case: stainless steel, ø 41 mm, height 11.5 mm; bidirectional bezel with 0-60 scale; sapphire crystal; transparent case back; screw-in crown; water-resistant to 20 atm
Band: calfskin, buckle
Price: $1,330
Variations: black dial ($1,330); black dial and Arabic numerals ($1,330)

SINN

U212 (EZM 16)
Reference number: 212.040
Movement: automatic, Sellita Caliber SW300-1; ø 25.6 mm, height 3.6 mm; 25 jewels; 28,800 vph; shock-resistant and antimagnetic according to German Industrial Norm (DIN); 42-hour power reserve
Functions: hours, minutes, sweep seconds; date
Case: stainless steel, ø 47 mm, height 14.5 mm; unidirectional bezel, with 0-60 scale; sapphire crystal; screw-in crown; water-resistant to 100 atm
Band: stainless steel, folding clasp with extension link
Remarks: European diving gear certification; dehumidifying technology (protective gas)
Price: $2,950
Variations: silicon strap ($2,870); textile strap ($2,680)

U1 S
Reference number: 1010.020
Movement: automatic, Sellita Caliber SW200-1; ø 25.6 mm, height 4.6 mm; 26 jewels; 28,800 vph; shock-resistant and antimagnetic according to German Industrial Norm (DIN); 38-hour power reserve
Functions: hours, minutes, sweep seconds; date
Case: tegimented stainless steel (submarine steel), with black hard coating, ø 44 mm, height 14 mm; unidirectional bezel, with 0-60 scale; sapphire crystal; screw-in crown; water-resistant to 100 atm
Band: silicon, folding clasp, with extension link
Remarks: European diving gear certification
Price: $2,530
Variations: without black hard coating ($2,080)

T1 B (EZM 14)
Reference number: 1014.011
Movement: automatic, Caliber SOP A10-2A (base Soprod A10); ø 25.6 mm, height 3.6 mm; 25 jewels; 28,800 vph; shockproof and antimagnetic (DIN-norm); 42-hour power reserve
Functions: hours, minutes, sweep seconds; date
Case: pearl-blasted titanium, ø 45 mm, height 12.5 mm; unidirectional bezel, with 0-60 scale; sapphire crystal; screw-in crown; water-resistant to 100 atm
Band: silicon, folding clasp with safety lock
Remarks: EU diving certified; dehumidifying technology (protective gas)
Price: $3,440
Variations: 41 mm case ($3,240)

240 St GZ
Reference number: 240.011
Movement: automatic, Sellita Caliber SW220-1; ø 25.6 mm, height 5.05 mm; 26 jewels; 28,800 vph; shock-resistant and antimagnetic according to German Industrial Norm (DIN); 38-hour power reserve
Functions: hours, minutes, sweep seconds; date, weekday
Case: stainless steel, ø 43 mm, height 11 mm; crown-controlled inner bezel with tides indication; power reserve indicator; sapphire crystal; crown with D3 seal; water-resistant to 10 atm
Band: stainless steel, folding clasp, with safety lock and extension link
Price: $1,930
Variations: 0-60 scale ring ($1,930)

EZM 9 TESTAF
Reference number: 949.010
Movement: automatic, Sellita Caliber SW200-1; ø 25.6 mm, height 4.6 mm; 26 jewels; 28,800 vph; shock-resistant and antimagnetic according to German Industrial Norm (DIN)
Functions: hours, minutes, sweep seconds; date
Case: tegimented titanium, ø 44 mm, height 12 mm; bidirectional bezel with 0-60 scale; sapphire crystal; screw-in crown; water-resistant to 20 atm
Band: tegimented titanium, folding clasp with safety lock and extension link
Remarks: certified according to Technical Standard for Aviator Watches (TESTAF)
Price: $3,970
Variations: leather strap ($3,540)

900 Diapal
Reference number: 900.013
Movement: automatic, Sellita Caliber SW500; ø 30 mm, height 7.9 mm; 25 jewels; 28,800 vph; lubrication-free escapement (Diapal), shock-resistant and antimagnetic according to German Industrial Norm (DIN); 48-hour power reserve
Functions: hours, minutes, subsidiary seconds; additional 24-hour display (2nd time zone); chronograph; date
Case: tegimented stainless steel, ø 44 mm, height 15.5 mm; crown-controlled scale ring; sapphire crystal; screw-in crown; water-resistant to 20 atm
Band: calfskin, buckle
Remarks: dehumidifying technology (protective gas); magnetic field protection to 80,000 A/m
Price: $4,310

6000 Rose Gold
Reference number: 6000.040
Movement: automatic, Sellita SW500 (modified); ø 30.4 mm, height 7.9 mm; 26 jewels; 28,800 vph; lubrication-free escapement (Diapal), finely finished, shock-resistant, and antimagnetic according to German Industrial Norm (DIN); 42-hour power reserve
Functions: hours, minutes, subsidiary seconds; additional 12-hour display (2nd time zone); chronograph; date
Case: rose gold, ø 38.5 mm, height 16.5 mm; crown-controlled inner bezel with 0-12 scale; sapphire crystal; transparent case back; water-resistant to 10 atm
Band: reptile skin, buckle
Price: $13,750
Variations: stainless steel ($3,980)

6096
Reference number: 6096.010
Movement: automatic, ETA Caliber 2893-2; ø 25.6 mm, height 4.1 mm; 21 jewels; 28,800 vph; shock-resistant and antimagnetic according to German Industrial Norm (DIN); 42-hour power reserve
Functions: hours, minutes, sweep seconds; additional 24-hour display (2nd time zone); date
Case: stainless steel, ø 41.5 mm, height 10 mm; crown-activated scale ring, with 0-12 scale; sapphire crystal; transparent case back; water-resistant to 10 atm
Band: calfskin, buckle
Remarks: comes with additional stainless steel bracelet
Price: $2,880

1746 Heimat
Reference number: 1746.012
Movement: automatic, ETA Caliber 2892-A2; ø 25.6 mm, height 3.6 mm; 21 jewels; 28,800 vph; shock-resistant and antimagnetic according to German Industrial Norm (DIN); 42-hour power reserve
Functions: hours, minutes
Case: stainless steel, ø 42 mm, height 9.4 mm; sapphire crystal; transparent case back; water-resistant to 10 atm
Band: Alcantara leather, buckle
Price: $2,180

6200 WG Meisterbund I
Reference number: 6200.020
Movement: manually wound, Caliber UWD 33.1; ø 33 mm, height 4.2 mm; 19 jewels; 21,600 vph; weighted balance; flying spring barrel, antimagnetic according to German Industrial Norm (DIN); 55-hour power reserve
Functions: hours, minutes, subsidiary seconds
Case: white gold, ø 40 mm, height 9.3 mm; sapphire crystal; transparent case back; water-resistant to 10 atm
Band: calfskin, buckle
Price: $16,460; limited to 55 pieces

434 TW68 WG
Reference number: 434.030
Movement: quartz; shock-resistant and antimagnetic according to German Industrial Norm (DIN)
Functions: hours, minutes, sweep seconds
Case: stainless steel, ø 34 mm, height 7.8 mm; bezel in white gold, with 68 diamonds; sapphire crystal; water-resistant to 10 atm
Band: calfskin, buckle
Remarks: Q-technology (protection from electromagnetic impulses); mother-of-pearl dial
Price: $4,930
Variations: without diamonds ($1,540)

556 I B
Reference number: 556.0104
Movement: automatic, ETA Caliber 2824-2; ø 25.6 mm, height 4.6 mm; 25 jewels; 28,800 vph; shock-resistant and antimagnetic according to German Industrial Norm (DIN); 38-hour power reserve
Functions: hours, minutes, sweep seconds
Case: stainless steel, ø 38.5 mm, height 11 mm; sapphire crystal; transparent case back; screw-in crown; water-resistant to 20 atm
Band: calfskin, buckle
Price: $1,160
Variations: black dial ($1,080); mocha-colored dial ($1,160)

STOWA

When a watch brand organizes a museum for itself, it is usually with good reason. The firm Stowa may not be the biggest fish in the horological pond, but it has been around for more than eighty years, and its products are well worth taking a look at as expressions of German watchmaking culture. Stowa began in Pforzheim, then moved to the little industrial town of Rheinfelden, and now operates in Engelsbrand, a "suburb" of Pforzheim. After a history as a family-owned company, today the brand is headed by Jörg Schauer, who has maintained the goal and vision of original founder Walter Storz: delivering quality watches at a reasonable price.

Stowa is one of the few German brands to have operated without interruption since the start of the twentieth century, albeit with a new owner as of 1990. Besides all the political upheavals, it survived the quartz crisis of the 1970s, during which Europe was flooded with cheap watches from Asia and many traditional German watchmakers were put out of business. Storz managed to keep Stowa going, but even a quality fanatic has to pay a price during times of trouble: With huge input from his son, Werner, Storz restructured the company so that it was able to begin encasing reasonably priced quartz movements rather than being strictly an assembler of mechanical ones.

Schauer bought the brand in 1996. Spurred on by the success of his own eponymous line, he also steered Stowa back toward mechanical watches, taking inspiration from older Stowa timepieces but using Swiss ETA movements. But the way out of the retro trap was about to become apparent: In 2015, Schauer joined forces with Hartmut Esslinger to create the Rana (frog) model, with an almost ethereal case and a modern dial, whose dot markers (DynaDots they are called) grow larger by the hour. These new shapes are above all expressed in the new Flieger (pilot) watches.

Stowa GmbH & Co. KG
Gewerbepark 16
D-75331 Engelsbrand
Germany

Tel.:
+49-7082-942630

E-mail:
info@stowa.com

Website:
www.stowa.com

Founded:
1927

Number of employees:
20

Annual production:
around 4,500 watches

Distribution:
direct sales; please contact company in Germany; orders taken by phone Monday–Friday 9 a.m.–5 p.m. European time. Note: prices are determined according to daily exchange rate.

Flyer Verus 40

Reference number: FliegerVerus40
Movement: automatic, ETA Caliber 2824-2; ø 25.6 mm, height 4.6 mm; 25 jewels; 28,800 vph; 40-hour power reserve
Functions: hours, minutes, sweep seconds; date
Case: stainless steel, ø 40 mm, height 10.2 mm; sapphire crystal; water-resistant to 5 atm
Band: calfskin, buckle
Price: $750
Variations: high-end movement ($900); manually wound movement ($920); without date

Flier Classic 40 White

Reference number: FliegerKlassik40weiss
Movement: automatic, ETA Caliber 2824-2; ø 25.6 mm, height 4.6 mm; 25 jewels; 28,800 vph; 40-hour power reserve
Functions: hours, minutes, sweep seconds; date
Case: stainless steel, ø 40 mm, height 10.2 mm; sapphire crystal; water-resistant to 5 atm
Band: calfskin, buckle
Price: $780
Variations: high-end movement ($920); manually wound movement ($950); blued hands ($830)

Antea 390 with Date "Back to Bauhaus"

Reference number: Antea390b2bdatumweiss
Movement: automatic, ETA Caliber 2824-2; ø 25.6 mm, height 4.6 mm; 25 jewels; 28,800 vph; blued screws, hand-made rotor; 40-hour power reserve
Functions: hours, minutes, sweep seconds; date
Case: stainless steel, ø 39 mm, height 9.2 mm; sapphire crystal; transparent case back; water-resistant to 5 atm
Band: calfskin, buckle
Price: $1,090
Variations: reptile skin strap ($1,170)

Marine Classic 40 Roman White

Reference number: MarineKlassik40Römisch
Movement: automatic, ETA Caliber 2824-2; ø 25.6 mm, height 4.6 mm; 25 jewels; 28,800 vph; 40-hour power reserve
Functions: hours, minutes, sweep seconds
Case: stainless steel, ø 40 mm, height 10.2 mm; sapphire crystal; water-resistant to 5 atm
Band: calfskin, buckle
Price: $750
Variations: high-end movement ($900); manually wound movement ($925); date function

Marine Original

Reference number: MarineOriginalpolweissarabisch
Movement: manually wound, ETA Caliber 6498-1; ø 36.6 mm, height 4.5 mm; 17 jewels; 18,000 vph; screw balance, swan-neck fine adjustment, côtes de Genève, blued screws; 46-hour power reserve
Functions: hours, minutes, subsidiary seconds
Case: stainless steel, ø 41 mm, height 12 mm; sapphire crystal; transparent case back; water-resistant to 5 atm
Band: calfskin, buckle
Price: $1,570
Variations: reptile skin strap ($1,700); stainless steel bracelet ($1,700)

Prodiver White

Reference number: Prodiverweiss
Movement: automatic, ETA Caliber 2824-2; ø 25.6 mm, height 4.6 mm; 25 jewels; 28,800 vph; 40-hour power reserve
Functions: hours, minutes, sweep seconds; date
Case: titanium, ø 42 mm, height 15.6 mm; unidirectional bezel, with 0-60 scale; sapphire crystal; screw-in crown; water-resistant to 100 atm
Band: rubber, double folding clasp, with safety lock and extension link
Price: $1,500
Variations: titanium bracelet ($1,725)

Antea 1919 White

Reference number: Antea1919weiss
Movement: automatic, ETA Caliber 2824-2; ø 25.6 mm, height 4.6 mm; 25 jewels; 28,800 vph; 40-hour power reserve
Functions: hours, minutes, sweep seconds
Case: stainless steel, ø 39 mm, height 9.2 mm; sapphire crystal; transparent case back; water-resistant to 5 atm
Band: calfskin, buckle
Price: $750
Variations: high-end movement ($900); manually wound movement ($925); black dial

Chronograph 1938 Bronze

Reference number: chronograph1938bronzepoliert
Movement: automatic, ETA Caliber 7753; ø 30 mm, height 7.9 mm; 27 jewels; 28,800 vph; 48-hour power reserve
Functions: hours, minutes, subsidiary seconds; chronograph
Case: stainless steel, ø 41 mm, height 14.7 mm; sapphire crystal; transparent case back; water-resistant to 5 atm
Band: calfskin, buckle
Price: $2,200
Variations: manually wound movement ($2,550)

Partitio Black Red Second

Reference number: PartitioschwarzroteSek
Movement: manually wound, ETA Caliber 2804-2; ø 25.6 mm, height 3.35 mm; 17 jewels; 28,800 vph; 42-hour power reserve
Functions: hours, minutes, sweep seconds
Case: stainless steel, ø 37 mm, height 9.8 mm; sapphire crystal; transparent case back; water-resistant to 5 atm
Band: calfskin, buckle
Price: $1,000
Variations: white dial; automatic movement

TAG HEUER

Measuring speed accurately in ever greater detail was always the goal of TAG Heuer. The brand also established numerous technical milestones, including the first automatic chronograph caliber with a microrotor (created in 1969 with Hamilton-Büren, Breitling, and Dubois Dépraz). Of more recent vintage is the fascinating mechanical movement V4 with its belt-driven transmission, unveiled in a limited edition. At the same time TAG Heuer released its first chronograph with an in-house movement, Caliber 1887, the basis of which was an existing chronograph movement by Seiko. Some of the components are made by the company itself in Switzerland, while assembly is done entirely in-house.

Lately, TAG Heuer has increased its manufacturing capacities to meet the strong and growing demand and to maintain its independence. It also serves as an extended workbench for companion brands Zenith and Hublot, also part of the LVMH Group.

TAG Heuer has continued to break world speed records for mechanical escapements. The Caliber 360 combined a standard movement with a 360,000-vph (50-Hz) chronograph mechanism able to measure 100ths of a second. In 2011, the Mikrograph 1/100th brought time display and measurement on a single plate. Shortly after, the Mikrotimer Flying 1000 broke the 1,000th of a second barrier. A year later, the Mikrogirder 2000 doubled the frequency using a vibrating metal strip instead of a balance wheel. The MikrotourbillonS features a separate chronograph escapement driven at a record-breaking 360,000 vph.

But these flights of fancy slowed when, at the end of 2014, the new LVMH coordinator Jean-Claude Biver devised a new strategy for TAG Heuer to cut back on the top end of the pricing scale. Guy Sémon, the new CEO, prescribed a return to former pricings, with a broader choice of models including a connected watch. The new portfolio includes a tourbillon at under $16,000 in the Carrera family, and a revived classic, the Autovia, which was the first chronograph by Heuer with a rotating bezel for remembering times or as a second time zone. The Monaco Caliber 11 is now in a charming re-release in the Gulf stable colors. The company has also started cooperations with the watch personalization specialist George Bamford and Aston Martin, or rather its Formula-One team Red Bull Racing.

TAG Heuer
Branch of LVMH SA
6a, rue L.-J.-Chevrolet
CH-2300 La Chaux-de-Fonds
Switzerland

Tel.:
+41-32-919-8164

Fax:
+41-32-919-9000

E-mail:
info@tagheuer.com

Website:
www.tagheuer.com

Founded:
1860

Number of employees:
1,600 employees internationally

U.S. distributor:
TAG Heuer/LVMH Watch & Jewelry USA
966 South Springfield Avenue
Springfield, NJ 07081
973-467-1890

Most important collections/price range:
TAG Heuer Formula 1, Aquaracer, Link, Carrera, Connected, Monaco, Heritage / from approx. $1,300 to $20,000

Formula 1 Caliber 16
Reference number: CAZ2010.BA0876
Movement: automatic, TAG Heuer Caliber 16 (base ETA 7750); ø 30.4 mm, height 7.9 mm; 25 jewels; 28,800 vph
Functions: hours, minutes, subsidiary seconds; chronograph; date
Case: stainless steel, ø 44 mm, height 15 mm; sapphire crystal; screw-in crown; water-resistant to 20 atm
Band: stainless steel, folding clasp
Price: $2,800

Aquaracer 300M Caliber 5
Reference number: WAY208D.FC8221
Movement: automatic, TAG Heuer Caliber 5 (base ETA 2824-2); ø 26 mm, height 4.6 mm; 25 jewels; 28,800 vph
Functions: hours, minutes, sweep seconds; date
Case: titanium with black PVD coating, ø 43 mm; unidirectional bezel with ceramic insert, 0-60 scale; sapphire crystal; screw-in crown; water-resistant to 30 atm
Band: textile, buckle
Price: $3,550

Link Caliber 5
Reference number: WBC2111.BA0603
Movement: automatic, TAG Heuer Caliber 5 (base ETA 2824-2); ø 26 mm, height 4.6 mm; 25 jewels; 28,800 vph
Functions: hours, minutes, sweep seconds; date
Case: stainless steel, ø 41 mm; sapphire crystal; water-resistant to 10 atm
Band: stainless steel, folding clasp
Price: $2,900

Link Caliber 17

Reference number: CBC2110.BA0603
Movement: automatic, TAG Heuer Caliber 17 (base ETA 2894-2); ø 28.6 mm, height 6.1 mm; 37 jewels; 28,800 vph; 42-hour power reserve
Functions: hours, minutes, subsidiary seconds; chronograph; date
Case: stainless steel, ø 41 mm; sapphire crystal; transparent case back; water-resistant to 10 atm
Band: stainless steel, folding clasp
Price: $4,500
Variations: blue dial

Carrera Caliber 5 Date

Reference number: WAR211A.BA0782
Movement: automatic, TAG Heuer Caliber 5 (base ETA 2824-2); ø 26 mm, height 4.6 mm; 25 jewels; 28,800 vph
Functions: hours, minutes, sweep seconds; date
Case: stainless steel, ø 39 mm, height 12 mm; sapphire crystal; transparent case back; water-resistant to 10 atm
Band: stainless steel, folding clasp
Price: $2,400
Variations: various dial colors; reptile skin strap ($2,400)

Carrera Caliber 5 Day-Date

Reference number: WAR201E.FC6292
Movement: automatic, TAG Heuer Caliber 5 (base ETA 2836-2); ø 26 mm, height 5.05 mm; 25 jewels; 28,800 vph
Functions: hours, minutes, sweep seconds; date, weekday
Case: stainless steel, ø 41 mm, height 13 mm; sapphire crystal; transparent case back; water-resistant to 10 atm
Band: reptile skin, folding clasp
Price: $2,600
Variations: various dials; stainless steel bracelet ($2,600)

Carrera Caliber 16

Reference number: CV201AP.FC6429
Movement: automatic, TAG Heuer Caliber 16 (base ETA 7750); ø 30 mm, height 7.9 mm; 25 jewels; 28,800 vph; 42-hour power reserve
Functions: hours, minutes, subsidiary seconds; chronograph; date
Case: stainless steel, ø 41 mm; bezel with ceramic inlay; sapphire crystal; screw-in crown; water-resistant to 10 atm
Band: calfskin, folding clasp
Price: $4,350

Carrera Caliber Heuer 01

Reference number: CAR201Z.BA0714
Movement: automatic, TAG Heuer Caliber Heuer 01; ø 29.3 mm, height 7.13 mm; 39 jewels; 28,800 vph
Functions: hours, minutes, subsidiary seconds; chronograph; date
Case: stainless steel, ø 43 mm, height 16 mm; ceramic bezel; sapphire crystal; transparent case back; water-resistant to 10 atm
Band: stainless steel, folding clasp
Price: $5,150

Carrera Caliber Heuer 01

Reference number: CAR2A1T.FT6052
Movement: automatic, TAG Heuer Caliber Heuer 01; ø 29.3 mm, height 7.13 mm; 39 jewels; 28,800 vph
Functions: hours, minutes, subsidiary seconds; chronograph; date
Case: stainless steel, ø 45 mm, height 16 mm; ceramic bezel; sapphire crystal; transparent case back; water-resistant to 10 atm
Band: rubber, folding clasp
Price: $5,300

TAG HEUER

Carrera Caliber Heuer 01
Reference number: CAR2A1W.BA0703
Movement: automatic, TAG Heuer Caliber Heuer 01; ø 29.3 mm, height 7.13 mm; 39 jewels; 28,800 vph
Functions: hours, minutes, subsidiary seconds; chronograph; date
Case: stainless steel, ø 45 mm, height 16 mm; ceramic bezel; sapphire crystal; transparent case back; water-resistant to 10 atm
Band: stainless steel, folding clasp
Price: $5,450

Carrera Caliber Heuer 01 Black Matte Ceramic
Reference number: CAR2A91.BH0742
Movement: automatic, TAG Heuer Caliber Heuer 01; ø 29.3 mm, height 7.13 mm; 39 jewels; 28,800 vph
Functions: hours, minutes, subsidiary seconds; chronograph; date
Case: ceramic, ø 45 mm, height 16 mm; ceramic bezel; sapphire crystal; transparent case back; water-resistant to 10 atm
Band: ceramic, folding clasp
Price: $6,300

Carrera Caliber Heuer 01 Special Edition Aston Martin
Reference number: CAR2A1AB.FT6163
Movement: automatic, TAG Heuer Caliber Heuer 01; ø 29.3 mm, height 7.13 mm; 39 jewels; 28,800 vph
Functions: hours, minutes, subsidiary seconds; chronograph; date
Case: stainless steel, ø 45 mm, height 16 mm; ceramic bezel; sapphire crystal; transparent case back; water-resistant to 10 atm
Band: calfskin, folding clasp
Price: $5,900

Carrera Caliber Heuer 01 Aston Martin Red Bull Racing
Reference number: CAR2A1N.FT6100
Movement: automatic, TAG Heuer Caliber Heuer 01; ø 29.3 mm, height 7.13 mm; 39 jewels; 28,800 vph
Functions: hours, minutes, subsidiary seconds; chronograph; date
Case: stainless steel, ø 45 mm, height 16 mm; ceramic bezel; sapphire crystal; transparent case back; water-resistant to 10 atm
Band: calfskin, folding clasp
Price: $5,850
Variations: stainless steel bracelet ($5,950)

Carrera Caliber Heuer 02 GMT
Reference number: CBG2A1Z.BA0658
Movement: automatic, TAG Heuer Caliber Heuer 02; ø 31 mm, height 6.9 mm; 33 jewels; 28,800 vph; 80-hour power reserve
Functions: hours, minutes, subsidiary seconds; additional 24-hour display (2nd time zone); chronograph; date
Case: stainless steel, ø 45 mm; bezel with ceramic inlay; sapphire crystal; transparent case back; water-resistant to 10 atm
Band: stainless steel, folding clasp
Price: $6,150
Variations: rubber strap ($5,950)

Carrera Caliber Heuer 02 Tourbillon C.O.S.C. Black Titanium
Reference number: CAR5A8Y.FC6377
Movement: automatic, TAG Heuer Caliber Heuer 02 T; ø 31 mm, height 6.9 mm; 33 jewels; 28,800 vph; 1-minute tourbillon; COSC-certified chronometer
Functions: hours, minutes; chronograph; date
Case: titanium with black titanium carbide coating, ø 45 mm; sapphire crystal; water-resistant to 10 atm
Band: reptile skin, folding clasp
Price: $15,900

TAG HEUER

Carrera Calibre Heuer 02 Tourbillon "Tête de Vipère"
Reference number: CAR5A93.FC6442
Movement: automatic, TAG Heuer Caliber Heuer 02 T; ø 31 mm, height 6.9 mm; 33 jewels; 28,800 vph; 1-minute tourbillon; certified chronometer (Besançon observatory)
Functions: hours, minutes; chronograph
Case: ceramic, ø 45 mm; sapphire crystal; transparent case back; water-resistant to 10 atm
Band: reptile skin, folding clasp
Remarks: "viper head" is seal of recently revived chronometer testing site at Besançon observatory
Price: $20,400; limited to 155 pieces

Connected Modular 41
Reference number: SBF818000.10BF0609
Movement: quartz, Intel Core Duo microprocessor; Android OS; near-field communication with smartphone for data sharing and reciprocal function control
Functions: hours, minutes, sweep seconds; other displays and functions through smartphone connection; chronograph; date
Case: titanium, ø 41 mm, height 13.2 mm; unidirectional ceramic bezel, 0-60 scale; sapphire crystal; water-resistant to 5 atm
Band: stainless steel, folding clasp
Remarks: removable smartwatch module
Price: $2,100

Connected Modular 45 Aston Martin Red Bull Racing
Reference number: SBF8A8028.11EB0147
Movement: quartz, Intel Core Duo microprocessor; Android OS; near-field communication with smartphone for data sharing and reciprocal function control
Functions: hours, minutes, sweep seconds; more displays and functions through smartphone connection; chronograph; date
Case: titanium, ø 45 mm, height 13.7 mm; unidirectional ceramic bezel, 0-60 scale; sapphire crystal; water-resistant to 5 atm
Band: calfskin, folding clasp
Price: $1,850

Heuer Monaco Caliber 11
Reference number: CAW211P.FC6356
Movement: automatic, TAG Heuer Caliber 11 (base Sellita SW300 with 2006 module of Dubois Dépraz); ø 30 mm, height 7.3 mm; 59 jewels; 28,800 vph
Functions: hours, minutes, subsidiary seconds; chronograph; date
Case: stainless steel, 39 × 39 mm, height 14.5 mm; sapphire crystal; transparent case back; water-resistant to 10 atm
Band: calfskin, folding clasp
Price: $5,900

Heuer Monaco Caliber 11 Special Edition Gulf
Reference number: CAW211R.FC6401
Movement: automatic, TAG Heuer Caliber 11 (base Sellita SW300 with 2006 module of Dubois Dépraz); ø 30 mm, height 7.3 mm; 59 jewels; 28,800 vph
Functions: hours, minutes, subsidiary seconds; chronograph; date
Case: stainless steel, 39 × 39 mm, height 14.5 mm; sapphire crystal; transparent case back; water-resistant to 10 atm
Band: calfskin, folding clasp
Price: $5,900

Heuer Autavia Caliber 02
Reference number: CBE2110.BA0687
Movement: automatic, TAG Heuer Caliber Heuer 02; ø 31 mm, height 6.9 mm; 33 jewels; 28,800 vph; 80-hour power reserve
Functions: hours, minutes, subsidiary seconds; chronograph; date
Case: stainless steel, ø 42 mm, height 15.5 mm; bidirectional bezel with aluminum insert and 0-12 scale; sapphire crystal; transparent case back; water-resistant to 10 atm
Band: stainless steel, folding clasp
Price: $5,150
Variations: calfskin strap ($5,300)

TEMPTION

Temption has been operating under the leadership of Klaus Ulbrich since 1997. Ulbrich is an engineer with special training in the construction of watches and movements, and right from the start, he intended to develop timekeepers that were modern in their esthetics but not subject to the whims of zeitgeist. Retro watches would have no place in his collections. The design behind all Temption models is inspired more by the Bauhaus or the Japanese concept of wabi sabi. Reduction to what is absolutely necessary is the golden rule here. Beauty emerges from clarity, or in other words, less is more.

Ulbrich sketches all the watches himself. Some of the components are even made in-house, but all the pieces are assembled in the company facility in Herrenberg, a town just to the east of the Black Forest. The primary functions are always easy to read, even in low light. The company logo is discreetly included on the dial.

Ulbrich works according to a model he calls the "information pyramid." Hours and minutes are at the tip, with all other functions subordinated. To maintain this hierarchy, the dials are dark, the date windows are in the same hue, and all subdials are not framed in any way. The most unimportant information for reading time comes at the end of the "pyramid"; it is shiny black on black: the logo can only be identified in lateral light.

The Cameo rectangular model is a perfect example of Ulbrich's esthetic ideas and his consistent technological approach: Because rectangular sapphire crystals can hardly be made water-resistant, the Cameo's crystal is chemically bonded to the case and water-resistant to 10 atm. The frame for the sapphire was metalized inside to hide the bonded edge. The overall look is one of stunning simplicity and elegance. With the CGK205 chronograph, Ulbrich took the concept out of the case. Whether it be the leather strap or the stainless steel bracelet, the watch's attachment is seamlessly integrated into the case, without any visible split.

Temption GmbH
Raistinger Str. 46
D-71083 Herrenberg
Germany

Tel.:
+49-7032-977-954

Fax:
+49-7032-977-955

E-mail:
ftemption@aol.com

Website:
www.temption.info

Founded:
1997

Number of employees:
4

Annual production:
700 watches

U.S. distributor:
TemptionUSA
Debby Gordon
2053 North Bridgeport Drive
Fayetteville, AR 72704
888-400-4293
temptionusa@sbcglobal.net

Most important collections/price range:
automatics (three-hand), GMT, chronographs, and chronographs with complications / approx. $1,900 to $4,200

CM05
Reference number: CM05A10SST
Movement: automatic, Temption Caliber T15.1 (base Soprod A10, or on request with a Caliber ETA 2892-A2); ø 25.6 mm, height 3.6 mm; 21 jewels; 28,800 vph; finely finished movement; 42-hour power reserve
Functions: hours, minutes, sweep seconds; date
Case: stainless steel, ø 42 mm, height 10.8 mm; sapphire crystal; transparent case back; screw-in crown; water-resistant to 10 atm
Band: stainless steel, double folding clasp with safety lock
Price: $2,400

Chronograph CGK205 V2
Reference number: 205V2316BSST
Movement: automatic, Temption Caliber T18.1 (base ETA 7751); ø 30 mm, height 7.8 mm; 25 jewels; 28,800 vph; finely finished movement; 42-hour power reserve
Functions: hours, minutes; second 24-hour display; chronograph; full calendar with date, weekday, month, moon phase
Case: stainless steel, ø 43 mm, height 14 mm; sapphire crystal; transparent case back; screw-down crown and pusher with colored cabochons; water-resistant to 10 atm
Band: stainless steel, double folding clasp
Remarks: comes with additional textile strap
Price: $3,540

Cameo-B
Reference number: CAMBLBFS151
Movement: automatic, Temption Caliber T15.1 (base Soprod A10); ø 25.6 mm, height 3.6 mm; 21 jewels; 28,800 vph; finely finished movement; 42-hour power reserve
Functions: hours, minutes, sweep seconds; date
Case: stainless steel, 37 × 41 mm, height 9.9 mm; sapphire crystal; transparent case back; screw-in crown; water-resistant to 10 atm
Band: calfskin, double folding clasp
Price: $1,750

CLAUDE MEYLAN
VALLÉE DE JOUX

Tortue de Joux

With its pure and timeless curves associated with a proven mechanical, "Tortue de Joux" introduces a new sculpture of time.

TISSOT

The Swiss watchmaker Tissot was founded in 1853 in the town of Le Locle in the Jura mountains. In the century that followed, it gained international recognition for its Savonnette pocket watch. And even when the wristwatch became popular in the early twentieth century, time and again Tissot managed to attract attention to its products. To this day, the Banana Watch of 1916 and its first watches in the art deco style (1919) remain design icons of that epoch. The watchmaker has always been at the top of its technical game as well: The first antimagnetic watch (1930), the first mechanical plastic watch (Astrolon, 1971), and its touch-screen T-Touch (1999) all bear witness to Tissot's remarkable capacity for finding unusual and modern solutions.

Today, Tissot belongs to the Swatch Group and, with its wide selection of quartz and inexpensive mechanical watches, serves as the group's entry-level brand. Within this price segment, Tissot offers something special for the buyer who values traditional watchmaking but is not of limitless financial means. The brand has been cultivating a sportive image of late, expanding into everything from basketball to superbike racing, from ice hockey to fencing—and water sports, of course. Partnerships with several NBA teams have been signed, notably with the Houston Rockets, Chicago Bulls, and Washington Wizards in October 2018. The chronograph Couturier line is outfitted with the new ETA chronograph caliber C01.211. This caliber features a number of plastic parts: another step in simplifying, and lowering the cost of, mechanical movements.

Increasingly, as well, a number of Tissot models are being equipped with silicon hairsprings, which are notorious for outstanding isochronous oscillation as well as imperviousness to magnetic fields and changes in temperature. And for the buyer, it means only a slight increase in pricing.

Tissot SA
Chemin des Tourelles, 17
CH-2400 Le Locle
Switzerland

Tel.:
+41-32-933-3111

Fax:
+41-32-933-3311

E-mail:
info@tissot.ch

Website:
www.tissot.ch

Founded:
1853

U.S. distributor:
Tissot
The Swatch Group (U.S.), Inc.
703 Waterford Way
Suite 450
Miami, FL 33126
www.us.tissotshop.com

Most important collections/price range:
Ballade / from $925; T-Touch / from $575; NBA Collection / from $395; Chemin des Tourelles / from $795; Seastar from $695; Swissmatic from $395

Seastar Automatic

Reference number: T120.407.17.041.00
Movement: automatic, Tissot Powermatic 80 (base ETA 2824-2); ø 25.6 mm, height 4.6 mm; 23 jewels; 21,600 vph; 80-hour power reserve
Functions: hours, minutes, sweep seconds; date
Case: stainless steel, ø 43 mm, height 12.7 mm; unidirectional bezel with ceramic inlay, 0-60 scale; sapphire crystal; water-resistant to 30 atm
Band: rubber, buckle
Price: $695

Chrono XL

Reference number: T116.617.36.057.00
Movement: quartz
Functions: hours, minutes, subsidiary seconds; chronograph; date
Case: stainless steel with black PVD coating, ø 45 mm, height 11.02 mm; sapphire crystal; water-resistant to 10 atm
Band: calfskin, buckle
Price: $350

Everytime Swissmatic

Reference number: T109.407.16.031.00
Movement: automatic, ETA Caliber C15.111; ø 31.9 mm, height 5.77 mm; 19 jewels; 21,600 vph; 72-hour power reserve
Functions: hours, minutes, sweep seconds; date
Case: stainless steel, ø 40 mm, height 11.62 mm; sapphire crystal; water-resistant to 3 atm
Band: calfskin, buckle
Price: $395
Variations: black dial

TISSOT

Heritage Petite Seconde
Reference number: T119.405.16.037.01
Movement: manually wound, ETA Caliber 6498-1; ø 36.6 mm, height 4.5 mm; 17 jewels; 21,600 vph; 38-hour power reserve
Functions: hours, minutes, subsidiary seconds
Case: stainless steel, ø 42 mm, height 11.35 mm; sapphire crystal; water-resistant to 5 atm
Band: calfskin, double folding clasp
Price: $995

Le Locle
Reference number: T006.407.11.033.00
Movement: automatic, Tissot Powermatic 80 (base ETA 2824-2); ø 25.6 mm, height 4.6 mm; 23 jewels; 21,600 vph; 80-hour power reserve
Functions: hours, minutes, sweep seconds; date
Case: stainless steel, ø 39.3 mm, height 9.75 mm; sapphire crystal; transparent case back; water-resistant to 3 atm
Band: stainless steel, double folding clasp
Price: $630
Variations: black dial; calfskin strap ($575)

Chemin des Tourelles Powermatic 80
Reference number: T099.407.16.048.00
Movement: automatic, Tissot Powermatic 80 (base ETA 2824-2); ø 25.6 mm, height 4.6 mm; 23 jewels; 21,600 vph; 80-hour power reserve
Functions: hours, minutes, sweep seconds; date
Case: stainless steel, ø 42 mm, height 10.89 mm; sapphire crystal; transparent case back; water-resistant to 5 atm
Band: calfskin, double folding clasp
Price: $795

V8 Swissmatic
Reference number: T106.407.11.031.00
Movement: automatic, ETA Caliber C15.111; ø 27.4 mm, height 5.77 mm; 19 jewels; 21,600 vph; 72-hour power reserve
Functions: hours, minutes, sweep seconds; date
Case: stainless steel, ø 42.5 mm; sapphire crystal; transparent case back; water-resistant to 10 atm
Band: stainless steel, folding clasp
Price: $450

Heritage Visodate
Reference number: T118.410.11.057.00
Movement: quartz
Functions: hours, minutes, subsidiary seconds; date
Case: stainless steel, ø 40 mm, height 10.66 mm; sapphire crystal; water-resistant to 3 atm
Band: stainless steel, jewelry clasp
Price: $350

T-Touch Expert Solar II
Reference number: T110.420.47.041.00
Movement: quartz, multifunctional movement with LCD display and solar cell for separate energy source
Functions: hours, minutes, additional 12-hour display (2nd time zone), barometer, altimeter, and altitude difference meter, compass, regatta function, 2 alarms; chronograph with countdown timer; perpetual calendar with date, weekday, weeks of the year
Case: titanium, ø 45 mm, height 13.1 mm; ceramic bezel; sapphire crystal; water-resistant to 10 atm
Band: silicon, buckle
Remarks: displays and functions controlled by touching the sapphire crystal
Price: $1,195

TOWSON WATCH COMPANY

After over forty years repairing high-grade watches, repeaters, and chronographs, and making his own tourbillons, George Thomas, a master watchmaker, met Hartwig Balke, a graduate in mechanical engineering and also a talented watchmaker, by chance in a bar in Annapolis. The two men, each well on their way to retirement, decided to turn their passion into a business and, in 2000, founded the Towson Watch Company.

Thomas's first tourbillon pocket watches are displayed at the National Watch and Clock Museum in Columbia, Pennsylvania. In 1999, Balke made his first wrist chronograph, the STS-99 Mission, for a NASA astronaut and mission specialist. It was worn during the first shuttle mission in the new millennium, in the year 2000. The two also restored one of the world's oldest watches, one belonging to Philip Melanchton. In 2009, Thomas was invited to open up a pocket watch belonging to President Lincoln, and revealed a secret message engraved by a servicing watchmaker and Union supporter working in Maryland: "Jonathan Dillon April 13-1861 Fort Sumpter [sic] was attacked by the rebels on the above date J Dillon."

Towson timepieces pay tribute to local sites, like the Choptank or Potomac rivers. The timepieces are imaginative, a touch retro, a bit nostalgic perhaps, and very personal—not to mention affordable. A number of chronographs give the brand a sportive look. For the Dress Chronograph, Towson recruited the German watchmaker and dial specialist Jochen Benzinger.

Their local commitment is also shared by entrepreneur and former University of Maryland football captain Kevin Plank, who launched the technological sports apparel company Under Armour. In early 2016, the company announced it had bought a 25 percent stake in Towson, to boost its market presence and ensure its future. The two founders had been thinking of succession. Those concerns have now been laid to rest: "The brand will continue to grow and thrive for a long time to come," they told the *Baltimore Sun*.

Towson Watch Co.
502 Dogwood Lane
Towson, MD 21286

Tel.:
410-823-1823

Fax:
410-823-8581

E-mail:
towsonwatchco@aol.com

Website:
www.twcwatches.com

Founded:
2000

Number of employees:
4

Annual production:
200 watches

Distribution:
retail

Most important collections/price range:
Skipjack GMT / approx. $2,950; Mission / approx. $2,500; Potomac / approx. $2,000; Choptank / approx. $4,500; Martin / approx. $3,950 / custom design / $10,000 to $35,000

14-kt Gold Potomac

Reference number: GP 001-14K
Movement: manually wound, Soprod Unitas Caliber 6497; diameter 37.2 mm; height 4.5 mm; 17 jewels; 18,000 vph; swan-neck fine adjustment; barley and solar guilloché on the dial, rhodium-plated dial; skeletonized movement
Functions: hours, minutes, subsidiary seconds
Case: rose gold; ø 42 mm, height 12.5 mm; sapphire crystal; transparent case back; water-resistant to 3 atm
Band: reptile skin, 14k rose gold buckle
Price: $23,500.

Gold Dress Watch

Reference number: TWC001-WG
Movement: manually wound, Soprod Unitas Caliber 6498; ø 37.2 mm; height 4.5 mm; 17 jewels; 18,000 vph; skeletonized, engraved and coated bridges: swan-neck fine adjustment
Functions: hours, minutes, subsidiary seconds
Case: white gold, 45 × 45 mm, height 12.5 mm; sapphire crystal, screw-down transparent case back, water-resistant to 3 atm
Band: reptile skin, white gold buckle
Price: $35,000

Potomac

Reference number: PO250-S
Movement: manually wound, Soprod Unitas Caliber 6498; ø 37.2 mm; height 4.5 mm; 17 jewels; 18,000 vph
Functions: hours, minutes, subsidiary seconds
Case: stainless steel, ø 42 mm, height 12.5 mm; domed sapphire crystal; screw-down transparent back; water-resistant to 3 atm
Band: calfskin, buckle
Price: $1,995
Variations: black dial with gold numerals and black calfskin band; stainless steel mesh bracelet ($2,345)

TOWSON WATCH COMPANY

Martin M-130
Reference number: CC100
Movement: automatic, ETA Caliber 7750 Valjoux, ø 30 mm; height 7.9 mm; 25 jewels; 28,800 vph; fine finishing with côtes de Genève.
Functions: hours, minutes, subsidiary seconds, chronograph, date
Case: stainless steel, ø 42 mm, height 13.5 mm, sapphire crystal, screw-down back with engraving, water-resistant to 5 atm
Band: leather, folding clasp
Price: $3,950
Variations: mesh stainless steel bracelet ($4,250)

Dress Chronograph
Reference number: BCH 25
Movement: automatic, Caliber 7750 Valjoux; diameter ø 30 mm; height 7.9 mm; 21 jewels; 28,800 vph; finely finished with côtes de Genève
Functions: hours, minutes, subsidiary seconds; chronograph; date
Case: stainless steel, ø 42 mm, height 15.8 mm; sapphire crystal; screw-down transparent case back; water-resistant to 5 atm
Band: reptile skin, folding clasp
Remarks: elaborate silver dial with guilloché by Jochen Benzinger
Price: $8,250
Variations: mesh stainless steel bracelet ($8,600)

Mission Moon SC
Reference number: MM250-CS
Movement: automatic ETA Caliber 7751; ø 25.6 mm, height 3.6 mm; 21 jewels; 28,800 vph; fine finishing with côtes de Genève.
Functions: hours, minutes, subsidiary seconds; weekday, month, date; moon phase; 24-hour display; chronograph
Case: stainless steel, 40 mm, height 13.5 mm, sapphire crystal, screw-down back with engraving, water-resistant to 5 atm
Band: calfskin, orange stitching, folding clasp
Price: $4,160
Variations: stainless steel bracelet ($4,460)

Pride II
Reference number: PR250-S
Movement: Automatic Caliber ETA 2892A2; ø 25.6 mm, height 3.6 mm; 21 jewels; 28,800 vph; fine finish with côtes de Genève.
Functions: hours, minutes, sweep seconds; date
Case: stainless steel; shield shape; ø 39 mm × 44 mm; sapphire crystal; screw-down back with engraving of Pride of Baltimore II; water-resistant to 5 atm
Band: calfskin, folding clasp
Price: $4,150
Variations: white dial and calfskin strap with folding clasp; stainless steel mesh bracelet ($3,850)

Skipjack GMT
Reference number: SKJ100-S
Movement: Automatic Caliber ETA 2893-2; ø 25.6 mm; height 4.1 mm; 21 jewels; 28,800 vph; fine finish with côtes de Genève
Functions: hours, minutes, sweep seconds; date; 24-hour adjustable hand
Case: stainless steel, cannelage; ø 41.5 mm; sapphire crystal; screw-down transparent back with sapphire crystal; water-resistant to 5 atm
Band: calfskin, folding clasp
Price: $2,950
Variations: black dial with rhodium-plated numerals and leather band with deployment clasp; stainless steel bracelet ($3,250)

Choptank Moon Chrono Special
Reference number: CT025-G
Movement: automatic, ETA Caliber 7751 Valjoux; ø 30 mm; height 7.9 mm; 25 jewels; 28,800 vph; fine finishing with côtes de Genève
Functions: hours, minutes, subsidiary seconds; weekday, month, date; moon phase; 24-hour display; chronograph
Case: stainless steel; 40 mm × 44 mm; height 13.5 mm; sapphire crystal at front; transparent screw-down back; water-resistant to 5 atm
Band: reptile skin with folding clasp
Price: $8,500
Variations: mesh stainless steel bracelet ($8,850)

TUDOR

Montres Tudor SA
Rue François-Dussaud 3-7
Case postale 1755
CH-1211 Geneva
Switzerland

Tel.:
+41-22-302-2200

Fax:
+41-22-300-2255

E-mail:
info@tudorwatch.com

Website:
www.tudorwatch.com

Founded:
1946

U.S. distributor:
Tudor Watch U.S.A., LLC
665 Fifth Avenue
New York, NY 10022
212-897-9900; 212-371-0371 (fax)
www.tudorwatch.com

Most important collections/price range:
Black Bay / $3,350 to $4,975; Heritage / $2,625 to $6,075; Pelagos / $4,400; 1926 / $1,675 to $3,400

The Tudor brand came out of the shadow cast by its "big sister" Rolex in 2007 and worked hard to develop its own personality. The strategy focuses on distinctive models that draw inspiration from the brand's rich past but remain in the "affordable quality watch segment."

Rolex founder Hans Wilsdorf started Tudor in 1946 as a second brand in order to offer the legendary reliability of his watches to a broader public at a more affordable price. To this day, Tudor still benefits from the same industrial platform as Rolex, especially in the area of cases and bracelets, assembly, and quality assurance, not to mention distribution and after-sales. However, the movements themselves are usually delivered by ETA and "Tudorized" according to the company's own esthetic and technical criteria.

In the era of vintage and retro, it's no wonder that the brand has started tapping into its own treasure trove of icons. Following the success of the Heritage Black Bay diver's watch, based on a 1954 model, came the turn of the blue-highlighted 1973 Chronograph Montecarlo. In 2014, Tudor completed the Heritage collection with the Ranger, a sports watch with an urban-adventurer feel, inspired by the same "tool watch" from the 1960s. In a bit of sibling rivalry, the brand has now come out with its own caliber, designed and built in-house. The MT-5621 made its debut in the simple North Flag and was built as a three-hander (MT-5612) for the Pelagos models. Two other caliber iterations are used for the new Black Bay models.

The M5601/5602 calibers, with three hands and a date was followed by an attractive automatic chronograph using Breitling's B01 Caliber in exchange for the three-hand MT5912. Meanwhile, the Tudor engineers came up with the automatic MT5652, which powers a brand-new GMT. These exchanges between the two brands give both independence from the large suppliers of movements.

Black Bay GMT

Reference number: 79830RB
Movement: automatic, Tudor Caliber MT5652; ø 31.8 mm, height 7.52 mm; 28 jewels; 28,800 vph; silicon hairspring; approx. 70-hour power reserve; COSC-certified chronometer
Functions: hours (crown-activated jumping GMT hand), minutes, sweep seconds; additional 24-hour display (2nd time zone); date
Case: stainless steel, ø 41 mm; bidirectional bezel with 0-24 scale; sapphire crystal; screw-in crown; waterproof to 20 atm
Band: stainless steel, folding clasp, with safety lock
Price: $3,900
Variations: calfskin or textile strap ($3,575)

Black Bay 41

Reference number: 79540
Movement: automatic, Tudor Caliber 2824 (base ETA 2824-2); ø 25.6 mm, height 4.6 mm; 25 jewels; 28,800 vph; approx. 38-hour power reserve
Functions: hours, minutes, sweep seconds
Case: stainless steel, ø 41 mm; sapphire crystal; screw-in crown; waterproof to 15 atm
Band: calfskin, folding clasp
Price: $2,625
Variations: stainless steel bracelet ($2,950); textile strap ($2,625)

Black Bay Fifty-Eight

Reference number: 79030N
Movement: automatic, Tudor Caliber MT5402; ø 26 mm, height 4.99 mm; 27 jewels; 28,800 vph; silicon hairspring; approx. 70-hour power reserve; COSC-certified chronometer
Functions: hours, minutes, sweep seconds
Case: stainless steel, ø 39 mm; unidirectional bezel with aluminum insert, with 0-60 scale; sapphire crystal; screw-in crown; waterproof to 20 atm
Band: textile, buckle
Price: $3,250
Variations: stainless steel bracelet ($3,575); calfskin strap ($3,250)

TUDOR

Heritage Black Bay S&G
Reference number: 79733N
Movement: automatic, Tudor Caliber MT5612; ø 31.8 mm, height 6.5 mm; 26 jewels; 28,800 vph; silicon hairspring, variable inertia balance; approx. 70-hour power reserve; COSC-certified chronometer
Functions: hours, minutes, sweep seconds; date
Case: stainless steel, ø 41 mm; unidirectional yellow gold bezel, with 0-60 scale; sapphire crystal; screw-in crown in yellow gold; waterproof to 20 atm
Band: stainless steel with yellow gold elements, folding clasp with safety lock
Price: $4,975
Variations: aged-leather strap ($3,775)

Heritage Black Bay Steel
Reference number: 79730
Movement: automatic, Tudor Caliber MT5612; ø 31.8 mm, height 6.5 mm; 26 jewels; 28,800 vph; silicon hairspring, variable inertia balance; approx. 70-hour power reserve; COSC-certified chronometer
Functions: hours, minutes, sweep seconds; date
Case: stainless steel, ø 41 mm; unidirectional bezel, with 0-60 scale; sapphire crystal; screw-in crown; waterproof to 20 atm
Band: calfskin, folding clasp with safety lock
Price: $3,475
Variations: stainless steel bracelet ($3,800)

Heritage Black Bay Chrono
Reference number: 79350
Movement: automatic, Tudor Caliber MT5813; ø 30.4 mm, height 7.23 mm; 41 jewels; 28,800 vph; silicon hairspring, variable inertia balance; approx. 70-hour power reserve; COSC-certified chronometer
Functions: hours, minutes, subsidiary seconds; chronograph; date
Case: stainless steel, ø 41 mm; sapphire crystal; screw-in crown; waterproof to 20 atm
Band: stainless steel, folding clasp
Price: $5,050
Variations: aged-leather strap ($4,725)

Heritage Black Bay Bronze
Reference number: 79250BM
Movement: automatic, Tudor Caliber MT5601; ø 33.8 mm, height 6.5 mm; 25 jewels; 28,800 vph; silicon hairspring; approx. 70-hour power reserve; COSC-certified chronometer
Functions: hours, minutes, sweep seconds
Case: bronze, ø 43 mm; unidirectional bezel, with 0-60 scale; sapphire crystal; screw-in crown; waterproof to 20 atm
Band: textile, buckle
Price: $3,975
Variation: leather strap ($3,975)

Heritage Black Bay Dark
Reference number: 79230DK
Movement: automatic, Tudor Caliber MT5602; ø 31.8 mm, height 6.5 mm; 25 jewels; 28,800 vph; silicon hairspring,; approx. 70-hour power reserve; COSC-certified chronometer
Functions: hours, minutes, sweep seconds
Case: stainless steel with black PVD coating, ø 41 mm; unidirectional bezel, with 0-60 scale; sapphire crystal; screw-in crown; waterproof to 20 atm
Band: stainless steel with black PVD coating, folding clasp with safety lock
Price: $4,475
Variation: aged-leather strap ($4,150)

Black Bay 41
Reference number: 79540
Movement: automatic, Tudor Caliber 2824 (base ETA 2824-2); ø 25.6 mm, height 4.6 mm; 25 jewels; 28,800 vph; approx. 38-hour power reserve
Functions: hours, minutes, sweep seconds
Case: stainless steel, ø 41 mm; sapphire crystal; screw-in crown; waterproof to 15 atm
Band: textile, buckle
Price: $2,625
Variations: calfskin strap ($2,625); steel bracelet ($2,950)

TUDOR

Heritage Advisor
Reference number: 79620TC
Movement: automatic, Tudor Caliber 2892 with module (base ETA 2892-A2); ø 25.6 mm; 21 jewels; 28,800 vph; approx. 42-hour power reserve
Functions: hours, minutes, sweep seconds; alarm; date
Case: stainless steel, titanium, ø 42 mm; sapphire crystal; waterproof to 10 atm
Band: reptile skin, folding clasp, with safety lock
Price: $5,850
Variations: stainless steel bracelet ($6,075); silk strap ($5,750)

Heritage Chrono
Reference number: 70330N
Movement: automatic, Tudor Caliber 2892 with module (base ETA 2892-A2); ø 25.6 mm; 21 jewels; 28,800 vph; approx. 42-hour power reserve
Functions: hours, minutes, subsidiary seconds; chronograph; date
Case: stainless steel, ø 42 mm; bidirectional 12-hour bezel; sapphire crystal; screw-in crown; waterproof to 15 atm
Band: textile, buckle
Price: $4,100
Variations: stainless steel bracelet ($4,425)

Heritage Ranger
Reference number: 79910
Movement: automatic, Tudor Caliber 2824 (base ETA 2824-2); ø 25.6 mm, height 4.6 mm; 25 jewels; 28,800 vph; approx. 38-hour power reserve
Functions: hours, minutes, sweep seconds
Case: stainless steel, ø 41 mm; sapphire crystal; screw-in crown; waterproof to 15 atm
Band: calfskin, buckle
Price: $2,625
Variations: textile strap ($2,625); stainless steel bracelet ($2,950)

Caliber MT5601
Automatic; single spring barrel, approx. 70-hour power reserve; COSC-certified chronometer
Functions: hours, minutes, sweep seconds
Diameter: 33.8 mm
Height: 6.5 mm
Jewels: 25
Balance: glucydur with weighted screws
Frequency: 28,800 vph
Balance spring: silicon
Relating caliber: MT5602 (with smaller encasement diameter: 31.8 mm)

Caliber MT5402
Automatic; single spring barrel, approx. 70-hour power reserve; COSC-certified chronometer
Functions: hours, minutes, sweep seconds
Diameter: 26 mm
Height: 4.99 mm
Jewels: 27
Balance: glucydur with weighted screws
Frequency: 28,800 vph
Balance spring: silicon

Caliber MT5813
Automatic; single spring barrel, approx. 70-hour power reserve; COSC-certified chronometer
Functions: hours, minutes, subsidiary seconds; chronograph; date
Diameter: 30.4 mm
Height: 7.23 mm
Jewels: 41
Balance: glucydur with weighted screws
Frequency: 28,800 vph
Balance spring: silicon

Tutima Uhrenfabrik GmbH Ndl. Glashütte
Altenberger Strasse 6
D-01768 Glashütte
Germany

Tel.:
+49-35053-320-20

Fax:
+49-35053-320-222

E-mail:
info@tutima.com

Website:
www.tutima.com

Founded:
1927

Number of employees:
approx. 60

U.S. distributor:
Tutima USA, Inc.
P.O. Box 983
Torrance, CA 90508
1-TUTIMA-1927
info@tutimausa.com
www.tutima.com

Most important collections/price range:
Patria, Saxon One, M2, Grand Flieger, Hommage / approx. $1,650 to $29,500

TUTIMA

The name Glashütte is synonymous with watches in Germany. The area, known also for precision engineering, already had quite a watchmaking industry going when World War I closed off markets, followed by the hyperinflation of the early twenties. To rebuild the local economy, a conglomerate was created to produce finished watches under the leadership of jurist Dr. Ernst Kurtz consisting of the movement manufacturer UROFA Glashütte AG and UFAG. The top watches were given the name Tutima, derived from the Latin *tutus*, meaning "whole" or "sound." Among the brand's most famous timepieces was a pilot's watch that set standards in terms of esthetics and functionality.

A few days before World War II ended, Kurtz left Glashütte and founded Uhrenfabrik Kurtz in southern Germany. A young businessman and former employee of Kurtz by the name of Dieter Delecate is credited with keeping the manufacturing facilities and the name Tutima going even as the company sailed through troubled waters. In founding Tutima Uhrenfabrik GmbH in Ganderkesee, this young, resolute entrepreneur prepared the company's strategy for the coming decades.

Delecate has had the joy of seeing Tutima return to its old home and vertically integrated operations, meaning it is once again a genuine *manufacture*. Under renowned designer Rolf Lang, it has developed an in-house minute repeater. In 2013, Tutima proudly announced a genuine made-in-Glashütte movement (at least 50 percent must be produced in the town), Caliber 617.

In addition to technically advanced and sportive watches, Tutima Glashütte has started reviving the great watchmaking crafts that have made the region world famous. There is the Hommage minute repeater and the three-hand Patria. In 2017, the brand introduced the Tempostopp, a flyback chronograph run on the Caliber 659, a replica of the legendary Urofa Caliber 59 from the 1940s with a few necessary improvements in the details.

Saxon One Chronograph Royal Blue

Reference number: 6420-05
Movement: automatic, Tutima Caliber 521 (base ETA 7750); ø 30 mm, height 7.9 mm; 25 jewels; 28,800 vph; sweep minute counter, rotor with gold seal; 48-hour power reserve
Functions: hours, minutes, subsidiary seconds; additional 24-hour display (2nd time zone); chronograph; date
Case: stainless steel, ø 43 mm, height 15.7 mm; bidirectional bezel with reference markers; sapphire crystal; transparent case back; screw-in crown; water-resistant to 20 atm
Band: stainless steel, folding clasp
Price: $6,500

Saxon One Lady Diamonds

Reference number: 6701-01
Movement: automatic, Tutima Caliber 340 (base ETA 2892-A2); ø 25.6 mm, height 3.6 mm; 21 jewels; 28,800 vph; rotor with gold seal; 42-hour power reserve
Functions: hours, minutes, sweep seconds; date
Case: stainless steel, ø 36 mm, height 10.7 mm; bezel set with 48 diamonds; sapphire crystal; transparent case back; water-resistant to 10 atm
Band: stainless steel, folding clasp
Remarks: mother-of-pearl dial
Price: $6,500

Saxon One M

Reference number: 6121-08
Movement: automatic, Tutima Caliber 330 (base ETA 2836-2); ø 25.6 mm, height 5.05 mm; 25 jewels; 28,800 vph; rotor with gold seal; 38-hour power reserve
Functions: hours, minutes, sweep seconds; date
Case: stainless steel, ø 40 mm, height 13 mm; sapphire crystal; transparent case back; screw-in crown; water-resistant to 10 atm
Band: calfskin, folding clasp
Price: $1,950
Variations: blue dial; stainless steel bracelet

TUTIMA

M2
Reference number: 6450-03
Movement: automatic, Tutima Caliber 521 (base ETA 7750); ø 30 mm, height 7.9 mm; 25 jewels; 28,800 vph; sweep minute counter; 48-hour power reserve
Functions: hours, minutes, subsidiary seconds; additional 24-hour display; chronograph; date
Case: titanium, ø 46 mm, height 15.5 mm; sapphire crystal; screw-in crown; water-resistant to 30 atm
Band: titanium, folding clasp
Remarks: soft iron inner case for antimagnetic protection (includes kit and Kevlar strap)
Price: $6,500
Variations: Kevlar strap ($5,900)

M2 Coastline
Reference number: 6150-04
Movement: automatic, Tutima Caliber 330 (ETA Caliber 2836-2); ø 25.6 mm, height 5.05 mm; 25 jewels; 28,800 vph; 42-hour power reserve
Functions: hours, minutes, sweep seconds; date, weekday
Case: titanium, ø 43 mm, height 12.9 mm; sapphire crystal; screw-in crown; water-resistant to 30 atm
Band: stainless steel, folding clasp
Price: $1,950
Variations: blue dial

M2 Seven Seas
Reference number: 6151-03
Movement: automatic, Tutima Caliber 330 (base ETA 2836-2); ø 25.6 mm, height 5.05 mm; 25 jewels; 28,800 vph; rotor with gold seal; 38-hour power reserve
Functions: hours, minutes, sweep seconds; date, weekday
Case: titanium, ø 44 mm, height 13 mm; unidirectional bezel, 0-60 scale; sapphire crystal; screw-in crown; water-resistant to 50 atm
Band: Kevlar, folding clasp
Price: $1,900

Grand Flieger Classic Chronograph
Reference number: 6402-01
Movement: automatic, Tutima Caliber 320 (base ETA 7750); ø 30 mm, height 7.9 mm; 25 jewels; 28,800 vph; rotor with gold seal; 48-hour power reserve; certified chronometer (DIN)
Functions: hours, minutes, subsidiary seconds; chronograph; date
Case: stainless steel, ø 43 mm, height 16 mm; bidirectional bezel with reference marker; sapphire crystal; transparent case back; screw-in crown; water-resistant to 20 atm
Band: calfskin, folding clasp
Price: $5,100
Variations: stainless steel bracelet ($5,500)

Flieger
Reference number: 6105-03
Movement: automatic, Tutima Caliber 330 (base ETA 2836-2); ø 25.6 mm, height 5.05 mm; 25 jewels; 28,800 vph; rotor with gold seal; 38-hour power reserve
Functions: hours, minutes, sweep seconds; date
Case: stainless steel, ø 41 mm, height 12.8 mm; sapphire crystal; transparent case back; screw-in crown; water-resistant to 10 atm
Band: calfskin, folding clasp
Price: $1,650

Grand Flieger Airport Automatic
Reference number: 6101-02
Movement: automatic, Tutima Caliber 330 (base ETA 2836-2); ø 25.6 mm, height 5.05 mm; 25 jewels; 28,800 vph; rotor with gold seal; 38-hour power reserve
Functions: hours, minutes, sweep seconds; date, weekday
Case: stainless steel, ø 43 mm, height 13 mm; bidirectional bezel with reference marker; sapphire crystal; transparent case back; screw-in crown; water-resistant to 20 atm
Band: stainless steel, folding clasp
Price: $2,900

Patria Power Reserve
Reference number: 6602-01
Movement: manually wound, Tutima Caliber 618; ø 31 mm, height 4.78 mm; 27 jewels; 21,600 vph; screw balance with weighted screws and Breguet hairspring; Glashütte three-quarter plate; winding wheels with click; gold-plated and finely finished movement; 65-hour power reserve
Functions: hours, minutes, subsidiary seconds; power reserve indicator
Case: rose gold, ø 43 mm, height 11.2 mm; sapphire crystal; transparent case back; water-resistant to 5 atm
Band: reptile skin, buckle
Price: $19,500

Tempostopp
Reference number: 6650-01
Movement: manually wound, Tutima Caliber 659; ø 33.7 mm, height 6.6 mm; 28 jewels; 21,600 vph; screw balance with gold weight screws and Breguet hairspring; winding wheels with click; hand-engraved balance cock, gold-plated and finely finished movement; 65-hour power reserve
Functions: hours, minutes, subsidiary seconds; flyback chronograph
Case: rose gold, ø 43 mm, height 12.95 mm; sapphire crystal; transparent case back
Band: reptile skin, buckle
Remarks: optimized replica of legendary UROFA Caliber 59 from 1940s
Price: $29,500; limited to 90 pieces

Hommage
Reference number: 6800-02
Movement: manually wound, Tutima Caliber 800; ø 32 mm, height 7.2 mm; 42 jewels; 21,600 vph; screw balance with gold weight screws and Breguet hairspring; Glashütte three-quarter plate; winding wheels with click; hand-engraved balance cock, gold-plated and finely finished movement; 65-hour power reserve
Functions: hours, minutes, subsidiary seconds; minute repeater
Case: rose gold, ø 43 mm, height 13.4 mm; sapphire crystal; transparent case back
Band: reptile skin, buckle
Price: on request; limited to 20 pieces

Caliber Tutima 618
Manually wound; 3 screw-mounted gold chatons, Glashütte three-quarter plate; winding wheels with click; single spring barrel, 65-hour power reserve
Functions: hours, minutes, subsidiary seconds; power reserve indicator
Diameter: 31 mm
Height: 4.78 mm
Jewels: 27
Balance: screw balance with gold weight screws
Frequency: 21,600 vph
Balance spring: Breguet hairspring
Remarks: gold-plated and finely finished movement

Caliber Tutima 659
Manually wound; column wheel control of chronograph functions; single spring barrel, 65-hour power reserve
Functions: hours, minutes, subsidiary seconds; flyback chronograph
Diameter: 33.7 mm
Height: 6.6 mm
Jewels: 28
Balance: screw balance with gold weight screws
Frequency: 21,600 vph
Balance spring: Breguet hairspring
Remarks: optimized replica of legendary UROFA Caliber 59; gold-plated and finely finished movement

Caliber Tutima 800
Manually wound; 4 screw-mounted gold chatons, Glashütte three-quarter plate, two gongs, winding wheels with click; single spring barrel, 65-hour power reserve
Functions: hours, minutes, subsidiary seconds; minute repeater
Diameter: 32 mm
Height: 7.2 mm
Jewels: 42
Balance: screw balance with gold weight screws
Frequency: 21,600 vph
Balance spring: Breguet hairspring
Remarks: gold-plated and finely finished movement

ULYSSE NARDIN

At the beginning of the 1980s, following the infamous quartz crisis, Rolf Schnyder revived the venerable Ulysse Nardin brand, which once upon a time had a reputation for marine chronometers and precision watches. He had the luck to meet the multitalented Dr. Ludwig Oechslin, who realized Schnyder's vision of astronomical wristwatches in the Trilogy of Time. Overnight, Ulysse Nardin became a name to be reckoned with in the world of fine watchmaking. Oechslin developed a host of innovations for Ulysse Nardin, from intelligent calendar movements to escapement systems. He was the first to use silicon and synthetic diamonds and thus gave the entire industry a great deal of food for thought. Just about every Ulysse Nardin has become famous for some spectacular technical innovation, be it the Moonstruck with its stunning moon phase accuracy or the outlandish Freak series that more or less does away with the dial.

After Schnyder's death in 2011, the brand developed a strategy of partnerships and acquisitions, notably of the enameler Donzé Cadrans SA, which gave rise to the Marine Chronometer Manufacture, powered by the Caliber UN-118. In 2014, the French luxury group Kering, owner of Girard-Perregaux, purchased Ulysse Nardin. The two companies are neighbors in La Chaux-de-Fonds, Switzerland, and this has created synergies. Ulysse Nardin's creative power remains strong, with such innovations as a new blade-driven anchor escapement, or the regatta countdown watch with a second hand that runs counterclockwise first before running clockwise like a conventional chronograph once the race has started. The popular and unique "Freak" also appears in new guises every now and then.

Ulysse Nardin SA
3, rue du Jardin
CH-2400 Le Locle
Switzerland

Tel.:
+41-32-930-7400

Fax:
+41-32-930-7419

Website:
www.ulysse-nardin.com

Founded:
1846

U.S. distributor:
Ulysse Nardin Inc.
7900 Glades Rd., Suite 200
Boca Raton, FL 33434
561-988-8600; 561-988-0123 (fax)
usa@ulysse-nardin.com

Most important collections:
Marine chronometers and diver's watches; Dual Time (also ladies' watches); complications (alarm clocks, perpetual calendar, tourbillons, minute repeaters, jacquemarts, astronomical watches)

Freak Vision

Reference number: 2505-250
Movement: automatic, Caliber UN-250; ø 31 mm; 19 jewels; 18,000 vph; flying 1-minute tourbillon on rotating carousel, automatic "grinder" winding system with pawl and flexible control, constant force lever escapement; silicon escapement and hairspring; movement components are used as hands, time-setting via bezel; 50-hour power reserve
Functions: hours, minutes
Case: platinum, ø 45 mm, height 13.5 mm; bidirectional bezel to set hands; sapphire crystal; transparent case back
Band: reptile skin, folding clasp
Price: $95,000

Freak Out

Reference number: 2053-132/BLACK
Movement: manually wound, Caliber UN-205; ø 35 mm; 28 jewels; 28,800 vph; 1-minute tourbillon on rotating carousel, silicon escapement and hairspring; movement components are used as hands, time-setting via bezel, winding by turning case back; 170-hour power reserve
Functions: hours, minutes, subsidiary seconds
Case: titanium with black PVD coating, ø 45 mm, height 13.5 mm; bidirectional bezel to set hands; sapphire crystal; transparent case back
Band: textile, folding clasp
Price: $48,000

Freak Out

Reference number: 2053-132/03
Movement: manually wound, Caliber UN-205; ø 35 mm; 28 jewels; 28,800 vph; 1-minute tourbillon on rotating carousel, silicon escapement and hairspring; movement components are used as hands, time-setting via bezel, winding by turning case back; 170-hour power reserve
Functions: hours, minutes, subsidiary seconds
Case: titanium, ø 45 mm, height 13.5 mm; bidirectional bezel for time setting; sapphire crystal; transparent case back
Band: textile, folding clasp
Price: $48,000

ULYSSE NARDIN

Executive Moonstruck Worldtimer

Reference number: 1062-113/01
Movement: automatic, Caliber UN-106; ø 35.4 mm, height 8.17 mm; 42 jewels; 28,800 vph; silicon escape wheel, pallet, balance wheel, and hairspring
Functions: hours (24h), switchable by pusher forward and backward, minutes; world time indicator, sun and moon position, moon phase, tides; date
Case: rose gold, ø 46 mm, height 14.2 mm; sapphire crystal; transparent case back
Band: reptile skin, folding clasp
Price: $75,000; limited to 100 pieces
Variations: platinum (limited to 100 pieces, $95,000)

Executive Skeleton Tourbillon

Reference number: 1713-139/MAGIC BLACK
Movement: manually wound, Caliber UN-171; ø 37 mm, height 5.86 mm; 23 jewels; 18,000 vph; 1-minute tourbillon; skeletonized movement; double spring barrel; silicon escape wheel and hairspring; 170-hour power reserve
Functions: hours, minutes
Case: titanium with black PVD coating, ø 45 mm, height 12.5 mm; ceramic bezel; sapphire crystal; transparent case back; water-resistant to 3 atm
Band: calfskin, folding clasp
Remarks: numerals of 46 baguette-cut diamonds
Price: $69,000

Executive Skeleton Tourbillon

Reference number: 1712-139
Movement: manually wound, Caliber UN-171; ø 37 mm, height 5.86 mm; 23 jewels; 18,000 vph; 1-minute tourbillon; skeletonized movement; double spring barrel; silicon escape wheel and hairspring; 170-hour power reserve
Functions: hours, minutes
Case: rose gold, ø 45 mm, height 12.5 mm; ceramic bezel; sapphire crystal; transparent case back; water-resistant to 3 atm
Band: reptile skin, folding clasp
Price: $49,000

Hour Striker "Samurai"

Reference number: 6109-130/SAMOURAI
Movement: automatic, Caliber UN-610; ø 33 mm, height 8.3 mm; 41 jewels; 28,800 vph; 40-hour power reserve
Functions: hours, minutes; hour and half-hour repeater with mobile jaquemarts
Case: platinum, ø 43 mm, height 13.2 mm; sapphire crystal
Band: reptile skin, folding clasp
Price: $119,000; limited to 18 pieces
Variations: rose gold (limited to 18 pieces, $99,000)

Classico Cloisonné "America Aurora Borealis"

Reference number: 8152-111-2/AMERICA V4
Movement: automatic, Caliber UN-815; ø 25.6 mm, height 3.6 mm; 21 jewels; 28,800 vph; 42-hour power reserve; COSC-certified chronometer
Functions: hours, minutes, sweep seconds
Case: rose gold, ø 40 mm, height 10.7 mm; sapphire crystal
Band: reptile skin, buckle
Remarks: enamel dial; unique piece
Price: on request; not available in U.S. market

Classico Small Second Manufacture

Reference number: 3203-136-2/E2
Movement: automatic, Caliber UN-320; ø 26.4 mm, height 4.6 mm; 39 jewels; 28,800 vph; silicon escapement and hairspring; 48-hour power reserve; COSC-certified chronometer
Functions: hours, minutes, subsidiary seconds; date
Case: stainless steel, ø 40 mm, height 9.5 mm; sapphire crystal; transparent case back
Band: reptile skin, buckle
Remarks: enamel dial
Price: $8,800
Variations: white dial ($8,800); stainless steel bracelet ($7,500)

ULYSSE NARDIN

Ulysse Nardin Diver Chronometer
Reference number: 1183-170-3/93
Movement: automatic, Caliber UN-118; ø 31.6 mm, height 6.45 mm; 50 jewels; 28,800 vph; "DIAMonSIL" escapement, silicon hairspring; 60-hour power reserve; COSC-certified chronometer
Functions: hours, minutes, subsidiary seconds; power reserve indicator; date
Case: titanium, ø 44 mm, height 12.3 mm; sapphire crystal; transparent case back; water-resistant to 30 atm
Band: rubber, buckle
Price: $7,900
Variations: titanium and rose gold, (limited to 100 pieces, $12,000)

Marine Chronograph Manufacture Regatta
Reference number: 1553-155-3/43
Movement: automatic, Caliber UN-155; ø 34 mm, height 8.28 mm; 67 jewels; 28,800 vph; silicon escapement and hairspring
Functions: hours, minutes, subsidiary seconds; chronograph with integrated 10-minute countdown function; date
Case: stainless steel, ø 44 mm; sapphire crystal; transparent case back; screw-in crown; water-resistant to 10 atm
Band: rubber and titanium, folding clasp
Remarks: countdown totalizer; second hand turns counterclockwise, then clockwise when reaching 0
Price: $15,900

Marine Tourbillon Manufacture
Reference number: 1283-181/E3
Movement: manually wound, Caliber UN-128; ø 31 mm, height 6.45 mm; 50 jewels; 28,800 vph; flying 1-minute tourbillon; 60-hour power reserve
Functions: hours, minutes; power reserve indicator
Case: stainless steel, ø 43 mm, height 12.2 mm; sapphire crystal; transparent case back; screw-in crown, with rubber overlay; water-resistant to 10 atm
Band: reptile skin, double folding clasp
Remarks: enamel dial
Price: $28,000
Variations: rubber strap ($27,900); stainless steel bracelet ($28,700)

Marine Torpilleur
Reference number: 1182-310/40
Movement: automatic, Caliber UN-118; ø 31.6 mm, height 6.45 mm; 50 jewels; 28,800 vph; "DIAMonSIL" escapement, silicon hairspring; 60-hour power reserve; COSC-certified chronometer
Functions: hours, minutes, subsidiary seconds; power reserve indicator; date
Case: pink gold, ø 42 mm, height 13 mm; sapphire crystal; transparent case back; screw-in crown; water-resistant to 10 atm
Band: reptile skin, folding clasp
Price: $17,900
Variations: black dial; reptile skin strap ($17,900)

Marine Torpilleur
Reference number: 1183-310/43
Movement: automatic, Caliber UN-118; ø 31.6 mm, height 6.45 mm; 50 jewels; 28,800 vph; "DIAMonSIL" escapement, silicon hairspring; 60-hour power reserve; COSC-certified chronometer
Functions: hours, minutes, subsidiary seconds; power reserve indicator; date
Case: stainless steel, ø 42 mm, height 13 mm; sapphire crystal; transparent case back; screw-in crown; water-resistant to 10 atm
Band: reptile skin, folding clasp
Price: $6,900
Variations: white dial; rubber strap ($6,900); stainless steel bracelet ($7,600)

Marine Torpilleur Military
Reference number: 1183-320LE/62
Movement: automatic, Caliber UN-118; ø 31.6 mm, height 6.45 mm; 50 jewels; 28,800 vph; "DIAMonSIL" escapement, silicon hairspring; 60-hour power reserve; COSC-certified chronometer
Functions: hours, minutes, subsidiary seconds
Case: stainless steel (sandblasted), ø 44 mm, height 13 mm; sapphire crystal; transparent case back; screw-in crown; water-resistant to 10 atm
Band: reptile skin, folding clasp
Price: $7,900; limited to 300 pieces
Variations: white dial (limited to 300 pieces)

Diver Deep Dive

Reference number: 3203-500LE-3/93-HAMMER
Movement: automatic, Caliber UN-320; ø 26.4 mm, height 4.6 mm; 39 jewels; 28,800 vph; silicon escapement and hairspring; 48-hour power reserve
Functions: hours, minutes, subsidiary seconds; date
Case: titanium, ø 46 mm, height 12.7 mm; unidirectional bezel, with 0-60 scale; sapphire crystal; screw-in crown, with guard; automatic helium valve; water-resistant to 100 atm
Band: rubber with titanium, folding clasp
Price: $12,000; limited to 300 pieces

Diver Lady

Reference number: 3203-190-3C/10.13
Movement: automatic, Caliber UN-320; ø 26.4 mm, height 4.6 mm; 39 jewels; 28,800 vph; silicon escapement and hairspring; 48-hour power reserve
Functions: hours, minutes, subsidiary seconds; date
Case: stainless steel, ø 40 mm, height 12 mm; unidirectional bezel, with 0-60 scale; sapphire crystal; water-resistant to 3 atm
Band: rubber with titanium, folding clasp
Remarks: case and strap set with 107 diamonds
Price: $10,800
Variations: black or white; without diamonds ($8,800)

Classic Dual Time Lady

Reference number: 3243-222B/93
Movement: automatic, Caliber UN-324; ø 26.4 mm, height 6.15 mm; 53 jewels; 28,800 vph; silicon escapement and hairspring; 42-hour power reserve
Functions: hours (quick forward and backward setting), minutes, subsidiary seconds; additional 24-hour display (2nd time zone), crown position display; large date
Case: stainless steel, ø 37 mm, height 10.2 mm; bezel with 60 diamonds; sapphire crystal
Band: reptile skin, double folding clasp
Price: $13,500
Variations: without diamonds ($9,700); white dial

Caliber UN-118

Automatic; DIAMonSIL (patented) escapement; single spring barrel, 60-hour power reserve
Functions: hours, minutes, subsidiary seconds; power reserve indicator; date
Diameter: 31.6 mm
Height: 6.45 mm
Jewels: 50
Balance: with variable inertia
Frequency: 28,800 vph
Balance spring: silicon
Shock protection: Incabloc
Remarks: perlage on mainplate, bridges with concentric côtes de Genève ("côtes circulaire")

Caliber UN-178

Manually wound; 1-minute tourbillon, Ulysse pallet lever escapement made of DIAMonSII (pivotless anchor between 2 blades) with new geometry; double spring barrel, 168-hour power reserve
Functions: hours, minutes; power reserve indicator
Diameter: 37 mm
Height: 6 mm
Jewels: 29
Balance: glucydur
Frequency: 18,000 vph
Balance spring: silicon
Remarks: image shows only the escapement

Caliber UN-334

Automatic; silicon escapement; single spring barrel, 48-hour power reserve
Functions: hours, minutes, subsidiary seconds; additional 24-hour display (2nd time zone); large date
Jewels: 49
Balance: with variable inertia
Frequency: 28,800 vph
Balance spring: silicon
Shock protection: Incabloc
Remarks: patented rapid time adjustment for 2nd time zone; perlage on mainplate, bridges with concentric côtes de Genève ("côtes circulaire")

URBAN JÜRGENSEN & SØNNER

For all aficionados and collectors of fine timekeepers, the name Urban Jürgensen & Sønner is synonymous with outstanding watches. The company was founded in 1773 and has always strived for the highest rungs of the horological art. Technical perfection consistently combines with imaginative cases. A lot of attention is given to dials and hands.

Today, Urban Jürgensen & Sønner—originally a Danish firm—manufactures watches in Switzerland, where a team of eight superbly qualified watchmakers do the work in three ateliers. For over a quarter century now, they have been making highly complicated unique pieces and very upmarket wristwatches in small editions of 50 to 300 pieces. The series were based mostly on *ébauches* by Frédéric Piguet. Like all keen watchmakers, those at Urban Jürgensen have also sought to make their own movements, which would meet the highest standards of precision and reliability and not require too much servicing.

In 2003, a team began collaborating with a well-known external design engineer to construct a base movement. The new UJS-P8 was conceived with both a traditional Swiss lever escapement and in a special variation featuring a pivoting chronometer escapement, available for the first time in a wristwatch.

The esthetic concept behind the brand's watches is clearly vintage. Urban Jürgensen & Sønner timepieces have the broad open face of old pocket watches and classic hands, including a Breguet-type hour hand. The lugs on the new 1741 recall the link to the watch chain. CEO Søren Jenry Petersen, an industrialist and watch lover, has kept up that watchmaking concept. The 1140 series is composed of classical watches with a modernized eighteenth-century feel. A detail worth noting is the complex "grenage" technique used to create that grainy look on the dial.

Urban Jürgensen & Sønner
Chemin Creux 18
CH-2503 Biel-Bienne
Switzerland

Tel.:
+41-32-365-1526

Fax:
+41-32-365-2266

E-mail:
info@urbanjurgensen.com

Website:
www.urbanjurgensen.com

Founded:
1773/1980

Annual production:
max. 200 watches

U.S. distributor:
Martin Pulli
4337 Main Street
Philadelphia, PA 19127
215-508-4610
martin@martinpulli.com
www.martinpulli.com

Most important collections:
High-end references with in-house movements

1741
Reference number: 1741 RT
Movement: manually wound, Urban Jürgensen Caliber P4; ø 32 mm, height 6.85 mm; 24 jewels; 21,600 vph; 2 spring barrels, 60-hour power reserve
Functions: hours, minutes, sweep seconds; perpetual calendar with date, weekday, month, moon phase, leap year
Case: platinum, ø 41 mm, height 12.3 mm; sapphire crystal; transparent case back; water-resistant to 3 atm
Band: reptile skin, buckle
Price: $98,600

The Alfred
Reference number: 1142 SS
Movement: manually wound, Urban Jürgensen Caliber UJS-P4; ø 32 mm, height 5.2 mm; 23 jewels; 21,600 vph; 2 spring barrels, 72-hour power reserve
Functions: hours, minutes, subsidiary seconds
Case: stainless steel, ø 42 mm, height 11.5 mm; sapphire crystal; transparent case back; water-resistant to 3 atm
Band: reptile skin, buckle
Remarks: solid silver "grenage" dial
Price: $15,200

1140 RG Brown
Reference number: 1140 PT
Movement: manually wound, Urban Jürgensen Caliber UJS-P4; ø 32 mm, height 5.2 mm; 23 jewels; 21,600 vph; 2 spring barrels, 72-hour power reserve
Functions: hours, minutes, subsidiary seconds
Case: red gold, ø 40 mm, height 10.5 mm; sapphire crystal; transparent case back; water-resistant to 3 atm
Band: reptile skin, buckle
Price: $38,100; limited to 20 pieces
Variations: platinum case and blue dial ($53,700, limited to 20 pieces)

URWERK

Urwerk SA
114, rue du Rhône
CH-1204 Geneva
Switzerland

Tel:
+41-22-900-2027

Fax:
+41-22-900-2026

E-mail:
info@urwerk.com

Website:
www.urwerk.com

Founded:
1995

Annual production:
150 watches

U.S. distributor:
Ildico Inc.
8701 Wilshire Blvd.
Beverly Hills, CA 90211
310-205-5555

Felix Baumgartner and designer Martin Frei count among the living legends of innovative horology. They founded their company, Urwerk, in 1997 with a name that is a play on the words *Uhrwerk*, for movement, and *Urwerk*, meaning a sort of primal mechanism. Their specialty is inventing surprising time indicators featuring digital numerals that rotate like satellites and display the time in a relatively linear depiction on a small "dial" at the front of the flattened case, which could almost—but not quite—be described as oval. Their inspiration goes back to the so-called night clock of the eighteenth-century Campanus brothers, but the realization is purely *2001: A Space Odyssey*.

Urwerk's debut was with the Harry Winston Opus 5. Later, they created the Black Cobra, which displays time using cylinders and other clever ways to recoup energy for driving rather heavy components. The Torpedo is another example of high-tech watchmaking, again based on the satellite system of revolving and turning hands. These pieces remind one of the frenetic engineering that has transformed the planet since the eighteenth century. And with each return to the drawing board, Baumgartner and Frei find new ways to explore what has now become an unmistakable form, using high-tech materials, like aluminum titanium nitride (AlTiN), or finding new functions for the owner to play with.

In 2017, Urwerk celebrated its twentieth anniversary. For the occasion, it gathered all its creative high points into a single watch. The minute hand hovers along an arc and bears the hour indices. The "Transformator" adds a rotatable, pivotable case to the watch. If you're not interested in the time, you can turn the watch around and admire the automatic mechanism through the back. The latest model is the UR-105 Kryptonite. The name is a reference to the indestructible material from the Marvel comics. In the case of the watch, it's a coating of aluminum-titanium nitrate.

UR-105 CT Kryptonite

Movement: automatic, Caliber UR 5.03; 52 jewels; 28,800 vph; revolving hour satellites with Maltese cross control and planetary transmission (all numerals always remain in vertical position), winding system regulated by fluid dynamics decoupling; 48-hour power reserve
Functions: hours (digital, rotating), minutes (segment display), subsidiary seconds (perforated disk display); power reserve indicator
Case: titanium with AlTiN coating, 39.5 × 53 mm, height 17.8 mm; sapphire crystal; water-resistant to 3 atm
Band: nylon, buckle
Price: $69,000

UR-210 "Clous de Paris"

Movement: automatic, Caliber UR 7.10; 51 jewels; 28,800 vph; revolving hours cube with retrograde minutes, arc minutes, winding system regulated by fluid dynamics decoupling and adjustable efficiency; 39-hour power reserve
Functions: hours (digital, rotating), minutes (retrograde); power reserve indicator, winding efficiency
Case: titanium and AlTin coated steel, 43.8 × 53.6 mm, height 17.8 mm; sapphire crystal
Band: textile, buckle
Remarks: hour cubes travel across semicircular minute scale, skeletonized minute hand jumps back at end of scale and "picks up" next satellite
Price: $150,000

UR-T8 Transformer

Movement: automatic, Caliber UR 8.01; 28,800 vph; revolving hour satellites with Maltese cross control and planetary transmission (all numerals always remain in vertical position), winding system regulated by fluid dynamics decoupling; 50-hour power reserve
Functions: hours (digital, rotating), minutes
Case: titanium with black PVD coating, 48.3 × 60.2 mm, height 20 mm; sapphire crystal; water-resistant to 3 atm
Band: reptile skin, buckle
Remarks: case on carrier frame can be pivoted and turned 180°
Price: $100,000

UTS

UTS, or "Uhren Technik Spinner," was the natural outgrowth of a company based in Munich and manufacturing CNC tools and machines for the watch industry. Nicolaus Spinner, a mechanical engineer and aficionado in his own right, learned the nitty-gritty of watchmaking by the age-old system of taking watches apart. From there to making robust diver's watches was just a short step. The collection has grown considerably since he started production in 1999. The watches are built mainly around ETA calibers. Some, like the new 4000M, feature a unique locking bezel using a stem, a bolt, and a ceramic ball bearing system invented by Spinner. Another specialty is the 6 mm sapphire crystal, which guarantees significant water resistance. Spinner's longtime friend and business partner, Stephen Newman, is the owner of the UTS trademark in the United States. He not only has worked on product development, but has also contributed his own design ideas and handles sales and marketing for the small brand. A new watch released in 2014, the 4000M Diver, boasts an extreme depth rating even without the need for a helium escape valve and is available in a GMT version. The collection is small, but UTS has a faithful following in Germany and the United States. The key for the fan club is a unique appearance coupled with mastery of the technology. These are pure muscle watches with no steroids.

UTS Watches, Inc.
P.O. Box 6293
Los Osos, CA 93412

Tel.:
877-887-0123 or 805-528-9800

E-mail:
info@utswatches.com

Website:
www.utswatches.com

Founded:
1999

Number of employees:
2

Annual production:
fewer than 500

Distribution:
direct sales only

Most important collections/price range:
sports and diver's watches, chronographs / from $2,500 to $7,000

Diver 4000M

Movement: automatic, ETA Caliber 2824-2; ø 25.6 mm, height 4.6 mm; 25 jewels; 28,800 vph; 42-hour power reserve
Functions: hours, minutes, sweep seconds; date
Case: stainless steel, ø 45 mm, height 17.5 mm; bidirectional bezel with 0-60 scale; 6 mm sapphire crystal, antireflective on back; screwed-down case back; screw-in crown and buttons; locking bezel; water-resistant to 400 atm
Band: stainless steel with diver's extension folding clasp or rubber or leather strap
Price: $6,800

Diver 4000M GMT

Movement: automatic, ETA Caliber 2893-2; ø 25.6 mm, height 4.6 mm; 25 jewels; 28,800 vph; 42-hour power reserve
Functions: hours, minutes, sweep seconds; date; 2nd time zone
Case: stainless steel, ø 45 mm, height 17.5 mm; bidirectional bezel with 0-60 scale; 6 mm sapphire crystal, antireflective on back; screwed-down case back; screw-in crown and buttons; locking bezel; water-resistant to 400 atm
Band: stainless steel with diver's extension folding clasp or rubber or leather strap
Price: $6,800

2000M

Movement: automatic, ETA Caliber 2824-2; ø 25.6 mm, height 4.6 mm; 25 jewels; 28,800 vph; 42-hour power reserve
Functions: hours, minutes, sweep seconds; date
Case: stainless steel, ø 44 mm, height 16.5 mm; unidirectional bezel with 0-60 scale; automatic helium escape valve; sapphire crystal, antireflective on back; screwed-down case back; screw-in crown and buttons; water-resistant to 200 atm
Band: stainless steel with diver's extension folding clasp, comes with rubber leather strap
Price: $3,950

1000M V2

Movement: automatic, ETA Caliber 2824-2; ø 25.6 mm, height 4.6 mm; 25 jewels; 28,800 vph; 42-hour power reserve
Functions: hours, minutes, sweep seconds; date
Case: stainless steel, ø 43 mm, height 14 mm; unidirectional bezel with 0-60 scale; sapphire crystal, antireflective on back; screwed-down sapphire (optional) crystal case back; screw-in crown and buttons; water-resistant to 10 atm
Band: stainless steel with diver's extension folding clasp or rubber or leather strap
Price: $3,390

1000 GMT

Movement: automatic, ETA Caliber ETA 2893-2; ø 25.6 mm, height 4.1 mm; 21 jewels; 28,800 vph; 42-hour power reserve
Functions: hours, minutes, sweep seconds; 2nd time zone; date; quick set GMT hand
Case: stainless steel, ø 43 mm, height 14 mm; unidirectional bezel with 0-60 scale; sapphire crystal, antireflective on back; screwed-down case back with optional transparent back (sapphire crystal); screw-in crown and buttons; water-resistant to 100 atm
Band: stainless steel with diver's extension folding clasp or rubber or leather strap
Price: $3,950

Adventure Automatic

Movement: automatic, ETA Valgranges Caliber A07.111; ø 36.6 mm, height 7.9 mm; 24 jewels; 28,800 vph; 46-hour power reserve
Functions: hours, minutes, sweep seconds; date
Case: stainless steel, ø 46 mm, height 15.5 mm; screw-in crown; antireflective sapphire crystal; screwed-down sapphire crystal case back; water-resistant to 50 atm
Band: rubber, buckle
Price: $3,950
Variations: leather strap; stainless steel bracelet with folding clasp and diver's extension

Adventure Automatic GMT

Movement: automatic, ETA Valgranges Caliber A07.171; ø 36.6 mm, height 7.9 mm; 24 jewels; 28,800 vph; 46-hour power reserve
Functions: hours, minutes, sweep seconds; date; 2nd time zone
Case: stainless steel, ø 46 mm, height 15.5 mm; screw-in crown; antireflective sapphire crystal; screwed-down sapphire crystal case back; water-resistant to 50 atm
Band: rubber, buckle
Price: $4,550
Variations: leather strap; stainless steel bracelet with folding clasp and diver's extension

Chrono Diver

Movement: automatic, ETA Valjoux Caliber 7750; ø 30 mm, height 7.9 mm; 25 jewels; 28,800 vph; 45-hour power reserve
Functions: hours, minutes, subsidiary seconds; date; chronograph
Case: stainless steel, ø 46 mm, height 16.5 mm; unidirectional bezel with 0-60 scale; screw-in crown and buttons; antireflective sapphire crystal; screwed-down case back; water-resistant to 600 m
Band: stainless steel, folding clasp
Price: $4,550
Variations: leather strap

Adventure Manual Wind

Movement: manually wound, ETA Unitas Caliber 6497; ø 36.6 mm, height 5.4 mm; 18 jewels; 18,000 vph; 48-hour power reserve
Functions: hours, minutes, subsidiary seconds
Case: stainless steel, ø 46 mm, height 14 mm; screw-in crown; antireflective sapphire crystal; screwed-down sapphire crystal case back; water-resistant to 50 atm
Band: leather, buckle
Price: $3,400
Variations: rubber strap

VACHERON CONSTANTIN

The origins of this oldest continuously operating watch *manufacture* can be traced back to 1755 when Jean-Marc Vacheron opened his workshop in Geneva. His highly complex watches were particularly appreciated by clients in Paris. The development of such an important outlet for horological works there had a lot to do with the emergence of a wealthy class around the powerful French court. The Revolution put an end to all the financial excesses of that market, however, and the Vacheron company suffered as well . . . until the arrival of marketing wizard François Constantin in 1819.

Fast-forward to the late twentieth century: The brand with the Maltese cross logo had evolved into a tradition-conscious keeper of *haute horlogerie* under the aegis, starting in the mid-1990s, of the Vendôme Luxury Group (today's Richemont SA). Gradually, it began creating collections that combine modern shapes with traditional patterns. The company has been expanding steadily. In 2013 it opened boutiques in the United States as well as China, and it has become a leading sponsor of the New York City Ballet.

Vacheron Constantin is one of the last luxury brands to have abandoned the traditional way of dividing up labor. Today, most of its basic movements are made in-house at the production facilities and headquarters in Plan-les-Ouates and the workshops in Le Brassus in Switzerland's Jura region, which were expanded in the summer of 2013.

The wide scope of the brand has been demonstrated repeatedly over the years. Products range from the world's most complicated watch, the 57260, and the finely crafted Les Cabinotiers collection of unique pieces, to the rejuvenated Overseas models, and, now, a new entry-level collection, the Fiftysix, with a basic movement and no Geneva Seal.

Vacheron Constantin
Chemin du Tourbillon
CH-1228 Plan-les-Ouates
Switzerland

Tel.:
+41-22-930-2005

E-mail:
info@vacheron-constantin.com

Website:
www.vacheron-constantin.com

Founded:
1755

Number of employees:
approx. 800

Annual production:
over 20,000 watches (estimated)

U.S. distributor:
Vacheron Constantin
Richemont North America
645 Fifth Avenue
New York, NY 10022
877-701-1755

Most important collections:
Harmony, Patrimony, Traditionnelle, Historiques, Métiers d'Art, Malte, Overseas, Quai de l'Île

Fiftysix Automatic

Reference number: 4600E/000A-B442
Movement: automatic, Vacheron Constantin Caliber 1326; ø 26.2 mm, height 4.3 mm; 25 jewels; 28,800 vph; 48-hour power reserve
Functions: hours, minutes, sweep seconds; date
Case: stainless steel, ø 40 mm, height 9.6 mm; sapphire crystal; transparent case back; water-resistant to 3 atm
Band: reptile skin, folding clasp
Price: $17,900
Variations: pink gold ($19,900)

Fiftysix Day/Date

Reference number: 4400E/000A-B437
Movement: automatic, Vacheron Constantin Caliber 2475 SC/2; ø 26.2 mm, height 5.7 mm; 27 jewels; 28,800 vph; 40-hour power reserve; Geneva Seal
Functions: hours, minutes, sweep seconds; power reserve indicator; date, weekday
Case: stainless steel, ø 40 mm, height 11.6 mm; sapphire crystal; transparent case back; water-resistant to 3 atm
Band: reptile skin, folding clasp
Price: $17,900
Variations: pink gold ($33,400)

Fiftysix Complete Calendar

Reference number: 4000E/000R-B438
Movement: automatic, Vacheron Constantin Caliber 2460 QCL/1; ø 29 mm, height 5.4 mm; 27 jewels; 28,800 vph; 40-hour power reserve; Geneva Seal
Functions: hours, minutes, sweep seconds; full calendar with date, weekday, month
Case: pink gold, ø 40 mm, height 11.6 mm; sapphire crystal; transparent case back; water-resistant to 3 atm
Band: reptile skin, folding clasp
Price: $36,800
Variations: stainless steel ($23,500)

VACHERON CONSTANTIN

Overseas Perpetual Calendar Extra-Thin
Reference number: 4300V/000R-B064
Movement: automatic, Vacheron Constantin Caliber 1120 QP/1; ø 29.6 mm, height 4.05 mm; 36 jewels; 19,800 vph; 40-hour power reserve; Geneva Seal
Functions: hours, minutes; perpetual calendar with date, weekday, month, moon phase, leap year
Case: pink gold, ø 41.5 mm, height 8.1 mm; sapphire crystal; transparent case back
Band: reptile skin, folding clasp
Remarks: comes with additional rubber strap
Price: $76,000

Overseas Dual Time
Reference number: 7900V/110A-B333
Movement: automatic, Vacheron Constantin Caliber 5110 DT; ø 30.6 mm, height 6 mm; 37 jewels; 28,800 vph; 60-hour power reserve; Geneva Seal
Functions: hours, minutes, sweep seconds; additional 12-hour display (2nd time zone), day/night indicator; date
Case: stainless steel, ø 41 mm, height 12.8 mm; sapphire crystal; transparent case back; water-resistant to 10 atm
Band: stainless steel, folding clasp
Remarks: comes with additional rubber and reptile skin strap
Price: $24,700

Overseas Automatic
Reference number: 4500V/110A-B128
Movement: automatic, Vacheron Constantin Caliber 5100; ø 30.6 mm, height 4.7 mm; 37 jewels; 28,800 vph; gold rotor; 60-hour power reserve; Geneva Seal
Functions: hours, minutes, sweep seconds; date
Case: stainless steel, ø 41 mm, height 11 mm; sapphire crystal; transparent case back; water-resistant to 15 atm
Band: stainless steel, double folding clasp
Price: $20,400
Variations: reptile skin strap; rubber strap

Overseas Chronograph
Reference number: 5500V/110A-B481
Movement: automatic, Vacheron Constantin Caliber 5200; ø 30.6 mm, height 6.6 mm; 54 jewels; 28,800 vph; column wheel control of chronograph functions; gold rotor; 52-hour power reserve; Geneva Seal
Functions: hours, minutes, subsidiary seconds; chronograph; date
Case: stainless steel, ø 42.5 mm, height 13.7 mm; sapphire crystal; transparent case back; screw-in crown and pusher; water-resistant to 15 atm
Band: rubber, double folding clasp
Price: $29,300
Variations: reptile skin strap

Overseas Small Model
Reference number: 2305V-100R-B077
Movement: automatic, Vacheron Constantin Caliber 5300; ø 22.6 mm, height 4 mm; 31 jewels; 28,800 vph; gold rotor; 44-hour power reserve; Geneva Seal
Functions: hours, minutes, subsidiary seconds
Case: rose gold, ø 37 mm, height 10.8 mm; with 84 diamonds; sapphire crystal; transparent case back; water-resistant to 15 atm
Band: rose gold, double folding clasp
Price: $25,400
Variations: reptile skin and rubber strap; in stainless steel

Overseas World Time Watch
Reference number: 7700V/110A-B172
Movement: automatic, Vacheron Constantin Caliber 2460 WT; ø 36.6 mm, height 7.55 mm; 27 jewels; 28,800 vph; 40-hour power reserve; Geneva Seal
Functions: hours, minutes, sweep seconds; world time indicator (2nd time zone), day/night indicator
Case: stainless steel, ø 43.5 mm, height 12.6 mm; sapphire crystal; transparent case back; screw-in crown; water-resistant to 15 atm
Band: reptile skin, buckle
Remarks: soft iron cage for antimagnetic protection; comes with stainless steel bracelet and rubber strap
Price: $37,500

VACHERON CONSTANTIN

Traditionnelle Tourbillon
Reference number: 6000T/000R-B346
Movement: manually wound, Vacheron Constantin Caliber 2160; ø 31 mm, height 5.65 mm; 30 jewels; 18,000 vph; 1-minute tourbillon; 80-hour power reserve; Geneva Seal
Functions: hours, minutes, subsidiary seconds (on tourbillon cage)
Case: rose gold, ø 41 mm, height 10.4 mm; sapphire crystal; transparent case back; water-resistant to 3 atm
Band: reptile skin, folding clasp
Price: on request

Traditionnelle Full Calendar
Reference number: 4010T/000R-B344
Movement: manually wound, Vacheron Constantin Caliber 2460 QCL; ø 29 mm, height 5.4 mm; 27 jewels; 28,800 vph; 40-hour power reserve; Geneva Seal
Functions: hours, minutes, sweep seconds; full calendar with date, weekday, month, moon phase
Case: rose gold, ø 41 mm, height 10.7 mm; sapphire crystal; transparent case back; water-resistant to 3 atm
Band: reptile skin, folding clasp
Price: $39,300

Traditionnelle Moon Phase and Power Reserve Small Model
Reference number: 83570/000G-9916
Movement: manually wound, Vacheron Constantin Caliber 1410 AS; ø 26 mm, height 4.2 mm; 22 jewels; 28,800 vph; 40-hour power reserve; Geneva Seal
Functions: hours, minutes, subsidiary seconds; power reserve indicator; moon phase
Case: white gold, ø 36 mm, height 9.1 mm; bezel and lugs set with 81 diamonds; sapphire crystal; transparent case back; crown with diamond; water-resistant to 3 atm
Band: reptile skin, buckle
Remarks: mother-of-pearl dial
Price: $40,200

Historique American 1921
Reference number: 1100S/000R-B430
Movement: manually wound, Vacheron Constantin Caliber 4400 AS; ø 28.6 mm, height 2.8 mm; 21 jewels; 28,800 vph; 65-hour power reserve; Geneva Seal
Functions: hours, minutes, subsidiary seconds
Case: pink gold, ø 36.5 mm, height 8 mm; sapphire crystal; transparent case back; water-resistant to 3 atm
Band: reptile skin, buckle
Remarks: modeled after a 1921 watch
Price: $35,700

Historique 1948
Reference number: 3100V/000R-B359
Movement: manually wound, Vacheron Constantin Caliber 4400 QC; ø 29 mm, height 4.6 mm; 21 jewels; 28,800 vph; 65-hour power reserve; Geneva Seal
Functions: hours, minutes, subsidiary seconds; full calendar with date, weekday, month, moon phase
Case: rose gold, ø 40 mm, height 10.35 mm; sapphire crystal; transparent case back; water-resistant to 3 atm
Band: reptile skin, buckle
Price: $35,300

Historique 1942
Reference number: 3110V/000A-B426
Movement: manually wound, Vacheron Constantin Caliber 4400 QC; ø 29 mm, height 4.6 mm; 21 jewels; 28,800 vph; 65-hour power reserve; Geneva Seal
Functions: hours, minutes, subsidiary seconds; full calendar with date, weekday, month
Case: stainless steel, ø 40 mm, height 10.35 mm; sapphire crystal; transparent case back; water-resistant to 3 atm
Band: reptile skin, buckle
Price: $19,700

VACHERON CONSTANTIN

Patrimony Moon Phase and Retrograde Date
Reference number: 4010U/000G-B330
Movement: automatic, Vacheron Constantin Caliber 2460 PDL; ø 27.2 mm, height 5.4 mm; 27 jewels; 28,800 vph; 40-hour power reserve; Geneva Seal
Functions: hours, minutes; date (retrograde), moon phase and age
Case: rose gold, ø 42.5 mm, height 9.7 mm; sapphire crystal; transparent case back; water-resistant to 3 atm
Band: reptile skin, buckle
Price: $40,600
Variations: white gold

Patrimony Automatic Small Model
Reference number: 4100U/000G-B181
Movement: automatic, Vacheron Constantin Caliber 2450 Q6; ø 26.2 mm, height 3.6 mm; 27 jewels; 28,800 vph; 40-hour power reserve; Geneva Seal
Functions: hours, minutes, sweep seconds; date
Case: white gold, ø 36 mm, height 8.1 mm; sapphire crystal; transparent case back; water-resistant to 3 atm
Band: reptile skin, buckle
Price: $25,400

Patrimony
Reference number: 81180/000R-9159
Movement: manually wound, Vacheron Constantin Caliber 1400; ø 20.35 mm, height 2.6 mm; 20 jewels; 28,800 vph; 40-hour power reserve; Geneva Seal
Functions: hours, minutes
Case: rose gold, ø 40 mm; sapphire crystal; transparent case back
Band: reptile skin, buckle
Price: $19,100
Variations: white gold or yellow gold

Malte Subsidiary Seconds
Reference number: 82230/000R-9963
Movement: manually wound, Vacheron Constantin Caliber 4400 AS; ø 28.6 mm, height 2.8 mm; 21 jewels; 28,800 vph; 65-hour power reserve; Geneva Seal
Functions: hours, minutes, subsidiary seconds
Case: rose gold, 36.7 × 47.61 mm, height 9.1 mm; sapphire crystal; water-resistant to 3 atm
Band: reptile skin, buckle
Price: $26,000
Variations: white gold ($26,000)

Les Cabinotiers Celestia Astronomica Grande Complication 3600
Reference number: 9720C/000G-B281
Movement: manually wound, Vacheron Constantin Caliber 3600; ø 36 mm, height 8.7 mm; 64 jewels; 18,000 vph; 1-minute tourbillon; 6 spring barrels, 21-day power reserve; Geneva Seal
Functions: hours, minutes, sweep seconds; power reserve indicator; perpetual calendar with date, weekday, month, moon phase, season, moon age, sunrise/sunset times; day and night duration; solar system conjunctions; celestial map of northern hemisphere
Case: white gold, ø 45 mm, height 13.6 mm; sapphire crystal; transparent case back; water-resistant to 3 atm
Price: on request

Les Cabinotiers Grande Complication Crocodile
Reference number: 9700C/001R-B187
Movement: manually wound; ø 33.9 mm, height 12.15 mm; 42 jewels; 18,000 vph; 1-minute tourbillon; 58-hour power reserve.; Geneva Seal
Functions: hours, minutes, subsidiary seconds (on tourbillon cage); power reserve indicator, minute repeater, time equation display; perpetual calendar with date, weekday, month, moon phase, leap year, season, moon age, sunrise/sunset times
Case: rose gold, ø 47 mm, height 19.1 mm; sapphire crystal; transparent case back
Band: reptile skin, buckle
Remarks: unique piece
Price: on request

VACHERON CONSTANTIN

Caliber 2460 QCL/1
Automatic; stop-seconds mechanism; single spring barrel, 40-hour power reserve; Geneva Seal
Functions: hours, minutes, sweep seconds; full calendar with date, weekday, month
Diameter: 29 mm
Height: 5.4 mm
Jewels: 27
Balance: glucydur
Frequency: 28,800 vph
Remarks: gold rotor; 308 parts

Caliber 2475 SC/2
Automatic; single spring barrel, 40-hour power reserve; Geneva Seal
Functions: hours, minutes, sweep seconds; power reserve indicator; date, weekday
Diameter: 26.2 mm
Height: 5.7 mm
Jewels: 27
Frequency: 28,800 vph
Remarks: gold rotor; 264 parts

Caliber 5110 DT
Automatic; single spring barrel, 60-hour power reserve; Geneva Seal
Functions: hours, minutes, sweep seconds; additional 12-hour display (2nd time zone), day/night indicator; date
Diameter: 30.6 mm
Height: 6 mm
Jewels: 37
Balance: glucydur
Frequency: 28,800 vph
Remarks: gold rotor; 234 parts

Caliber 2160
Manually wound; 1-minute tourbillon; double spring barrel, 80-hour power reserve; Geneva Seal
Functions: hours, minutes, subsidiary seconds (on tourbillon cage)
Diameter: 31 mm
Height: 5.65 mm
Jewels: 30
Balance: glucydur
Frequency: 18,000 vph
Remarks: 188 parts

Caliber 2460 G4/1
Automatic; single spring barrel, 40-hour power reserve; Geneva Seal
Functions: disk display for hours, minutes; disk display for date, weekday
Diameter: 31 mm
Height: 6.05 mm
Jewels: 27
Balance: glucydur
Frequency: 28,800 vph
Remarks: gold rotor; 237 parts

Caliber 3300
Manually wound; column wheel control of chronograph functions, horizontal clutch; single spring barrel, 65-hour power reserve; Geneva Seal
Functions: hours, minutes, subsidiary seconds; power reserve indicator; chronograph with crown pusher
Diameter: 32.8 mm
Height: 6.7 mm
Jewels: 35
Balance: glucydur
Frequency: 21,600 vph
Remarks: 252 parts

VACHERON CONSTANTIN

Caliber 5200
Automatic; column wheel control of chronograph functions; single spring barrel, 52-hour power reserve; Geneva Seal
Functions: hours, minutes, subsidiary seconds; chronograph; date
Diameter: 30.6 mm
Height: 6.6 mm
Jewels: 54
Balance: glucydur
Frequency: 28,800 vph
Remarks: gold rotor; 263 parts

Caliber 1120 QP
Automatic; extra-thin construction; winding rotor with supporting ring; single spring barrel, 40-hour power reserve; Geneva Seal
Functions: hours, minutes; perpetual calendar with date, weekday, month, moon phase, leap year
Diameter: 29.6 mm
Height: 4.05 mm
Jewels: 36
Balance: glucydur
Frequency: 19,800 vph
Remarks: skeletonized rotor with gold oscillating mass; 276 parts

Caliber 1731
Manually wound; single spring barrel, 65-hour power reserve; Geneva Seal
Functions: hours, minutes, subsidiary seconds; hour, quarter-hour, and minute repeater
Diameter: 32.8 mm
Height: 3.9 mm
Jewels: 36
Balance: glucydur
Frequency: 21,600 vph
Remarks: perlage on mainplate, beveled edges, bridges with côtes de Genève

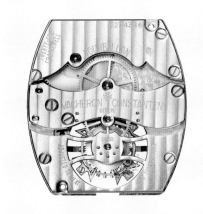

Caliber 2795
Automatic; 1-minute tourbillon; single spring barrel, 45-hour power reserve; Geneva Seal
Functions: hours, minutes, subsidiary seconds (on tourbillon cage)
Dimensions: 27.37 × 29.3 mm
Height: 6.1 mm
Jewels: 27
Balance: glucydur
Frequency: 18,000 vph
Remarks: tonneau-shaped

Caliber 2260
Manually wound; 1-minute tourbillon; quadruple spring barrel, 336-hour power reserve; Geneva Seal
Functions: hours, minutes, subsidiary seconds (on tourbillon cage); power reserve indicator
Diameter: 29.1 mm
Height: 6.8 mm
Jewels: 31
Balance: glucydur
Frequency: 18,000 vph
Remarks: 231 parts

Caliber 1003
Manually wound; single spring barrel, 31-hour power reserve; Geneva Seal
Functions: hours, minutes
Diameter: 21.1 mm
Height: 1.64 mm
Jewels: 18
Balance: glucydur
Frequency: 18,000 vph
Remarks: thinnest movement currently being manufactured

VAN CLEEF & ARPELS

Van Cleef & Arpels
2, rue du Quatre-Septembre
F-75002 Paris
France

Tel.:
+33-1-70-70-36-56

Website:
www.vancleefarpels.com

Founded:
1906

U.S. distributor:
1-877-VAN-CLEEF

Most important collections:
Charms; Pierre Arpels; Poetic Complications

In 1999, while shopping around for more companies to add to its roster of high-end jewelers, Richemont Group decided to purchase Van Cleef & Arpels. The venerable jewelry brand had a lot of name recognition, thanks in part to a host of internationally known customers, like Jacqueline Kennedy Onassis, whose two marriages each involved a Van Cleef & Arpels ring. It also had a reputation for the high quality of its workmanship. It was Van Cleef & Arpels that came up with the mystery setting using a special rail and cut totally hidden from the casual eye.

Van Cleef & Arpels was a family business that came to be when a young stonecutter, Alfred van Cleef, married Estelle Arpels in 1896, and ten years later opened a business on Place Vendôme in Paris with Estelle's brother Charles. More of Estelle's brothers joined the firm, which was soon booming and serving, quite literally, royalty.

Watches were always a part of the portfolio. But after joining Richemont, Van Cleef now had the support of a very complete industrial portfolio that would allow it to make stunning movements that could bring dials to life. A collaboration with Jean-Marc Wiederrecht and Agenhor produced outstanding combinations of artistry in design and crafts, with horological excellence that made the watch-loving public take notice. The brand has come up with some genuine innovations: The Midnight Nuit Lumineuse lights up six diamonds with a pusher using a ceramic band and the phenomenon of piezoelectricity. The Lady Arpels Planétarium won the Ladies' Complication Prize at the prestigious GPHG in 2018. The watch features an extraordinary complication: Mercury, Venus, and Earth rotating in real time around the sun, with the Moon rotating around Earth, and a shooting star fulfilling wishes all day long on an aventurine sky.

Lady Arpels Planétarium
Reference number: VCAROAR500
Movement: automatic, Valfleurier Q020 with exclusive module (van der Klaauw); 34 jewels; 40-hour power reserve
Functions: retrograde hours and minutes
Case: white gold, ø 38 mm, height 11.8 mm; white gold bezel with round diamonds; sculpted bridge; diamond on crown
Band: reptile skin, white gold buckle
Remarks: planetarium with aventurine dial, pink gold sun and white gold shooting star, pink mother-of-pearl Mercury, green enamel Venus, turquoise Earth, diamond Moon; planets rotate at actual speed
Price: $245,000
Variation: on diamond bracelet ($330,000)

Midnight Nuit Lumineuse
Reference number: VCARO5YB00
Movement: automatic, Valfleurier Q020, exclusive caliber developed for Van Cleef & Arpels; 50 jewels; 40-hour power reserve
Functions: retrograde hours; minutes
Case: white gold, ø 42 mm, height 12.1 mm; white gold bezel set with diamonds; round diamond on crown; water-resistant up to 3 atm
Remarks: six LEDs backlight diamonds on the dial to form the Unicorn constellation on the dial, lit by piezoelectricity
Band: reptile skin, buckle
Remarks: aventurine dial with miniature painting and diamonds
Price: on request

Pierre Arpels Heure d'ici & Heure d'ailleurs
Reference number: VCARO4II00
Movement: automatic, exclusive Agenhor caliber; 48-hour power reserve
Functions: double jumping hours and retrograde minute; dual time zone
Case: rose gold, ø 42 mm, height 7.97 mm; white gold bezel; crown with round diamond; sapphire case back; water-resistant up to 3 atm
Remarks: black lacquer dial with sunburst motif on the edge
Band: reptile skin, white gold buckle
Price: $28,300
Variations: in white gold and white lacquer dial with "piqué" motif ($40,900)

Vortic Watch Co.
517 N. Link Lane Unit A
Fort Collins, CO 80524

Tel.:
855-285-7884

E-mail:
info@vorticwatches.com

Website:
www.vorticwatches.com

Founded:
2014

Number of employees:
5

Most important collections/price range:
American Artisan, Railroad Edition, Journeyman Edition, "Convert your watch" / $1,450 to $5,000

VORTIC WATCH COMPANY

The U.S. watch industry produced some very fine timepieces back in the nineteenth century. Big names like Ball, Elgin, Hamilton, and Waltham have that ring of rugged individualism and the lonesome whistle of trains chugging back and forth across the country. So where did the millions of pocket watches go?

Enter R. T. Custer from Pennsylvania, and open the book on a tale of American entrepreneurial gumption. While golfing with a friend during his student days, Custer heard about companies gathering cases of old pocket watches for their gold and silver, and throwing out the movements, dials, hands, and anything deemed worthless.

His mental escapement started ticking. He took some classes in industrial design, learned all about 3D printing, graduated, and moved out to Colorado. He invited his golfing buddy to come along for the ride, crowdfunded seed money, and started printing cases of bronze for the movements. Restoration work was done by a professional watchmaker, who designed the uniform sleeve to fit the trademark gold-plated winding crowns at 12 o'clock.

Four years later, the business has been transformed and expanded. The 3D printed titanium cases are made for the American Artisan series, which also features a crystal of Corning's very robust Gorilla Glass. Potential customers can send in their old inherited pocket watches for a reconditioning and wrist conversion. The Railroad Edition offers vintage railroad watches featuring a lever under the removable bezel, a system that prevented the watches from being accidentally reset. Finally, there's the Journeyman Edition, which is in fact a watch customizer the consumer can use to create his or her own watch with a nice vintage look.

American Artisan Series "The Boston"

Movement: manually wound, antique Waltham pocket watch movement; 12size (ø 20.32 mm); 17 jewels
Functions: hours, minutes, seconds
Case: 3D printed titanium, ø 46 mm, height 12 mm; Gorilla Glass crystal; transparent case back; water-resistant to 1 atm
Band: distressed leather, buckle
Remarks: unique subsidiary seconds dial; original dial and hands
Price: $1,495 to $3,995 (all unique pieces)
Variations: comes in different cases, 36 mm, 46 mm, 49 mm; cases in machined titanium and machined bronze

American Artisan Series "The Chicago"

Movement: antique Elgin Pocket Watch movement; 12size (ø 20.32 mm); 17 jewels
Functions: hours, minutes, seconds
Case: bronze, ø 46 mm, height 12 mm; Gorilla Glass crystal; transparent case back; water-resistant to 1 atm
Band: distressed leather, buckle
Remarks: unique patina on dial; original antique dial and hands
Price: $1,495 to $3,995 (all unique pieces)
Variations: comes in different cases, 36 mm, 46 mm, 49 mm; cases in machined titanium and machined bronze

The Railroad Edition

Movement: antique Illinois railroad-grade pocket watch movement; 16size (ø 16.93 mm); 21 jewels
Functions: hours, minutes, seconds
Case: CNC machined titanium, ø 51 mm, height 15 mm; special bezel system allows access to railroad lever setting mechanism; Gorilla Glass crystal; transparent case back; water-resistant 1 atm
Band: distressed leather, buckle
Remarks: unique; original dial and hands
Price: $2,495 to $4,995 (all unique pieces)
Variations: cases in machined titanium and machined bronze

VOSTOK-EUROPE

Vostok-Europe is a young brand with old roots. In 2014, it celebrated its tenth anniversary.

What started as a joint venture between the original Vostok company—a wholly separate entity—deep in the heart of Russia and a start-up in the newly minted European Union member nation of Lithuania has grown into something altogether different over the years. Originally, every Vostok model had a proprietary Russian engine, a 32-jewel automatic built by Vostok in Russia. Over the years, demand and the need for alternative complications expanded the portfolio of movements to include Swiss and Japanese ones. While the heritage of the eighty-year Russian watch industry is still evident in the inspirations and designs of Vostok-Europe, the watches built today have become favorites of extreme athletes the world over.

"Real people doing real things" is the mantra that Igor Zubovskij, managing director of the company, often repeats. "We don't use models to market our watches. Only real people test our watches in many different conditions."

That community of "real people" includes cross-country drivers in the Dakar Rally, one of the most famous aerobatic pilots in the world, a team of spelunkers who literally went to the bottom of the world in the Krubera Cave, and world free-diving champions. Much of the Vostok-Europe line is of professional dive quality. For illumination, some models incorporate tritium tube technology, which offers about twenty-five years of constant lighting. The Lunokhod 2, the current flagship of the brand, incorporates vertical tubes in a "candleholder" design for full 360-degree illumination.

The watches are assembled in Vilnius, Lithuania, and Zubvoskij still personally oversees quality control operations. The Mriya, named after the world's largest cargo airplane, was the first watch in the world to carry the new Seiko NE88 column wheel chronograph movement.

Koliz Vostok Co. Ltd.
Naugarduko 41
LT-03227 Vilnius
Lithuania

Tel.:
+370-5-2106342

Fax:
+370-4-2130777

E-mail:
info@vostok-europe.com

Website:
www.vostok-europe.com

Founded:
2003

Number of employees:
24

Annual production:
30,000 watches

U.S. distributor:
Vostok-Europe
Détente Watch Group
244 Upton Road, Suite 4
Colchester, CT 06415
877-486-7865
www.detentewatches.com

Most important collections/price range:
Anchar collection / starting at $759; Mriya / starting at $649

Expedition Everest Underground Automatic

Reference number: YN84-597A543
Movement: automatic, Seiko Epson YN84; ø 29.36 mm; 22 jewels; 21,600 vph; 40-hour power reserve
Functions: hours, minutes, sweep seconds; power reserve indication, 24-hour display
Case: stainless steel, ø 48 mm, height 17.5 mm; unidirectional bezel with 0-60 scale, hardened anti-reflective K1 mineral glass; screw-in crown; water-resistant to 20 atm
Band: leather, buckle
Remarks: "Trigalight" constant illumination; comes with silicon strap, changing tool, and dry box
Price: $769; limited and numbered edition of 3,000 pieces

Expedition Everest Underground

Reference number: YM8J-597E546
Movement: quartz, Seiko Epson YM8J
Functions: hours, minutes, sweep seconds; world time with city references; weekday, date, 24-hour chronograph, 24-hour countdown, 24-hour sound alarm
Case: stainless steel, ø 47 mm, height 17.5 mm; unidirectional bezel with 0-60 scale, hardened antireflective K1 mineral glass; screw-in crown; water-resistant to 20 atm
Band: leather, buckle
Remarks: "Trigalight" constant illumination; comes with silicon strap, changing tool, and dry box
Price: $599; limited and numbered edition of 3,000 pieces

Lunokhod 2 "Tritium Gaslight"

Reference number: YM86-620A506
Movement: S. Epson YM86; ø 27 mm, height 3.7 mm
Functions: hours, minutes, subsidiary seconds; 24-hour chronograph; 24-hour sound alarm; perpetual calendar; days of week
Case: stainless steel, ø 49 mm, height 17.5 mm; unidirectional bezel with 0-60 scale, hardened antireflective K1 mineral glass; water-resistant to 30 atm, helium release valve
Band: calfskin, buckle
Remarks: 2nd silicon band, screwdriver and dry box; "Trigalight" constant tritium illumination
Price: $899

VOSTOK-EUROPE

Ekranoplan Automatic "Tritium Gaslight"
Reference number: NH35-546H515
Movement: automatic, Seiko Caliber NH35A; ø 27.4 mm, height 5.32 mm; 24 jewels; 21,600 vph; 40-hour power reserve
Functions: hours, minutes, sweep seconds; date
Case: titanium, ø 47 mm, height 16 mm; unidirectional bezel with 0-60 scale, K1 hardened antireflective mineral glass; screw-in crown; water-resistant to 20 atm
Band: leather, buckle
Variations: silicon band, nylon strap
Price: $759

Energia "Tritium Gaslight"
Reference number: NH35-575O286
Movement: automatic, Seiko Caliber NH35; ø 26.4 mm, height 5.32 mm; 24 jewels, 21,600 vph, 40-hour power reserve
Functions: hours, minutes, sweep seconds; date
Case: bronze, ø 48 mm, height 17.3 mm; unidirectional bezel with 0-60 scale, K1 hardened antireflective mineral glass; water-resistant to 30 atm, helium release valve
Band: leather, buckle
Remarks: 2nd silicon band, screwdriver, and dry box; "Trigalight" constant tritium illumination
Price: $999

Energia 2
Reference number: YN84/575O540
Movement: automatic, Seiko Caliber YN84; ø 26.6 mm, height 5.32 mm; 22 jewels, 21,600 vph, 41-hour power reserve
Functions: hours, minutes, sweep seconds; power reserve indicator; 24-hour indicator
Case: bronze, ø 48 mm, height 17 mm; unidirectional bezel with 0-60 scale, K1 hardened antireflective mineral glass; water-resistant to 30 atm, helium release valve; helium release valve
Band: leather, buckle
Remarks: 2nd silicon band, screwdriver, and dry box; "Trigalight" constant tritium illumination
Price: $1,079; limited and numbered edition of 3,000 pieces

Anchar Men's Diver Watch
Reference number: NH35A-510C530
Movement: automatic, Seiko Caliber NH35; ø 26.6 mm, height 5.32 mm; 24 jewels, 21,600 vph, 41-hour power reserve
Functions: hours, minutes, sweep seconds; date
Case: stainless steel, ø 48 mm, height 16 mm; unidirectional bezel with 0-60 scale, hardened antireflective K1 mineral glass; screw-in crown; water-resistant to 30 atm;
Band: calfskin, buckle
Remarks: 2nd silicon band, screwdriver, and dry box; "Trigalight" constant tritium illumination
Price: $639
Variations: various dial colors

Gaz-14 Automatic Dual Time
Reference number: 2426-5601059
Movement: automatic, Vostok Caliber 2426; ø 24 mm, 32 jewels; 18,000 vph; blued screws; 31-hour power reserve
Functions: hours, minutes, sweep seconds; additional 24-hour display (2nd time zone)
Case: stainless steel, ø 43 mm, height 13.8 mm; hardened antireflective K1 mineral crystal; transparent case back; water-resistant to 5 atm
Band: calfskin, buckle
Price: $549

GAZ-14 World Timer & Alarm "Tritium Gaslight"
Reference number: YM26-565B293
Movement: S. Epson YM26; ø 27 mm, height 3.7 mm
Functions: hours, minutes, subsidiary seconds; hours and minutes of 2nd time zone, hours and minutes of sound alarm; date
Case: stainless steel with RG PVD coating, ø 45 mm, height 15 mm; hardened antireflective K1 mineral crystal; water-resistant to 5 atm
Band: calfskin, buckle
Remarks: "Trigalight" constant tritium illumination
Price: $739

WEMPE GLASHÜTTE

Ever since 2005, the global jewelry chain Gerhard D. Wempe KG has been putting out watches under its own name again. It was probably inevitable: Gerhard D. Wempe, who founded the company in the late nineteenth century in Oldenburg, was himself a watchmaker. And in the 1930s, the company also owned the Hamburg chronometer works that made watches for seafarers and pilots.

Today, while Wempe remains formally in Hamburg, its manufacturing is done in Glashütte. The move to the fully renovated and expanded Urania observatory in the hills above town was engineered by Eva-Kim Wempe, great-granddaughter of the founder. There, the company does all its after-sales service and tests watches according to the strict German Industrial Norm (DIN 8319), with official blessings from the Saxon and Thuringian offices for measurement and calibration and accreditation from the German Calibration Service.

The move to Glashütte coincided with a push to verticalize by creating a line of in-house movements for the exclusive Chronometerwerke models, like the very retro Chronometerwerke Power Reserve, or the eminently noticeable Tonneau Tourbillon. The calibers, bearing the initials CW, are made in partnership with companies like Nomos in Glashütte or the Swiss workshop MHVJ. In 2016, Wempe released the CW4, an automatic that had its first "outing" in a classic three-hander. It has a promising future ahead of it.

The second Wempe line is called Zeitmeister, or Master of Time. This collection uses more standard, but reworked, ETA or Sellita calibers. It meets all the requirements of the high art of watchmaking and, thanks to its accessible pricing, is attractive for budding collectors. All models are in the middle price range, which the luxury watch industry has long shunned.

Gerhard D. Wempe KG
Steinstrasse 23
D-20095 Hamburg
Germany

Tel.:
+49-40-334-480

Fax:
+49-40-331-840

E-mail:
info@wempe.de

Website:
www.wempe.de

Founded:
1878

Number of employees:
717 worldwide; 24 at Wempe Glashütte I/SA

Annual production:
5,000 watches

U.S. distributor:
Wempe Timepieces
700 Fifth Avenue
New York, NY 10019
212-397-9000
www.wempe.com

Most important collections/price range:
Wempe Zeitmeister / approx. $1,000 to $4,500;
Wempe Chronometerwerke / approx. $5,000 to $95,000

Chronometerwerke Automatic
Reference number: WG 090002
Movement: automatic, Wempe Caliber CW4; ø 32.8 mm, height 6 mm; 35 jewels; 28,800 vph; 2 spring barrels, three-quarter plate, hand-engraved balance cock, 6 gold chatons, tungsten microrotor, finely finished with Glashütte ribbing; 92-hour power reserve; DIN-certified chronometer
Functions: hours, minutes, sweep seconds; date
Case: stainless steel, ø 41 mm, height 11.7 mm; sapphire crystal; transparent case back; water-resistant to 3 atm
Band: reptile skin, buckle
Price: $8,030
Variations: yellow gold ($17,270)

Chronometerwerke Power Reserve
Reference number: WG 080005
Movement: manually wound, Wempe Caliber CW3; ø 32 mm, height 6.1 mm; 40 jewels; 28,800 vph; three-quarter plate, 3 screw-mounted gold chatons, hand-engraved balance cock; 42-hour power reserve; DIN-certified chronometer
Functions: hours, minutes, subsidiary seconds; power reserve indicator
Case: yellow gold, ø 43 mm, height 12.5 mm; sapphire crystal; transparent case back; water-resistant to 3 atm
Band: reptile skin, buckle
Price: $16,120
Variations: stainless steel ($8,500)

Chronometerwerke Small Seconds
Reference number: WG 070002
Movement: manually wound, Wempe Caliber CW3.1; ø 32.8 mm, height 6.1 mm; 40 jewels; 28,800 vph; three-quarter plate, 3 screw-mounted gold chatons, swan-neck fine adjustment, hand-engraved balance cock; 42-hour power reserve; DIN-certified chronometer
Functions: hours, minutes, subsidiary seconds
Case: stainless steel, ø 41 mm, height 12.5 mm; sapphire crystal; transparent case back; water-resistant to 3 atm
Band: reptile skin, buckle
Price: $6,880
Variations: yellow gold ($17,000)

WEMPE GLASHÜTTE

Zeitmeister Annual Calendar

Reference number: WM 690003
Movement: automatic, ETA Caliber 2892-A2 with Dubois Dépraz module 5900; ø 25.6 mm, height 5.29 mm; 21 jewels; 28,800 vph; 42-hour power reserve; DIN-certified chronometer
Functions: hours, minutes, sweep seconds; annual calendar with date, weekday, month, moon phase
Case: stainless steel, ø 42 mm, height 14.28 mm; sapphire crystal; water-resistant to 5 atm
Band: reptile skin, folding clasp
Price: $9,215; limited to 100 pieces

Zeitmeister Chronograph with Moon Phase and Full Calendar

Reference number: WM 530001
Movement: automatic, ETA Caliber 7751; ø 30 mm, height 7.9 mm; 25 jewels; 28,800 vph; 48-hour power reserve; DIN-certified chronometer
Functions: hours, minutes, subsidiary seconds; additional 24-hour display; chronograph; full calendar with date, weekday, month, moon phase
Case: stainless steel, ø 42 mm, height 14.71 mm; sapphire crystal; water-resistant to 5 atm
Band: reptile skin, folding clasp
Price: $4,015

Zeitmeister Chronograph

Reference number: WM 540001
Movement: automatic, ETA Caliber 7753; ø 30 mm, height 7.9 mm; 27 jewels; 28,800 vph; 42-hour power reserve; DIN-certified chronometer
Functions: hours, minutes, subsidiary seconds; chronograph; date
Case: stainless steel, ø 42 mm, height 15 mm; sapphire crystal; water-resistant to 5 atm
Band: reptile skin, folding clasp
Price: $2,515

Zeitmeister Stahl I

Reference number: WM 690010
Movement: automatic, ETA Caliber 2000-1; ø 19.4 mm, height 3.6 mm; 20 jewels; 28,800 vph; 40-hour power reserve
Functions: hours, minutes, sweep seconds; date
Case: stainless steel with gray PVD coating, 31.8 × 49 mm, height 10 mm; sapphire crystal; water-resistant to 5 atm
Band: calfskin, buckle
Remarks: designed by German pop star Herbert Grönemeyer
Price: $3,400; limited to 250 pieces

Zeitmeister Stahl I

Reference number: WM 690011
Movement: automatic, ETA Caliber 2000-1; ø 19.4 mm, height 3.6 mm; 20 jewels; 28,800 vph; 40-hour power reserve
Functions: hours, minutes, sweep seconds; date
Case: stainless steel, 31.8 × 49 mm, height 10 mm; sapphire crystal; water-resistant to 5 atm
Band: calfskin, buckle
Remarks: designed by German pop star Herbert Grönemeyer
Price: $3,290; limited to 250 pieces

Zeitmeister Aviator Watch Chronograph XL

Reference number: WM 600005
Movement: automatic, ETA Caliber A07.211; ø 36.6 mm, height 7.9 mm; 25 jewels; 28,800 vph; 46-hour power reserve; DIN-certified chronometer
Functions: hours, minutes, subsidiary seconds; chronograph; date
Case: stainless steel, ø 45 mm, height 15.45 mm; sapphire crystal; water-resistant to 5 atm
Band: horse leather, folding clasp
Price: $2,890

ZEITWINKEL

Zeitwinkel turned ten in 2016, but that is not really important for this small, independent company based in St.-Imier, one of the hubs of the watch industry in Switzerland. The key attributes of the brand, ones that many watch manufacturers aspire to endow their creations with, are "timeless, simple, and sustainable." What are ten years compared to timelessness?

The models produced by Zeitwinkel (the name means "time angle") are deceptively classical: The simplest exemplar is a two-hand watch; the most complicated, the 273°, a three-hand timepiece with power reserve display and large date. The most decoration one will find on the dials is a spangling of stylized *W*s, for *Winkel* (angle). With cases designed by Jean-François Ruchonnet (TAG Heuer V4, Cabestan), the watches look fairly "German," which comes as no surprise, because Zeitwinkel's founders, Ivica Maksimovic and Peter Nikolaus, hail from there. Some details will catch the eye, notably the extra-large subsidiary seconds dial or the large date aperture, found beside the 11 o'clock marker.

The most valuable part of the watches is their veritable *manufacture* movements, the likes of which are very rare in the business. The calibers were developed by Laurent Besse and his *artisans horlogers*, or watchmaking craftspeople. All components come courtesy of independent suppliers—Zeitwinkel balance wheels, pallets, escape wheels, and Straumann spirals, for example, are produced by Precision Engineering, a company associated with watch brand H. Moser & Cie. The 273° comes with a smoked sapphire crystal dial; the new 083° is smaller (39 millimeters) as an epitome of discreetness.

In keeping with the company's ideals, you won't find any alligator in Zeitwinkel watch bands. Choices here are exclusively rubber, calfskin, or calfskin with an alligator-like pattern. Gold cases were also once taboo for the brand, but thanks to a partnership with the Alliance for Responsible Mining, the watches now come in "fair-mined" gold cases.

Zeitwinkel Montres SA
Rue Pierre-Jolissaint 35
CH-2610 Saint-Imier
Switzerland

Tel.:
+41-32-940-17-71

Fax:
+41-32-940-17-81

E-mail:
info@zeitwinkel.ch

Website:
www.zeitwinkel.ch

Founded:
2006

Annual production:
approx. 800 watches

U.S. distributor:
Tourneau
510 Madison Avenue
New York, NY 10022
212-758-5830
Right Time
7110 E. County Line Road
Highlands Ranch, CO 80126
303-862-3900; 303-862-3905 (fax)

Most important collections/price range:
mechanical wristwatches / starting at around $7,500

082° Email Grand Feu

Reference number: 082-3.S02-01-23
Movement: automatic, Caliber ZW0102; ø 30.4 mm, height 5.7 mm; 30 jewels; 28,800 vph; German silver three-quarter plate and bridges, côtes de Genève, polished screws and edges; 72-hour power reserve
Functions: hours, minutes, sweep seconds
Case: stainless steel, ø 39 mm, height 11.6 mm; sapphire crystal; transparent case back; water-resistant to 5 atm
Band: calfskin, folding clasp
Remarks: white enamel dial, grand feu
Price: $10,500
Variations: different bands

273° Saphir Fumé

Reference number: 273-4.S01-01-21
Movement: automatic, Caliber ZW0103; ø 30.4 mm, height 8 mm; 49 jewels; 28,800 vph; German silver three-quarter plate and bridges, côtes de Genève, polished screws and edges; perlage on dial side; 72-hour power reserve
Functions: hours, minutes, subsidiary seconds; power reserve indicator; patented big date mechanism
Case: stainless steel, ø 42.5 mm, height 13.8 mm; sapphire crystal; transparent back; water-resistant to 5 atm
Band: calfskin, folding clasp
Remarks: smoky black sapphire crystal dial
Price: $15,500
Variations: various dial colors; different bands

188° Galvano-blue

Reference number: 188-23-01-23
Movement: automatic, Caliber ZW0102; ø 30.4 mm, height 5.7 mm; 28 jewels; 28,800 vph; German silver three-quarter plate and bridges, côtes de Genève, polished screws and edges; 72-hour power reserve
Functions: hours, minutes, subsidiary seconds; date
Case: stainless steel, ø 39 mm, height 11.6 mm; sapphire crystal; transparent case back; water-resistant to 5 atm
Band: calfskin, folding clasp
Price: $7,490
Variations: various dial colors

ZENITH

Zenith Branch
LVMH Swiss Manufactures SA
34, rue des Billodes
CH-2400 Le Locle
Switzerland

Tel.:
+41-32-930-6262

Fax:
+41-32-930-6363

Website:
www.zenith-watches.com

Founded:
1865

Number of employees:
over 330 employees worldwide

U.S. distributor:
Zenith Watches
966 South Springfield Avenue
Springfield, NJ 07081
866-675-2079
contact.zenith@lvmhwatchjewelry.com

Most important collections/price range:
Academy / from $76,100; Elite / from $4,700; Chronomaster / from $6,700; Pilot / from $5,700; Defy / from $9,600

The tall, narrow building in Le Locle, with its closely spaced high windows to let in daylight, is a testimony to Zenith's history as a self-sufficient *manufacture* in the entrepreneurial spirit of the Industrial Revolution. The company, founded in 1865 by Georges Favre-Jacot as a small watch reassembly workshop, has produced and distributed every type of watch from the simple pocket watch to the most complicated calendar. But it remains primarily linked with the El Primero caliber, the first wristwatch chronograph movement boasting automatic winding and a frequency of 36,000 vph. Only a few watch manufacturers had risked such a high oscillation frequency—and none of them with such complexity as the integrated chronograph mechanism and bilaterally winding rotor of the El Primero.

The movement might have sunk into obscurity had it not been for the purchase of the brand by the LVMH Group in 1999. Zenith was dusted off and modernized—perhaps a little too much for some, but the eccentric creations did catapult the name into the world of *haute horlogerie*.

Also dusted was the historic complex in Le Locle, which was put on UNESCO's World Heritage list in 2009. Over eighty different crafts are practiced here, from watchmaking to design, from art to prototyping. Synergies with the Group companion Hublot and TAG Heuer produced the Defy 21, a complex chronograph movement based on the 36,000-vph El Primero. It features two separate gear trains and escapements for time and chronograph functions, respectively. The chronograph movement beats at 360,000 vph, allowing the hundredths of a second to be displayed. The other technical feat is the Zero G that keeps the escapement system in the horizontal position. To keep lovers of classical watches happy, though, Zenith also offers a growing collection of classic-nostalgic pilots' watches.

Defy El Primero 21

Reference number: 95.9000.9004/78.R582
Movement: automatic, Zenith Caliber 9004 "El Primero"; ø 32 mm, height 7.9 mm; 53 jewels; 36,000 vph; independent chronograph mechanism with separate escapement (360,000 vph) and power management; 50-hour power reserve; COSC-certified chronometer
Functions: hours, minutes, subsidiary seconds; power reserve indicator (for chronograph functions); chronograph (1/100th of a second display)
Case: titanium, ø 44 mm, height 14.5 mm; sapphire crystal; transparent case back; water-resistant to 10 atm
Band: rubber with reptile skin overlay, double folding clasp
Price: $11,200

Defy El Primero 21

Reference number: 95.9002.9004/78.M9000
Movement: automatic, Zenith Caliber 9004 "El Primero"; ø 32 mm, height 7.9 mm; 53 jewels; 36,000 vph; independent chronograph mechanism with separate escapement (360,000 vph) and power management; 50-hour power reserve; COSC-certified chronometer
Functions: hours, minutes, subsidiary seconds; power reserve indicator (for chronograph functions); chronograph (1/100th of a second display)
Case: titanium, ø 44 mm, height 14.5 mm; sapphire crystal; transparent case back; water-resistant to 10 atm
Band: titanium, double folding clasp
Price: $12,200

Defy El Primero 21

Reference number: 49.9000.9004/78.R782
Movement: automatic, Zenith Caliber 9004 "El Primero"; ø 32 mm, height 7.9 mm; 53 jewels; 36,000 vph; independent chronograph mechanism with separate escapement (360,000 vph) and power management; COSC-certified chronometer
Functions: hours, minutes, subsidiary seconds; power reserve indicator (for chronograph functions); chronograph (1/100th of a second display)
Case: ceramic, ø 44 mm, height 14.5 mm; sapphire crystal; transparent case back; water-resistant to 10 atm
Band: rubber, double folding clasp
Price: $12,200

ZENITH

Defy El Primero 21
Reference number: 95.9005.9004/01.M9000
Movement: automatic, Zenith Caliber 9004 "El Primero"; ø 32 mm, height 7.9 mm; 53 jewels; 36,000 vph; independent chronograph mechanism with separate escapement (360,000 vph) and power management; 50-hour power reserve; COSC-certified chronometer
Functions: hours, minutes, subsidiary seconds; power reserve indicator (for chronograph functions); chronograph (1/100th of a second display)
Case: titanium, ø 44 mm, height 14.5 mm; sapphire crystal; transparent case back; water-resistant to 10 atm
Band: titanium, double folding clasp
Price: $11,300

Defy El Primero 21
Reference number: 32.9000.9004/78.R582
Movement: automatic, Zenith Caliber 9004 "El Primero"; ø 32 mm, height 7.9 mm; 53 jewels; 36,000 vph; independent chronograph mechanism with separate escapement (360,000 vph); 50-hour power reserve; COSC-certified chronometer
Functions: hours, minutes, subsidiary seconds; power reserve indicator (for chronograph functions); chronograph (1/100th of a second display)
Case: titanium, ø 44 mm, height 14.5 mm; bezel in white gold, with 44 baguette diamonds; sapphire crystal; transparent case back; water-resistant to 10 atm
Band: rubber with reptile skin overlay, double folding clasp
Remarks: case set with 288 diamonds
Price: $29,500

Defy Zero G
Reference number: 18.9000.8812/79.R584
Movement: manually wound, Zenith Caliber 8812 "El Primero"; ø 38.5 mm, height 7.85 mm; 41 jewels; 36,000 vph; gyroscopic "gravity control" module that keeps the regulating organ horizontal irrespective of the watch's position; skeletonized movement; 50-hour power reserve
Functions: hours, minutes, subsidiary seconds; power reserve indicator
Case: rose gold, ø 44 mm, height 14.85 mm; sapphire crystal; transparent case back; water-resistant to 10 atm
Band: rubber with reptile skin overlay, double folding clasp
Price: $115,900
Variations: titanium ($99,800)

Defy Classic
Reference number: 95.9000.670/78.R584
Movement: automatic, Zenith Caliber 670 SK "Elite"; ø 25.6 mm, height 3.88 mm; 27 jewels; 28,800 vph; skeletonized movement; 48-hour power reserve
Functions: hours, minutes, sweep seconds; date
Case: titanium, ø 41 mm, height 10.75 mm; sapphire crystal; transparent case back; water-resistant to 10 atm
Band: rubber with reptile skin overlay, double folding clasp
Price: $6,500
Variations: titanium bracelet ($7,500); rubber strap ($6,500)

Defy Classic
Reference number: 95.9000.670/78.M9000
Movement: automatic, Zenith Caliber 670 SK "Elite"; ø 25.6 mm, height 3.88 mm; 27 jewels; 28,800 vph; skeletonized movement; 48-hour power reserve
Functions: hours, minutes, sweep seconds; date
Case: titanium, ø 41 mm, height 10.75 mm; sapphire crystal; transparent case back; water-resistant to 10 atm
Band: titanium, double folding clasp
Price: $7,500
Variations: reptile skin strap ($6,500); rubber strap ($6,500)

Defy Classic
Reference number: 95.9000.670/51.R790
Movement: automatic, Zenith Caliber 670 SK "Elite"; ø 25.6 mm, height 3.88 mm; 27 jewels; 28,800 vph; 48-hour power reserve
Functions: hours, minutes, sweep seconds; date
Case: titanium, ø 41 mm, height 10.75 mm; sapphire crystal; transparent case back; water-resistant to 10 atm
Band: rubber, double folding clasp
Price: $5,900
Variations: titanium bracelet ($6,900)

Chronomaster El Primero Grande Date Full Open

Reference number: 03.2530.4047/78.C813
Movement: automatic, Zenith Caliber 4047B "El Primero"; ø 30.5 mm, height 9.05 mm; 32 jewels; 36,000 vph; partially skeletonized; 50-hour power reserve
Functions: hours, minutes, subsidiary seconds; chronograph; large date, moon phase
Case: stainless steel, ø 45 mm, height 15.6 mm; sapphire crystal; transparent case back; water-resistant to 10 atm
Band: calfskin, double folding clasp
Price: $10,600

Chronomaster El Primero

Reference number: 03.2040.4061/69.C496
Movement: automatic, Zenith Caliber 4061 "El Primero"; ø 30 mm, height 6.6 mm; 31 jewels; 36,000 vph; partially skeletonized under regulating organ; 50-hour power reserve
Functions: hours, minutes; chronograph
Case: stainless steel, ø 42 mm, height 14.05 mm; sapphire crystal; transparent case back; water-resistant to 10 atm
Band: reptile skin, double folding clasp
Price: $8,600
Variations: rose gold ($16,300)

Chronomaster El Primero

Reference number: 03.2040.4061/21.M2040
Movement: automatic, Zenith Caliber 4061 "El Primero"; ø 30 mm, height 6.6 mm; 31 jewels; 36,000 vph; partially skeletonized under regulating organ; 50-hour power reserve
Functions: hours, minutes; chronograph
Case: stainless steel, ø 42 mm, height 14.05 mm; sapphire crystal; transparent case back; water-resistant to 10 atm
Band: stainless steel, double folding clasp
Price: $9,100

Chronomaster El Primero

Reference number: 51.2080.400/69.C494
Movement: automatic, Zenith Caliber 400B "El Primero"; ø 30 mm, height 6.6 mm; 31 jewels; 36,000 vph; 50-hour power reserve
Functions: hours, minutes, subsidiary seconds; chronograph
Case: stainless steel, ø 42 mm, height 12.75 mm; bezel in rose gold; sapphire crystal; transparent case back; water-resistant to 10 atm
Band: reptile skin, double folding clasp
Price: $8,200

Chronomaster El Primero Classic Cars

Reference number: 03.2046.400/25.C771
Movement: automatic, Zenith Caliber 400B "El Primero"; ø 30 mm, height 6.6 mm; 31 jewels; 36,000 vph; 50-hour power reserve
Functions: hours, minutes, subsidiary seconds; chronograph; date
Case: stainless steel, ø 42 mm, height 12.75 mm; sapphire crystal; transparent case back; water-resistant to 10 atm
Band: calfskin, double folding clasp
Price: $6,700

Chronomaster El Primero

Reference number: 03.2150.400/53.C700
Movement: automatic, Zenith Caliber 400 "El Primero"; ø 30 mm, height 6.6 mm; 31 jewels; 36,000 vph; 50-hour power reserve
Functions: hours, minutes, subsidiary seconds; chronograph; date
Case: stainless steel, ø 38 mm, height 12.45 mm; sapphire crystal; transparent case back; water-resistant to 10 atm
Band: reptile skin, double folding clasp
Price: $6,700

ZENITH

Pilot Type 20 Chronograph Ton-Up

Reference number: 11.2432.4069/21.C900
Movement: automatic, Zenith Caliber 4069 "El Primero"; ø 30 mm, height 6.6 mm; 35 jewels; 36,000 vph; 50-hour power reserve
Functions: hours, minutes, subsidiary seconds; chronograph
Case: stainless steel with black PVD coating, ø 45 mm, height 14.25 mm; sapphire crystal; water-resistant to 10 atm
Band: calfskin, buckle
Price: $7,100

Pilot Type 20 Chronograph Extra Special

Reference number: 29.2430.4069/57.C808
Movement: automatic, Zenith Caliber 4069 "El Primero"; ø 30 mm, height 6.6 mm; 35 jewels; 36,000 vph; 50-hour power reserve
Functions: hours, minutes, subsidiary seconds; chronograph
Case: bronze, ø 45 mm, height 14.25 mm; sapphire crystal; water-resistant to 10 atm
Band: calfskin, buckle
Price: $7,100

Pilot Type 20 Cohiba Edition

Reference number: 18.2430.679/27.C721
Movement: automatic, Zenith Caliber 679 "Elite"; ø 25.6 mm, height 3.9 mm; 27 jewels; 28,800 vph; 50-hour power reserve
Functions: hours, minutes, sweep seconds
Case: rose gold, ø 45 mm, height 14.25 mm; sapphire crystal; water-resistant to 10 atm
Band: reptile skin, buckle
Remarks: not available in the U.S.
Price: on request; limited to 50 pieces

Pilot Type 20 Extra Special

Reference number: 29.1940.679/21.C800
Movement: automatic, Zenith Caliber 679 "Elite"; ø 25.6 mm, height 3.9 mm; 27 jewels; 28,800 vph; 50-hour power reserve
Functions: hours, minutes, sweep seconds
Case: bronze, ø 40 mm, height 12.95 mm; sapphire crystal; water-resistant to 10 atm
Band: calfskin, buckle
Price: $5,700

Pilot Cronometro Tipo CP-2 Flyback

Reference number: 29.2240.405/18.C801
Movement: automatic, Zenith Caliber 405B "El Primero"; ø 30 mm, height 6.6 mm; 31 jewels; 36,000 vph; column wheel control of chronograph functions; 50-hour power reserve
Functions: hours, minutes, subsidiary seconds; flyback chronograph
Case: bronze, ø 43 mm, height 12.85 mm; unidirectional bezel, with 0-60 scale; sapphire crystal; water-resistant to 10 atm
Band: calfskin, double folding clasp
Price: $7,700

Pilot Cronometro Tipo CP-2 Flyback

Reference number: 11.2240.405/21.C773
Movement: automatic, Zenith Caliber 406B "El Primero"; ø 30 mm, height 6.6 mm; 31 jewels; 36,000 vph; column wheel control of chronograph functions; 50-hour power reserve
Functions: hours, minutes, subsidiary seconds; flyback chronograph
Case: stainless steel, ø 43 mm, height 12.85 mm; unidirectional bezel, with 0-60 scale; sapphire crystal; water-resistant to 10 atm
Band: calfskin, double folding clasp
Price: $7,700

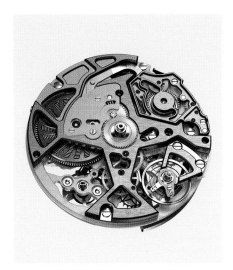

Caliber 9004 El Primero

Automatic; independent chronograph mechanism with separate escapement (360,000 vph) and power management; hairsprings of carbon nanotube matrix; single spring barrel, 50-hour power reserve; COSC-certified chronometer
Functions: hours, minutes, subsidiary seconds; power reserve indicator (for chronograph functions); chronograph with 1/100th of a second display
Diameter: 32.8 mm
Height: 7.9 mm
Jewels: 53
Balance: glucydur
Frequency: 36,000 vph
Shock protection: Kif
Remarks: 203 components; finely finished with côtes de Genève

Caliber 4054 El Primero

Automatic; single spring barrel, 50-hour power reserve
Functions: hours, minutes; chronograph; annual calendar with date, weekday, month
Diameter: 30 mm
Height: 8.3 mm
Jewels: 29
Balance: glucydur
Frequency: 36,000 vph
Balance spring: self-compensating flat hairspring
Shock protection: Kif

Caliber 400 El Primero

Automatic; column wheel control of chronograph functions; single spring barrel, 50-hour power reserve
Functions: hours, minutes, subsidiary seconds; chronograph; date
Diameter: 30 mm
Height: 6.5 mm
Jewels: 31
Balance: glucydur
Frequency: 36,000 vph
Balance spring: self-compensating flat hairspring
Shock protection: Kif
Remarks: 278 parts

Caliber 4047 El Primero

Automatic; single spring barrel, 50-hour power reserve
Functions: hours, minutes; chronograph; large date, sun- and moon-phase display (integrated day/night indicator)
Diameter: 30.5 mm
Height: 9.05 mm
Jewels: 41
Balance: glucydur
Frequency: 36,000 vph
Balance spring: auto-compensating flat hairspring
Shock protection: Kif

Caliber 400B El Primero

Automatic; single spring barrel, 50-hour power reserve
Functions: hours, minutes, subsidiary seconds; chronograph; date
Diameter: 30 mm
Height: 6.6 mm
Jewels: 31
Balance: glucydur
Frequency: 36,000 vph
Balance spring: self-compensating flat hairspring
Shock protection: Kif

Caliber 681 Elite

Automatic; single spring barrel, 50-hour power reserve
Functions: hours, minutes, subsidiary seconds
Diameter: 25.6 mm
Height: 3.81 mm
Jewels: 27
Balance: glucydur
Frequency: 28,800 vph
Balance spring: self-compensating flat hairspring
Shock protection: Kif

CONCEPTO

The Concepto Watch Factory, founded in 2006 in La Chaux-de-Fonds, is the successor to the family-run company Jaquet SA, which changed its name to La Joux-Perret a little while ago and then moved to a different location on the other side of the hub of watchmaking. In 2008, Valérien Jaquet, son of the company founder Pierre Jaquet, began systematically building up a modern movement and watch component factory on an empty floor of the building.

Today, the Concepto Watch Factory employs eighty people in various departments, such as Development/Prototyping, Decoparts (partial manufacturing using lathes, machining, or wire erosion), Artisia (production of movements and complications in large series), as well as Optimo (escapements). In addition to the standard family of calibers, the C2000 (based on the Valjoux) and the vintage chronograph movement C7000 (the evolution of the Venus Caliber), the company's product portfolio includes various tourbillon movements (Caliber C8000) and several modules for adding onto ETA movements (Caliber C1000). A brand-new caliber series, the C3000, features a retrograde calendar and seconds, a power reserve indicator, and a chronograph. The C4000 chronograph caliber with automatic winding is currently in pre-series testing.

One of Concepto's greatest assets is its flexibility. Most of the company's movements are not sold off the shelf, as it were, but rather designed according to the specific requirements of the customer with regard to form or technical DNA. Complicated movements are assembled entirely and tested by the company's watchmakers, while others are sold as kits for assembly by the watchmakers. Annual production is somewhere between 30,000 and 40,000 units, with additional hundreds of thousands of components made for contract manufacturing.

Caliber 1053
Automatic; inverted construction with dial-side escapement; bidirectional off-center winding rotor; single spring barrel; 42-hour power reserve
Functions: hours, minutes, subsidiary seconds (all off-center)
Diameter: 33 mm
Height: 3.75 mm
Jewels: 31
Balance: glucydur
Frequency: 28,800 vph
Balance spring: flat hairspring
Remarks: black finishing on movement

Caliber 2904 (dial side)
Inverted construction with dial-side escapement; single spring barrel; 48-hour power reserve
Functions: hours, minutes, subsidiary seconds
Diameter: 30.4 mm
Height: 4.6 mm
Jewels: 31
Balance: screw balance
Frequency: 28,800 vph
Balance spring: flat hairspring

Caliber 3041 Skeleton (dial side)
Manually wound; skeletonized symmetrical construction; single spring barrel; 48-hour power reserve
Functions: hours, minutes
Diameter: 32.6 mm
Height: 5.5 mm
Jewels: 21
Balance: screw balance
Frequency: 28,800 vph
Balance spring: flat hairspring
Remarks: extensive personalization options for finishing and accessories

Caliber 2000-RAC

Automatic; column wheel control of chronograph functions; stop-second system; single spring barrel; 48-hour power reserve
Functions: hours, minutes, subsidiary seconds; chronograph
Diameter: 30.4 mm; **Height:** 8.4 mm
Jewels: 26; **Balance:** screw balance
Frequency: 28,800 vph
Balance spring: flat hairspring
Shock protection: Incabloc
Remarks: related calibers: 2000 (without control wheel); with two or three totalizers ("tricompax") with or without date; various additional displays (moon phase, retrograde date hand, additional 24-hour sweep hand, power reserve indicator)

Caliber 8500

Manually wound; 1-minute tourbillon; column wheel control of chronograph functions; single spring barrel; 50-hour power reserve
Functions: hours, minutes, subsidiary seconds; split-seconds chronograph
Diameter: 31.3 mm
Height: 7.2 mm
Jewels: 31
Balance: screw balance
Frequency: 21,600 vph
Balance spring: flat hairspring
Remarks: very fine movement finishing

Caliber 8950-A

Automatic; 1-minute tourbillon; single spring barrel; 60-hour power reserve
Functions: hours, minutes
Diameter: 30.4 mm
Height: 6.7 mm
Jewels: 27
Balance: glucydur
Frequency: 28,800 vph
Balance spring: flat hairspring
Remarks: related caliber: 8950-M (manual winding); extensive personalization options for the finishing, accessories, and functions

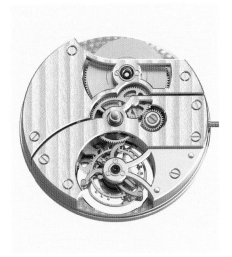

Caliber 8000 (dial side)

Manually wound; 1-minute tourbillon; single spring barrel; 72-hour power reserve
Functions: hours, minutes
Diameter: 32.6 mm
Height: 5.7 mm
Jewels: 19
Balance: screw balance
Frequency: 21,600 vph
Balance spring: flat hairspring
Remarks: extensive personalization options for the finishing, accessories, and functions

Caliber 8152

Automatic; 1-minute tourbillon; bridges and plate made of sapphire crystal; off-center, bidirectional rotor; single spring barrel; 72-hour power reserve
Functions: hours, minutes
Diameter: 32.6 mm
Height: 8.5 mm
Jewels: 25
Balance: screw balance
Frequency: 21,600 vph
Balance spring: flat hairspring
Remarks: extensive personalization options for the finishing, accessories, and functions

Caliber 8908-M (dial side)

Manually wound; flying 1-minute tourbillon; single spring barrel; 42-hour power reserve
Functions: hours, minutes
Diameter: 34.6 mm
Height: 6.6 mm
Jewels: 21
Balance: screw balance
Frequency: 28,800 vph
Balance spring: flat hairspring
Remarks: extensive personalization options for the finishing, accessories, and functions

ETA

This Swatch Group movement manufacturer produces more than five million movements a year. And after the withdrawal of Richemont's Jaeger-LeCoultre as well as Swatch Group sisters Nouvelle Lémania and Frédéric Piguet from the business of selling movements on the free market, most watch brands can hardly help but beat down the door of this full-service manufacturer.

ETA offers a broad spectrum of automatic movements in various dimensions with different functions, chronograph mechanisms in varying configurations, pocket watch classics (Calibers 6497 and 98), and manually wound calibers of days gone by (Calibers 1727 and 7001). This company truly offers everything that a manufacturer's heart could desire—not to mention the sheer variety of quartz technology from inexpensive three-hand mechanisms to highly complicated multifunctional movements and futuristic ETA-quartz featuring autonomous energy creation using a rotor and generator.

The almost stereotypical accusation of ETA being "mass goods" is not justified, however, for it is a real art to manufacture filigreed micromechanical technology in consistently high quality. This is certainly one of the reasons why there have been very few movement factories in Europe that can compete with ETA, or that would want to. Since the success of Swatch—a pure ETA product—millions of Swiss francs have been invested in new development and manufacturing technologies. ETA today owns more than twenty production locales in Switzerland, France, Germany, Malaysia, and Thailand.

In 2002, ETA's management announced it would discontinue providing half-completed component kits for reassembly and/or embellishment to specialized workshops, and from 2010 only offer completely assembled and finished movements for sale. The Swiss Competition Commission, however, studied the issue, and a new deal was struck in 2013, phasing out sales to customers over a period of six years. ETA is already somewhat of a competitor of independent reassemblers such as Soprod, Sellita, La Joux-Perret, Dubois Dépraz, and others thanks to its diversification of available calibers, which has led the rest to counter by creating their own base movements.

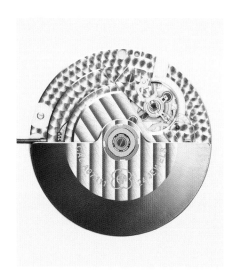

Caliber A07.111
Automatic; ETACHRON regulating system with fine-timing device, rotor on ball bearings, stop-second system; single spring barrel; 48-hour power reserve
Functions: hours, minutes, sweep seconds
Diameter: 37.2
Height: 7.9 mm
Jewels: 24
Frequency: 28,800 vph
Balance spring: flat hairspring
Shock protection: Incabloc
Remarks: related calibers: A07.161 (with power reserve display)

Caliber A07.171 (dial side)
Automatic; ETACHRON regulating system with fine-timing device, rotor on ball bearings, stop-second system; single spring barrel; 48-hour power reserve
Functions: hours, minutes, sweep seconds; 2nd time zone, additional 24-hour display (2nd time zone); quick-set date window
Diameter: 37.2 mm
Height: 7.9 mm
Jewels: 24
Frequency: 28,800 vph
Balance spring: flat hairspring
Shock protection: Incabloc

Caliber A07.211 (dial side)
Automatic; ETACHRON regulating system with fine-timing device, rotor on ball bearings, stop-second system; single spring barrel; 48-hour power reserve
Functions: hours, minutes, subsidiary seconds; chronograph; quick-set date window
Diameter: 37.2 mm
Height: 7.9 mm
Jewels: 25
Frequency: 28,800 vph
Balance spring: flat hairspring
Shock protection: Incabloc

Caliber 2000-1

Automatic; ball bearing–mounted rotor; stop-seconds, ETACHRON regulating system; single spring barrel; 40-hour power reserve
Functions: hours, minutes, sweep seconds; quick-set date window
Diameter: 20 mm
Height: 3.6 mm
Jewels: 20
Balance: glucydur
Frequency: 28,800 vph
Balance spring: flat hairspring
Shock protection: Incabloc

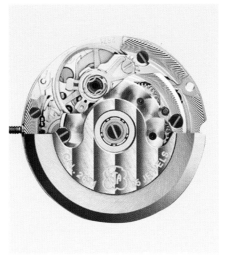

Caliber 2671

Automatic; ball bearing–mounted rotor; stop-seconds, ETACHRON regulating system; single spring barrel; 38-hour power reserve
Functions: hours, minutes, sweep seconds; date window
Diameter: 17.5 mm
Height: 4.8 mm
Jewels: 25
Balance: glucydur
Frequency: 28,800 vph
Balance spring: flat hairspring
Shock protection: Incabloc
Remarks: related calibers: 2678 (additional weekday window, height 5.35 mm)

Caliber 2681 (dial side)

Automatic; ball bearing–mounted rotor; stop-seconds, ETACHRON regulating system; single spring barrel; 38-hour power reserve
Functions: hours, minutes, sweep seconds; quick-set date window
Diameter: 20 mm
Height: 4.8 mm
Jewels: 25
Balance: glucydur
Frequency: 28,800 vph
Balance spring: flat hairspring
Shock protection: Incabloc

Caliber 2801-2

Manually wound; ETACHRON regulating system; 42-hour power reserve
Functions: hours, minutes, sweep seconds
Diameter: 26 mm
Height: 3.35 mm
Jewels: 17
Frequency: 28,800 vph
Related caliber: 2804-2 (with date window and quick set)

Caliber 2824-2

Automatic; ball bearing–mounted rotor; stop-seconds, ETACHRON regulating system; 38-hour power reserve
Functions: hours, minutes, sweep seconds; quick-set date window at 3 o'clock
Diameter: 26 mm
Height: 4.6 mm
Jewels: 25
Frequency: 28,800 vph
Related calibers: 2836-2 (additional day window at 3 o'clock, height 5.05 mm); 2826-2 (with large date, height 6.2 mm)

Caliber 2834-2 (dial side)

Automatic; ball bearing–mounted rotor; stop-seconds, ETACHRON regulating system; single spring barrel; 38-hour power reserve
Functions: hours, minutes, sweep seconds; quick-set date window, quick-set weekday
Diameter: 29.4 mm
Height: 5.05 mm
Jewels: 25
Balance: glucydur
Frequency: 28,800 vph
Balance spring: flat hairspring
Shock protection: Incabloc

ETA

Caliber 2892-A2

Automatic; ball bearing–mounted rotor; stop-seconds, ETACHRON regulating system; single spring barrel; 42-hour power reserve
Functions: hours, minutes, sweep seconds; quick-set date window
Diameter: 26.2 mm
Height: 3.6 mm
Jewels: 21
Balance: glucydur
Frequency: 28,800 vph
Balance spring: flat hairspring
Shock protection: Incabloc

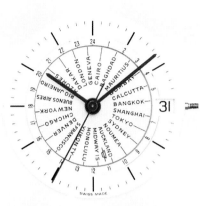

Caliber 2893-1 (dial side)

Automatic; ball bearing rotor; stop-seconds, ETACHRON regulating system; 42-hour power reserve
Functions: hours, minutes, sweep seconds; quick-set date window at 3 o'clock; world time display via central disk
Diameter: 25.6 mm
Height: 4.1 mm
Jewels: 21
Frequency: 28,800 vph
Related calibers: 2893-2 (24-hour hand; 2nd time zone instead of world time disk); 2893-3 (only world time disk without date window)

Caliber 2894-2

Automatic; ball bearing–mounted rotor; stop-seconds, ETACHRON regulating system; single spring barrel; 42-hour power reserve
Functions: hours, minutes, subsidiary seconds; chronograph; quick-set date window
Diameter: 28.6 mm
Height: 6.1 mm
Jewels: 37
Balance: glucydur
Frequency: 28,800 vph
Balance spring: flat hairspring
Shock protection: Incabloc
Related caliber: 2094 (diameter 23.9 mm, height 5.5 mm, 33 jewels)

Caliber 2895-2 (dial side)

Automatic; ball bearing–mounted rotor; stop-seconds, ETACHRON regulating system; single spring barrel; 42-hour power reserve
Functions: hours, minutes, subsidiary seconds, at 6 o'clock; quick-set date window
Diameter: 26.2 mm
Height: 4.35 mm
Jewels: 27
Balance: glucydur
Frequency: 28,800 vph
Balance spring: flat hairspring
Shock protection: Incabloc

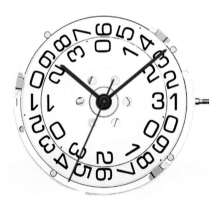

Caliber 2896 (dial side)

Automatic; ball bearing rotor; stop-seconds, ETACHRON regulating system; 42-hour power reserve
Functions: hours, minutes, sweep seconds; power reserve display at 3 o'clock
Diameter: 25.6 mm
Height: 4.85 mm
Jewels: 21
Frequency: 28,800 vph

Caliber 2897 (dial side)

Automatic; ball bearing–mounted rotor; stop-seconds, ETACHRON regulating system; single spring barrel; 42-hour power reserve
Functions: hours, minutes, sweep seconds; power reserve indicator; quick-set date window
Diameter: 26.2 mm
Height: 4.85 mm
Jewels: 21
Balance: glucydur
Frequency: 28,800 vph
Balance spring: flat hairspring
Shock protection: Incabloc

ETA

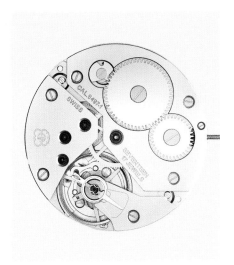

Caliber 6497-1

Manually wound; ETACHRON regulating system; single spring barrel; 46-hour power reserve
Functions: hours, minutes, subsidiary seconds
Diameter: 37.2 mm
Height: 4.5 mm
Jewels: 17
Frequency: 18,000 vph
Balance spring: flat hairspring
Remarks: pocket watch movement (Unitas model) in Lépine version with subsidiary seconds extending from the winding stem); as Caliber 6497-2 with 21,600 vph and 53-hour power reserve

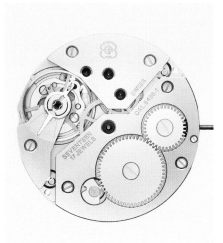

Caliber 6498-1

Manually wound; ETACHRON regulating system; single spring barrel; 46-hour power reserve
Functions: hours, minutes, subsidiary seconds
Diameter: 37.2 mm
Height: 4.5 mm
Jewels: 17
Frequency: 18,000 vph
Balance spring: flat hairspring
Remarks: pocket watch movement (Unitas model) in savonette version (subsidiary seconds at right angle to the winding stem); as Caliber 6498-2 with 21,600 vph and 53-hour power reserve)

Caliber 7001

Manually wound; ultrathin construction; single spring barrel; 42-hour power reserve
Functions: hours, minutes, subsidiary seconds
Diameter: 23.7 mm
Height: 2.5 mm
Jewels: 17
Frequency: 21,600 vph
Balance spring: flat hairspring

Caliber 7750 (dial side)

Automatic; stop-second system; single spring barrel; 42-hour power reserve
Functions: hours, minutes, subsidiary seconds; chronograph; quick-set date and weekday window
Diameter: 30.4 mm
Height: 7.9 mm
Jewels: 25
Balance: glucydur
Frequency: 28,800 vph
Balance spring: flat hairspring
Shock protection: Incabloc

Caliber 7751 (dial side)

Automatic; stop-second system; single spring barrel; 42-hour power reserve
Functions: hours, minutes, subsidiary seconds; additional 24-hour display; chronograph; full calendar with date, weekday, month, moon phase
Diameter: 30.4 mm
Height: 7.9 mm
Jewels: 25
Balance: glucydur
Frequency: 28,800 vph
Balance spring: flat hairspring
Shock protection: Incabloc
Remarks: related caliber: 7754 with sweep 24-hour hand (2nd time zone)

Caliber 7753

Automatic; stop-second system; single spring barrel; 42-hour power reserve
Functions: hours, minutes, subsidiary seconds; chronograph; quick-set date window with pusher
Diameter: 30.4
Height: 7.9 mm
Jewels: 25
Balance: glucydur
Frequency: 28,800 vph
Balance spring: flat hairspring
Shock protection: Incabloc
Remarks: variation of the Valjoux chronograph caliber with symmetrical "tricompax" layout of the totalizers

RONDA

Ronda is a Swiss company with a long tradition. It was founded by William Mosset, born in 1909 in the village of Hölstein, a man whose gift for micro-engineering declared itself early on when he invented a way to drill thirty-two holes in a metal plate in one operation and with great accuracy. The company was founded in 1946 in Lausen, a little town in the hinterlands of German-speaking Switzerland near Basel, where the first factory was built.

In the meantime the company has turned into a group with five subsidiaries: There are two production sites in Ticino, one in the Jura mountains, one operation in Thailand, and sales offices in Hong Kong. Overall, Ronda employs around 1,800 people in Switzerland and Asia.

The shareholders of the family enterprise, which is now in its second generation, value the company's absolute independence. This is undoubtedly a key advantage for the customer, since Ronda can continue defining its own strategy and can react decisively to customer needs.

That is why the company, which had already made a name for itself with quartz movements, decided to add a portfolio of automatic mechanical movements. The first product batches arrived on the market in early 2017; in the medium term, the mechanical Ronda Caliber R150 is to be produced in batches of six figures per year.

Caliber R150
Automatic; ball bearing–mounted rotor; stop-seconds, index for fine adjustment; single spring barrel; 40-hour power reserve
Functions: hours, minutes, sweep seconds; quick-set date
Diameter: 25.6 mm
Height: 4.4 mm
Jewels: 25
Frequency: 28,800 vph
Balance spring: flat hairspring
Shock protection: Incabloc

Caliber 5040.B
Quartz; 54-month power reserve; single spring barrel
Functions: hours, minutes, subsidiary seconds; chronograph, with add and split function; large date
Diameter: 28.6 mm
Height: 4.4 mm
Jewels: 13

Caliber 7004.P
Quartz; 48-month power reserve; single spring barrel
Functions: hours, minutes, subsidiary seconds; large date and weekday (retrograde)
Diameter: 34.6 mm
Height: 5.6 mm
Jewels: 6

FULL MONTH
2018 COLLECTION

 Independent Timepiece Maker | www.itay-noy.com | studio@itay-noy.com

SELLITA

Sellita, founded in 1950 by Pierre Grandjean in La Chaux-de-Fonds, is one of the biggest reassemblers and embellishers in the mechanical watch industry. On average, Sellita embellishes and finishes about one million automatic and hand-wound movements annually—a figure that represents about 25 percent of Switzerland's mechanical movement production, according to Miguel García, Sellita's president.

Reassembly can be defined as the assembly and regulation of components to make a functioning movement. This is the type of work that ETA loved to give to outside companies back in the day in order to concentrate on manufacturing complete quartz movements and individual components for them.

Reassembly workshops like Sellita refine and embellish components purchased from ETA according to their customers' wishes and can even successfully fulfill smaller orders made by the company's estimated 350 clients.

When ETA announced that it would only sell ébauches to companies outside the Swatch Group until the end of 2010, García, who has owned Sellita since 2003, reacted by shifting production to the development and manufacturing of new in-house products.

He planned and implemented a new line of movements based on the dimensions of the most popular ETA calibers, whose patents had expired. The company now has a line of manually wound or automatic movements with little complications, like a date, weekday, GMT, or a second time zone, as well as chronographs with different display constellations. The new design types are all based on mature models. A whole new line of automatic movements was launched, the Caliber SW1000, which has no ETA parts at all. The way to identify these movements will simply be the four digits. The price range will be similar to that of other products offered by the company.

Caliber SW200-1

Automatic; ball bearing–mounted rotor; stop-second system; single spring barrel; 38-hour power reserve
Functions: hours, minutes, sweep seconds; quick-set date
Diameter: 25.6 mm
Height: 4.6 mm
Jewels: 26
Balance: nickel or glucydur
Frequency: 28,800 vph
Balance spring: Nivaflex
Shock protection: Novodiac or Incabloc

Caliber SW210-1

Manually wound; stop-second system; single spring barrel; 42-hour power reserve
Functions: hours, minutes, sweep seconds
Diameter: 25.6 mm
Height: 3.35 mm
Jewels: 19
Balance: nickel
Frequency: 28,800 vph
Balance spring: Nivaflex
Shock protection: Novodiac or Incabloc
Remarks: related caliber: SW215 (with window date)

Caliber SW220-1

Automatic; ball bearing–mounted rotor; stop-second system; single spring barrel; 38-hour power reserve
Functions: hours, minutes, sweep seconds; quick-set date and weekday
Diameter: 25.6 mm
Height: 5.05 mm
Jewels: 26
Balance: nickel or glucydur
Frequency: 28,800 vph
Balance spring: flat hairspring, Nivaflex
Shock protection: Novodiac or Incabloc
Remarks: related calibers: SW221-1 (with hand date); SW240-1 with larger mainplate (ø 29 mm)

SELLITA

Caliber SW260-1

Automatic; ball bearing–mounted rotor; stop-second system; single spring barrel; 38-hour power reserve
Functions: hours, minutes, subsidiary seconds at 6 o'clock; quick-set date
Diameter: 25.6 mm
Height: 5.6 mm
Jewels: 31
Balance: nickel or glucydur
Frequency: 28,800 vph
Balance spring: flat hairspring, Nivaflex
Shock protection: Novodiac or Incabloc
Remarks: related caliber: SW290-1 (subsidiary seconds at 9 o'clock)

Caliber SW300-1

Automatic; ball bearing–mounted rotor; stop-second system; single spring barrel; 42-hour power reserve
Functions: hours, minutes, sweep seconds; quick-set date
Diameter: 25.6 mm
Height: 3.6 mm
Jewels: 25
Balance: glucydur
Frequency: 28,800 vph
Balance spring: flat hairspring, Nivaflex
Shock protection: Incabloc
Remarks: related caliber: SW360-1 (with subsidiary seconds, height 4.35 mm, 31 jewels)

Caliber SW330-1

Automatic; ball bearing–mounted rotor; stop-second system; single spring barrel, 42-hour power reserve
Functions: hours, minutes, sweep seconds; additional 24-hour display (2nd time zone); quick-set date
Diameter: 25.6 mm
Height: 4.1 mm
Jewels: 25
Balance: glucydur
Frequency: 28,800 vph
Balance spring: flat hairspring, Nivaflex
Shock protection: Incabloc

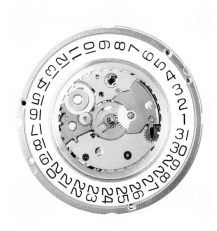

Caliber SW400-1

Automatic; ball bearing–mounted rotor; stop-second system; single spring barrel; 38-hour power reserve
Functions: hours, minutes, sweep seconds; quick-set date
Diameter: 31 mm
Height: 4.67 mm
Jewels: 26
Balance: nickel or glucydur
Frequency: 28,800 vph
Balance spring: Nivaflex
Shock protection: Novodiac or Incabloc

Caliber SW500-1

Automatic; ball bearing–mounted rotor; stop-second system; single spring barrel; 48-hour power reserve
Functions: hours, minutes, subsidiary seconds; chronograph quick-set date and weekday
Diameter: 30 mm
Height: 7.9 mm
Jewels: 25
Balance: nickel or glucydur
Frequency: 28,800 vph
Balance spring: Nivaflex
Shock protection: Incabloc

Caliber SW1000-1

Automatic; ball bearing–mounted rotor; stop-second system; single spring barrel; 38-hour power reserve
Functions: hours, minutes, sweep seconds; quick-set date
Diameter: 20 mm
Height: 3.9 mm
Jewels: 18
Balance: nickel or glucydur
Frequency: 28,800 vph
Balance spring: Nivaflex
Shock protection: Incabloc

Watch Your Watch

Mechanical watches are not only by and large more expensive and complex than quartzes, they are also a little high-maintenance, as it were. The mechanism within does need servicing occasionally—perhaps a touch of oil and an adjustment. Worse yet, the complexity of all those wheels and pinions engaged in reproducing the galaxy means that a user will occasionally do something perfectly harmless like wind his or her watch up only to find everything grinding to a halt. Here are some tips for dealing with these mechanical beauties for new watch owners and reminders for the old hands.

1. DATE CHANGES

Do not change the date manually (via the crown or pusher) on any mechanical watch—whether manual wind or automatic—when the time indicated on the dial reads between 10 and 2 o'clock. Although some better watches are protected against this horological quirk, most mechanical watches with a date indicator are engaged in the process of automatically changing the date between the hours of 10 p.m. and 2 a.m. Intervening with a forced manual change while the automatic date shift is engaged can damage the movement. Of course, you can make the adjustment between 10 a.m. and 2 p.m. in most cases—but this is just not a good habit to get into. When in doubt, roll the time past 12 o'clock and look for an automatic date change before you set the time and date. The Ulysse Nardin brand is notable, among a very few others, for in-house mechanical movements immune to this effect.

2. CHRONOGRAPH USE

On a simple chronograph, start and stop are almost always the same button. Normally located above the crown, the start/stop actuator can be pressed at will to initiate and end the interval timing. The reset button, normally below the crown, is only used for resetting the chronograph to zero, but only when the chronograph is stopped—never while engaged. Only a "flyback" chronograph allows safe resetting to zero while running. With the chronograph engaged, you simply hit the reset button and all the chronograph indicators (seconds, minutes, and hours) snap back to zero and the chronograph begins to accumulate the interval time once again. In the early days of air travel this was a valuable complication as pilots would reset their chronographs when taking on a new heading—without having to fumble about with a three-step procedure with gloved hands.

Nota bene: Don't actuate or reset your chronograph while your watch is submerged—even if you have one of those that are built for such usage, like Omega, IWC, and a few other brands. Feel free to hit the buttons before submersion and jump in and swim while they run; just don't push anything while in the water.

3. CHANGING TIME BACKWARD

Don't adjust the time on your watch in a counterclockwise direction—especially if the watch has calendar functions. A few watches can tolerate the abuse, but it's better to avoid the possibility of damage altogether. Change the dates as needed (remembering the 10 and 2 rule above).

4. SHOCKS

Almost all modern watches are equipped with some level of shock protection. Best practices for the Swiss brands allow for a three-foot fall onto a hard wood surface. But if your watch is running poorly—or even worse has stopped entirely after an impact—do not shake, wind, or bang it again to get it running; take it to an expert for service as you may do even more damage. Sports like tennis, squash, or golf can have a deleterious effect on your watch, including flattening the pivots, overbanking, or even bending or breaking a pivot.

5. OVERWINDING

Most modern watches are fitted with a mechanism that allows the mainspring to slide inside the barrel—or stops it completely once the spring is fully wound—for protection against overwinding. The best advice here is just don't force it. Over the years, a winding crown may start to get "stickier" and more difficult to turn even when unwound. That's a sure sign it is due for service.

6. JACUZZI TEMPERATURE

Don't jump into the Jacuzzi—or even a steaming hot shower—with your watch on. Better-built watches with a deeper water-resistance rating typically have no problem with this scenario. However, take a 3 or 5 atm water-resistant watch into the Jacuzzi, and there's a chance the different rates of expansion and contraction of the metals and sapphire or mineral crystals may allow moisture into the case.

Bovet's barrier to pressing the wrong pusher.

WATCH YOUR WATCH

Panerai makes sure you think before touching the crown.

Do it yourself at your own risk.

7. SCREW THAT CROWN DOWN (AND THOSE PUSHERS)!

Always check and double-check to ensure a watch fitted with a screwed-down crown is closed tightly. Screwed-down pushers for a chronograph—or any other functions—deserve the same attention. This one oversight has cost quite a few owners their watches. If a screwed-down crown is not secured, water will likely get into the case and start oxidizing the metal. In time, the problem can destroy the watch.

8. MAGNETISM

If your watch is acting up, running faster or slower, it may have become magnetized. This can happen if you leave your timepiece near a computer, cell phone, or some other electronic device. Many service centers have a so-called degausser to take care of the problem. A number of brands also make watches with a soft iron core to deflect magnetic fields, though this might not work with the stronger ones.

9. TRIBOLOGY

Keeping a mechanical timepiece hidden away in a box for extended lengths of time is not the best way to care for it. Even if you don't wear a watch every day, it is a good idea to run your watch at regular intervals to keep its lubricating oils and greases viscous. Think about a can of house paint: Keep it stirred and it stays liquid almost indefinitely; leave it still for too long and a skin develops. On a smaller level the same thing can happen to the lubricants inside a mechanical watch.

10. SERVICE

Most mechanical watches call for a three- to five-year service cycle for cleaning, oiling, and maintenance. Some mechanical watches can run twice that long and have functioned within acceptable parameters, but if you're not going to have your watch serviced at regular intervals, you do run the risk of having timing issues. Always have your watch serviced by a qualified watchmaker (see box), not at the kiosk in the local mall. The best you can expect there is a quick battery change.

Gary Girdvainis is the founder of Isochron Media LLC, publishers of WristWatch *and* AboutTime *magazines.*

WATCH REPAIR SERVICE CENTERS

RGM
www.rgmwatches.com/repair

Stoll & Co.
www.americaswatchmaker.com

Swiss Watchmakers & Company
www.swisswatchland.com

Universal Watch Repair
www.universalwatch.net

Watch Repairs USA
www.watchrepairsusa.com

Glossary

ANNUAL CALENDAR

The automatic allowances for the different lengths of each month of a year in the calendar module of a watch. This type of watch usually shows the month and date, and sometimes the day of the week (like this one by Patek Philippe) and the phases of the moon.

ANTIMAGNETIC

Magnetic fields found in common everyday places affect mechanical movements, hence the use of anti- or non-magnetic components in the movement. Some companies encase movements in antimagnetic cores such as Sinn's Model 756, the Duograph, shown here.

ANTIREFLECTION

A film created by steaming the crystal to eliminate light reflection and improve legibility. Antireflection functions best when applied to both sides of the crystal, but because it scratches, some manufacturers prefer to have it only on the interior of the crystal. It is mainly used on synthetic sapphire crystals. Dubey & Schaldenbrand applies antireflection on both sides for all of the company's wristwatches, such as this Aquadyn model.

AUTOMATIC WINDING

A rotating weight set into motion by moving the wrist winds the spring barrel via the gear train of a mechanical watch movement. Automatic winding was invented during the pocket watch era in 1770, but the breakthrough automatic winding movement via rotor began with the ball bearing Eterna-Matic in the late 1940s. Today we speak of unidirectional winding and bidirectionally winding rotors, depending on the type of gear train used. Shown is IWC's automatic Caliber 50611.

BALANCE

The beating heart of a mechanical watch movement is the balance. Fed by the energy of the mainspring, a tirelessly oscillating little wheel, just a few millimeters in diameter and possessing a spiral-shaped balance spring, sets the rhythm for the escape wheel and pallets with its vibration frequency. Today the balance is usually made of one piece of antimagnetic glucydur, an alloy that expands very little when exposed to heat.

BEVELING

To uniformly file down the sharp edges of a plate, bridge, or bar and give it a high polish. The process is also called *anglage*. Edges are usually beveled at a 45° angle. As the picture shows, this is painstaking work that needs the skilled hands and eyes of an experienced watchmaker or *angleur*.

BAR OR COCK

A metal plate fastened to the base plate at one point, leaving room for a gear wheel or pinion. The balance is usually attached to a bar called the balance cock. Glashütte tradition dictates that the balance cock be decoratively engraved by hand like this one by Glashütte Original.

GLOSSARY

BRIDGE

A metal plate fastened to the base plate at two points leaving room for a gear wheel or pinion. This vintage Favre-Leuba movement illustrates the point with three individual bridges.

CALIBER

A term, similar to type or model, that refers to different watch movements. Pictured here is Heuer's Caliber 11, the legendary automatic chronograph caliber from 1969. This movement was a coproduction jointly researched and developed for four years by Heuer-Leonidas, Breitling, and Hamilton-Büren. Each company gave the movement a different name after serial production began.

CARBON FIBER

A very light, tough composite material, carbon fiber is composed of filaments comprised of several thousand seven-micron carbon fibers held together by resin. The arrangement of the filaments determines the quality of a component, making each unique. Carbon fiber is currently being used for dials, cases, and even movement components.

CHAMPLEVÉ

A dial decoration technique, whereby the metal is engraved, filled with enamel, and baked, as in this cockatoo on a Cartier Tortue, enhanced with mother-of-pearl slivers.

CERAMIC

An inorganic, nonmetallic material formed by the action of heat and practically unscratchable. Pioneered by Rado, ceramic is a high-tech material generally made from aluminum and zirconia oxide. Today, it is used generally for cases and bezels and now comes in many colors.

CHRONOGRAPH

From the Greek *chronos* (time) and *graphein* (to write). Originally a chronograph literally wrote, inscribing the time elapsed on a piece of paper with the help of a pencil attached to a type of hand. Today this term is used for watches that show not only the time of day, but also certain time intervals via independent hands that may be started or stopped at will. Stopwatches differ from chronographs because they do not show the time of day. This exploded illustration shows the complexity of a Breitling chronograph.

CHRONOMETER

Literally, "measurer of time." As the term is used today, a chronometer denotes an especially accurate watch (one with a deviation of no more than 5 seconds a day for mechanical movements). Chronometers are usually supplied with an official certificate from an independent testing office such as the COSC. The largest producer of chronometers in 2008 was Rolex, with 769,850 officially certified movements. Chopard came in sixth with more than 22,000 certified L.U.C mechanisms, like the 4.96 in the Pro One model shown here.

COLUMN WHEEL

The component used to control chronograph functions within a true chronograph movement. The presence of a column wheel indicates that the chronograph is fully integrated into the movement. In the modern era, modules are generally used that are attached to a base caliber movement. This particular column wheel is made of blued steel.

GLOSSARY

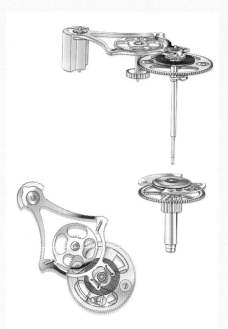

CONSTANT FORCE MECHANISM

Sometimes called a constant force escapement, it isn't really: in most cases this mechanism is "simply" an initial tension spring. It is also known in English by part of its French name, the *remontoir*, which actually means "winding mechanism." This mechanism regulates and portions the energy that is passed on through the escapement, making the rate as even and precise as possible. Shown here is the constant force escapement from A. Lange & Söhne's Lange 31—a mechanism that gets as close to its name as possible.

COSC

The Contrôle Officiel Suisse de Chronomètrage, the official Swiss testing office for chronometers. The COSC is the world's largest issuer of so-called chronometer certificates, which are only otherwise given out individually by certain observatories (such as the one in Neuchâtel, Switzerland). For a fee, the COSC tests the rate of movements that have been adjusted by watchmakers. These are usually mechanical movements, but the office also tests some high-precision quartz movements. Those that meet the specifications for being a chronometer are awarded an official certificate as shown here.

CÔTES DE GENÈVE

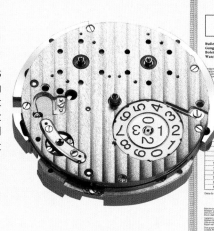

Also called *vagues de Genève* and Geneva stripes. This is a traditional Swiss surface decoration comprising an even pattern of parallel stripes, applied to flat movement components with a quickly rotating plastic or wooden peg. Glashütte watchmakers have devised their own version of *côtes de Genève* that is applied at a slightly different angle, called Glashütte ribbing.

CROWN

The crown is used to wind and set a watch. A few simple turns of the crown will get an automatic movement started, while a manually wound watch is completely wound by the crown. The crown is also used for the setting of various functions, almost always including at least the hours, minutes, seconds, and date. A screwed-down crown like the one on the TAG Heuer Aquagraph pictured here can be tightened to prevent water entering the case or any mishaps while performing extreme sports such as diving.

EQUATION OF TIME

The mean time that we use to keep track of the passing of the day (24 hours evenly divided into minutes and seconds) is not equal to true solar time. The equation of time is a complication devised to show the difference between the mean time shown on one's wristwatch and the time the sun dictates. The Équation Marchante by Blancpain very distinctly indicates this difference via the golden suntipped hand that also rotates around the dial in a manner known to watch connoisseurs as *marchant*. Other wristwatch models, such as the Boreas by Martin Braun, display the difference on an extra scale on the dial.

GLOSSARY

ESCAPEMENT

The combination of the balance, balance spring, pallets, and escape wheel, a subgroup which divides the impulses coming from the spring barrel into small, accurately portioned doses. It guarantees that the gear train runs smoothly and efficiently. The pictured escapement is one newly invented by Parmigiani, containing pallet stones of varying colors, though they are generally red synthetic rubies. Here one of them is a colorless sapphire, or corundum, the same geological material that ruby is made of.

FLINQUÉ

A dial decoration in which a guilloché design is given a coat of enamel, softening the pattern and creating special effects, as shown here on a unique Bovet.

FLYBACK CHRONOGRAPH

A chronograph with a special dial train switch that makes the immediate reuse of the chronograph movement possible after resetting the hands. It was developed for special timekeeping duties such as those found in aviation, which require the measurement of time intervals in quick succession. A flyback may also be called a *retour en vol*. An elegant example of this type of chronograph is Corum's Classical Flyback Large Date shown here.

GEAR TRAIN

A mechanical watch's gear train transmits energy from the mainspring to the escapement. The gear train comprises the minute wheel, the third wheel, the fourth wheel, and the escape wheel.

GUILLOCHÉ

A surface decoration usually applied to the dial and the rotor using a grooving tool with a sharp tip, such as a rose engine, to cut an even pattern onto a level surface. The exact adjustment of the tool for each new path is controlled by a device similar to a pantograph, and the movement of the tool can be controlled either manually or mechanically. Real *guillochis* (the correct term used by a master of guilloché) are very intricate and expensive to produce, which is why most dials decorated in this fashion are produced by stamping machines. Breguet is one of the very few companies to use real guilloché on every one of its dials.

GLUCYDUR

Glucydur is a functional alloy of copper, beryllium, and iron that has been used to make balances in watches since the 1930s. Its hardness and stability allow watchmakers to use balances that were poised at the factory and no longer required adjustment screws.

INDEX

A regulating mechanism found on the balance cock and used by the watchmaker to adjust the movement's rate. The index changes the effective length of the balance spring, thus making it move more quickly or slowly. This is the standard index found on an ETA Valjoux 7750.

JEWEL

To minimize friction, the hardened steel tips of a movement's rotating gear wheels (called pinions) are lodged in synthetic rubies (fashioned as polished stones with a hole) and lubricated with a very thin layer of special oil. These synthetic rubies are produced in exactly the same way as sapphire crystal using the same material. During the pocket watch era, real rubies with hand-drilled holes were still used, but because of the high costs involved, they were only used in movements with especially quickly rotating gears. The jewel shown here on a bridge from A. Lange & Söhne's Double Split is additionally embedded in a gold chaton secured with three blued screws.

LIGA

The word LIGA is actually a German acronym that stands for lithography (*Lithografie*), electroplating (*Galvanisierung*), and plastic molding (*Abformung*). It is a lithographic process exposed by UV or X-ray light that literally "grows" perfect micro components made of nickel, nickel-phosphorus, or 23.5-karat gold in a plating bath. The components need no finishing or trimming after manufacture.

GLOSSARY

LUMINOUS SUBSTANCE

Tritium paint is a slightly radioactive substance that replaced radium as a luminous coating for hands, numerals, and hour markers on watch dials. Watches bearing tritium must be marked as such, with the letter T on the dial near 6 o'clock. It has now for the most part been replaced by nonradioactive materials such as Superluminova. Traser technology (as seen on these Ball timepieces) uses tritium gas enclosed in tiny silicate glass tubes coated on the inside with a phosphorescing substance. The luminescence is constant and will hold around twenty-five years.

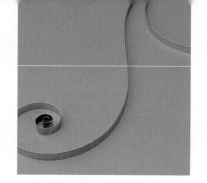

MAINSPRING

The mainspring, located in the spring barrel, stores energy when tensioned and passes it on to the escapement via the gear train as the tension relaxes. Today, mainsprings are generally made of Nivaflex, an alloy invented by Swiss engineer Max Straumann at the beginning of the 1950s. This alloy basically comprises iron, nickel, chrome, cobalt, and beryllium.

MINUTE REPEATER

A striking mechanism with hammers and gongs for acoustically signaling the hours, quarter hours, and minutes elapsed since noon or midnight. The wearer pushes a slide, which winds the spring. Normally a repeater uses two different gongs to signal hours (low tone), quarter hours (high and low tones in succession), and minutes (high tone). Some watches have three gongs, called a carillon. The Chronoswiss Répétition à Quarts is a prominent repeating introduction of recent years.

PERPETUAL CALENDAR

The calendar module for this type of timepiece automatically makes allowances for the different lengths of each month as well as leap years until the next secular year, which will occur in 2100. A perpetual calendar usually shows the date, month, and four-year cycle, and may show the day of the week and moon phase as well, as does this one introduced by George J von Burg at Baselworld 2005. Perpetual calendars need much skill to complete.

PERLAGE

Surface decoration comprising an even pattern of partially overlapping dots, applied with a quickly rotating plastic or wooden peg, as shown here on the plates of Frédérique Constant's *manufacture* Caliber FC 910-1.

PLATE

A metal platform having several tiers for the gear train. The base plate of a movement usually incorporates the dial and carries the bearings for the primary pinions of the "first floor" of a gear train. The gear wheels are made complete by tightly fitting screwed-in bridges and bars on the back side of the plate. A specialty of the so-called Glashütte school, as opposed to the Swiss school, is the reverse completion of a movement not via different bridges and bars, but rather with a three-quarter plate. Glashütte Original's Caliber 65 (shown) displays a beautifully decorated three-quarter plate.

GLOSSARY

POWER RESERVE DISPLAY

A mechanical watch contains only a certain amount of power reserve. A fully wound modern automatic watch usually possesses between 36 and 42 hours of energy before it needs to be wound again. The power reserve display keeps the wearer informed about how much energy his or her watch still has in reserve, a function that is especially practical on manually wound watches with several days of possible reserve. The Nomos Tangente Power Reserve pictured here represents an especially creative way to illustrate the state of the mainspring's tension. On some German watches the power reserve is also displayed with the words "auf" and "ab."

PULSOMETER

A scale on the dial, flange, or bezel that, in conjunction with the second hand, may be used to measure a pulse rate. A pulsometer is always marked with a reference number—if it is marked with *gradué pour 15 pulsations*, for example, then the wearer counts fifteen pulse beats. At the last beat, the second hand will show what the pulse rate is in beats per minute on the pulsometer scale. The scale on Sinn's World Time Chronograph (shown) is marked simply with the German world *Puls* (pulse), but the function remains the same.

QUALITÉ FLEURIER

This certification of quality was established by Chopard, Parmigiani Fleurier, Vaucher, and Bovet Fleurier in 2004. Watches bearing the seal must fulfill five criteria, including COSC certification, passing several tests for robustness and precision, top-notch finishing, and being 100 percent Swiss-made (except for the raw materials). The seal appears here on the dial of the Parmigiani Fleurier Tonda 39.

RETROGRADE DISPLAY

A retrograde display shows the time linearly instead of circularly. The hand continues along an arc until it reaches the end of its scale, at which precise moment it jumps back to the beginning instantaneously. This Nienaber model not only shows the minutes in retrograde form, it is also a regulator display.

ROTOR

The rotor is the component that keeps an automatic watch wound. The kinetic motion of this part, which contains a heavy metal weight around its outer edge, winds the mainspring. It can either wind unilaterally or bilaterally (to one or both sides) depending on the caliber. The rotor from this Temption timepiece belongs to an ETA Valjoux 7750.

SAPPHIRE CRYSTAL

Synthetic sapphire crystal is known to gemologists as aluminum oxide (Al_2O_3) or corundum. It can be colorless (corundum), red (ruby), blue (sapphire), or green (emerald). It is virtually scratchproof; only a diamond is harder. The innovative Royal Blue Tourbillon by Ulysse Nardin pictured here features not only sapphire crystals on the front and back of the watch, but also actual plates made of both colorless and blue corundum within the movement.

SCREW BALANCE

Before the invention of the perfectly weighted balance using a smooth ring, balances were fitted with weighted screws to get the exact impetus desired. Today a screw balance is a subtle sign of quality in a movement due to its costly construction and assembly utilizing minuscule weighted screws.

GLOSSARY

SEAL OF GENEVA

Since 1886 the official seal of this canton has been awarded to Genevan watch *manufactures* who must follow a defined set of high-quality criteria that include the following: polished jewel bed drillings, jewels with olive drillings, polished winding wheels, quality balances and balance springs, steel levers and springs with beveling of 45 degrees and *côtes de Genève* decoration, and polished stems and pinions. The list was updated in 2012 to include the entire watch and newer components. Testing is done on the finished piece. The Seal consists of two, one on the movement, one on the case. The pictured seal was awarded to Vacheron Constantin, a traditional Genevan *manufacture*.

SILICIUM/SILICON

Silicon is an element relatively new to mechanical watches. It is currently being used in the manufacture of precision escapements. Ulysse Nardin's Freak has lubrication-free silicon wheels, and Breguet has successfully used flat silicon balance springs.

SKELETONIZATION

The technique of cutting a movement's components down to their weight-bearing basic substance. This is generally done by hand in painstaking hours of microscopic work with a small handheld saw, though machines can skeletonize parts to a certain degree, such as the version of the Valjoux 7750 that was created for Chronoswiss's Opus and Pathos models. This tourbillon created by Christophe Schaffo is additionally—and masterfully—hand-engraved.

SONNERIE

A variety of minute repeater that—like a tower clock—sounds the time not at the will of the wearer, but rather automatically *(en passant)* every hour *(petite sonnerie)* or quarter hour *(grande sonnerie)*. Gérald Genta designed the most complicated sonnerie back in the early nineties. Shown is a recent model from the front and back.

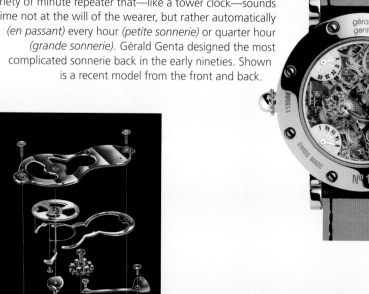

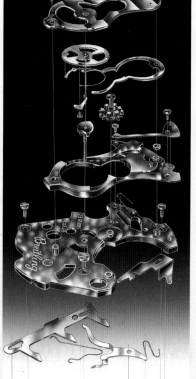

SPLIT-SECONDS CHRONOGRAPH

Also known in the watch industry by its French name, the *rattrapante* (exploded view at left). A watch with two second hands, one of which can be blocked with a special dial train lever to indicate an intermediate time while the other continues to run. When released, the split-seconds hand jumps ahead to the position of the other second hand. The PTC by Porsche Design illustrates this nicely.